Multivariate Statistics

MULTIVARIATE STATISTICS
A Vector Space Approach

MORRIS L. EATON
Department of Theoretical Statistics
University of Minnesota, Minneapolis

JOHN WILEY & SONS
New York Chichester Brisbane Toronto Singapore

Library of Congress Cataloging in Publication Data:

Eaton, Morris L.
Multivariate statistics.

(Wiley series in probability and mathematical statistics. Probability and mathematical statistics, ISSN 0271-6232)
Includes bibliographical references and index.
1. Multivariate analysis. 2. Vector spaces.
I. Title. II. Series.
QA278.E373 1983 519.5′35 83-1215
ISBN 0-471-02776-6

Printed in the United States of America

10 9 8 7 6 5 4 3 2 1

Dedicated to
Marcia

Preface

The purpose of this book is to present a version of multivariate statistical theory in which vector space and invariance methods replace, to a large extent, more traditional multivariate methods. The book is a text. Over the past ten years, various versions have been used for graduate multivariate courses at the University of Chicago, the University of Copenhagen, and the University of Minnesota. Designed for a one year lecture course or for independent study, the book contains a full complement of problems and problem solutions.

My interest in using vector space methods in multivariate analysis was aroused by William Kruskal's success with such methods in univariate linear model theory. In the late 1960s, I had the privilege of teaching from Kruskal's lecture notes where a coordinate free (vector space) approach to univariate analysis of variance was developed. (Unfortunately, Kruskal's notes have not been published.) This approach provided an elegant unification of linear model theory together with many useful geometric insights. In addition, I found the pedagogical advantages of the approach far outweighed the extra effort needed to develop the vector space machinery. Extending the vector space approach to multivariate situations became a goal, which is realized here. Basic material on vector spaces, random vectors, the normal distribution, and linear models take up most of the first half of this book.

Invariance (group theoretic) arguments have long been an important research tool in multivariate analysis as well as in other areas of statistics. In fact, invariance considerations shed light on most multivariate hypothesis testing, estimation, and distribution theory problems. When coupled with vector space methods, invariance provides an important complement to the traditional distribution theory–likelihood approach to multivariate analysis. Applications of invariance to multivariate problems occur throughout the second half of this book.

A brief summary of the contents and flavor of the ten chapters herein follows. In Chapter 1, the elements of vector space theory are presented. Since my approach to the subject is geometric rather than algebraic, there is an emphasis on inner product spaces where the notions of length, angle, and orthogonal projection make sense. Geometric topics of particular importance in multivariate analysis include singular value decompositions and angles between subspaces. Random vectors taking values in inner product spaces is the general topic of Chapter 2. Here, induced distributions, means, covariances, and independence are introduced in the inner product space setting. These results are then used to establish many traditional properties of the multivariate normal distribution in Chapter 3. In Chapter 4, a theory of linear models is given that applies directly to multivariate problems. This development, suggested by Kruskal's treatment of univariate linear models, contains results that identify all the linear models to which the Gauss–Markov Theorem applies.

Chapter 5 contains some standard matrix factorizations and some elementary Jacobians that are used in later chapters. In Chapter 6, the theory of invariant integrals (measures) is outlined. The many examples here were chosen to illustrate the theory and prepare the reader for the statistical applications to follow. A host of statistical applications of invariance, ranging from the invariance of likelihood methods to the use of invariance in deriving distributions and establishing independence, are given in Chapter 7. Invariance arguments are used throughout the remainder of the book.

The last three chapters are devoted to a discussion of some traditional and not so traditional problems in multivariate analysis. Here, I have stressed the connections between classical likelihood methods, linear model considerations, and invariance arguments. In Chapter 8, the Wishart distribution is defined via its representation in terms of normal random vectors. This representation, rather than the form of the Wishart density, is used to derive properties of the Wishart distribution. Chapter 9 begins with a thorough discussion of the multivariate analysis of variance (MANOVA) model. Variations on the MANOVA model including multivariate linear models with structured covariances are the main topic of the rest of Chapter 9. An invariance argument that leads to the relationship between canonical correlations and angles between subspaces is the lead topic in Chapter 10. After a discussion of some distribution theory, the chapter closes with the connection between testing for independence and testing in multivariate regression models.

Throughout the book, I have assumed that the reader is familiar with the basic ideas of matrix and vector algebra in coordinate spaces and has some knowledge of measure and integration theory. As for statistical prerequisites, a solid first year graduate course in mathematical statistics should suffice. The book is probably best read and used as it was written—from front to back. However, I have taught short (one quarter) courses on topics in MANOVA using the material in Chapters 1, 2, 3, 4, 8, and 9 as a basis.

It is very difficult to compare this text with others on multivariate analysis. Although there may be a moderate amount of overlap with other texts, the approach here is sufficiently different to make a direct comparison inappropriate. Upon reflection, my attraction to vector space and invariance methods was, I think, motivated by a desire for a more complete understanding of multivariate statistical models and techniques. Over the years, I have found vector space ideas and invariance arguments have served me well in this regard. There are many multivariate topics not even mentioned here. These include discrimination and classification, factor analysis, Bayesian multivariate analysis, asymptotic results and decision theory results. Discussions of these topics can be found in one or more of the books listed in the Bibliography.

As multivariate analysis is a relatively old subject within statistics, a bibliography of the subject is very large. For example, the entries in *A Bibliography of Multivariate Analysis* by T. W. Anderson, S. Das Gupta, and G. H. P. Styan, published in 1972, number over 6000. The condensed bibliography here contains a few of the important early papers plus a sample of some recent work that reflects my bias. A more balanced view of the subject as a whole can be obtained by perusing the bibliographies of the multivariate texts listed in the Bibliography.

My special thanks go to the staff of the Institute of Mathematical Statistics at the University of Copenhagen for support and encouragement. It was at their invitation that I spent the 1971–1972 academic year at the University of Copenhagen lecturing on multivariate analysis. These lectures led to *Multivariate Statistical Analysis*, which contains some of the ideas and the flavor of this book. Much of the work herein was completed during a second visit to Copenhagen in 1977–1978. Portions of the work have been supported by the National Science Foundation and the University of Minnesota. This generous support is gratefully acknowledged.

A number of people have read different versions of my manuscript and have made a host of constructive suggestions. Particular thanks go to Michael Meyer, whose good sense of pedagogy led to major revisions in a number of places. Others whose help I would like to acknowledge are Murray Clayton, Siu Chuen Ho, and Takeaki Kariya.

Most of the typing of the manuscript was done by Hanne Hansen. Her efforts are very much appreciated. For their typing of various corrections, addenda, changes, and so on, I would like to thank Melinda Hutson, Catherine Stepnes, and Victoria Wagner.

MORRIS L. EATON

Minneapolis, Minnesota
May 1983

Contents

Notation

$(V, (\cdot, \cdot))$	an inner product space, vector space V and inner product $(\cdot, \cdot)$
$\mathcal{L}(V, W)$	the vector space of linear transformations on V to W
$Gl(V)$	the group of nonsingular linear transformations on V to V
$\mathcal{O}(V)$	the orthogonal group of the inner product space $(V, (\cdot, \cdot))$
R^n	Euclidean coordinate space of all n-dimensional column vectors
$\mathcal{L}_{p,n}$	the linear space of all $n \times p$ real matrices
Gl_n	the group of $n \times n$ nonsingular matrices
$\mathcal{O}_n$	the group of $n \times n$ orthogonal matrices
$\mathcal{F}_{p,n}$	the space of $n \times p$ real matrices whose p columns form an orthonormal set in R^n
G_T^+	the group of lower triangular matrices with positive diagonal elements—dimension implied by context
G_U^+	the group of upper triangular matrices with positive diagonal elements—dimension implied by context
$\mathcal{S}_p^+$	the set of $p \times p$ real symmetric positive definite matrices
$A > 0$	the matrix or linear transformation A is positive definite
$A \geqslant 0$	A is positive semidefinite (non-negative definite)
det	determinant
tr	trace
$x \Box y$	the outer product of the vectors x and y
$A \otimes B$	the Kronecker product of the linear transformations A and B
Δ_r	the right-hand modulus of a locally compact topological group
$\mathcal{L}(\cdot)$	the distributional law of "$\cdot$"

$N(\mu, \Sigma)$	the normal distribution with mean μ and covariance Σ on an inner product space
$W(\Sigma, p, n)$	the Wishart distribution with n degrees of freedom and $p \times p$ parameter matrix Σ

CHAPTER 1

Vector Space Theory

In order to understand the structure and geometry of multivariate distributions and associated statistical problems, it is essential that we be able to distinguish those aspects of multivariate distributions that can be described without reference to a coordinate system and those that cannot. Finite dimensional vector space theory provides us with a framework in which it becomes relatively easy to distinguish between coordinate free and coordinate concepts. It is fair to say that the material presented in this chapter furnishes the language we use in the rest of this book to describe many of the geometric (coordinate free) and coordinate properties of multivariate probability models. The treatment of vector spaces here is far from complete, but those aspects of the theory that arise in later chapters are covered. Halmos (1958) has been followed quite closely in the first two sections of this chapter, and because of space limitations, proofs sometimes read "see Halmos (1958)."

The material in this chapter runs from the elementary notions of basis, dimension, linear transformation, and matrix to inner product space, orthogonal projection, and the spectral theorem for self-adjoint linear transformations. In particular, the linear space of linear transformations is studied in detail, and the chapter ends with a discussion of what is commonly known as the singular value decomposition theorem. Most of the vector spaces here are finite dimensional real vector spaces, although excursions into infinite dimensions occur via applications of the Cauchy–Schwarz Inequality. As might be expected, we introduce complex coordinate spaces in the discussion of determinants and eigenvalues.

Multilinear algebra and tensors are not covered systematically, although the outer product of vectors and the Kronecker product of linear transformations are covered. It was felt that the simplifications and generality obtained by introducing tensors were not worth the price in terms of added notation, vocabulary, and abstractness.

1.1. VECTOR SPACES

Let R denote the set of real numbers. Elements of R, called scalars, are denoted by $\alpha, \beta, \ldots$.

Definition 1.1. A set V, whose elements are called vectors, is called a real vector space if:

(I) to each pair of vectors $x, y \in V$, there is a vector $x + y \in V$, called the sum of x and y, and for all vectors in V,

(i) $x + y = y + x$.
(ii) $(x + y) + z = x + (y + z)$.
(iii) There exists a unique vector $0 \in V$ such that $x + 0 = x$ for all x.
(iv) For each $x \in V$, there is a unique vector $-x$ such that $x + (-x) = 0$.

(II) For each $\alpha \in R$ and $x \in V$, there is a vector denoted by $\alpha x \in V$, called the product of α and x, and for all scalars and vectors,

(i) $\alpha(\beta x) = (\alpha\beta)x$.
(ii) $1x = x$.
(iii) $(\alpha + \beta)x = \alpha x + \beta x$.
(iv) $\alpha(x + y) = \alpha x + \alpha y$.

In II(iii), $(\alpha + \beta)x$ means the sum of the two scalars, α and β, times x, while $\alpha x + \beta x$ means the sum of the two vectors, αx and βx. This multiple use of the plus sign should not cause any confusion. The reason for calling V a real vector space is that multiplication of vectors by real numbers is permitted.

A classical example of a real vector space is the set R^n of all ordered n-tuples of real numbers. An element of R^n, say x, is represented as

$$x = \begin{pmatrix} x_1 \\ x_2 \\ \vdots \\ x_n \end{pmatrix}, \qquad x_i \in R, \quad i = 1, \ldots, n,$$

and x_i is called the ith coordinate of x. The vector $x + y$ has ith coordinate $x_i + y_i$ and αx, $\alpha \in R$, is the vector with coordinates αx_i, $i = 1, \ldots, n$. With

$0 \in R^n$ representing the vector of all zeroes, it is routine to check that R^n is a real vector space. Vectors in the coordinate space R^n are always represented by a column of n real numbers as indicated above. For typographical convenience, a vector is often written as a row and appears as $x' = (x_1, \ldots, x_n)$. The prime denotes the *transpose* of the vector $x \in R^n$.

The following example provides a method of constructing real vector spaces and yields the space R^n as a special case.

◆ **Example 1.1.** Let $\mathfrak{X}$ be a set. The set V is the collection of all the real-valued functions defined on $\mathfrak{X}$. For any two elements $x_1, x_2 \in V$, define $x_1 + x_2$ as the function on $\mathfrak{X}$ whose value at t is $x_1(t) + x_2(t)$. Also, if $\alpha \in R$ and $x \in V$, αx is the function on $\mathfrak{X}$ given by $(\alpha x)(t) \equiv \alpha x(t)$. The symbol $0 \in V$ is the zero function. It is easy to verify that V is a real vector space with these definitions of addition and scalar multiplication. When $\mathfrak{X} = \{1, 2, \ldots, n\}$, then V is just the real vector space R^n and $x \in R^n$ has as its ith coordinate the value of x at $i \in \mathfrak{X}$. Every vector space discussed in the sequel is either V (for some set $\mathfrak{X}$) or a linear subspace (to be defined in a moment) of some V. ◆

Before defining the dimension of a vector space, we need to discuss linear dependence and independence. The treatment here follows Halmos (1958, Sections 5–9). Let V be a real vector space.

Definition 1.2. A finite set of vectors $\{x_i | i = 1, \ldots, k\}$ is *linearly dependent* if there exist real numbers $\alpha_1, \ldots, \alpha_k$, not all zero, such that $\Sigma \alpha_i x_i = 0$. Otherwise, $\{x_i | i = 1, \ldots, k\}$ is *linearly independent*.

A brief word about summation notation. Ordinarily, we do not indicate indices of summation on a summation sign when the range of summation is clear from the context. For example, in Definition 1.2, the index i was specified to range between 1 and k before the summation on i appeared; hence, no range was indicated on the summation sign.

An arbitrary subset $S \subseteq V$ is linearly independent if every finite subset of S is linearly independent. Otherwise, S is linearly dependent.

Definition 1.3. A *basis* for a vector space V is a linearly independent set S such that every vector in V is a linear combination of elements of S. V is *finite dimensional* if it has a finite set S that is a basis.

◆ **Example 1.2.** Take $V = R^n$ and let $\varepsilon_i' = (0, \ldots, 0, 1, 0, \ldots, 0)$ where the one occurs as the ith coordinate of ε_i, $i = 1, \ldots, n$. For $x \in R^n$, it is clear that $x = \Sigma x_i \varepsilon_i$ where x_i is the ith coordinate of x. Thus every vector in R^n is a linear combination of $\varepsilon_1, \ldots, \varepsilon_n$. To show that $\{\varepsilon_i | i = 1, \ldots, n\}$ is a linearly independent set, suppose $\Sigma \alpha_i \varepsilon_i = 0$ for some scalars α_i, $i = 1, \ldots, n$. Then $x = \Sigma \alpha_i \varepsilon_i = 0$ has α_i as its ith coordinate, so $\alpha_i = 0$, $i = 1, \ldots, n$. Thus $\{\varepsilon_i | i = 1, \ldots, n\}$ is a basis for R^n and R^n is finite dimensional. The basis $\{\varepsilon_i | i = 1, \ldots, n\}$ is called the *standard basis* for R^n. ◆

Let V be a finite dimensional real vector space. The basic properties of linearly independent sets and bases are:

(i) If $\{x_1, \ldots, x_m\}$ is a linearly independent set in V, then there exist vectors $x_{m+1}, \ldots, x_{m+k}$ such that $\{x_1, \ldots, x_{m+k}\}$ is a basis for V.

(ii) All bases for V have the same number of elements. The *dimension* of V is defined to be the number of elements in any basis.

(iii) Every set of $n + 1$ vectors in an n-dimensional vector space is linearly dependent.

Proofs of the above assertions can be found in Halmos (1958, Sections 5–8). The dimension of a finite dimensional vector space is denoted by $\dim(V)$. If $\{x_1, \ldots, x_n\}$ is a basis for V, then every $x \in V$ is a unique linear combination of $\{x_1, \ldots, x_n\}$—say $x = \Sigma \alpha_i x_i$. That every x can be so expressed follows from the definition of a basis and the uniqueness follows from the linear independence of $\{x_1, \ldots, x_n\}$. The numbers $\alpha_1, \ldots, \alpha_n$ are called the *coordinates* of x in the basis $\{x_1, \ldots, x_n\}$. Clearly, the coordinates of x depend on the order in which we write the basis. Thus by a basis we always mean an ordered basis.

We now introduce the notion of a subspace of a vector space.

Definition 1.4. A nonempty subset $M \subseteq V$ is a *subspace* (or *linear manifold*) of V if, for each $x, y \in M$ and $\alpha, \beta \in R$, $\alpha x + \beta y \in M$.

A subspace M of a real vector space V is easily shown to satisfy the vector space axioms (with addition and scalar multiplication inherited from V), so subspaces are real vector spaces. It is not difficult to verify the following assertions (Halmos, 1958, Sections 10–12):

(i) The intersection of subspaces is a subspace.

(ii) If M is a subspace of a finite dimensional vector space V, then $\dim(M) \leqslant \dim(V)$.

(iii) Given an m-dimensional subspace M of an n-dimensional vector space V, there is a basis $\{x_1, \ldots, x_m, \ldots, x_n\}$ for V such that $\{x_1, \ldots, x_m\}$ is a basis for M.

Given any set $S \subseteq V$, span(S) is defined to be the intersection of all the subspaces that contain S—that is, span(S) is the smallest subspace that contains S. It is routine to show that span(S) is equal to the set of all linear combinations of elements of S. The subspace span(S) is often called the subspace spanned by the set S.

If M and N are subspaces of V, then span$(M \cup N)$ is the set of all vectors of the form $x + y$ where $x \in M$ and $y \in N$. The suggestive notation $M + N \equiv \{z | z = x + y, x \in M, y \in N\}$ is used for span$(M \cup N)$ when M and N are subspaces. Using the fact that a linearly independent set can be extended to a basis in a finite dimensional vector space, we have the following. Let V be finite dimensional and suppose M and N are subspaces of V.

(i) Let $m = \dim(M)$, $n = \dim(N)$, and $k = \dim(M \cap N)$. Then there exist vectors $x_1, \ldots, x_k, y_{k+1}, \ldots, y_m$, and $z_{k+1}, \ldots, z_n$ such that $\{x_1, \ldots, x_k\}$ is a basis for $M \cap N$, $\{x_1, \ldots, x_k, y_{k+1}, \ldots, y_m\}$ is a basis for M, $\{x_1, \ldots, x_k, z_{k+1}, \ldots, z_n\}$ is a basis for N, and $\{x_1, \ldots, x_k, y_{k+1}, \ldots, y_m, z_{k+1}, \ldots, z_n\}$ is a basis for $M + N$. If $k = 0$, then $\{x_1, \ldots, x_k\}$ is interpreted as the empty set.

(ii) $\dim(M + N) = \dim(M) + \dim(N) - \dim(M \cap N)$.

(iii) There exists a subspace $M_1 \subseteq V$ such that $M \cap M_1 = \{0\}$ and $M + M_1 = V$.

Definition 1.5. If M and N are subspaces of V that satisfy $M \cap N = \{0\}$ and $M + N = V$, then M and N are *complementary* subspaces.

The technique of decomposing a vector space into two (or more) complementary subspaces arises again and again in the sequel. The basic property of such a decomposition is given in the following proposition.

Proposition 1.1. Suppose M and N are complementary subspaces in V. Then each $x \in V$ has a unique representation $x = y + z$ with $y \in M$ and $z \in N$.

Proof. Since $M + N = V$, each $x \in V$ can be written $x = y_1 + z_1$ with $y_1 \in M$ and $z_1 \in N$. If $x = y_2 + z_2$ with $y_2 \in M$ and $z_2 \in N$, then $0 = x -$

$x = (y_1 - y_2) + (z_1 - z_2)$. Hence $(y_2 - y_1) = (z_1 - z_2)$ so $(y_2 - y_1) \in M \cap N = \{0\}$. Thus $y_1 = y_2$. Similarly, $z_1 = z_2$. □

The above proposition shows that we can decompose the vector space V into two vector spaces M and N and each x in V has a unique piece in M and in N. Thus x can be represented as (y, z) with $y \in M$ and $z \in N$. Also, note that if $x_1, x_2 \in V$ and have the representations (y_1, z_1), (y_2, z_2), then $\alpha x_1 + \beta x_2$ has the representation $(\alpha y_1 + \beta y_2, \alpha z_1 + \beta z_2)$, for $\alpha, \beta \in R$. In other words the function that maps x into its decomposition (y, z) is linear. To make this a bit more precise, we now define the direct sum of two vector spaces.

Definition 1.6. Let V_1 and V_2 be two real vector spaces. The *direct sum* of V_1 and V_2, denoted by $V_1 \oplus V_2$, is the set of all ordered pairs $\{x, y\}$, $x \in V_1$, $y \in V_2$, with the linear operations defined by $\alpha_1\{x_1, y_1\} + \alpha_2\{x_2, y_2\} \equiv \{\alpha_1 x_1 + \alpha_2 x_2, \alpha_1 y_1 + \alpha_2 y_2\}$.

That $V_1 \oplus V_2$ is a real vector space with the above operations can easily be verified. Further, identifying V_1 with $\{\{x_1, 0\} | x \in V_1\} \equiv \tilde{V}_1$ and V_2 with $\{\{0, y\} | y \in V_2\} \equiv \tilde{V}_2$, we can think of V_1 and V_2 as complementary subspaces of $V_1 \oplus V_2$, since $\tilde{V}_1 + \tilde{V}_2 = V_1 \oplus V_2$ and $\tilde{V}_1 \cap \tilde{V}_2 = \{0, 0\}$, which is the zero element in $V_1 \oplus V_2$. The relation of the direct sum to our previous decomposition of a vector space should be clear.

◆ **Example 1.3.** Consider $V = R^n$, $n \geqslant 2$, and let p and q be positive integers such that $p + q = n$. Then R^p and R^q are both real vector spaces. Each element of R^n is a n-tuple of real numbers, and we can construct subspaces of R^n by setting some of these coordinates equal to zero. For example, consider $M = \{x \in R^n | x = \begin{pmatrix} y \\ 0 \end{pmatrix}$ with $y \in R^p, 0 \in R^q\}$ and $N = \{x \in R^n | x = \begin{pmatrix} 0 \\ z \end{pmatrix}$ with $0 \in R^p$ and $z \in R^q\}$. It is clear that $\dim(M) = p$, $\dim(N) = q$, $M \cap N = \{0\}$, and $M + N = R^n$. The identification of R^p with M and R^q with N shows that it is reasonable to write $R^p \oplus R^q = R^{p+q}$. ◆

1.2. LINEAR TRANSFORMATIONS

Linear transformations occupy a central position, both in vector space theory and in multivariate analysis. In this section, we discuss the basic properties of linear transforms, leaving the deeper results for consideration after the introduction of inner products. Let V and W be real vector spaces.

Definition 1.7. Any function A defined on V and taking values in W is called a *linear transformation* if $A(\alpha_1 x_1 + \alpha_2 x_2) = \alpha_1 A(x_1) + \alpha_2 A(x_2)$ for all $x_1, x_2 \in V$ and $\alpha_1, \alpha_2 \in R$.

Frequently, $A(x)$ is written Ax when there is no danger of confusion. Let $\mathcal{L}(V, W)$ be the set of all linear transformations on V to W. For two linear transformations A_1 and A_2 in $\mathcal{L}(V, W)$, $A_1 + A_2$ is defined by $(A_1 + A_2)(x) = A_1 x + A_2 x$ and $(\alpha A)(x) = \alpha A x$ for $\alpha \in R$. The zero linear transformation is denoted by 0. It should be clear that $\mathcal{L}(V, W)$ is a real vector space with these definitions of addition and scalar multiplication.

◆ **Example 1.4.** Suppose $\dim(V) = m$ and let $x_1, \ldots, x_m$ be a basis for V. Also, let $y_1, \ldots, y_m$ be arbitrary vectors in W. The claim is that there is a unique linear transformation A such that $Ax_i = y_i$, $i = 1, \ldots, m$. To see this, consider $x \in V$ and express x as a unique linear combination of the basis vectors, $x = \Sigma \alpha_i x_i$. Define A by

$$Ax = \sum_{i}^{n} \alpha_i A x_i = \sum_{1}^{n} \alpha_i y_i.$$

The linearity of A is easy to check. To show that A is unique, let B be another linear transformation with $Bx_i = y_i$, $i = 1, \ldots, n$. Then $(A - B)(x_i) = 0$ for $i = 1, \ldots, n$, and $(A - B)(x) = (A - B)(\Sigma \alpha_i x_i) = \Sigma \alpha_i (A - B)(x_i) = 0$ for all $x \in V$. Thus $A = B$. ◆

The above example illustrates a general principle—namely, a linear transformation is completely determined by its values on a basis. This principle is used often to construct linear transformations with specified properties. A modification of the construction in Example 1.4 yields a basis for $\mathcal{L}(V, W)$ when V and W are both finite dimensional. This basis is given in the proof of the following proposition.

Proposition 1.2. If $\dim(V) = m$ and $\dim(W) = n$, then $\dim(\mathcal{L}(V, W)) = mn$.

Proof. Let $x_1, \ldots, x_m$ be a basis for V and let $y_1, \ldots, y_n$ be a basis for W. Define a linear transformation A_{ji}, $i = 1, \ldots, m$ and $j = 1, \ldots, n$, by

$$A_{ji}(x_k) = \begin{cases} 0 & \text{if } k \neq i \\ y_j & \text{if } k = i \end{cases}.$$

For each (j, i), A_{ji} has been defined on a basis in V so the linear transformation A_{ji} is uniquely determined. We now claim that $\{A_{ji}|i = 1,\dots, m; j = 1,\dots, n\}$ is a basis for $\mathcal{L}(V, W)$. To show linear independence, suppose $\Sigma\Sigma\alpha_{ji}A_{ji} = 0$. Then for each $k = 1,\dots, m$,

$$0 = \sum_j \sum_i \alpha_{ji} A_{ji}(x_k) = \sum_j \alpha_{jk} y_j.$$

Since $\{y_1,\dots, y_m\}$ is a linearly independent set, this implies that $\alpha_{jk} = 0$ for all j and k. Thus linear independence holds. To show every $A \in \mathcal{L}(V, W)$ is a linear combination of the A_{ji}, first note that Ax_k is a vector in W and thus is a unique linear combination of $y_1,\dots, y_n$, say $Ax_k = \Sigma_j a_{jk} y_j$ where $a_{jk} \in R$. However, the linear transformation $\Sigma\Sigma a_{ji}A_{ji}$ evaluated at x_k is

$$\sum_j \sum_i a_{ji} A_{ji}(x_k) = \sum_j a_{jk} y_j.$$

Since A and $\Sigma\Sigma a_{ji}A_{ji}$ agree on a basis in V, they are equal. This completes the proof since there are mn elements in the basis $\{A_{ji}|i = 1,\dots, m; j = 1,\dots, n\}$ for $\mathcal{L}(V, W\}$. □

Since $\mathcal{L}(V, W)$ is a vector space, general results about vector spaces, of course, apply to $\mathcal{L}(V, W)$. However, linear transformations have many interesting properties not possessed by vectors in general. For example, consider vector spaces V_i, $i = 1, 2, 3$. If $A \in \mathcal{L}(V_1, V_2)$ and $B \in \mathcal{L}(V_2, V_3)$, then we can compose the functions B and A by defining $(BA)(x) = B(A(x))$. The linearity of A and B implies that BA is a linear transformation on V_1 to V_3—that is, $BA \in \mathcal{L}(V_1, V_3)$. Usually, BA is called the product of B and A.

There are two special cases of $\mathcal{L}(V, W)$ that are of particular interest. First, if $A, B \in \mathcal{L}(V, V)$, then $AB \in \mathcal{L}(V, V)$ and $BA \in \mathcal{L}(V, V)$, so we have a multiplication defined in $\mathcal{L}(V, V)$. However, this multiplication is not commutative—that is, AB is not, in general, equal to BA. Clearly, $A(B + C) = AB + AC$ for $A, B, C \in \mathcal{L}(V, V)$. The identity linear transformation in $\mathcal{L}(V, V)$, usually denoted by I, satisfies $AI = IA = A$ for all $A \in \mathcal{L}(V, V)$, since $Ix = x$ for all $x \in V$. Thus $\mathcal{L}(V, V)$ is not only a vector space, but there is a multiplication defined in $\mathcal{L}(V, V)$.

The second special case of $\mathcal{L}(V, W)$ we wish to consider is when $W = R$—that is, W is the one-dimensional real vector space R. The space $\mathcal{L}(V, R)$ is called the *dual space* of V and, if $\dim(V) = n$, then $\dim(\mathcal{L}(V, R)) = n$. Clearly, $\mathcal{L}(V, R)$ is the vector space of all real-valued linear functions defined on V. We have more to say about $\mathcal{L}(V, R)$ after the introduction of inner products on V.

Understanding the geometry of linear transformations usually begins with a specification of the range and null space of the transformation. These objects are now defined. Let $A \in \mathcal{L}(V, W)$ where V and W are finite dimensional.

Definition 1.8. The *range of A*, denoted by $\mathcal{R}(A)$, is

$$\mathcal{R}(A) \equiv \{u | u \in W, Ax = u \text{ for some } x \in V\}.$$

The *null space of A*, denoted by $\mathcal{N}(A)$, is

$$\mathcal{N}(A) \equiv \{x | x \in V, Ax = 0\}.$$

It is routine to verify that $\mathcal{R}(A)$ is a subspace of W and $\mathcal{N}(A)$ is a subspace of V. The *rank of A*, denoted by $r(A)$, is the dimension of $\mathcal{R}(A)$.

Proposition 1.3. If $A \in \mathcal{L}(V, W)$ and $n = \dim(V)$, then $r(A) + \dim(\mathcal{N}(A)) = n$.

Proof. Let M be a subspace of V such that $M \oplus \mathcal{N}(A) = V$, and consider a basis $\{x_1, \ldots, x_k\}$ for M. Since $\dim(M) + \dim(\mathcal{N}(A)) = n$, we need to show that $k = r(A)$. To do this, it is sufficient to show that $\{Ax_1, \ldots, Ax_k\}$ is a basis for $\mathcal{R}(A)$. If $0 = \Sigma\alpha_i Ax_i = A(\Sigma\alpha_i x_i)$, then $\Sigma\alpha_i x_i \in M \cap \mathcal{N}(A)$ so $\Sigma\alpha_i x_i = 0$. Hence $\alpha_1 = \cdots = \alpha_k = 0$ as $\{x_1, \ldots, x_k\}$ is a basis for M. Thus $\{Ax_1, \ldots, Ax_k\}$ is a linearly independent set. To verify that $\{Ax_1, \ldots, Ax_k\}$ spans $\mathcal{R}(A)$, suppose $w \in \mathcal{R}(A)$. Then $w = Ax$ for some $x \in V$. Write $x = y + z$ where $y \in M$ and $z \in \mathcal{N}(A)$. Then $w = A(y + z) = Ay$. Since $y \in M$, $y = \Sigma\alpha_i x_i$ for some scalars $\alpha_1, \ldots, \alpha_k$. Therefore, $w = A(\Sigma\alpha_i x_i) = \Sigma\alpha_i Ax_i$. □

Definition 1.9. A linear transformation $A \in \mathcal{L}(V, V)$ is called *invertible* if there exists a linear transformation, denoted by A^{-1}, such that $AA^{-1} = A^{-1}A = I$.

The following assertions hold; see Halmos (1958, Section 36):

(i) A is invertible iff $\mathcal{R}(A) = V$ iff $Ax = 0$ implies $x = 0$.

(ii) If $A, B, C \in \mathcal{L}(V, V)$ and if $AB = CA = I$, then A is invertible and $B = C = A^{-1}$.

(iii) If A and B are invertible, then AB is invertible and $(AB)^{-1} = B^{-1}A^{-1}$. If A is invertible and $\alpha \neq 0$, then $(\alpha A)^{-1} = \alpha^{-1}A^{-1}$ and $(A^{-1})^{-1} = A$.

In terms of bases, invertible transformations are characterized by the following.

Proposition 1.4. Let $A \in \mathcal{L}(V, V)$ and suppose $\{x_1, \dots, x_n\}$ is a basis for V. The following are equivalent:

(i) A is invertible.
(ii) $\{Ax_1, \dots, Ax_n\}$ is a basis for V.

Proof. Suppose A is invertible. Since $\dim(V) = n$, we must show $\{Ax_1, \dots, Ax_n\}$ is a linearly independent set. Thus if $0 = \Sigma\alpha_i Ax_i = A(\Sigma\alpha_i x_i)$, then $\Sigma\alpha_i x_i = 0$ since A is invertible. Hence $\alpha_i = 0$, $i = 1, \dots, n$, as $\{x_1, \dots, x_n\}$ is a basis for V. Therefore, $\{Ax_1, \dots, Ax_n\}$ is a basis.

Conversely, suppose $\{Ax_1, \dots, Ax_n\}$ is a basis. We show that $Ax = 0$ implies $x = 0$. First, write $x = \Sigma\alpha_i x_i$ so $Ax = 0$ implies $\Sigma\alpha_i Ax_i = 0$. Hence $\alpha_i = 0$, $i = 1, \dots, n$, as $\{Ax_1, \dots, Ax_n\}$ is a basis. Thus $x = 0$, so A is invertible. □

We now introduce real matrices and consider their relation to linear transformations. Consider vector spaces V and W of dimension m and n, respectively, and bases $\{x_1, \dots, x_m\}$ and $\{y_1, \dots, y_n\}$ for V and W. Each $x \in V$ has a unique representation $x = \Sigma\alpha_i x_i$. Let $[x]$ denote the column vector of coordinates of x in the given basis. Thus $[x] \in R^m$ and the ith coordinate of $[x]$ is α_i, $i = 1, \dots, m$. Similarly, $[y] \in R^n$ is the column vector of $y \in W$ in the basis $\{y_1, \dots, y_n\}$. Consider $A \in \mathcal{L}(V, W)$ and express Ax_j in the given basis of W; $Ax_j = \Sigma_i a_{ij} y_i$ for unique scalars a_{ij}, $i = 1, \dots, n$, $j = 1, \dots, m$. The $n \times m$ rectangular array of real scalars

$$[A] = \begin{bmatrix} a_{11} & a_{12} & \cdots & a_{1m} \\ a_{21} & a_{22} & \cdots & a_{2m} \\ \vdots & & & \\ a_{n1} & & \cdots & a_{nm} \end{bmatrix} \equiv \{a_{ij}\}$$

is called the *matrix* of A relative to the two given bases. Conversely, given any $n \times m$ rectangular array of real scalars $\{a_{ij}\}$, $i = 1, \dots, n$, $j = 1, \dots, m$, the linear transformation A defined by $Ax_j = \Sigma_i a_{ij} y_i$ has as its matrix $[A] = \{a_{ij}\}$.

Definition 1.10. A rectangular array $\{a_{ij}\} : m \times n$ of real scalars is called an $m \times n$ *matrix*. If $A = \{a_{ij}\} : m \times n$ is a matrix and $B = \{b_{ij}\} : n \times p$ is a matrix, then $C = AB$, called the *matrix product* of A and B (in that order) is defined to be the matrix $\{c_{ij}\} : m \times p$ with $c_{ij} = \Sigma_k a_{ik} b_{kj}$.

In this book, the distinction between linear transformations, matrices, and the matrix of a linear transformation is always made. The notation $[A]$ means the matrix of a linear transformation with respect to two given bases. However, symbols like A, B, or C may represent either linear transformations or real matrices; care is taken to clearly indicate which case is under consideration.

Each matrix $A = \{a_{ij}\}: m \times n$ defines a linear transformation on R^n to R^m as follows. For $x \in R^n$ with coordinates $x_1, \ldots, x_n$, Ax is the vector y in R^m with coordinates $y_i = \Sigma_j a_{ij} x_j$, $i = 1, \ldots, m$. Of course, this is the usual row by column rule of a matrix operating on a vector. The matrix of this linear transformation in the standard bases for R^n and R^m is just the matrix A. However, if the bases are changed, then the matrix of the linear transformation changes. When $m = n$, the matrix $A = \{a_{ij}\}$ determines a linear transformation on R^n to R^n via the above definition of a matrix times a vector. The matrix A is called nonsingular (or invertible) if there exists a matrix, denoted by A^{-1}, such that $AA^{-1} = A^{-1}A = I_n$ where I_n is the $n \times n$ identity matrix consisting of ones on the diagonal and zeroes off the diagonal. As with linear transformations, A^{-1} is unique and exists iff $Ax = 0$ implies $x = 0$.

The symbol $\mathcal{L}_{n,m}$ denotes the real vector space of $m \times n$ real matrices with the usual operations of addition and scalar multiplication. In other words, if $A = \{a_{ij}\}$ and $B = \{b_{ij}\}$ are elements of $\mathcal{L}_{n,m}$, then $A + B = \{a_{ij} + b_{ij}\}$ and $\alpha A = \{\alpha a_{ij}\}$. Notice that $\mathcal{L}_{n,m}$ is the set of $m \times n$ matrices (m and n are in reverse order). The reason for writing $\mathcal{L}_{n,m}$ is that an $m \times n$ matrix determines a linear transformation from R^n to R^m. We have made the choice of writing $\mathcal{L}(V, W)$ for linear transformations from V to W, and it is an unpleasant fact that the dimensions of a matrix occur in reverse order to the dimensions of the spaces V and W. The next result summarizes the relations between linear transformations and matrices.

Proposition 1.5. Consider vector spaces V_1, V_2, and V_3 with bases $\{x_1, \ldots, x_{n_1}\}$, $\{y_1, \ldots, y_{n_2}\}$, and $\{z_1, \ldots, z_{n_3}\}$, respectively. For $x \in V_1$, $y \in V_2$, and $z \in V_3$, let $[x]$, $[y]$, and $[z]$ denote the vector of coordinates of x, y, and z in the given bases, so $[x] \in R^{n_1}$, $[y] \in R^{n_2}$, and $[z] \in R^{n_3}$. For $A \in \mathcal{L}(V_1, V_2)$ and $B \in \mathcal{L}(V_2, V_3)$ let $[A]$ ($[B]$) denote the matrix of $A(B)$ relative to the bases $\{x_1, \ldots, x_{n_1}\}$ and $\{y_1, \ldots, y_{n_2}\}$ ($\{y_1, \ldots, y_{n_2}\}$ and $\{z_1, \ldots, z_{n_3}\}$). Then:

(i) $[Ax] = [A][x]$.

(ii) $[BA] = [B][A]$.

(iii) If $V_1 = V_2$ and A is invertible, $[A^{-1}] = [A]^{-1}$. Here, $[A^{-1}]$ and $[A]$ are matrices in the bases $\{x_1, \ldots, x_{n_1}\}$ and $\{x_1, \ldots, x_{n_1}\}$.

Proof. A few words are in order concerning the notation in (i), (ii), and (iii). In (i), $[Ax]$ is the vector of coordinates of $Ax \in V_2$ with respect to the basis $\{y_1, \ldots, y_{n_2}\}$ and $[A][x]$ means the matrix $[A]$ times the coordinate vector $[x]$ as defined previously. Since both sides of (i) are linear in x, it suffices to verify (i) for $x = x_j$, $j = 1, \ldots, n_1$. But $[A][x_j]$ is just the column vector with coordinates a_{ij}, $i = 1, \ldots, n_2$, and $Ax_j = \Sigma_i a_{ij} y_i$, so $[Ax_j]$ is the column vector with coordinates a_{ij}, $i = 1, \ldots, n_2$. Hence (i) holds.

For (ii), $[B][A]$ is just the matrix product of $[B]$ and $[A]$. Also, $[BA]$ is the matrix of the linear transformation $BA \in \mathcal{L}(V_1, V_3)$ with respect to the bases $\{x_1, \ldots, x_{n_1}\}$ and $\{z_1, \ldots, z_{n_3}\}$. To show that $[BA] = [B][A]$, we must verify that, for all $x \in V$, $[BA][x] = [B][A][x]$. But by (i), $[BA][x] = [BAx]$ and, using (i) twice, $[B][A][x] = [B][Ax] = [BAx]$. Thus (ii) is established.

In (iii), $[A]^{-1}$ denotes the inverse of the matrix $[A]$. Since A is invertible, $AA^{-1} = A^{-1}A = I$ where I is the identity linear transformation on V_1 to V_1. Thus by (ii), with I_n denoting the $n \times n$ identity matrix, $I_n = [I] = [AA^{-1}] = [A][A^{-1}] = [A^{-1}A] = [A^{-1}][A]$. By the uniqueness of the matrix inverse, $[A^{-1}] = [A]^{-1}$. □

Projections are the final topic in this section. If V is a finite dimensional vector space and M and N are subspaces of V such that $M \oplus N = V$, we have seen that each $x \in V$ has a unique piece in M and a unique piece in N. In other words, $x = y + z$ where $y \in M$, $z \in N$, and y and z are unique.

Definition 1.11. Given subspaces M and N in V such that $M \oplus N = V$, if $x = y + z$ with $y \in M$ and $z \in N$, then y is called the *projection* of x on M along N and z is called the projection of x on N along M.

Since M and N play symmetric roles in the above definition, we concentrate on the projection on M.

Proposition 1.6. The function P mapping V into V whose value at x is the projection of x on M along N is a linear transformation that satisfies

(i) $\mathcal{R}(P) = M$, $\mathcal{N}(P) = N$.
(ii) $P^2 = P$.

Proof. We first show that P is linear. If $x = y + z$ with $y \in M$, $z \in N$, then by definition, $Px = y$. Also, if $x_1 = y_1 + z_1$ and $x_2 = y_2 + z_2$ are the decompositions of x_1 and x_2, respectively, then $\alpha_1 x_1 + \alpha_2 x_2 = (\alpha_1 y_1 + \alpha_2 y_2) + (\alpha_1 z_1 + \alpha_2 z_2)$ is the decomposition of $\alpha_1 x_1 + \alpha_2 x_2$. Thus $P(\alpha_1 x_1 + \alpha_2 x_2) = \alpha_1 P x_1 + \alpha_2 P x_2$ so P is linear. By definition $Px \in M$, so $\mathcal{R}(P)$

$\subseteq M$. But if $x \in M$, $Px = x$ and $\mathcal{R}(P) = M$. Also, if $x \in N$, $Px = 0$ so $\mathcal{N}(P) \supseteq N$. However, if $Px = 0$, then $x = 0 + x$, and therefore $x \in N$. Thus $\mathcal{N}(P) = N$. To show $P^2 = P$, note that $Px \in M$ and $Px = x$ for $x \in M$. Hence, $Px = P(Px) = P^2x$, which implies that $P = P^2$. □

A converse to Proposition 1.6 gives a complete description of all linear transformations on V to V that satisfy $A^2 = A$.

Proposition 1.7. If $A \in \mathcal{L}(V, V)$ and satisfies $A^2 = A$, then $\mathcal{R}(A) \oplus \mathcal{N}(A) = V$ and A is the projection on $\mathcal{R}(A)$ along $\mathcal{N}(A)$.

Proof. To show $\mathcal{R}(A) \oplus \mathcal{N}(A) = V$, we must verify that $\mathcal{R}(A) \cap \mathcal{N}(A) = \{0\}$ and that each $x \in V$ is the sum of a vector in $\mathcal{R}(A)$ and a vector in $\mathcal{N}(A)$. If $x \in \mathcal{R}(A) \cap \mathcal{N}(A)$, then $x = Ay$ for some $y \in V$ and $Ax = 0$. Since $A^2 = A$, $0 = Ax = A^2y = Ay = x$ and $\mathcal{R}(A) \cap \mathcal{N}(A) = \{0\}$. For $x \in V$, write $x = Ax + (I - A)x$ and let $y = Ax$ and $z = (I - A)x$. Then $y \in \mathcal{R}(A)$ by definition and $Az = A(I - A)x = (A - A^2)x = 0$, so $z \in \mathcal{N}(A)$. Thus $\mathcal{R}(A) \oplus \mathcal{N}(A) = V$.

The verification that A is the projection on $\mathcal{R}(A)$ along $\mathcal{N}(A)$ goes as follows. A is zero on $\mathcal{N}(A)$ by definition. Also, for $x \in \mathcal{R}(A)$, $x = Ay$ for some $y \in V$. Thus $Ax = A^2y = Ay = x$, so $Ax = x$ and $x \in \mathcal{R}(A)$. However, the projection on $\mathcal{R}(A)$ along $\mathcal{N}(A)$, say P, also satisfies $Px = x$ for $x \in \mathcal{R}(A)$ and $Px = 0$ for $x \in \mathcal{N}(A)$. This implies that $P = A$ since $\mathcal{R}(A) \oplus \mathcal{N}(A) = V$. □

The above proof shows that the projection on M along N is the unique linear transformation that is the identity on M and zero on N. Also, it is clear that P is the projection on M along N iff $I - P$ is the projection on N along M.

1.3. INNER PRODUCT SPACES

The discussion of the previous section was concerned mainly with the linear aspects of vector spaces. Here, we introduce inner products on vector spaces so that the geometric notions of length, angle, and orthogonality become meaningful. Let us begin with an example.

◆ **Example 1.5.** Consider coordinate space R^n with the standard basis $\{\varepsilon_1, \ldots, \varepsilon_n\}$. For $x, y \in R^n$, define $x'y \equiv \sum x_i y_i$ where x and y have coordinates $x_1, \ldots, x_n$ and $y_1, \ldots, y_n$. Of course, x' is the transpose of the vector x and $x'y$ can be thought of as the $1 \times n$

matrix x' times the $n \times 1$ matrix y. The real number $x'y$ is sometimes called the scalar product (or inner product) of x and y. Some properties of the scalar product are:

(i) $x'y = y'x$ (symmetry).
(ii) $x'y$ is linear in y for fixed x and linear in x for fixed y.
(iii) $x'x = \Sigma_1^n x_i^2 \geqslant 0$ and is zero iff $x = 0$.

The *norm* of x, defined by $\|x\| = (x'x)^{1/2}$, can be thought of as the distance between x and $0 \in R^n$. Hence, $\|x - y\| = (\Sigma(x_i - y_i)^2)^{1/2}$ is usually called the distance between x and y. When x and y are both not zero, then the cosine of the angle between x and y is $x'y/\|x\|\|y\|$ (see Halmos, 1958, p. 118). Thus we have a geometric interpretation of the scalar product. In particular, the angle between x and y is $\pi/2(\cos \pi/2 = 0)$ iff $x'y = 0$. Thus we say x and y are orthogonal (perpendicular) iff $x'y = 0$. ◆

Let V be a real vector space. An inner product on V is obtained by simply abstracting the properties of the scalar product on R^n.

Definition 1.12. An *inner product* on a real vector space V is a real valued function on $V \times V$, denoted by $(\cdot, \cdot)$, with the following properties:

(i) $(x, y) = (y, x)$ (symmetry).
(ii) $(\alpha_1 x_1 + \alpha_2 x_2, y) = \alpha_1(x_1, y) + \alpha_2(x_2, y)$ (linearity).
(iii) $(x, x) \geqslant 0$ and $(x, x) = 0$ only if $x = 0$ (positivity).

From (i) and (ii) it follows that $(x, \alpha_1 y_1 + \alpha_2 y_2) = \alpha_1(x, y_1) + \alpha_2(x, y_2)$. In other words, inner products are linear in each variable when the other variable is fixed. The *norm* of x, denoted by $\|x\|$, is defined to be $\|x\| = (x, x)^{1/2}$ and the *distance* between x and y is $\|x - y\|$. Hence geometrically meaningful names and properties related to the scalar product on R^n have become definitions on V. To establish the existence of inner products on finite dimensional vector spaces, we have the following proposition.

Proposition 1.8. Suppose $\{x_1, \ldots, x_n\}$ is a basis for the real vector space V. The function $(\cdot, \cdot)$ defined on $V \times V$ by $(x, y) = \Sigma_1^n \alpha_i \beta_i$, where $x = \Sigma \alpha_i x_i$ and $y = \Sigma \beta_i x_i$, is an inner product on V.

Proof. Clearly $(x, y) = (y, x)$. If $x = \Sigma \alpha_i x_i$ and $z = \Sigma \gamma_i x_i$, then $(\alpha x + \gamma z, y) = \Sigma(\alpha \alpha_i + \gamma \gamma_i)\beta_i = \alpha \Sigma \alpha_i \beta_i + \gamma \Sigma \gamma_i \beta_i = \alpha(x, y) + \gamma(z, y)$. This establishes the linearity. Also, $(x, x) = \Sigma\alpha_i^2$, which is zero iff all the α_i are zero and this is equivalent to x being zero. Thus $(\cdot, \cdot)$ is an inner product on V. □

A vector space V with a given inner product $(\cdot, \cdot)$ is called an *inner product space*.

Definition 1.13. Two vectors x and y in an inner product space $(V, (\cdot, \cdot))$ are *orthogonal*, written $x \perp y$, if $(x, y) = 0$. Two subsets S_1 and S_2 of V are *orthogonal*, written $S_1 \perp S_2$, if $x \perp y$ for all $x \in S_1$ and $y \in S_2$.

Definition 1.14. Let $(V, (\cdot, \cdot))$ be a finite dimensional inner product space. A set of vectors $\{x_1, \ldots, x_k\}$ is called an *orthonormal set* if $(x_i, x_j) = \delta_{ij}$ for $i, j = 1, \ldots, k$ where $\delta_{ij} = 1$ if $i = j$ and 0 if $i \neq j$. A set $\{x_1, \ldots, x_k\}$ is called an *orthonormal basis* if the set is both a basis and is orthonormal.

First note that an orthonormal set $\{x_1, \ldots, x_k\}$ is linearly independent. To see this, suppose $0 = \Sigma\alpha_i x_i$. Then $0 = (0, x_j) = (\Sigma\alpha_i x_i, x_j) = \Sigma\alpha_i(x_i, x_j) = \Sigma_i \alpha_i \delta_{ij} = \alpha_j$. Hence $\alpha_j = 0$ for $j = 1, \ldots, k$ and the set $\{x_1, \ldots, x_k\}$ is linearly independent.

In Proposition 1.8, the basis used to define the inner product is, in fact, an orthonormal basis for the inner product. Also, the standard basis for R^n is an orthonormal basis for the scalar product on R^n—this scalar product is called the *standard inner product* on R^n. An algorithm for constructing orthonormal sets from linearly independent sets is now given. It is known as the *Gram–Schmidt orthogonalization procedure*.

Proposition 1.9. Let $\{x_1, \ldots, x_k\}$ be a linearly independent set in the inner product space $(V, (\cdot, \cdot))$. Define vectors $y_1, \ldots, y_k$ as follows:

$$y_1 = \frac{x_1}{\|x_1\|}$$

and

$$y_{i+1} = \frac{x_{i+1} - \sum_{j=1}^{i} (x_{i+1}, y_j) y_j}{\|x_{i+1} - \sum_{j=1}^{i} (x_{i+1}, y_j) y_j\|},$$

for $i = 1, \ldots, k - 1$. Then $\{y_1, \ldots, y_k\}$ is an orthonormal set and $\text{span}\{x_1, \ldots, x_i\} = \text{span}\{y_1, \ldots, y_i\}$, $i = 1, \ldots, k$.

Proof. See Halmos (1958, Section 65). □

An immediate consequence of Proposition 1.9 is that if $\{x_1, \ldots, x_n\}$ is a basis for V, then $\{y_1, \ldots, y_n\}$ constructed above is an orthonormal basis for $(V, (\cdot, \cdot))$. If $\{y_1, \ldots, y_n\}$ is an orthonormal basis for $(V, (\cdot, \cdot))$, then each x in V has the representation $x = \Sigma(x, y_i)y_i$ in the given basis. To see this, we know $x = \Sigma \alpha_i y_i$ for unique scalars $\alpha_1, \ldots, \alpha_n$. Thus

$$(x, y_j) = \sum_i \alpha_i (y_i, y_j) = \sum_i \alpha_i \delta_{ij} = \alpha_j.$$

Therefore, the coordinates of x in the orthonormal basis are (x, y_i), $i = 1, \ldots, n$. Also, it follows that $(x, x) = \Sigma(x, y_i)^2$.

Recall that the dual space of V was defined to be the set of all real-valued linear functions on V and was denoted by $\mathcal{L}(V, R)$. Also $\dim(V) = \dim(\mathcal{L}(V, R))$ when V is finite dimensional. The identification of V with $\mathcal{L}(V, R)$ via a given inner product is described in the following proposition.

Proposition 1.10. If $(V, (\cdot, \cdot))$ is a finite dimensional inner product space and if $f \in \mathcal{L}(V, R)$, then there exists a vector $x_0 \in V$ such that $f(x) = (x_0, x)$ for $x \in V$. Conversely, $(x_0, \cdot)$ is a linear function on V for each $x_0 \in V$.

Proof. Let $x_1, \ldots, x_n$ be an orthonormal basis for V and set $\alpha_i = f(x_i)$ for $i = 1, \ldots, n$. For $x_0 = \Sigma \alpha_i x_i$, it is clear that $(x_0, x_j) = \alpha_j = f(x_j)$. Since the two linear functions f and $(x_0, \cdot)$ agree on a basis, they are the same function. Thus $f(x) = (x_0, x)$ for $x \in V$. The converse is clear. □

Definition 1.15. If S is a subset of V, the *orthogonal complement* of S, denoted by $S^\perp$, is $S^\perp = \{x | x \perp y \text{ for all } y \in S\}$.

It is easily verified that $S^\perp$ is a subspace of V for any set S, and $S \perp S^\perp$. The next result provides a basic decomposition for a finite dimensional inner product space.

Proposition 1.11. Suppose M is a k-dimensional subspace of an n-dimensional inner product space $(V, (\cdot, \cdot))$. Then

(i) $M \cap M^\perp = \{0\}$.
(ii) $M \oplus M^\perp = V$.
(iii) $(M^\perp)^\perp = M$.

Proof. Let $\{x_1, \ldots, x_n\}$ be a basis for V such that $\{x_1, \ldots, x_k\}$ is a basis for M. Applying the Gram–Schmidt process to $\{x_1, \ldots, x_n\}$, we get an orthonormal basis $\{y_1, \dots, y_n\}$ such that $\{y_1, \dots, y_k\}$ is a basis for M. Let $N = \text{span}\{y_{k+1}, \dots, y_n\}$. We claim that $N = M^\perp$. It is clear that $N \subseteq M^\perp$ since $y_j \perp M$ for $j = k+1, \dots, n$. But if $x \in M^\perp$, then $x = \Sigma_1^n (x, y_i) y_i$ and $(x, y_i) = 0$ for $i = 1, \dots, k$ since $x \in M^\perp$, that is, $x = \Sigma_{k+1}^n (x, y_i) y_i \in N$. Therefore, $M = N^\perp$. Assertions (i) and (ii) now follow easily. For (iii), $M^\perp$ is spanned by $\{y_{k+1}, \dots, y_n\}$ and, arguing as above, $(M^\perp)^\perp$ must be spanned by $y_1, \dots, y_k$, which is just M. □

The decomposition, $V = M \oplus M^\perp$, of an inner product space is called an *orthogonal direct sum decomposition*. More generally, if $M_1, \dots, M_k$ are subspaces of V such that $M_i \perp M_j$ for $i \neq j$ and $V = M_1 \oplus M_2 \oplus \cdots \oplus M_k$, we also speak of the orthogonal direct sum decomposition of V. As we have seen, every direct sum decomposition of a finite dimensional vector space has associated with it two projections. When V is an inner product space and $V = M \oplus M^\perp$, then the projection on M along $M^\perp$ is called the *orthogonal projection* onto M. If P is the orthogonal projection onto M, then $I - P$ is the orthogonal projection onto $M^\perp$. The thing that makes a projection an orthogonal projection is that its null space must be the orthogonal complement of its range. After introducing adjoints of linear transformations, a useful characterization of orthogonal projections is given.

When $(V, (\cdot, \cdot))$ is an inner product space, a number of special types of linear transformations in $\mathcal{L}(V, V)$ arise. First, we discuss the adjoint of a linear transformation. For $A \in \mathcal{L}(V, V)$, consider (x, Ay). For x fixed, (x, Ay) is a linear function of y, and, by Proposition 1.9, there exists a unique vector (which depends on x) $z(x) \in V$ such that $(x, Ay) = (z(x), y)$ for all $y \in V$. Thus z defines a function from V to V that takes x into $z(x)$. However, the verification that $z(\alpha_1 x_1 + \alpha_2 x_2) = \alpha_1 z(x_1) + \alpha_2 z(x_2)$ is routine. Thus the function z is a linear transformation on V to V, and this leads to the following definition.

Definition 1.16. For $A \in \mathcal{L}(V, V)$, the unique linear transformation in $\mathcal{L}(V, V)$, denoted by A', which satisfies $(x, Ay) = (A'x, y)$, for all $x, y \in V$, is called the *adjoint* (or *transpose*) of A.

The uniqueness of A' in Definition 1.16 follows from the observation that if $(Bx, y) = (Cx, y)$ for all $x, y \in V$, then $((B - C)x, y) = 0$. Taking $y = (B - C)x$ yields $((B - C)x, (B - C)x) = 0$ for all x, so $(B - C)x = 0$ for all x. Hence $B = C$.

Proposition 1.12. If $A, B \in \mathcal{L}(V, V)$, then $(AB)' = B'A'$, and if A is invertible, then $(A^{-1})' = (A')^{-1}$. Also, $(A')' = A$.

Proof. $(AB)'$ is the transformation in $\mathcal{L}(V, V)$ that satisfies $((AB)'x, y) = (x, ABy)$. Using the definition of A' and B', $(x, ABy) = (A'x, By) = (B'A'x, y)$. Thus $(AB)' = B'A'$. The other assertions are proved similarly. □

Definition 1.17. A linear transformation in $\mathcal{L}(V, V)$ is called:

(i) *Self-adjoint* (or symmetric) if $A = A'$.
(ii) *Skew symmetric* if $A' = -A$.
(iii) *Orthogonal* if $(Ax, Ay) = (x, y)$ for $x, y \in V$.

For self-adjoint transformations, A is:

(iv) *Non-negative definite* (or positive semidefinite) if $(x, Ax) \geqslant 0$ for $x \in V$.
(v) *Positive definite* if $(x, Ax) > 0$ for all $x \neq 0$.

The remainder of this section is concerned with a variety of descriptions and characterizations of the classes of transformations defined above.

Proposition 1.13. Let $A \in \mathcal{L}(V, V)$. Then

(i) $\mathcal{R}(A) = (\mathcal{N}(A'))^{\perp}$.
(ii) $\mathcal{R}(A) = \mathcal{R}(AA')$.
(iii) $\mathcal{N}(A) = \mathcal{N}(A'A)$.
(iv) $r(A) = r(A')$

Proof. Assertion (i) is equivalent to $(\mathcal{R}(A))^{\perp} = \mathcal{N}(A')$. But $x \in \mathcal{N}(A')$ means that $0 = (y, A'x)$ for all $y \in V$, and this is equivalent to $x \perp \mathcal{R}(A)$ since $(y, A'x) = (Ay, x)$. This proves (i). For (ii), it is clear that $\mathcal{R}(AA') \subseteq \mathcal{R}(A)$. If $x \in \mathcal{R}(A)$, then $x = Ay$ for some $y \in V$. Write $y = y_1 + y_2$ where $y_1 \in \mathcal{R}(A')$ and $y_2 \in (\mathcal{R}(A'))^{\perp}$. From (i), $(\mathcal{R}(A'))^{\perp} = \mathcal{N}(A)$, so $Ay_2 = 0$. Since $y_1 \in \mathcal{R}(A')$, $y_1 = A'z$ for some $z \in V$. Thus $x = Ay = Ay_1 = AA'z$, so $x \in \mathcal{R}(AA')$.

To prove (iii), if $Ax = 0$, then $A'Ax = 0$, so $\mathcal{N}(A) \subseteq \mathcal{N}(A'A)$. Conversely, if $A'Ax = 0$, then $0 = (x, A'Ax) = (Ax, Ax)$, so $Ax = 0$, and $\mathcal{N}(A'A) \subseteq \mathcal{N}(A)$.

Since $\dim(\mathcal{R}(A)) + \dim(\mathcal{N}(A)) = \dim(V)$, $\dim(\mathcal{R}(A')) + \dim(\mathcal{N}(A')) = \dim(V)$, and $\mathcal{R}(A) = (\mathcal{N}(A'))^{\perp}$, it follows that $r(A) = r(A')$. □

If $A \in \mathcal{L}(V, V)$ and $r(A) = 0$, then $A = 0$ since A must map everything into $0 \in V$. We now discuss the rank one linear transformations and show that these can be thought of as the "building blocks" for $\mathcal{L}(V, V)$.

Proposition 1.14. For $A \in \mathcal{L}(V, V)$, the following are equivalent:

(i) $r(A) = 1$.
(ii) There exist $x_0 \neq 0$ and $y_0 \neq 0$ in V such that $Ax = (y_0, x)x_0$ for $x \in V$.

Proof. That (ii) implies (i) is clear since, if $Ax = (y_0, x)x_0$, then $\mathfrak{R}(A) = \text{span}\{x_0\}$, which is one-dimensional. Thus suppose $r(A) = 1$. Since $\mathfrak{R}(A)$ is one-dimensional, there exists $x_0 \in \mathfrak{R}(A)$ with $x_0 \neq 0$ and $\mathfrak{R}(A) = \text{span}\{x_0\}$. As $Ax \in \mathfrak{R}(A)$ for all x, $Ax = \alpha(x)x_0$ where $\alpha(x)$ is some scalar that depends on x. The linearity of A implies that $\alpha(\beta_1 x_1 + \beta_2 x_2) = \beta_1\alpha(x_1) + \beta_2\alpha(x_2)$. Thus α is a linear function on V and, by Proposition 1.10, $\alpha(x) = (y_0, x)$ for some $y_0 \in V$. Since $\alpha(x) \neq 0$ for some $x \in V$, $y_0 \neq 0$. Therefore, (i) implies (ii). □

This description of the rank one linear transformations leads to the following definition.

Definition 1.18. Given $x, y \in V$, the *outer product* of x and y, denoted by $x \,\square\, y$, is the linear transformation on V to V whose value at z is $(x \,\square\, y)z = (y, z)x$.

Thus $x \,\square\, y \in \mathcal{L}(V, V)$ and $x \,\square\, y = 0$ iff x or y is zero. When $x \neq 0$ and $y \neq 0$, $\mathfrak{R}(x \,\square\, y) = \text{span}\{x\}$ and $\mathfrak{N}(x \,\square\, y) = (\text{span}\{y\})^{\perp}$. The result of Proposition 1.14 shows that every rank one transformation is an outer product of two nonzero vectors. The following properties of outer products are easily verified:

(i) $x \,\square\, (\alpha_1 y_1 + \alpha_2 y_2) = \alpha_1 x \,\square\, y_1 + \alpha_2 x \,\square\, y_2$.
(ii) $(\alpha_1 x_1 + \alpha_2 x_2) \,\square\, y = \alpha_1 x_1 \,\square\, y + \alpha_2 x_2 \,\square\, y$.
(iii) $(x \,\square\, y)' = y \,\square\, x$.
(iv) $(x_1 \,\square\, y_1)(x_2 \,\square\, y_2) = (y_1, x_2) x_1 \,\square\, y_2$.

One word of caution: the definition of the outer product depends on the inner product on V. When there is more than one inner product for V, care must be taken to indicate which inner product is being used to define the outer product. The claim that rank one linear transformations are the building blocks for $\mathcal{L}(V, V)$ is partially justified by the following proposition.

Proposition 1.15. Let $\{x_1, \dots, x_n\}$ be an orthonormal basis for $(V, (\cdot, \cdot))$. Then $\{x_i \square x_j;\ i, j = 1, \dots, n\}$ is a basis for $\mathcal{L}(V, V)$.

Proof. If $A \in \mathcal{L}(V, V)$, A is determined by the n^2 numbers $\alpha_{ij} = (x_i, Ax_j)$. But the linear transformation $B = \sum\sum \alpha_{ij} x_i \square x_j$ satisfies

$$(x_i, Bx_j) = \left(x_i, \left(\sum_k \sum_l \alpha_{kl} x_k \square x_l\right) x_j\right) = \sum_k \sum_l \alpha_{kl}(x_l, x_j)(x_i, x_k) = \alpha_{ij}.$$

Thus $B = A$ so every $A \in \mathcal{L}(V, V)$ is a linear combination of $\{x_i \square x_j | i, j = 1, \dots, n\}$. Since $\dim(\mathcal{L}(V, V)) = n^2$, the result follows. □

Using outer products, it is easy to give examples of self-adjoint linear transformations. First, since linear combinations of self-adjoint linear transformations are again self-adjoint, the set M of self-adjoint transformations is a subspace of $\mathcal{L}(V, V)$. Also, the set N of skew symmetric transformations is a subspace of $\mathcal{L}(V, V)$. It is clear that the only transformation that is both self-adjoint and skew symmetric is 0, so $M \cap N = \{0\}$. But if $A \in \mathcal{L}(V, V)$, then

$$A = \frac{A + A'}{2} + \frac{A - A'}{2}, \qquad \frac{A + A'}{2} \in M, \qquad \text{and} \qquad \frac{A - A'}{2} \in N.$$

This shows that $\mathcal{L}(V, V) = M \oplus N$. To give examples of elements of M, let $x_1, \dots, x_n$ be an orthonormal basis for $(V, (\cdot, \cdot))$. For each i, $x_i \square x_i$ is self-adjoint, so for scalars α_i, $B = \sum \alpha_i x_i \square x_i$ is self-adjoint. The geometry associated with the transformation B is interesting and easy to describe. Since $\|x_i\| = 1$, $(x_i \square x_i)^2 = x_i \square x_i$, so $x_i \square x_i$ is a projection on span$\{x_i\}$ along $(\text{span}\{x_i\})^\perp$ —that is, $x_i \square x_i$ is the orthogonal projection on span$\{x_i\}$ as the null space of $x_i \square x_i$ is the orthogonal complement of its range. Let $M_i = \text{span}\{x_i\}$, $i = 1, \dots, k$. Each M_i is a one-dimensional subspace of $(V, (\cdot, \cdot))$, $M_i \perp M_j$ if $i \neq j$, and $M_1 \oplus M_2 \oplus \cdots \oplus M_n = V$. Hence, V is the direct sum of n mutually orthogonal subspaces and each $x \in V$ has the unique representation $x = \sum(x, x_i)x_i$ where $(x, x_i)x_i = (x_i \square x_i)x$ is the projection of x onto M_i, $i = 1, \dots, n$. Since B is linear, the value of Bx is completely determined by the value of B on each M_i, $i = 1, \dots, n$. However, if $y \in M_j$, then $y = \alpha x_j$ for some $\alpha \in R$ and $By = \alpha B x_j = \alpha \sum \alpha_i (x_i \square x_i) x_j = \alpha \alpha_j x_j = \alpha_j y$. Thus when B is restricted to M_j, B is α_j times the identity transformation, and understanding how B transforms vectors has become particularly simple. In summary, take $x \in V$ and write $x = \sum(x, x_i)x_i$; then $Bx = \sum \alpha_i (x, x_i) x_i$. What is especially fascinating and useful is that every self-adjoint transformation in $\mathcal{L}(V, V)$ has the representation $\sum \alpha_i x_i \square x_i$ for some orthonormal basis for V and some scalars $\alpha_1, \dots, \alpha_n$. This fact is known as the spectral theorem and is discussed in more detail later in this chapter. For the time being, we are content with the following observation about the self-adjoint transformation $B = \Sigma \alpha_i x_i \square x_i$: B is positive definite iff $\alpha_i > 0$, $i = 1, \ldots, n$. This follows since $(x, Bx) = \Sigma \alpha_i (x, x_i)^2$ and $x = 0$ iff $(x, x_i)^2 = 0$ for all $i = 1, \ldots, n$. For exactly the same reasons, B is non-negative definite iff $\alpha_i \geqslant 0$ for $i = 1, \ldots, n$. Proposition 1.16 introduces a useful property of self-adjoint transformations.

Proposition 1.16. If A_1 and A_2 are self-adjoint linear transformations in $\mathcal{L}(V, V)$ such that $(x, A_1 x) = (x, A_2 x)$ for all x, then $A_1 = A_2$.

Proof. It suffices to show that $(x, A_1 y) = (x, A_2 y)$ for all $x, y \in V$. But

$$
\begin{aligned}
(x + y, A_1(x + y)) &= (x, A_1 x) + (y, A_1 y) + 2(x, A_1 y) \\
&= (x + y, A_2(x + y)) \\
&= (x, A_2 x) + (y, A_2 y) + 2(x, A_2 y).
\end{aligned}
$$

Since $(z, A_1 z) = (z, A_2 z)$ for all $z \in V$, we see that $(x, A_1 y) = (x, A_2 y)$. □

In the above discussion, it has been observed that, if $x \in V$ and $\|x\| = 1$, then $x \square x$ is the orthogonal projection onto the one-dimensional subspace span$\{x\}$. Recall that $P \in \mathcal{L}(V, V)$ is an orthogonal projection if P is a projection (i.e., $P^2 = P$) and if $\mathcal{N}(P) = (\mathcal{R}(P))^\perp$. The next result characterizes orthogonal projections as those projections that are self-adjoint.

Proposition 1.17. If $P \in \mathcal{L}(V, V)$, the following are equivalent:

(i) P is an orthogonal projection.
(ii) $P^2 = P = P'$.

Proof. If (ii) holds, then P is a projection and P is self-adjoint. By Proposition 1.13, $\mathcal{N}(P) = (\mathcal{R}(P'))^\perp = (\mathcal{R}(P))^\perp$ since $P = P'$. Thus P is an orthogonal projection. Conversely, if (i) holds, then $P^2 = P$ since P is a projection. We must show that if P is a projection and $\mathcal{N}(P) = (\mathcal{R}(P))^\perp$, then $P = P'$. Since $V = \mathcal{R}(P) \oplus \mathcal{N}(P)$, consider $x, y \in V$ and write $x = x_1 + x_2$, $y = y_1 + y_2$ with $x_1, y_1 \in \mathcal{R}(P)$ and $x_2, y_2 \in \mathcal{N}(P) = (\mathcal{R}(P))^\perp$. Using the fact that P is the identity on $\mathcal{R}(P)$, compute as follows:

$$
\begin{aligned}
(P'x, y) &= (x, Py) = (x_1 + x_2, Py_1) = (x_1, y_1) = (Px_1, y_1) \\
&= (P(x_1 + x_2), y_1 + y_2) = (Px, y).
\end{aligned}
$$

Since P' is the unique linear transformation that satisfies $(x, Py) = (P'x, y)$, we have $P = P'$. $\square$

It is sometimes convenient to represent an orthogonal projection in terms of outer products. If P is the orthogonal projection onto M, let $\{x_1, \ldots, x_k\}$ be an orthonormal basis for M in $(V, (\cdot, \cdot))$. Set $A = \Sigma x_i \square x_i$ so A is self-adjoint. If $x \in M$, then $x = \Sigma(x, x_i)x_i$ and $Ax = (\Sigma x_i \square x_i)x = \Sigma(x, x_i)x_i = x$. If $x \in M^\perp$, then $Ax = 0$. Since A agrees with P on M and $M^\perp$, $A = P = \Sigma x_i \square x_i$. Thus all orthogonal projections are sums of rank one orthogonal projections (given by outer products) and different terms in the sum are orthogonal to each other (i.e., $(x_i \square x_i)(x_j \square x_j) = 0$ if $i \neq j$). Generalizing this a little bit, two orthogonal projections P_1 and P_2 are called *orthogonal* if $P_1 P_2 = 0$. It is not hard to show that P_1 and P_2 are orthogonal to each other iff the range of P_1 and the range of P_2 are orthogonal to each other, as subspaces. The next result shows that a sum of orthogonal projections is an orthogonal projection iff each pair of summands is orthogonal.

Proposition 1.18. Let $P_1, \ldots, P_k$ be orthogonal projections on $(V, (\cdot, \cdot))$. Then $P = P_1 + \cdots + P_k$ is an orthogonal projection iff $P_i P_j = 0$ for $i \neq j$.

Proof. See Halmos (1958, Section 76). $\square$

We now turn to a discussion of orthogonal linear transformations on an inner product space $(V, (\cdot, \cdot))$. Basically, an orthogonal transformation is one that preserves the geometric structure (distance and angles) of the inner product. A variety of characterizations of orthogonal transformations is possible.

Proposition 1.19. If $(V, (\cdot, \cdot))$ is a finite dimensional inner product space and if $A \in \mathcal{L}(V, V)$, then the following are equivalent:

(i) $(Ax, Ay) = (x, y)$ for all $x, y \in V$.
(ii) $\|Ax\| = \|x\|$ for all $x \in V$.
(iii) $AA' = A'A = I$.
(iv) If $\{x_1, \ldots, x_n\}$ is an orthonormal basis for $(V, (\cdot, \cdot))$, then $\{Ax_1, \ldots, Ax_n\}$ is also an orthonormal basis for $(V, (\cdot, \cdot))$.

Proof. Recall that (i) is our definition of an orthogonal transformation. We prove that (i) implies (ii), (ii) implies (iii), (iii) implies (i), and then show that (i) implies (iv) and (iv) implies (ii). That (i) implies (ii) is clear since

$\|Ax\|^2 = (Ax, Ax)$. For (ii) implies (iii), $(x, x) = (Ax, Ax) = (x, A'Ax)$ implies that $A'A = I$ since $A'A$ and I are self-adjoint (see Proposition 1.16). But, by the uniqueness of inverses, this shows that $A' = A^{-1}$ so $I = AA^{-1} = AA'$ and (iii) holds. Assuming (iii), we have $(x, y) = (x, A'Ay) = (Ax, Ay)$ and (i) holds. If (i) holds and $\{x_1, \dots, x_n\}$ is an orthonormal basis for $(V, (\cdot, \cdot))$, then $\delta_{ij} = (x_i, x_j) = (Ax_i, Ax_j)$, which implies that $\{Ax_1, \dots, Ax_n\}$ is an orthonormal basis. Now, assume (iv) holds. For $x \in V$, we have $x = \Sigma(x, x_i)x_i$ and $\|x\|^2 = \Sigma(x, x_i)^2$. Thus

$$\|Ax\|^2 = (Ax, Ax) = \left(\sum_i (x, x_i)Ax_i, \sum_j (x, x_j)Ax_j\right)$$
$$= \sum_i \sum_j (x, x_i)(x, x_j)(Ax_i, Ax_j) = \sum_i \sum_j (x, x_i)(x, x_j)\delta_{ij}$$
$$= \sum_i (x, x_i)^2 = \|x\|^2.$$

Therefore (ii) holds. □

Some immediate consequences of the preceding proposition are: if A is orthogonal, so is $A^{-1} = A'$ and if A_1 and A_2 are orthogonal, then $A_1 A_2$ is orthogonal. Let $\mathcal{O}(V)$ denote all the orthogonal transformations on the inner product space $(V, (\cdot, \cdot))$. Then $\mathcal{O}(V)$ is closed under inverses, $I \in \mathcal{O}(V)$, and $\mathcal{O}(V)$ is closed under products of linear transformations. In other words, $\mathcal{O}(V)$ is a group of linear transformations on $(V, (\cdot, \cdot))$ and $\mathcal{O}(V)$ is called the *orthogonal group* of $(V, (\cdot, \cdot))$. This and many other groups of linear transformations are studied in later chapters.

One characterization of orthogonal transformations on $(V, (\cdot, \cdot))$ is that they map orthonormal bases into orthonormal bases. Thus given two orthonormal bases, there exists a unique orthogonal transformation that maps one basis onto the other. This leads to the following question. Suppose $\{x_1, \dots, x_k\}$ and $\{y_1, \dots, y_k\}$ are two finite sets of vectors in $(V(\cdot, \cdot))$. Under what conditions will there exist an orthogonal transformation A such that $Ax_i = y_i$ for $i = 1, \dots, k$? If such an $A \in \mathcal{O}(V)$ exists, then $(x_i, x_j) = (Ax_i, Ax_j) = (y_i, y_j)$ for all $i, j = 1, \dots, k$. That this condition is also sufficient for the existence of an $A \in \mathcal{O}(V)$ that maps x_i to y_i, $i = 1, \dots, k$, is the content of the next result.

Proposition 1.20. Let $\{x_1, \dots, x_k\}$ and $\{y_1, \dots, y_k\}$ be finite sets in $(V, (\cdot, \cdot))$. The following are equivalent:

(i) $(x_i, x_j) = (y_i, y_j)$ for $i, j = 1, \dots, k$.
(ii) There exists an $A \in \mathcal{O}(V)$ such that $Ax_i = y_i$ for $i = 1, \dots, k$.

Proof. That (ii) implies (i) is clear, so assume that (i) holds. Let $M = \text{span}\{x_1, \dots, x_k\}$. The idea of the proof is to define A on M using linearity and then extend the definition of A to V using linearity again. Of course, it must be verified that all this makes sense and that the A so defined is orthogonal. The details of this, which are primarily computational, follow. First, by (i), $\Sigma\alpha_i x_i = 0$ iff $\Sigma\alpha_i y_i = 0$ since $(\Sigma\alpha_i x_i, \Sigma\alpha_j x_j) = \Sigma\Sigma\alpha_i\alpha_j(x_i, x_j) = \Sigma\Sigma\alpha_i\alpha_j(y_i, y_j) = (\Sigma\alpha_i y_i, \Sigma\alpha_j y_j)$. Let $N = \text{span}\{y_1, \dots, y_k\}$ and define B on M to N by $B(\Sigma\alpha_i x_i) \equiv \Sigma\alpha_i y_i$. B is well defined since $\Sigma\alpha_i x_i = \Sigma\beta_i x_i$ implies that $\Sigma\alpha_i y_i = \Sigma\beta_i y_i$ and the linearity of B on M is easy to check. Since B maps M onto N, $\dim(N) \leqslant \dim(M)$. But if $B(\Sigma\alpha_i x_i) = 0$, then $\Sigma\alpha_i y_i = 0$, so $\Sigma\alpha_i x_i = 0$. Therefore the null space of B is $\{0\} \subseteq M$ and $\dim(M) = \dim(N)$. Let $M^\perp$ and $N^\perp$ be the orthogonal complements of M and N, respectively, and let $\{u_1, \dots, u_s\}$ and $\{v_1, \dots, v_s\}$ be orthonormal bases for $M^\perp$ and $N^\perp$, respectively. Extend the definition of B to V by first defining $B(u_i) = v_i$ for $i = 1, \dots, s$ and then extend by linearity. Let A be the linear transformation so defined. We now claim that $\|Aw\|^2 = \|w\|^2$ for all $w \in V$. To see this write $w = w_1 + w_2$ where $w_1 \in M$ and $w_2 \in M^\perp$. Then $Aw_1 \in N$ and $Aw_2 \in N^\perp$. Thus $\|Aw\|^2 = \|Aw_1 + Aw_2\|^2 = \|Aw_1\|^2 + \|Aw_2\|^2$. But $w_1 = \Sigma\alpha_i x_i$ for some scalars α_i. Thus

$$\begin{aligned}\|Aw_1\|^2 &= \left(A\left(\sum\alpha_i x_i\right), A\left(\sum\alpha_j x_j\right)\right) = \sum\sum\alpha_i\alpha_j(Ax_i, Ax_j)\\ &= \sum\sum\alpha_i\alpha_j(y_i, y_j) = \sum\sum\alpha_i\alpha_j(x_i, x_j)\\ &= \left(\sum\alpha_i x_i, \sum\alpha_j x_j\right) = \|w_1\|^2.\end{aligned}$$

Similarly, $\|Aw_2\|^2 = \|w_2\|^2$. Since $\|w\|^2 = \|w_1\|^2 + \|w_2\|^2$, the claim that $\|Aw\|^2 = \|w\|^2$ is established. By Proposition 1.19, A is orthogonal. □

◆ **Example 1.6.** Consider the real vector space R^n with the standard basis and the usual inner product. Also, let $\mathcal{L}_{n,n}$ be the real vector space of all $n \times n$ real matrices. Thus each element of $\mathcal{L}_{n,n}$ determines a linear transformation on R^n and vice versa. More precisely, if A is a linear transformation on R^n to R^n and $[A]$ denotes the matrix of A in the standard basis on both the range and domain of A, then $[Ax] = [A]x$ for $x \in R^n$. Here, $[Ax] \in R^n$ is the vector of coordinates of Ax in the standard basis and $[A]x$ means the matrix $[A] = \{a_{ij}\}$ times the coordinate vector $x \in R^n$. Conversely, if $[A] \in \mathcal{L}_{n,n}$ and we define a linear transformation A by $Ax = [A]x$, then the matrix of A is $[A]$. It is easy to show that if A is a linear transformation on R^n to R^n with the standard inner product, then $[A'] = [A]'$ where A' denotes the adjoint of A and $[A]'$ denotes the transpose of the matrix $[A]$. Now, we are in a position to relate the notions of self-adjointness and skew symmetry of linear transformations to properties of matrices. Proofs of the following two assertions are straightforward and are left to the reader. Let A be a linear transformation on R^n to R^n with matrix $[A]$.

(i) A is self-adjoint iff $[A] = [A]'$.
(ii) A is skew-symmetric iff $[A]' = -[A]$.

Elements of $\mathcal{L}_{n,n}$ that satisfy $B = B'$ are usually called *symmetric* matrices, while the term *skew-symmetric* is used if $B' = -B$, $B \in \mathcal{L}_{n,n}$. Also, the matrix B is called *positive definite* if $x'Bx > 0$ for all $x \in R^n$, $x \neq 0$. Of course $x'Bx$ is just the standard inner product of x with Bx. Clearly, B is positive definite iff the linear transformation it defines is positive definite.

If A is an orthogonal transformation on R^n to R^n, then $[A]$ must satisfy $[A][A]' = [A]'[A] = I_n$ where I_n is the $n \times n$ identity matrix. Thus a matrix $B \in \mathcal{L}_{n,n}$ is called *orthogonal* if $BB' = B'B = I_n$. An interesting geometric interpretation of the condition $BB' = B'B = I_n$ follows. If $B = \{b_{ij}\}$, the vectors $b_j \in R^n$ with coordinates b_{ij}, $i = 1, \ldots, n$, are the *column vectors* of B and the vectors $c_i \in R^n$ with coordinates b_{ij}, $j = 1, \ldots, n$, are the *row vectors* of B. The matrix BB' has elements $c_i'c_j$ and the condition $BB' = I_n$ means that $c_i'c_j = \delta_{ij}$—that is, the vectors $c_1, \ldots, c_n$ form an orthonormal basis for R^n in the usual inner product. Similarly, the condition $B'B = I_n$ holds iff the vectors $b_1, \ldots, b_n$ form an orthonormal basis for R^n. Hence a matrix B is orthogonal iff both its rows and columns determine an orthonormal basis for R^n with the standard inner product. ◆

1.4. THE CAUCHY–SCHWARZ INEQUALITY

The form of the Cauchy–Schwarz Inequality given here is general enough to be applicable to both finite and infinite dimensional vector spaces. The examples below illustrate that the generality is needed to treat some standard situations that arise in analysis and in the study of random variables. In a finite dimensional inner product space $(V, (\cdot, \cdot))$, the inequality established in this section shows that $|(x, y)| \leqslant \|x\| \|y\|$ where $\|x\|^2 = (x, x)$. Thus $-1 \leqslant (x, y)/\|x\| \|y\| \leqslant 1$ and the quantity $(x, y)/\|x\| \|y\|$ is defined to be the cosine of the angle between the vectors x and y. A variety of applications of the Cauchy–Schwarz Inequality arise in later chapters. We now proceed with the technical discussion.

Suppose that V is a real vector space, not necessarily finite dimensional. Let $[\cdot, \cdot]$ denote a non-negative definite symmetric bilinear function on $V \times V$—that is, $[\cdot, \cdot]$ is a real-valued function on $V \times V$ that satisfies (i) $[x, y] = [y, x]$, (ii) $[\alpha_1 x_1 + \alpha_2 x_2, y] = \alpha_1 [x_1, y] + \alpha_2 [x_2, y]$, and (iii) $[x, x] \geqslant 0$. It is clear that (i) and (ii) imply that $[x, \alpha_1 y_1 + \alpha_2 y_2] = \alpha_1 [x, y_1] + \alpha_2 [x, y_2]$. The Cauchy–Schwarz Inequality states that $[x, y]^2 \leqslant [x, x][y, y]$. We also give necessary and sufficient conditions for equality to hold in this inequality. First, a preliminary result.

Proposition 1.21. Let $M = \{x | [x, x] = 0\}$. Then M is a subspace of V.

Proof. If $x \in M$ and $\alpha \in R$, then $[\alpha x, \alpha x] = \alpha^2 [x, x] = 0$ so $\alpha x \in M$. Thus we must show that if $x_1, x_2 \in M$, then $x_1 + x_2 \in M$. For $\alpha \in R$, $0 \leqslant [x_1 + \alpha x_2, x_1 + \alpha x_2] = [x_1, x_1] + 2\alpha [x_1, x_2] + \alpha^2 [x_2, x_2] = 2\alpha [x_1, x_2]$ since $x_1, x_2 \in M$. But if $2\alpha [x_1, x_2] \geqslant 0$ for all $\alpha \in R$, $[x_1, x_2] = 0$, and this implies that $0 = [x_1 + \alpha x_2, x_1 + \alpha x_2]$ for all $\alpha \in R$ by the above equality. Therefore, $x_1 + \alpha x_2 \in M$ for all α when $x_1, x_2 \in M$ and thus M is a subspace. □

Theorem 1.1. (Cauchy–Schwarz Inequality). Let $[\cdot, \cdot]$ be a non-negative definite symmetric bilinear function on $V \times V$ and set $M = \{x | [x, x] = 0\}$. Then:

(i) $[x, y]^2 \leqslant [x, x][y, y]$ for $x, y \in V$.

(ii) $[x, y]^2 = [x, x][y, y]$ iff $\alpha x + \beta y \in M$ for some real α and β not both zero.

Proof. To prove (i), we consider two cases. If $x \in M$, then $0 \leqslant [y + \alpha x, y + \alpha x] = [y, y] + 2\alpha [x, y]$ for all $\alpha \in R$, so $[x, y] = 0$ and (i) holds. Similarly, if $y \in M$, (i) holds. If $x \notin M$ and $y \notin M$, let $x_1 = x/[x, x]^{1/2}$ and let $y_1 = y/[y, y]^{1/2}$. Then we must show that $|[x_1, y_1]| \leqslant 1$. This follows from the two inequalities

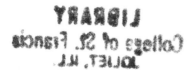

$$0 \leqslant [x_1 - y_1, x_1 - y_1] = 2 - 2[x_1, y_1]$$

and

$$0 \leqslant [x_1 + y_1, x_1 + y_1] = 2 + 2[x_1, y_1].$$

The proof of (i) is now complete.

To prove (ii), first assume that $[x, y]^2 = [x, x][y, y]$. If either $x \in M$ or $y \in M$, then $\alpha x + \beta y \in M$ for some α, β not both zero. Thus consider $x \notin M$ and $y \notin M$. An examination of the proof of (i) shows that we can have equality in (i) iff either $0 = [x_1 - y_1, x_1 - y_1]$ or $0 = [x_1 + y_1, x_1 + y_1]$ and, in either case, this implies that $\alpha x + \beta y \in M$ for some real α, β not both zero. Now, assume $\alpha x + \beta y \in M$ for some real α, β not both zero. If $\alpha = 0$ or $\beta = 0$ or $x \in M$ or $y \in M$, we clearly have equality in (i). For the case when $\alpha\beta \neq 0$, $x \notin M$, and $y \notin M$, our assumption implies that $x_1 + \gamma y_1 \in M$ for some $\gamma \neq 0$, since M is a subspace. Thus there is a real $\gamma \neq 0$ such that $0 = [x_1 + \gamma y_1, x_1 + \gamma y_1] = 1 + 2\gamma[x_1, y_1] + \gamma^2$. The equation for the roots of a quadratic shows that this can hold only if $|[x_1, y_1]| = 1$. Hence equality in (i) holds. □

◆ **Example 1.7.** Let $(V, (\cdot, \cdot))$ be a finite dimensional inner product space and suppose A is a non-negative definite linear transformation on V to V. Then $[x, y] \equiv (x, Ay)$ is a non-negative definite symmetric bilinear function. The set $M = \{x | (x, Ax) = 0\}$ is equal to $\mathfrak{N}(A)$—this follows easily from Theorem 1.1(i). Theorem 1.1 shows that $(x, Ay)^2 \leqslant (x, Ax)(y, Ay)$ and provides conditions for equality. In particular, when A is nonsingular, $M = \{0\}$ and equality holds iff x and y are linearly dependent. Of course, if $A = I$, then we have $(x, y)^2 \leqslant \|x\|^2\|y\|^2$, which is one classical form of the Cauchy–Schwarz Inequality. ◆

◆ **Example 1.8.** In this example, take V to be the set of all continuous real-valued functions defined on a closed bounded interval, say a to b, of the real line. It is easily verified that

$$[x_1, x_2] \equiv \int_a^b x_1(t)x_2(t)\,dt$$

is symmetric, bilinear, and non-negative definite. Also $[x, x] > 0$ unless $x = 0$ since x is continuous. Hence $M = \{0\}$. The Cauchy–Schwarz Inequality yields

$$\left(\int_a^b x_1(t)x_2(t)\,dt\right)^2 \leqslant \int_a^b x_1^2(t)\,dt \int_a^b x_2^2(t)\,dt.$$ ◆

◆ **Example 1.9.** The following example has its origins in the study of the covariance between two real-valued random variables. Consider a probability space $(\Omega, \mathcal{F}, P_0)$ where Ω is a set, $\mathcal{F}$ is a σ-algebra of subsets of Ω, and P_0 is a probability measure on $\mathcal{F}$. A *random variable* X is a real-valued function defined on Ω such that the inverse image of each Borel set in R is an element of $\mathcal{F}$; symbolically, $X^{-1}(B) \in \mathcal{F}$ for each Borel set B of R. Sums and products of random variables are random variables and the constant functions on Ω are random variables. If X is a random variable such that $\int |X(\omega)| P_0(d\omega) < +\infty$, then X is *integrable* and we write $\mathcal{E}X$ for $\int X(\omega) P_0(d\omega)$.

Now, let V be the collection of all real-valued random variables X, such that $\mathcal{E}X^2 < +\infty$. It is clear that if $X \in V$, then $\alpha X \in V$ for all real α. Since $(X_1 + X_2)^2 \leqslant 2(X_1^2 + X_2^2)$, if X_1 and X_2 are in V, then $X_1 + X_2$ is in V. Thus V is a real vector space with addition being the pointwise addition of random variables and scalar multiplication being pointwise multiplication of random variables by scalars. For $X_1, X_2 \in V$, the inequality $|X_1 X_2| \leqslant X_1^2 + X_2^2$ implies that $X_1 X_2$ is integrable. In particular, setting $X_2 = 1$, X_1 is integrable. Define $[\cdot, \cdot]$ on $V \times V$ by $[X_1, X_2] = \mathcal{E}(X_1 X_2)$. That $[\cdot, \cdot]$ is symmetric and bilinear is clear. Since $[X_1, X_1] = \mathcal{E}X_1^2 \geqslant 0$, $[\cdot, \cdot]$ is non-negative definite. The Cauchy–Schwarz Inequality yields $(\mathcal{E}X_1 X_2)^2 \leqslant \mathcal{E}X_1^2 \mathcal{E}X_2^2$, and setting $X_2 = 1$, this gives $(\mathcal{E}X_1)^2 \leqslant \mathcal{E}X_1^2$. Of course, this is just a verification that the variance of a random variable is non-negative. For future use, let $\text{var}(X_1) \equiv \mathcal{E}X_1^2 - (\mathcal{E}X_1)^2$. To discuss conditions for equality in the Cauchy–Schwarz Inequality, the subspace $M = \{X | [X, X] = 0\}$ needs to be described. Since $[X, X] = \mathcal{E}X^2$, $X \in M$ iff X is zero, except on set of P_0 measure zero—that is, $X = 0$ a.e. (P_0). Therefore, $(\mathcal{E}X_1 X_2)^2 = \mathcal{E}X_1^2 \mathcal{E}X_2^2$ iff $\alpha X_1 + \beta X_2 = 0$ a.e. (P_0) for some real α, β not both zero. In particular, $\text{var}(X_1) = 0$ iff $X_1 - \mathcal{E}X_1 = 0$ a.e. (P_0).

A somewhat more interesting non-negative definite symmetric bilinear function on $V \times V$ is

$$\text{cov}\{X_1, X_2\} \equiv \mathcal{E}X_1 X_2 - \mathcal{E}X_1 \mathcal{E}X_2,$$

and is called the *covariance* between X_1 and X_2. Symmetry is clear and bilinearity is easily checked. Since $\text{cov}\{X_1, X_1\} = \mathcal{E}X_1^2 - (\mathcal{E}X_1)^2 = \text{var}(X_1)$, $\text{cov}\{\cdot, \cdot\}$ is non-negative definite and $M_1 = \{X | \text{cov}\{X, X\} = 0\}$ is just the set of random variables in V that have variance zero. For this case, the Cauchy–Schwarz Inequality is

$$(\text{cov}\{X_1, X_2\})^2 \leqslant \text{var}(X_1)\text{var}(X_2).$$

Equality holds iff there exist α, β, not both zero, such that $\text{var}(\alpha X_1 + \beta X_2) = 0$; or equivalently, $\alpha(X_1 - \mathcal{E}X_1) + \beta(X_2 - \mathcal{E}X_2) = 0$ a.e. (P_0) for some α, β not both zero. The properties of $\text{cov}\{\cdot, \cdot\}$ given here are used in the next chapter to define the covariance of a random vector. ◆

1.5. THE SPACE $\mathcal{L}(V, W)$

When $(V, (\cdot, \cdot))$ is an inner product space, the adjoint of a linear transformation in $\mathcal{L}(V, V)$ was introduced in Section 1.3 and used to define some special linear transformations in $\mathcal{L}(V, V)$. Here, some of the notions discussed in relation to $\mathcal{L}(V, V)$ are extended to the case of linear transformations in $\mathcal{L}(V, W)$ where $(V, (\cdot, \cdot))$ and $(W, [\cdot, \cdot])$ are two inner product spaces. In particular, adjoints and outer products are defined, bilinear functions on $V \times W$ are characterized, and Kronecker products are introduced. Of course, all the results in this section apply to $\mathcal{L}(V, V)$ by taking $(W, [\cdot, \cdot]) = (V, (\cdot, \cdot))$ and the reader should take particular notice of this special case. There is one point that needs some clarification. Given $(V, (\cdot, \cdot))$ and $(W, [\cdot, \cdot])$, the adjoint of $A \in \mathcal{L}(V, W)$, to be defined below, depends on both the inner products $(\cdot, \cdot)$ and $[\cdot, \cdot]$. However, in the previous discussion of adjoints in $\mathcal{L}(V, V)$, it was assumed that the inner product was the same on both the range and the domain of the linear transformation (i.e., V is the domain and range). Whenever we discuss adjoints of $A \in \mathcal{L}(V, V)$ it is assumed that only one inner product is involved, unless the contrary is explicitly stated—that is, when specializing results from $\mathcal{L}(V, W)$ to $\mathcal{L}(V, V)$, we take $W = V$ and $[\cdot, \cdot] = (\cdot, \cdot)$.

The first order of business is to define the adjoint of $A \in \mathcal{L}(V, W)$ where $(V, (\cdot, \cdot))$ and $(W, [\cdot, \cdot])$ are inner product spaces. For a fixed $w \in W$, $[w, Ax]$ is a linear function of $x \in V$ and, by Proposition 1.10, there exists a unique vector $y(w) \in V$ such that $[w, Ax] = (y(w), x)$ for all $x \in V$. It is easy to verify that $y(\alpha_1 w_1 + \alpha_2 w_2) = \alpha_1 y(w_1) + \alpha_2 y(w_2)$. Hence $y(\cdot)$ determines a linear transformation on W to V, say A', which satisfies $[w, Ax] = (A'w, x)$ for all $w \in W$ and $x \in V$.

Definition 1.19. Given inner product spaces $(V, (\cdot, \cdot))$ and $(W, [\cdot, \cdot])$, if $A \in \mathcal{L}(V, W)$, the unique linear transformation $A' \in \mathcal{L}(W, V)$ that satisfies $[w, Ax] = (A'w, x)$ for all $w \in W$ and $x \in V$ is called the *adjoint* of A.

The existence and uniqueness of A' was demonstrated in the discussion preceeding Definition 1.19. It is not hard to show that $(A + B)' = A' + B'$, $(A')' = A$, and $(\alpha A)' = \alpha A'$. In the present context, Proposition 1.13 becomes Proposition 1.22.

Proposition 1.22. Suppose $A \in \mathcal{L}(V, W)$. Then:

(i) $\mathcal{R}(A) = (\mathcal{N}(A'))^{\perp}$.
(ii) $\mathcal{R}(A) = \mathcal{R}(AA')$.
(iii) $\mathcal{N}(A) = \mathcal{N}(A'A)$.
(iv) $r(A) = r(A')$.

Proof. The proof here is essentially the same as that given for Proposition 1.13 and is left to the reader. □

The notion of an outer product has a natural extension to $\mathcal{L}(V, W)$.

Definition 1.20. For $x \in (V, (\cdot, \cdot))$ and $w \in (W, [\cdot, \cdot])$, the *outer product*, $w \square x$ is that linear transformation in $\mathcal{L}(V, W)$ given by $(w \square x)(y) \equiv (x, y)w$ for all $y \in V$.

If $w = 0$ or $x = 0$, then $w \square x = 0$. When both w and x are not zero, then $w \square x$ has rank one, $\mathcal{R}(w \square x) = \text{span}\{w\}$, and $\mathcal{N}(w \square x) = (\text{span}\{x\})^{\perp}$. Also, a minor modification of the proof of Proposition 1.14 shows that, if $A \in \mathcal{L}(V, W)$, then $r(A) = 1$ iff $A = w \square x$ for some nonzero w and x.

Proposition 1.23. The outer product has the following properties:

(i) $(\alpha_1 w_1 + \alpha_2 w_2) \square x = \alpha_1 w_1 \square x + \alpha_2 w_2 \square x$.
(ii) $w \square (\alpha_1 x_1 + \alpha_2 x_2) = \alpha_1 w \square x_1 + \alpha_2 w \square x_2$.
(iii) $(w \square x)' = x \square w \in \mathcal{L}(W, V)$.

If $(V_1, (\cdot, \cdot)_1)$, $(V_2, (\cdot, \cdot)_2)$, and $(V_3, (\cdot, \cdot)_3)$ are inner product spaces with $x_1 \in V_1$, $x_2, y_2 \in V_2$, and $y_3 \in V_3$, then

(iv) $(y_3 \square y_2)(x_2 \square x_1) = (x_2, y_2)_2 y_3 \square x_1 \in \mathcal{L}(V_1, V_3)$.

Proof. Assertions (i), (ii), and (iii) follow easily. For (iv), consider $x \in V_1$. Then $(x_2 \square x_1)x = (x_1, x)_1 x_2$, so $(y_3 \square y_2)(x_2 \square x_1)x = (x_1, x)_1 (y_3 \square y_2)x_2 = (x_1, x)_1 (y_2, x_2)_2 y_3 \in V_3$. However, $(x_2, y_2)_2 (y_3 \square x_1)x = (x_2, y_2)_2 (x_1, x)_1 y_3$. Thus (iv) holds. □

There is a natural way to construct an inner product on $\mathcal{L}(V, W)$ from inner products on V and W. This construction and its relation to outer products are described in the next proposition.

Proposition 1.24. Let $\{x_1, \ldots, x_m\}$ be an orthonormal basis for $(V, (\cdot, \cdot))$ and let $\{w_1, \ldots, w_n\}$ be an orthonormal basis for $(W, [\cdot, \cdot])$. Then:

(i) $\{w_i \square x_j | i = 1, \ldots, n,\ j = 1, \ldots, m\}$ is a basis for $\mathcal{L}(V, W)$.

Let $a_{ij} = [w_i, Ax_j]$. Then:

(ii) $A = \Sigma\Sigma a_{ij} w_i \square x_j$ and the matrix of A is $[A] = \{a_{ij}\}$ in the given bases.

If $A = \Sigma\Sigma a_{ij} w_i \square x_j$ and $B = \Sigma\Sigma b_{ij} w_i \square x_j$, define $\langle A, B\rangle \equiv \Sigma\Sigma a_{ij} b_{ij}$. Then:

(iii) $\langle \cdot , \cdot \rangle$ is an inner product on $\mathcal{L}(V, W)$ and $\{w_i \square x_j | i = 1, \ldots, n,\ j = 1, \ldots, m\}$ is an orthonormal basis for $(\mathcal{L}(V, W), \langle \cdot , \cdot \rangle)$.

Proof. Since $\dim(\mathcal{L}(V, W)) = mn$, to prove (i) it suffices to prove (ii). Let $B = \Sigma\Sigma a_{ij} w_i \square x_j$. Then

$$[w_k, Bx_l] = \sum_i \sum_j a_{ij} [w_k, (w_i \square x_j) x_l] = \sum_i \sum_j a_{ij} \delta_{ik} \delta_{jl} = a_{kl},$$

so $[w_i, Bx_j] = [w_i, Ax_j]$ for $i = 1, \ldots, n$ and $j = 1, \ldots, m$. Therefore, $[w, Bx] = [w, Ax]$ for all $w \in W$ and $x \in V$, which implies that $[w, (B - A)x] = 0$. Choosing $w = (B - A)x$, we see that $(B - A)x = 0$ for all $x \in V$ and, therefore, $B = A$. To show that the matrix of A is $[A] = \{a_{ij}\}$, recall that the matrix of A consists of the scalars b_{kj} defined by $Ax_j = \Sigma_k b_{kj} w_k$. The inner product of w_i with each side of this equation is

$$a_{ij} = [w_i, Ax_j] = \sum_k b_{kj} [w_i, w_k] = b_{ij}$$

and the proof of (ii) is complete.

For (iii), $\langle \cdot , \cdot \rangle$ is clearly symmetric and bilinear. Since $\langle A, A\rangle = \Sigma\Sigma a_{ij}^2$, the positivity of $\langle \cdot , \cdot \rangle$ follows. That $\{w_i \square x_j | i = 1, \ldots, n,\ j = 1, \ldots, m\}$ is an orthonormal basis for $(\mathcal{L}(V, W), \langle \cdot , \cdot \rangle)$ follows immediately from the definition of $\langle \cdot , \cdot \rangle$. □

A few words are in order concerning the inner product $\langle \cdot , \cdot \rangle$ on $\mathcal{L}(V, W)$. Since $\{w_i \square x_j | i = 1, \ldots, n,\ j = 1, \ldots, m\}$ is an orthonormal basis, we know that if $A \in \mathcal{L}(V, W)$, then

$$A = \sum \sum \langle A, w_i \square x_j \rangle w_i \square x_j,$$

since this is the unique expansion of a vector in any orthonormal basis. However, $A = \Sigma\Sigma[w_i, Ax_j]w_i \square x_j$ by (ii) of Proposition 1.24. Thus $\langle A, w_i \square x_j \rangle = [w_i, Ax_j]$ for $i = 1, \ldots, n$ and $j = 1, \ldots, m$. Since both sides of this relation are linear in w_i and x_j, we have $\langle A, w \square x \rangle = [w, Ax]$ for all $w \in W$ and $x \in V$. In particular, if $A = \tilde{w} \square \tilde{x}$, then

$$\langle \tilde{w} \square \tilde{x}, w \square x \rangle = [w, (\tilde{w} \square \tilde{x})x] = [w, (\tilde{x}, x)\tilde{w}] = [w, \tilde{w}](x, \tilde{x}).$$

This relation has some interesting implications.

Proposition 1.25. The inner product $\langle \cdot, \cdot \rangle$ on $\mathcal{L}(V, W)$ satisfies

(i) $\langle \tilde{w} \square \tilde{x}, w \square x \rangle = [\tilde{w}, w](\tilde{x}, x)$

for all $\tilde{w}, w \in W$ and $\tilde{x}, x \in V$, and $\langle \cdot, \cdot \rangle$ is the unique inner product with this property. Further, if $\{z_1, \ldots, z_n\}$ and $\{y_1, \ldots, y_m\}$ are any orthonormal bases for W and V, respectively, then $\{z_i \square y_j | i = 1, \ldots, n,\ j = 1, \ldots, m\}$ is an orthonormal basis for $(\mathcal{L}(V, W), \langle \cdot, \cdot \rangle)$.

Proof. Equation (i) has been verified. If $\{\cdot, \cdot\}$ is another inner product on $\mathcal{L}(V, W)$ that satisfies (i), then

$$\{w_{i_1} \square x_{j_1}, w_{i_2} \square x_{j_2}\} = \langle w_{i_1} \square x_{j_1}, w_{i_2} \square x_{j_2} \rangle$$

for all $i_1, i_2 = 1, \ldots, n$ and $j_1, j_2 = 1, \ldots, m$ where $\{x_1, \ldots, x_m\}$ and $\{w_1, \ldots, w_n\}$ are the orthonormal bases used to define $\langle \cdot, \cdot \rangle$. Using (i) of Proposition 1.24 and the bilinearity of inner products, this implies that $\{A, B\} = \langle A, B \rangle$ for all $A, B \in \mathcal{L}(V, W)$. Therefore, the two inner products are the same. The verification that $\{z_i \square y_j | i = 1, \ldots, n,\ j = 1, \ldots, m\}$ is an orthonormal basis follows easily from (i). □

The result of Proposition 1.25 is a formal statement of the fact that $\langle \cdot, \cdot \rangle$ does not depend on the particular orthonormal bases used to define it, but $\langle \cdot, \cdot \rangle$ is determined by the inner products on V and W. Whenever V and W are inner product spaces, the symbol $\langle \cdot, \cdot \rangle$ always means the inner product on $\mathcal{L}(V, W)$ as defined above.

◆ **Example 1.10.** Consider $V = R^m$ and $W = R^n$ with the usual inner products and the standard bases. Thus we have the inner product $\langle \cdot, \cdot \rangle$ on $\mathcal{L}_{m,n}$—the linear space of $n \times m$ real matrices. For $A = \{a_{ij}\}$ and $B = \{b_{ij}\}$ in $\mathcal{L}_{m,n}$,

$$\langle A, B \rangle = \sum_{i=1}^{n} \sum_{j=1}^{m} a_{ij} b_{ij}.$$

If $C = AB' : n \times n$, then

$$c_{ii} = \sum_j a_{ij} b_{ij}, \qquad i = 1, \ldots, n,$$

so $\langle A, B \rangle = \Sigma c_{ii}$. In other words, $\langle A, B \rangle$ is just the sum of the diagonal elements of the $n \times n$ matrix AB'. This observation leads to the definition of the trace of any square matrix. If $C : k \times k$ is a real matrix, the *trace of* C, denoted by $\operatorname{tr} C$, is the sum of the diagonal elements of C. The identity $\langle A, B \rangle = \langle B, A \rangle$ shows that $\operatorname{tr} AB' = \operatorname{tr} B'A$ for all $A, B \in \mathcal{L}_{m,n}$. In the present example, it is clear that $w \square x = wx'$ for $x \in R^m$ and $w \in R^n$, so $w \square x$ is just the $n \times 1$ matrix w times the $1 \times m$ matrix x'. Also, the identity in Proposition 1.25 is a reflection of the fact that

$$\operatorname{tr} \tilde{w}\tilde{x}'xw' = \tilde{w}'w\tilde{x}'x$$

for $w, \tilde{w} \in R^n$ and $x, \tilde{x} \in R^m$. ◆

If $(V, (\cdot, \cdot))$ and $(W, [\cdot, \cdot])$ are inner product spaces and $A \in \mathcal{L}(V, W)$, then $[Ax, w]$ is linear in x for fixed w and linear in w for fixed x. This observation leads to the following definition.

Definition 1.21. A function f defined on $V \times W$ to R is called *bilinear* if:

(i) $f(\alpha_1 x_1 + \alpha_2 x_2, w) = \alpha_1 f(x_1, w) + \alpha_2 f(x_2, w)$.
(ii) $f(x, \alpha_1 w_1 + \alpha_2 w_2) = \alpha_1 f(x, w_1) + \alpha_2 f(x, w_2)$.

These conditions apply for scalars α_1 and α_2; $x, x_1, x_2 \in V$ and $w, w_1, w_2 \in W$.

Our next result shows there is a natural one-to-one correspondence between bilinear functions and $\mathcal{L}(V, W)$.

Proposition 1.26. If f is a bilinear function on $V \times W$ to R, then there exists an $A \in \mathcal{L}(V, W)$ such that $f(x, w) = [Ax, w]$ for all $x \in V$ and $w \in W$. Conversely, each $A \in \mathcal{L}(V, W)$ determines the bilinear function $[Ax, w]$ on $V \times W$.

Proof. Let $\{x_1, \dots, x_m\}$ be an orthonormal basis for $(V, (\cdot, \cdot))$ and $\{w_1, \dots, w_n\}$ be an orthonormal basis for $(W, [\cdot, \cdot])$. Set $a_{ij} = f(x_j, w_i)$ for $i = 1, \dots, n$ and $j = 1, \dots, m$ and let $A = \Sigma\Sigma a_{ij} w_i \Box x_j$. By Proposition 1.24, we have

$$a_{ij} = \left[Ax_j, w_i\right] = f(x_j, w_i).$$

The bilinearity of f and of $[Ax, w]$ implies $[Ax, w] = f(x, w)$ for all $x \in V$ and $w \in W$. The converse is obvious. □

Thus far, we have seen that $\mathcal{L}(V, W)$ is a real vector space and that, if V and W have inner products $(\cdot, \cdot)$ and $[\cdot, \cdot]$, respectively, then $\mathcal{L}(V, W)$ has a natural inner product determined by $(\cdot, \cdot)$ and $[\cdot, \cdot]$. Since $\mathcal{L}(V, W)$ is a vector space, there are linear transformations on $\mathcal{L}(V, W)$ to other vector spaces and there is not much more to say in general. However, $\mathcal{L}(V, W)$ is built from outer products and it is natural to ask if there are special linear transformations on $\mathcal{L}(V, W)$ that transform outer products into outer products. For example, if $A \in \mathcal{L}(V, V)$ and $B \in \mathcal{L}(W, W)$, suppose we define $B \otimes A$ on $\mathcal{L}(V, W)$ by $(B \otimes A)C = BCA'$ where A' denotes the transpose of $A \in \mathcal{L}(V, V)$. Clearly, $B \otimes A$ is a linear transformation. If $C = w \Box x$, then $(B \otimes A)(w \Box x) = B(w \Box x)A' \in \mathcal{L}(V, W)$. But for $v \in V$,

$$\begin{aligned}(B(w \Box x)A')v &= B(w \Box x)(A'v) = B((x, A'v)w) \\ &= (Ax, v)Bw = ((Bw) \Box (Ax))v.\end{aligned}$$

This calculation shows that $(B \otimes A)(w \Box x) = (Bw) \Box (Ax)$, so outer products get mapped into outer products by $B \otimes A$. Generalizing this a bit, we have the following definition.

Definition 1.22. Let $(V_1, (\cdot, \cdot)_1)$, $(V_2, (\cdot, \cdot)_2)$, $(W_1, [\cdot, \cdot]_1)$, and $(W_2, [\cdot, \cdot]_2)$ be inner product spaces. For $A \in \mathcal{L}(V_1, V_2)$ and $B \in \mathcal{L}(W_1, W_2)$, the *Kronecker product* of B and A, denoted by $B \otimes A$, is a linear transformation on $\mathcal{L}(V_1, W_1)$ to $\mathcal{L}(V_2, W_2)$, defined by

$$(B \otimes A)C \equiv BCA'$$

for all $C \in \mathcal{L}(V_1, W_1)$.

In most applications of Kronecker products, $V_1 = V_2$ and $W_1 = W_2$, so $B \otimes A$ is a linear transformation on $\mathcal{L}(V_1, W_1)$ to $\mathcal{L}(V_1, W_1)$. It is not easy to say in a few words why the transpose of A should appear in the definition of the Kronecker product, but the result below should convince the reader that the definition is the "right" one. Of course, by A', we mean the linear transformation on V_2 to V_1, which satisfies $(x_2, Ax_1)_2 = (A'x_2, x_1)_1$ for $x_1 \in V_1$ and $x_2 \in V_2$.

Proposition 1.27. In the notation of Definition 1.22,

(i) $(B \otimes A)(w_1 \Box v_1) = (Bw_1)\Box(Av_1) \in \mathcal{L}(V_2, W_2)$.

Also,

(ii) $(B \otimes A)' = B' \otimes A'$,

where $(B \otimes A)'$ denotes the transpose of the linear transformation $B \otimes A$ on $(\mathcal{L}(V_1, W_1), \langle \cdot, \cdot \rangle_1)$ to $(\mathcal{L}(V_2, W_2), \langle \cdot, \cdot \rangle_2)$.

Proof. To verify (i), for $v_2 \in V_2$, compute as follows:

$$[(B \otimes A)(w_1 \Box v_1)](v_2) = B(w_1 \Box v_1)A'v_2 = B(v_1, A'v_2)_2 w_1$$
$$= (Av_1, v_2)_1 Bw_1 = [(Bw_1)\Box(Av_1)](v_2).$$

Since this holds for all $v_2 \in V_2$, assertion (i) holds. The proof of (ii) requires we show that $B' \otimes A'$ satisfies the defining equation of the adjoint—that is, for $C_1 \in \mathcal{L}(V_1, W_1)$ and $C_2 \in \mathcal{L}(V_2, W_2)$,

$$\langle C_2, (B \otimes A)C_1 \rangle_2 = \langle (B' \otimes A')C_2, C_1 \rangle_1.$$

Since outer products generate $\mathcal{L}(V_1, W_1)$, it is enough to show the above holds for $C_1 = w_1 \Box x_1$ with $w_1 \in W_1$ and $x_1 \in V_1$. But, by (i) and the definition of transpose,

$$\langle C_2, (B \otimes A)(w_1 \Box x_1) \rangle_2 = \langle C_2, Bw_1 \Box Ax_1 \rangle_2 = [C_2 A x_1, Bw_1]_2$$
$$= [B'C_2 A x_1, w_1]_1 = \langle B'C_2 A, w_1 \Box x_1 \rangle_1$$
$$= \langle (B' \otimes A')C_2, w_1 \Box x_1 \rangle_1,$$

and this completes the proof of (ii). □

We now turn to the case when $A \in \mathcal{L}(V, V)$ and $B \in \mathcal{L}(W, W)$ so $B \otimes A$ is a linear transformation on $\mathcal{L}(V, W)$ to $\mathcal{L}(V, W)$. First note that if A is self-adjoint relative to the inner product on V and B is self-adjoint relative to the inner product on W, then Proposition 1.27 shows that $B \otimes A$ is self-adjoint relative to the natural induced inner product on $\mathcal{L}(V, W)$.

Proposition 1.28. For $A_i \in \mathcal{L}(V, V)$, $i = 1, 2$, and $B_i \in \mathcal{L}(W, W)$, $i = 1, 2$, we have:

(i) $(B_1 \otimes A_1)(B_2 \otimes A_2) = (B_1 B_2) \otimes (A_1 A_2)$.
(ii) If A_1^{-1} and B_1^{-1} exist, then $(B_1 \otimes A_1)^{-1} = B_1^{-1} \otimes A_1^{-1}$.
(iii) If A_1 and B_1 are orthogonal projections, then $B_1 \otimes A_1$ is an orthogonal projection.

Proof. The proof of (i) goes as follows: For $C \in \mathcal{L}(V, W)$,

$$(B_1 \otimes A_1)(B_2 \otimes A_2)C = (B_1 \otimes A_1)(B_2 C A_2') = B_1 B_2 C A_2' A_1'$$
$$= B_1 B_2 C (A_1 A_2)' = ((B_1 B_2) \otimes (A_1 A_2))C.$$

Now, (ii) follows immediately from (i). For (iii), it needs to be shown that $(B_1 \otimes A_1)^2 = B_1 \otimes A_1 = (B_1 \otimes A_1)'$. The second equality has been verified. The first follows from (i) and the fact that $B_1^2 = B_1$ and $A_1^2 = A_1$. □

Other properties of Kronecker products are given as the need arises. One issue to think about is this: if $C \in \mathcal{L}(V, W)$ and $B \in \mathcal{L}(W, W)$, then BC can be thought of as the product of the two linear transformations B and C. However, BC can also be interpreted as $(B \otimes I)C$, $I \in \mathcal{L}(V, V)$—that is, BC is the value of the linear transformation $B \otimes I$ at C. Of course, the particular situation determines the appropriate way to think about BC.

Linear isometries are the final subject of discussion in this section, and are a natural generalization of orthogonal transformations on $(V, (\cdot, \cdot))$. Consider finite dimensional inner product spaces V and W with inner products $(\cdot, \cdot)$ and $[\cdot, \cdot]$ and assume that $\dim V \leqslant \dim W$. The reason for this assumption is made clear in a moment.

Definition 1.23. A linear transformation $A \in \mathcal{L}(V, W)$ is a *linear isometry* if $(v_1, v_2) = [Av_1, Av_2]$ for all $v_1, v_2 \in V$.

If A is a linear isometry and $v \in V$, $v \neq 0$, then $0 < (v, v) = [Av, Av]$. This implies that $\mathcal{N}(A) = \{0\}$, so necessarily $\dim V \leqslant \dim W$. When $W = V$

and $[\cdot,\cdot]=(\cdot,\cdot)$, then linear isometries are simply orthogonal transformations. As with orthogonal transformations, a number of equivalent descriptions of linear isometries are available.

Proposition 1.29. For $A \in \mathcal{L}(V, W)$ $(\dim V \leqslant \dim W)$, the following are equivalent:

(i) A is a linear isometry.
(ii) $A'A = I \in \mathcal{L}(V, V)$.
(iii) $[Av, Av] = (v, v)$, $v \in V$.

Proof. The proof is similar to the proof of Proposition 1.19 and is left to the reader. □

The next proposition is an analog of Proposition 1.20 that covers linear isometries and that has a number of applications.

Proposition 1.30. Let $v_1, \ldots, v_k$ be vectors in $(V, (\cdot,\cdot))$, let $w_1, \ldots, w_k$ be vectors in $(W, [\cdot,\cdot])$, and assume $\dim V \leqslant \dim W$. There exists a linear isometry $A \in \mathcal{L}(V, W)$ such that $Av_i = w_i$, $i = 1, \ldots, k$, iff $(v_i, v_j) = [w_i, w_j]$ for $i, j = 1, \ldots, k$.

Proof. The proof is a minor modification of that given for Proposition 1.20 and the details are left to the reader. □

Proposition 1.31. Suppose $A \in \mathcal{L}(V, W_1)$ and $B \in \mathcal{L}(V, W_2)$ where $\dim W_2 \leqslant \dim W_1$, and $(\cdot,\cdot)$, $[\cdot,\cdot]_1$, and $[\cdot,\cdot]_2$ are inner products on V, W_1, and W_2. Then $A'A = B'B$ iff there exists a linear isometry $\Psi \in \mathcal{L}(W_2, W_1)$ such that $A = \Psi B$.

Proof. If $A = \Psi B$, then $A'A = B'\Psi'\Psi B = B'B$, since $\Psi'\Psi = I \in \mathcal{L}(W_2, W_2)$. Conversely, suppose $A'A = B'B$ and let $\{v_1, \ldots, v_m\}$ be a basis for V. With $x_i = Av_i \in W_1$ and $y_i = Bv_i \in W_2$, $i = 1, \ldots, m$, we have $[x_i, x_j]_1 = [Av_i, Av_j]_1 = (v_i, A'Av_j) = (v_i, B'Bv_j) = [Bv_i, Bv_j]_2 = [y_i, y_j]_2$ for $i, j = 1, \ldots, m$. Applying Proposition 1.30, there exists a linear isometry $\Psi \in \mathcal{L}(W_2, W_1)$ such that $\Psi y_i = x_i$ for $i = 1, \ldots, m$. Therefore, $\Psi B v_i = A v_i$ for $i = 1, \ldots, m$ and, since $\{v_1, \ldots, v_m\}$ is a basis for V, $\Psi B = A$. □

◆ **Example 1.11.** Take $V = R^m$ and $W = R^n$ with the usual inner products and assume $m \leqslant n$. Then a matrix $A = \{a_{ij}\} : n \times m$ is a linear isometry iff $A'A = I_m$ where I_m is the $m \times m$ identity matrix. If $a_1, \ldots, a_m$ denote the columns of the matrix A, then $A'A$ is just the $m \times m$ matrix with elements $a_i'a_j$, $i, j = 1, \ldots, m$. Thus the condition $A'A = I_m$ means that $a_i'a_j = \delta_{ij}$ so A is a linear isometry on R^m to R^n iff the columns of A are an orthonormal set of vectors in R^n. Now, let $\mathfrak{F}_{m,n}$ be the set of all $n \times m$ real matrices that are linear isometries—that is, $A \in \mathfrak{F}_{m,n}$ iff $A'A = I_m$. The set $\mathfrak{F}_{m,n}$ is sometimes called the space of *m-frames in* R^n as the columns of A form an m-dimensional orthonormal "frame" in R^n. When $m = 1$, $\mathfrak{F}_{1,n}$ is just the set of vectors in R^n of length one, and when $m = n$, $\mathfrak{F}_{n,n}$ is the set of all $n \times n$ orthogonal matrices. We have much more to say about $\mathfrak{F}_{m,n}$ in later chapters.

An immediate application of Proposition 1.31 shows that, if $A : n_1 \times m$ and $B : n_2 \times m$ are real matrices with $n_2 \leqslant n_1$, then $A'A = B'B$ iff $A = \Psi B$ where $\Psi : n_1 \times n_2$ satisfies $\Psi'\Psi = I_{n_2}$. In particular, when $n_1 = n_2$, $A'A = B'B$ iff there exists an orthogonal matrix $\Psi : n_1 \times n_1$ such that $A = \Psi B$. ◆

1.6. DETERMINANTS AND EIGENVALUES

At this point in our discussion, we are forced, by mathematical necessity, to introduce complex numbers and complex matrices. Eigenvalues are defined as the roots of a certain polynomial and, to insure the existence of roots, complex numbers arise. This section begins with complex matrices, determinants, and their basic properties. After defining eigenvalues, the properties of the eigenvalues of linear transformations on real vector spaces are described.

In what follows, $\mathbb{C}$ denotes the field of complex numbers and the symbol i is reserved for $\sqrt{-1}$. If $\alpha \in \mathbb{C}$, say $\alpha = a + ib$, then $\bar{\alpha} = a - ib$ is the complex conjugate of α. Let $\mathbb{C}^n$ be the set of all n-tuples (henceforth called vectors) of complex numbers—that is, $x \in \mathbb{C}^n$ iff

$$x = \begin{pmatrix} x_1 \\ \vdots \\ x_n \end{pmatrix}; \qquad x_j \in \mathbb{C}, \quad j = 1, \ldots, n.$$

The number x_j is called the jth coordinate of x, $j = 1, \ldots, n$. For $x, y \in \mathbb{C}^n$, $x + y$ is defined to be the vector with coordinates $x_j + y_j$, $j = 1, \ldots, n$, and for $\alpha \in \mathbb{C}$, αx is the vector with coordinates αx_j, $j = 1, \ldots, n$. Replacing R by $\mathbb{C}$ in Definition 1.1, we see that $\mathbb{C}^n$ satisfies all the axioms of a vector space where scalars are now taken to be complex numbers, rather than real numbers. More generally, if we replace R by $\mathbb{C}$ in (II) of Definition 1.1, we have the definition of a *complex vector space*. All of the definitions, results, and proofs in Sections 1.1 and 1.2 are valid, without change, for complex vector spaces. In particular, $\mathbb{C}^n$ is an n-dimensional complex vector space and the *standard basis* for $\mathbb{C}^n$ is $\{\varepsilon_1, \dots, \varepsilon_n\}$ where ε_j has its jth coordinate equal to one and the remaining coordinates are zero.

As with real matrices, an $m \times n$ array $A = \{a_{jk}\}$ for $j = 1, \dots, m$, and $k = 1, \dots, n$ where $a_{jk} \in \mathbb{C}$ is called an $m \times n$ complex matrix. If $A = \{a_{jk}\} : m \times n$ and $B = \{b_{kl}\} : n \times p$ are complex matrices, then $C = AB$ is the $m \times p$ complex matrix with with entries $c_{jl} = \sum_k a_{jk} b_{kl}$ for $j = 1, \dots, m$ and $l = 1, \dots, p$. The matrix C is called the *product* of A and B (in that order). In particular, when $p = 1$, the matrix B is $n \times 1$ so B is an element of $\mathbb{C}^n$. Thus if $x \in \mathbb{C}^n$ (x now plays the role of B) and $A : m \times n$ is a complex matrix, $Ax \in \mathbb{C}^m$. Clearly, each $A : m \times n$ determines a linear transformation on $\mathbb{C}^n$ to $\mathbb{C}^m$ via the definition of Ax for $x \in \mathbb{C}^n$. For an $m \times n$ complex matrix $A = \{a_{jk}\}$, the *conjugate transpose* of A, denoted by A^*, is the $n \times m$ matrix $A^* = \{\bar{a}_{kj}\}$, $k = 1, \dots, n$, $j = 1, \dots, m$, where $\bar{a}_{kj}$ is the complex conjugate of $a_{kj} \in \mathbb{C}$. In particular, if $x \in \mathbb{C}^n$, x^* denotes the *conjugate transpose* of x. The following relation is easily verified:

$$\overline{y^*Ax} = x^*A^*y$$

where $y \in \mathbb{C}^m$, $x \in \mathbb{C}^n$, and A is an $m \times n$ complex matrix. Of course, the bar over $\overline{y^*Ax}$ denotes the complex conjugate of $y^*Ax \in \mathbb{C}$.

With the preliminaries out of the way, we now want to define determinant functions. Let $\mathcal{C}_n$ denote the set of all $n \times n$ complex matrices so $\mathcal{C}_n$ is an n^2-dimensional complex vector space. If $A \in \mathcal{C}_n$, write $A = (a_1, a_2, \dots, a_n)$ where a_j is the jth column of A.

Definition 1.24. A function D defined on $\mathcal{C}_n$ and taking values in $\mathbb{C}$ is called a *determinant function* if

(i) $D(A) = D(a_1, \dots, a_n)$ is linear in each column vector a_j when the other columns are held fixed. That is,

$$D(a_1, \dots, \alpha a_j + \beta b_j, \dots, a_n) = \alpha D(a_1, \dots, a_j, \dots, a_n) + \beta D(a_1, \dots, b_j, \dots, a_n)$$

for $\alpha, \beta \in \mathbb{C}$.

(ii) For any two indices j and k, $j < k$,

$$D(a_1, \dots, a_j, \dots, a_k, \dots, a_n) = -D(a_1, \dots, a_k, \dots, a_j, \dots, a_n).$$

Functions D on $\mathcal{C}_n$ to $\mathbb{C}$ that satisfy (i) are called *n-linear* since they are linear in each of the n vectors $a_1, \ldots, a_n$ when the remaining ones are held fixed. If D is n-linear and satisfies (ii), D is sometimes called an *alternating* n-linear function, since $D(A)$ changes sign if two columns of A are interchanged. The basic result that relates all determinant functions is the following.

Proposition 1.32. The set of determinant functions is a one-dimensional complex vector space. If D is a determinant function and $D \not\equiv 0$, then $D(I) \neq 0$ where I is the $n \times n$ identity matrix in $\mathcal{C}_n$.

Proof. We briefly outline the proof of this proposition since the proof is instructive and yields the classical formula defining the determinant of an $n \times n$ matrix. Suppose $D(A) = D(a_1, \ldots, a_n)$ is a determinant function. For each $k = 1, \ldots, n$, $a_k = \sum_j a_{jk} \varepsilon_j$ where $\{\varepsilon_1, \ldots, \varepsilon_n\}$ is the standard basis for $\mathbb{C}_n$ and $A = \{a_{jk}\} : n \times n$. Since D is n-linear and $a_1 = \sum a_{j1} \varepsilon_j$,

$$D(a_1, \ldots, a_n) = \sum_j a_{j1} D(\varepsilon_j, a_2, \ldots, a_n).$$

Applying this same argument for $a_2 = \sum a_{j2} \varepsilon_j$,

$$D(a_1, \ldots, a_n) = \sum_{j_1} \sum_{j_2} a_{j_1 1} a_{j_2 2} D(\varepsilon_{j_1}, \varepsilon_{j_2}, a_3, \ldots, a_n).$$

Continuing in the obvious way,

$$D(a_1, \ldots, a_n) = \sum_{j_1, \ldots, j_n} a_{j_1 1} a_{j_2 2} \cdots a_{j_n n} D(\varepsilon_{j_1}, \varepsilon_{j_2}, \ldots, \varepsilon_{j_n})$$

where the summation extends over all $j_1, \ldots, j_n$ with $1 \leqslant j_l \leqslant n$ for $l = 1, \ldots, n$. The above formula shows that a determinant function is determined by the n^n numbers $D(\varepsilon_{j_1}, \ldots, \varepsilon_{j_n})$ for $1 \leqslant j_l \leqslant n$, and this fact followed solely from the assumption that D is n-linear. But since D is alternating, it is clear that, if two columns of A are the same, then $D(A) = 0$. In particular, if two indices j_l and j_k are the same, then $D(\varepsilon_{j_1}, \ldots, \varepsilon_{j_n}) = 0$. Thus the summation above extends only over those indices where $j_1, \ldots, j_n$ are all distinct. In other words, the summation extends over all permutations of the set $\{1, 2, \ldots, n\}$. If π denotes a permutation of $1, 2, \ldots, n$, then

$$D(a_1, \ldots, a_n) = \sum_{\pi} a_{\pi(1)1} \cdots a_{\pi(n)n} D(\varepsilon_{\pi(1)}, \ldots, \varepsilon_{\pi(n)})$$

where the summation now extends over all $n!$ permutations. But for a fixed permutation $\pi(1),\ldots,\pi(n)$ of $1,\ldots,n$, there is a sequence of pairwise interchanges of the elements of $\pi(1),\ldots,\pi(n)$, which results in the order $1,2,\ldots,n$. In fact there are many such sequences of interchanges, but the number of interchanges is always odd or always even (see Hoffman and Kunze, 1971, Section 5.3). Using this, let $\text{sgn}(\pi)=1$ if the number of interchanges required to put $\pi(1),\ldots,\pi(n)$ into the order $1,2,\ldots,n$ is even and let $\text{sgn}(\pi)=-1$ otherwise. Now, since D is alternating, it is clear that

$$D(\varepsilon_{\pi(1)},\ldots,\varepsilon_{\pi(n)}) = \text{sgn}(\pi)D(\varepsilon_1,\ldots,\varepsilon_n).$$

Therefore, we have arrived at the formula $D(a_1,\ldots,a_n) = D(I)\Sigma_\pi \text{sgn}(\pi)a_{\pi(1)1}\cdots a_{\pi(n)n}$ since $D(I)=D(\varepsilon_1,\ldots,\varepsilon_n)$. It is routine to verify that, for any complex number α, the function defined by

$$D_\alpha(a_1,\ldots,a_n) \equiv \alpha\sum_\pi \text{sgn}(\pi)a_{\pi(1)1}\cdots a_{\pi(n)n}$$

is a determinant function and the argument given above shows that every determinant function is a D_α for some $\alpha \in \mathbb{C}$. This completes the proof; for more details, the reader is referred to Hoffman and Kunze (1971, Chapter 5). □

Definition 1.25. If $A \in \mathcal{C}_n$, the *determinant* of A, denoted by $\det(A)$ (or $\det A$), is defined to be $D_1(A)$ where D_1 is the unique determinant function with $D_1(I)=1$.

The proof of Proposition 1.32 gives the formula for $\det(A)$, but that is not of much concern to us. The properties of $\det(\cdot)$ given below are most easily established using the fact that $\det(\cdot)$ is an alternating n-linear function of the columns of A.

Proposition 1.33. For $A, B \in \mathcal{C}_n$:

(i) $\det(AB) = \det A \det B$.
(ii) $\det A^* = \overline{\det A}$.
(iii) $\det A \neq 0$ iff the columns of A are linear independent vectors in the complex vector space $\mathbb{C}^n$.

If $A_{11}: n_1 \times n_1$, $A_{12}: n_1 \times n_2$, $A_{21}: n_2 \times n_1$, and $A_{22}: n_2 \times n_2$ are complex matrices, then:

(iv) $\det\begin{pmatrix} A_{11} & 0 \\ A_{21} & A_{22} \end{pmatrix} = \det\begin{pmatrix} A_{11} & A_{12} \\ 0 & A_{22} \end{pmatrix} = \det A_{11} \det A_{22}$.

(v) If A is a real matrix, then $\det(A)$ is real and $\det(A) = 0$ iff the columns of A are linearly dependent vectors over the real vector space R^n.

Proof. The proofs of these assertions can be found in Hoffman and Kunze (1971, Chapter 5). □

These properties of $\det(\cdot)$ have a number of useful and interesting implications. If A has columns $a_1, \dots, a_n$, then the range of the linear transformation determined by A is just $\text{span}\{a_1, \dots, a_n\}$. Thus A is invertible iff $\text{span}\{a_1, \dots, a_n\} = \mathbb{C}^n$ iff $\det A \neq 0$. If $\det A \neq 0$, then $1 = \det AA^{-1} = \det A \det A^{-1}$, so $\det A^{-1} = 1/\det A$. Consider $B_{11} : n_1 \times n_1$, $B_{12} : n_1 \times n_2$, $B_{21} : n_2 \times n_1$, and $B_{22} : n_2 \times n_2$— complex matrices. Then it is easy to verify the identity:

$$\begin{pmatrix} A_{11} & A_{12} \\ A_{21} & A_{22} \end{pmatrix}\begin{pmatrix} B_{11} & B_{12} \\ B_{21} & B_{22} \end{pmatrix} = \begin{pmatrix} A_{11}B_{11} + A_{12}B_{21} & A_{11}B_{12} + A_{12}B_{22} \\ A_{21}B_{11} + A_{22}B_{21} & A_{21}B_{12} + A_{22}B_{22} \end{pmatrix}$$

where A_{11}, A_{12}, A_{21}, and A_{22} are defined in Proposition 1.33. This tells us how to multiply the two $(n_1 + n_2) \times (n_1 + n_2)$ complex matrices in terms of their blocks. Of course, such matrices are called partitioned matrices.

Proposition 1.34. Let A be a complex matrix, partitioned as above. If $\det A_{11} \neq 0$, then:

(i) $\det\begin{pmatrix} A_{11} & A_{12} \\ A_{21} & A_{22} \end{pmatrix} = \det A_{11} \det(A_{22} - A_{21}A_{11}^{-1}A_{12})$.

If $\det A_{22} \neq 0$, then:

(ii) $\det\begin{pmatrix} A_{11} & A_{12} \\ A_{22} & A_{22} \end{pmatrix} = \det A_{22} \det(A_{11} - A_{12}A_{22}^{-1}A_{21})$.

Proof. For (i), first note that

$$\det\begin{pmatrix} I_{n_1} & -A_{11}^{-1}A_{12} \\ 0 & I_{n_2} \end{pmatrix} = 1,$$

by Proposition 1.33, (iv). Therefore, by (i) of Proposition 1.33,

$$\det\begin{pmatrix} A_{11} & A_{12} \\ A_{21} & A_{22} \end{pmatrix} = \det\begin{pmatrix} A_{11} & A_{12} \\ A_{21} & A_{22} \end{pmatrix}\begin{pmatrix} I_{n_1} & -A_{11}^{-1}A_{12} \\ 0 & I_{n_2} \end{pmatrix}$$

$$= \det\begin{pmatrix} A_{11} & 0 \\ A_{21} & A_{22} - A_{21}A_{11}^{-1}A_{12} \end{pmatrix}$$

$$= \det A_{11}\det\left(A_{22} - A_{21}A_{11}^{-1}A_{12}\right).$$

The proof of (ii) is similar. □

Proposition 1.35. Let $A : n \times m$ and $B : m \times n$ be complex matrices. Then

$$\det(I_n + AB) = \det(I_m + BA).$$

Proof. Apply the previous proposition to

$$\begin{pmatrix} I_n & -A \\ B & I_m \end{pmatrix}.$$ □

We now turn to a discussion of the eigenvalues of an $n \times n$ complex matrix. The definition of an eigenvalue is motivated by the following considerations. Let $A \in \mathcal{C}_n$. To analyze the linear transformation determined by A, we would like to find a basis $x_1, \ldots, x_n$ of $\mathbb{C}_n$ such that $Ax_j = \lambda_j x_j$, $j = 1, \ldots, n$, where $\lambda_j \in \mathbb{C}$. If this were possible, then the matrix of the linear transformation in the basis $\{x_1, \ldots, x_n\}$ would simply be

$$\begin{pmatrix} \lambda_1 & & & \\ & \lambda_2 & & \\ & & \ddots & \\ & & & \lambda_n \end{pmatrix}$$

where the elements not indicated are zero. Of course, this says that the linear transformation is λ_j times the identity transformation when restricted to span$\{x_j\}$. Unfortunately, it is not possible to find such a basis for each linear transformation. However, the numbers $\lambda_1, \ldots, \lambda_n$, which are called eigenvalues after we have an appropriate definition, can be interpreted in another way. Given $\lambda \in \mathbb{C}$, $Ax = \lambda x$ for some nonzero vector x iff $(A - \lambda I)x = 0$, and this is equivalent to saying that $A - \lambda I$ is a singular matrix, that is, $\det(A - \lambda I) = 0$. In other words, $A - \lambda I$ is singular iff there exists $x \neq 0$ such that $Ax = \lambda x$. However, using the formula for $\det(\cdot)$, a bit of calculation shows that

$$\det(A - \lambda I) = (-1)^n \lambda^n + \alpha_{n-1}\lambda^{n-1} + \cdots + \alpha_1 \lambda + \alpha_0$$

where $\alpha_0, \alpha_1, \ldots, \alpha_{n-1}$ are complex numbers. Thus $\det(A - \lambda I)$ is a polynomial of degree n in the complex variable λ, and it has n roots (counting multiplicities). This leads to the following definition.

Definition 1.26. Let $A \in \mathcal{C}_n$ and set

$$p(\lambda) = \det(A - \lambda I).$$

Then nth degree polynomial p is called the *characteristic polynomial* of A and the n roots of the polynomial (counting multiplicities) are called the *eigenvalues* of A.

If $p(\lambda) = \det(A - \lambda I)$ has roots $\lambda_1, \ldots, \lambda_n$, then it is clear that

$$p(\lambda) = \prod_{j=1}^{n} (\lambda_j - \lambda)$$

since the right-hand side of the above equation is an nth degree polynomial with roots $\lambda_1, \ldots, \lambda_n$ and the coefficient of λ^n is $(-1)^n$. In particular,

$$p(0) = \prod_1^n \lambda_j = \det(A)$$

so the determinant of A is the product of its eigenvalues.

There is a particular case when the characteristic polynomial of A can be computed explicitly. If $A \in \mathcal{C}_n$, $A = \{a_{jk}\}$ is called *lower triangular* if $a_{jk} = 0$ when $k > j$. Thus A is lower triangular if all the elements above the diagonal of A are zero. An application of Proposition 1.33 (iv) shows that when A is lower triangular, then

$$\det(A) = \prod_1^n a_{jj}.$$

But when A is lower triangular with diagonal elements a_{jj}, $j = 1, \ldots, n$, then $A - \lambda I$ is lower triangular with diagonal elements $(a_{jj} - \lambda)$, $j = 1, \ldots, n$.

Thus

$$p(\lambda) = \det(A - \lambda I) = \prod_1^n (a_{jj} - \lambda),$$

so A has eigenvalues $a_{11}, \ldots, a_{nn}$.

Before returning to real vector spaces, we first establish the existence of eigenvectors (to be defined below).

Proposition 1.36. If λ is an eigenvalue of $A \in \mathcal{C}_n$, then there exists a nonzero vector $x \in \mathbb{C}^n$ such that $Ax = \lambda x$.

Proof. Since λ is an eigenvalue of A, the matrix $A - \lambda I$ is singular, so the dimension of the range of $A - \lambda I$ is less than n. Thus the dimension of the null space of $A - \lambda I$ is greater than 0. Hence there is a nonzero vector in the null space of $A - \lambda I$, say x, and $(A - \lambda I)x = 0$. □

Definition 1.27. If $A \in \mathcal{C}_n$, a nonzero vector $x \in C^n$ is called an *eigenvector* of A if there exists a complex number $\lambda \in \mathbb{C}$ such that $Ax = \lambda x$.

If $x \neq 0$ is an eigenvector of A and $Ax = \lambda x$, then $(A - \lambda I)x = 0$ so $A - \lambda I$ is singular. Therefore, λ must be an eigenvalue for A. Conversely, if $\lambda \in \mathbb{C}$ is an eigenvalue, Proposition 1.36 shows there is an eigenvector x such that $Ax = \lambda x$.

Now, suppose V is an n-dimensional real vector space and B is a linear transformation on V to V. We want to define the characteristic polynomial, and hence the eigenvalues of B. Let $\{v_1, \ldots, v_n\}$ be a basis for V so the matrix of B is $[B] = \{b_{jk}\}$ where the b_{jk}'s satisfy $Bv_k = \sum_j b_{jk} v_j$. The characteristic polynomial of $[B]$ is

$$p(\lambda) = \det([B] - \lambda I)$$

where I is the $n \times n$ identity matrix and $\lambda \in \mathbb{C}$. If we could show that $p(\lambda)$ does not depend on the particular basis for V, then we would have a reasonable definition of the characteristic polynomial of B.

Proposition 1.37. Suppose $\{v_1, \ldots, v_n\}$ and $\{y_1, \ldots, y_n\}$ are bases for the real vector space V, and let $B \in \mathcal{L}(V, V)$. Let $[B] = \{b_{jk}\}$ be the matrix of B in the basis $\{v_1, \ldots, v_n\}$ and let $[B]_1 = \{a_{jk}\}$ be the matrix of B in the basis $\{y_1, \ldots, y_n\}$. Then there exists a nonsingular real matrix $C = \{c_{jk}\}$ such that

$$[B]_1 = C^{-1}[B]C.$$

Proof. The numbers a_{jk} are uniquely determined by the relations

$$By_k = \sum_j a_{jk} y_j, \qquad k = 1, \ldots, n.$$

Define the linear transformation C_1 on V to V by $C_1 v_j = y_j$, $j = 1, \ldots, n$. Then C_1 is nonsingular since C_1 maps a basis onto a basis. Therefore,

$$BC_1 v_k = \sum_j a_{jk} C_1 v_j = C_1\left(\sum_j a_{jk} v_j\right)$$

and this yields

$$C_1^{-1} BC_1 v_k = \sum_j a_{jk} v_j.$$

Thus the matrix of $C_1^{-1}BC_1$ in the basis $\{v_1, \ldots, v_n\}$ is $\{a_{jk}\}$. From Proposition 1.5, we have

$$[B]_1 = \{a_{jk}\} = [C_1^{-1}BC_1] = [C_1^{-1}][B][C_1] = [C_1]^{-1}[B][C_1]$$

where $[C_1]$ is the matrix of C_1 in the basis $\{v_1, \ldots, v_n\}$. Setting $C = [C_1]$, the conclusion follows. □

The above proposition implies that

$$p(\lambda) = \det([B] - \lambda I) = \det\left(C^{-1}([B] - \lambda I)C\right)$$

$$= \det\left(C^{-1}[B]C - \lambda I\right) = \det([B]_1 - \lambda I).$$

Thus $p(\lambda)$ does not depend on the particular basis we use to represent B, and, therefore, we call p the characteristic polynomial of the linear transformation B. The suggestive notation

$$p(\lambda) = \det(B - \lambda I)$$

is often used. Notice that Proposition 1.37 also shows that it makes sense to define $\det(B)$ for $B \in \mathcal{L}(V, V)$ as the value of $\det[B]$ in any basis, since the value does not depend on the basis. Of course, the roots of the polynomial $p(\lambda) = \det(B - \lambda I)$ are called the eigenvalues of the linear transformation B. Even though $[B]$ is a real matrix in any basis for V, some or all of the eigenvalues of B may be complex numbers. Proposition 1.37 also allows us to define the trace of $A \in \mathcal{L}(V, V)$. If $\{v_1, \ldots, v_n\}$ is a basis for V, let $\operatorname{tr} A \equiv \operatorname{tr}[A]$ where $[A]$ is the matrix of A in the given basis. For any nonsingular matrix C,

$$\operatorname{tr}[A] = \operatorname{tr} CC^{-1}[A] = \operatorname{tr} C^{-1}[A]C,$$

which shows that our definition of $\operatorname{tr} A$ does not depend on the particular basis chosen.

The next result summarizes the properties of eigenvalues for linear transformations on a real inner product space.

Proposition 1.38. Suppose $(V, (\cdot, \cdot))$ is a finite dimensional real inner product space and let $A \in \mathcal{L}(V, V)$.

(i) If $\lambda \in \mathbb{C}$ is an eigenvalue of A, then $\bar{\lambda}$ is an eigenvalue of A.
(ii) If A is symmetric, the eigenvalues of A are real
(iii) If A is skew-symmetric, then the eigenvalues of A are pure imaginary
(iv) If A is orthogonal and λ is an eigenvalue of A, then $\lambda\bar{\lambda} = 1$.

Proof. If $A \in \mathcal{L}(V, V)$, then the characteristic polynomial of A is

$$p(\lambda) = \det([A] - \lambda I), \qquad \lambda \in \mathbb{C},$$

where $[A]$ is the matrix of A in a basis for V. An examination of the formula for $\det(\cdot)$ shows that

$$p(\lambda) = (-1)^n \lambda^n + \alpha_{n-1}\lambda^{n-1} + \cdots + \alpha_1 \lambda + \alpha_0$$

where $\alpha_0, \ldots, \alpha_{n-1}$ are real numbers since $[A]$ is a real matrix Thus if $p(\lambda) = 0$, then $p(\bar{\lambda}) = \overline{p(\lambda)} = 0$ so whenever $p(\lambda) = 0$, $p(\bar{\lambda}) = 0$. This establishes assertion (i).

For (ii), let λ be an eigenvalue of A, and let $\{v_1, \ldots, v_n\}$ be an orthonormal basis for $(V, (\cdot, \cdot))$. Thus the matrix of A, say $[A]$, is a real symmetric matrix and $[A] - \lambda I$ is singular as a matrix acting on $\mathbb{C}^n$. By Proposition 1.36, there exists a nonzero vector $x \in \mathbb{C}^n$ such that $[A]x = \lambda x$. Thus $x^*[A]x = \lambda x^* x$. But since $[A]$ is real and symmetric,

$$\overline{x^*[A]x} = x^*[A]^* x = x^*[A]x = \overline{\lambda x^* x} = \bar{\lambda} x^* x.$$

Thus $\bar{\lambda} x^* x = \lambda x^* x$ and, since $x \neq 0$, $\bar{\lambda} = \lambda$ so λ is real.

To prove (iii), again let $[A]$ be the matrix of A in the orthonormal basis $\{v_1,\ldots,v_n\}$ so $[A]' = [A]^* = -[A]$. If λ is an eigenvalue of A, then there exists $x \in \mathbb{C}^n$, $x \neq 0$, such that $[A]x = \lambda x$. Thus $x^*[A]x = \lambda x^*x$ and

$$\bar{\lambda}x^*x = \overline{x^*[A]x} = -x^*[A]x = -\lambda x^*x.$$

Since $x \neq 0$, $\bar{\lambda} = -\lambda$, which implies that $\lambda = ib$ for some real number b—that is, λ is pure imaginary and this proves (iii).

If A is orthogonal, then $[A]$ is an $n \times n$ orthogonal matrix in the orthonormal basis $\{v_1,\ldots,v_n\}$. Again, if λ is an eigenvalue of A, then $[A]x = \lambda x$ for some $x \in \mathbb{C}^n$, $x \neq 0$. Thus $\bar{\lambda}x^* = x^*[A]^* = x^*[A]'$ since $[A]$ is a real matrix. Therefore

$$\lambda\bar{\lambda}x^*x = x^*[A]'[A]x = x^*x$$

as $[A]'[A] = I$. Hence $\lambda\bar{\lambda} = 1$ and the proof of Proposition 1.38 is complete. □

It has just been shown that if $(V,(\cdot,\cdot))$ is a finite dimensional vector space and if $A \in \mathcal{L}(V,V)$ is self-adjoint, then the eigenvalues of A are real. The spectral theorem, to be established in the next section, provides much more useful information about self-adjoint transformations. For example, one application of the spectral theorem shows that a self-adjoint transformation is positive definite iff all its eigenvalues are positive.

If $A \in \mathcal{L}(V,W)$ and $B \in \mathcal{L}(W,V)$, the next result compares the eigenvalues of $AB \in \mathcal{L}(W,W)$ with those of $BA \in \mathcal{L}(V,V)$.

Proposition 1.39. The nonzero eigenvalues of AB are the same as the nonzero eigenvalues of BA, including multiplicities. If $W = V$, AB and BA have the same eigenvalues and multiplicities.

Proof. Let $m = \dim V$ and $n = \dim W$. The characteristic polynomial of BA is

$$p_1(\lambda) = \det(BA - \lambda I_m).$$

Now, for $\lambda \neq 0$, compute as follows:

$$\begin{aligned}\det(BA - \lambda I_m) &= \det(-\lambda)\left(\frac{BA}{-\lambda} + I_m\right)\\ &= (-\lambda)^m \det\left(\frac{BA}{-\lambda} + I_m\right) = (-\lambda)^m \det\left(\frac{AB}{-\lambda} + I_n\right)\\ &= (-\lambda)^m \det\left(\frac{1}{-\lambda}\right)(AB - \lambda I_n) = \frac{(-\lambda)^m}{(-\lambda)^n}\det(AB - \lambda I_n).\end{aligned}$$

Therefore, the characteristic polynomial of AB, say $p_2(\lambda) = \det(AB - \lambda I_n)$, is related to $p_1(\lambda)$ by

$$p_1(\lambda) = \frac{(-\lambda)^m}{(-\lambda)^n} p_2(\lambda), \qquad \lambda \in \mathbb{C}, \quad \lambda \neq 0.$$

Both of the assertions follow from this relationship. □

1.7. THE SPECTRAL THEOREM

The spectral theorem for self-adjoint linear transformations on a finite dimensional real inner product space provides a basic theoretical tool not only for understanding self-adjoint transformations but also for establishing a variety of useful facts about general linear transformations. The form of the spectral theorem given below is slightly weaker than that given in Halmos (1958, see Section 79), but it suffices for most of our purposes. Applications of this result include a necessary and sufficient condition that a self-adjoint transformation be positive definite and a demonstration that positive definite transformations possess square roots. The singular value decomposition theorem, which follows from the spectral theorem, provides a useful decomposition result for linear transformations on one inner product space to another. This section ends with a description of the relationship between the singular value decomposition theorem and angles between two subspaces of an inner product space.

Let $(V, (\cdot, \cdot))$ be a finite dimensional real inner product space. The spectral theorem follows from the two results below. If $A \in \mathcal{L}(V, V)$ and M is subspace of V, M is called *invariant* under A if $A(M) = \{Ax | x \in M\} \subseteq M$.

Proposition 1.40. Suppose $A \in \mathcal{L}(V, V)$ is self-adjoint and let M be a subspace of V. If $A(M) \subseteq M$, then $A(M^\perp) \subseteq M^\perp$.

Proof. Suppose $v \in A(M^\perp)$. It must be shown that $(v, x) = 0$ for all $x \in M$. Since $v \in A(M^\perp)$, $v = Av_1$ for $v_1 \in M^\perp$. Therefore,

$$(v, x) = (Av_1, x) = (v_1, Ax) = 0$$

since A is self-adjoint and $x \in M$ implies $Ax \in M$ by assumption. □

Proposition 1.41. Suppose $A \in \mathcal{L}(V, V)$ is self-adjoint and λ is an eigenvalue of A. Then there exists a $v \in V$, $v \neq 0$, such that $Av = \lambda v$.

Proof. Since A is self-adjoint, the eigenvalues of A are real. Let $\{v_1, \ldots, v_n\}$ be a basis for V and let $[A]$ be the matrix of A in this basis. By Proposition 1.36, there exists a nonzero vector $z \in \mathbb{C}^n$ such that $[A]z = \lambda z$. Write $z = z_1 + iz_2$ where $z_1 \in R^n$ is the real part of z and $z_2 \in R^n$ is the imaginary part of z. Since $[A]$ is real and λ is real, we have $[A]z_1 = \lambda z_1$ and $[A]z_2 = \lambda z_2$. But, z_1 and z_2 cannot both be zero as $z \neq 0$. For definiteness, say $z_1 \neq 0$ and let $v \in V$ be the vector whose coordinates in basis $\{v_1, \ldots, v_n\}$ are z_1. Then $v \neq 0$ and $[A][v] = \lambda[v]$. Therefore $Av = \lambda v$. □

Theorem 1.2 (Spectral Theorem). If $A \in \mathcal{L}(V, V)$ is self-adjoint, then there exists an orthonormal basis $\{x_1, \ldots, x_n\}$ for V and real numbers $\lambda_1, \ldots, \lambda_n$ such that

$$A = \sum_1^n \lambda_i x_i \square x_i.$$

Further, $\lambda_1, \ldots, \lambda_n$ are the eigenvalues of A and $Ax_i = \lambda_i x_i$, $i = 1, \ldots, n$.

Proof. The proof of the first assertion is by induction on dimension. For $n = 1$, the result is obvious. Assume the result is true for integers $1, 2, \ldots, n - 1$ and consider $A \in \mathcal{L}(V, V)$, which is self-adjoint on the inner product space $(V, (\cdot, \cdot))$, $n = \dim V$. Let λ be an eigenvalue of A. By Proposition 1.41, there exists $v \in V$, $v \neq 0$, such that $Av = \lambda v$. Set $x_n = v/\|v\|$ and $\lambda_n = \lambda$. Then $Ax_n = \lambda_n x_n$. With $M = \text{span}\{x_n\}$, it is clear that $A(M) \subseteq M$ so $A(M^\perp) \subseteq M^\perp$ by Proposition 1.40. However, if we let A_1 be the restriction of A to the $(n - 1)$-dimensional inner product space $(M^\perp, (\cdot, \cdot))$, then A_1 is clearly self-adjoint. By the induction hypothesis there is an orthonormal basis $\{x_1, \ldots, x_{n-1}\}$ for $M^\perp$ and real numbers $\lambda_1, \ldots, \lambda_{n-1}$ such that

$$A_1 = \sum_1^{n-1} \lambda_i x_i \square x_i.$$

It is clear that $\{x_1, \ldots, x_n\}$ is an orthonormal basis for V and we claim that

$$A = \sum_1^n \lambda_i x_i \square x_i.$$

To see this, consider $v_0 \in V$ and write $v_0 = v_1 + v_2$ with $v_1 \in M$ and $v_2 \in M^\perp$. Then

$$Av_0 = Av_1 + Av_2 = \lambda_n v_1 + A_1 v_2 = \lambda_n v_1 + \sum_1^{n-1} \lambda_i (x_i \square x_i) v_2.$$

However,

$$\left(\sum_1^n \lambda_i x_i \square x_i\right)(v_1 + v_2) = \lambda_n (v_1, x_n) x_n + \sum_1^{n-1} \lambda_i (x_i \square x_i) v_2$$

since $v_1 \in M$ and $v_2 \in M^{\perp}$. But $(v_1, x_n)x_n = v_1$ since $v_1 \in \text{span}\{x_n\}$. Therefore $A = \sum_1^n \lambda_i x_i \square x_i$, which establishes the first assertion.

For the second assertion, if $A = \sum_1^n \lambda_i x_i \square x_i$ where $\{x_1, \dots, x_n\}$ is an orthonormal basis for $(V, (\cdot, \cdot))$, then

$$Ax_j = \sum_i \lambda_i (x_i \square x_i) x_j = \sum_i \lambda_i (x_i, x_j) x_i = \lambda_j x_j.$$

Thus the matrix of A, say $[A]$, in this basis has diagonal elements $\lambda_1, \dots, \lambda_n$ and all other elements of $[A]$ are zero. Therefore the characteristic polynomial of A is

$$p(\lambda) = \det([A] - \lambda I) = \prod_1^n (\lambda_i - \lambda),$$

which has roots $\lambda_1, \dots, \lambda_n$. The proof of the spectral theorem is complete. □

When $A = \sum \lambda_i x_i \square x_i$, then A is particularly easy to understand. Namely, A is λ_i times the identity transformation when restricted to $\text{span}\{x_i\}$. Also, if $x \in V$, then $x = \sum (x_i, x) x_i$ so $Ax = \sum \lambda_i (x_i, x) x_i$. In the case when A is an orthogonal projection onto the subspace M, we know that $A = \sum_1^k x_i \square x_i$ where $k = \dim M$ and $\{x_1, \dots, x_k\}$ is an orthonormal basis for M. Thus A has eigenvalues of zero and one, and one occurs with multiplicity $k = \dim M$. Conversely, the spectral theorem implies that, if A is self-adjoint and has only zero and one as eigenvalues, then A is an orthogonal projection onto a subspace of dimensional equal to the multiplicity of the eigenvalue one.

We now begin to reap the benefits of the spectral theorem.

Proposition 1.42. If $A \in \mathcal{L}(V, V)$, then A is positive definite iff all the eigenvalues of A are strictly positive. Also, A is positive semidefinite iff the eigenvalues of A are non-negative.

Proof. Write A in spectral form:

$$A = \sum_1^n \lambda_i x_i \square x_i$$

where $\{x_1, \ldots, x_n\}$ is an orthonormal basis for $(V, (\cdot, \cdot))$. Then $(x, Ax) = \Sigma\lambda_i(x_i, x)^2$. If $\lambda_i > 0$ for $i = 1, \ldots, n$, then $x \neq 0$ implies that $\Sigma\lambda_i(x_i, x)^2 > 0$ and A is positive definite. Conversely, if A is positive definite, set $x = x_j$ and we have $0 < (x_j, Ax_j) = \lambda_j$. Thus all the eigenvalues of A are strictly positive. The other assertion is proved similarly. □

The representation of A in spectral form suggests a way to define various functions of A. If $A = \Sigma\lambda_i x_i \square x_i$, then

$$A^2 = \Big(\sum_i \lambda_i x_i \square x_i\Big)\Big(\sum_j \lambda_j x_j \square x_j\Big) = \sum_i\sum_j \lambda_i\lambda_j (x_i \square x_i)(x_j \square x_j)$$

$$= \sum_i\sum_j \lambda_i\lambda_j (x_i, x_j) x_i \square x_j = \sum_i \lambda_i^2 x_i \square x_i.$$

More generally, if k is a positive integer, a bit of calculation shows that

$$A^k = \sum_i \lambda_i^k x_i \square x_i, \qquad k = 1, 2, \ldots.$$

For $k = 0$, we adopt the convention that $A^0 = I$ since $\Sigma x_i \square x_i = I$. Now if p is any polynomial on R, the above equation forces us to define $p(A)$ by

$$p(A) = \sum_i p(\lambda_i) x_i \square x_i.$$

This suggests that, if f is any real-valued function that is defined at $\lambda_1, \ldots, \lambda_n$, we should define $f(A)$ by

$$f(A) = \sum_i f(\lambda_i) x_i \square x_i.$$

Adopting this suggestive definition shows that if $\lambda_1, \ldots, \lambda_n$ are the eigenvalues of A, then $f(\lambda_1), \ldots, f(\lambda_n)$ are the eigenvalues of $f(A)$. In particular, if $\lambda_i \neq 0$ for all i and $f(t) = t^{-1}$, $t \neq 0$, then it is clear that $f(A) = A^{-1}$. Another useful choice for f is given in the following proposition.

Proposition 1.43. If $A \in \mathcal{L}(V, V)$ is positive semidefinite, then there exists a $B \in \mathcal{L}(V, V)$ that is positive semidefinite and satisfies $B^2 = A$.

Proof. Choose $f(t) = t^{1/2}$, and let

$$B \equiv f(A) = \sum_1^n \lambda_i^{1/2} x_i \square x_i.$$

The square root is well defined since $\lambda_i \geqslant 0$ for $i = 1, \ldots, n$ as A is positive semidefinite. Since B has non-negative eigenvalues, B is positive definite. That $B^2 = A$ is clear. $\square$

There is a technical problem with our definition of $f(A)$ that is caused by the nonuniqueness of the representation

$$A = \sum_1^n \lambda_i x_i \square x_i$$

for self-adjoint transformations. For example, if the first n_1 λ_i's are equal and the last $n - n_1$ λ_i's are equal, then

$$A = \lambda_1 \left(\sum_1^{n_1} x_i \square x_i \right) + \lambda_n \left(\sum_{n_1+1}^{n} x_i \square x_i \right).$$

However, $\Sigma_1^{n_1} x_i \square x_i$ is the orthogonal projection onto $M_1 \equiv \text{span}(x_1, \ldots, x_{n_1}\}$. If $y_1, \ldots, y_n$ is any other orthonormal basis for $(V, (\cdot, \cdot))$ such that $\text{span}\{x_1, \ldots, x_{n_1}\} = \text{span}\{y_1, \ldots, y_{n_1}\}$, it is clear that

$$A = \lambda_1 \sum_1^{n_1} y_i \square y_i + \lambda_n \sum_{n_1+1}^{n} y_i \square y_i = \sum_1^n \lambda_i y_i \square y_i.$$

Obviously, $\lambda_1, \ldots, \lambda_n$ are uniquely defined as the eigenvalues for A (counting multiplicities), but the orthonormal basis $\{x_1, \ldots, x_n\}$ providing the spectral form for A is not unique. It is therefore necessary to verify that the definition of $f(A)$ does not depend on the particular orthonormal basis in the representation for A or to provide an alternative representation for A. It is this latter alternative that we follow. The result below is also called the spectral theorem.

Theorem 1.2a (Spectral Theorem). Suppose A is a self-adjoint linear transformation on V to V where $n = \dim V$. Let $\lambda_1 > \lambda_2 > \cdots > \lambda_r$ be the distinct eigenvalues of A and let n_i be the multiplicity of λ_i, $i = 1, \ldots, r$. Then there exists orthogonal projections $P_1, \ldots, P_r$ with $P_i P_j = 0$ for $i \neq j$, $n_i = \text{rank}(P_i)$, and $\Sigma_1^n P_i = I$ such that

$$A = \sum_1^r \lambda_i P_i.$$

Further, this decomposition is unique in the following sense. If $\mu_1 > \cdots >$

μ_k and $Q_1, \ldots, Q_k$ are orthogonal projections such that $Q_iQ_j = 0$ for $i \neq j$, $\Sigma Q_i = I$, and

$$A = \sum_1^k \mu_i Q_i,$$

then $k = r$, $\mu_i = \lambda_i$, and $Q_i = P_i$ for $i = 1, \ldots, k$.

Proof. The first assertion follows immediately from the spectral representation given in Theorem 1.2. For a proof of the uniqueness assertion, see Halmos (1958, Section 79). □

Now, our definition of $f(A)$ is

$$f(A) = \sum_1^r f(\lambda_i) P_i$$

when $A = \Sigma_1^r \lambda_i P_i$. Of course, it is assumed that f is defined at $\lambda_1, \ldots, \lambda_r$. This is exactly the same definition as before, but the problem about the nonuniqueness of the representation of A has disappeared. One application of the uniqueness part of the above theorem is that the positive semidefinite square root given in Proposition 1.43 is unique. The proof of this is left to the reader (see Halmos, 1958, Section 82).

Other functions of self-adjoint linear transformations come up later and we consider them as the need arises. Another application of the spectral theorem solves an interesting extremal problem. To motivate this problem, suppose A is self-adjoint on $(V, (\cdot, \cdot))$ with eigenvalues $\lambda_1 \geqslant \lambda_2 \geqslant \cdots \geqslant \lambda_n$. Thus $A = \Sigma \lambda_i x_i \Box x_i$ where $\{x_1, \ldots, x_n\}$ is an orthonormal basis for V. For $x \in V$ and $\|x\| = 1$, we ask how large (x, Ax) can be. To answer this, write $(x, Ax) = \Sigma \lambda_i (x, (x_i \Box x_i)x) = \Sigma \lambda_i (x, x_i)^2$, and note that $0 \leqslant (x, x_i)^2$ and $1 = \|x\|^2 = \Sigma_1^n (x_i, x)^2$. Therefore, $\lambda_1 \geqslant \Sigma \lambda_i (x, x_i)^2$ with equality for $x = x_1$. The conclusion is

$$\sup_{x, \|x\| = 1} (x, Ax) = \lambda_1$$

where λ_1 is the largest eigenvalue of A. This result also shows that $\lambda_1(A)$—the largest eigenvalue of the self-adjoint transformation A—is a convex function of A. In other words, if A_1 and A_2 are self-adjoint and $\alpha \in [0, 1]$, then $\lambda_1(\alpha A_1 + (1 - \alpha) A_2) \leqslant \alpha \lambda_1(A_1) + (1 - \alpha)\lambda_1(A_2)$. To prove this, first notice that for each $x \in V$, (x, Ax) is a linear, and hence convex, function of A. Since the supremum of a family of convex functions is a convex function, it follows that

$$\lambda_1(A) = \sup_{x,\|x\|=1} (x, Ax)$$

is a convex function defined on the real linear space of self-adjoint linear transformations. An interesting generalization of this is the following.

Proposition 1.44. Consider a self-adjoint transformation A defined on the n-dimensional inner product space $(V, (\cdot, \cdot))$ and let $\lambda_1 \geqslant \lambda_2 \geqslant \cdots \geqslant \lambda_n$ be the ordered eigenvalues of A. For $1 \leqslant k \leqslant n$, let $\mathfrak{B}_k$ be the collection of all k-tuples $\{v_1, \ldots, v_k\}$ such that $\{v_1, \ldots, v_k\}$ is an orthonormal set in $(V, (\cdot, \cdot))$. Then

$$\sup_{\{v_1,\ldots,v_k\}\in\mathfrak{B}_k} \sum_1^k (v_i, Av_i) = \sum_1^k \lambda_i.$$

Proof. Recall that $\langle \cdot, \cdot \rangle$ is the inner product on $\mathcal{L}(V, V)$ induced by the inner product $(\cdot, \cdot)$ on V, and $(x, Ax) = \langle x \square x, A\rangle$ for $x \in V$. Thus

$$\sum_1^k (v_i, Av_i) = \sum_1^k \langle v_i \square v_i, A\rangle = \left\langle \sum_1^k v_i \square v_i, A \right\rangle.$$

Write A in spectral form, $A = \sum_1^n \lambda_1 x_i \square x_i$. For $\{v_1, \ldots, v_k\} \in \mathfrak{B}_k$, $P_k = \sum_1^k v_i \square v_i$ is the orthogonal projection onto span$\{v_1, \ldots, v_k\}$. Thus for $\{v_1, \ldots, v_k\} \in \mathfrak{B}_k$,

$$\left\langle \sum_i^k v_i \square v_i, A \right\rangle = \left\langle P_k, \sum_1^n \lambda_i x_i \square x_i \right\rangle$$

$$= \sum_1^n \langle P_k, \lambda_i x_i \square x_i\rangle = \sum_1^n \lambda_i (x_i, P_k x_i).$$

Since P_k is an orthogonal projection and $\|x_i\| = 1$, $i = 1, \ldots, n$, $0 \leqslant (x_i, P_k x_i) \leqslant 1$. Also,

$$\sum_1^n (x_i, P_k x_i) = \sum_1^n \langle P_k, x_i \square x_i\rangle = \left\langle P_k, \sum_1^n x_i \square x_i \right\rangle = \langle P_k, I\rangle$$

because $\sum_1^n x_i \square x_i = I \in \mathcal{L}(V, V)$. But $P_k = \sum_1^k v_i \square v_i$, so

$$\langle P_k, I\rangle = \sum_1^k \langle v_i \square v_i, I\rangle = \sum_1^k (v_i, v_i) = k.$$

Therefore, the real numbers $\alpha_i = (x_i, P_k x_i)$, $i = 1,\ldots, n$, satisfy $0 \leqslant \alpha_i \leqslant 1$ and $\Sigma_1^n \alpha_i = k$. A moment's reflection shows that, for any numbers $\alpha_1,\ldots, \alpha_n$ satisfying these conditions, we have

$$\sum_1^n \lambda_i \alpha_i \leqslant \sum_1^k \lambda_i$$

since $\lambda_1 \geqslant \cdots \geqslant \lambda_n$. Therefore,

$$\left\langle \sum_1^k v_i \Box v_i, A \right\rangle \leqslant \sum_1^k \lambda_i$$

for $\{v_1,\ldots, v_k\} \in \mathfrak{B}_k$. However, setting $v_i = x_i$, $i = 1,\ldots, k$, yields equality in the above inequality. □

For $A \in \mathcal{L}(V, V)$, which is self-adjoint, define $\operatorname{tr}_k A = \Sigma_1^k \lambda_i$ where $\lambda_1 \geqslant \cdots \geqslant \lambda_n$ are the ordered eigenvalues of A. The symbol $\operatorname{tr}_k A$ is read "trace sub-k of A." Since $\langle \Sigma_1^k v_i \Box v_i, A\rangle$ is a linear function of A and $\operatorname{tr}_k A$ is the supremum over all $\{v_1,\ldots, v_k\} \in \mathfrak{B}_k$, it follows that $\operatorname{tr}_k A$ is a convex function of A. Of course, when $k = n$, $\operatorname{tr}_k A$ is just the trace of A.

For completeness, a statement of the spectral theorem for $n \times n$ symmetric matrices is in order.

Proposition 1.45. Suppose A is an $n \times n$ real symmetric matrix. Then there exists an $n \times n$ orthogonal matrix Γ and an $n \times n$ diagonal matrix D such that $A = \Gamma D \Gamma'$. The columns of Γ are the eigenvectors of A and the diagonal elements of D, say $\lambda_1,\ldots, \lambda_n$, are the eigenvalues of A.

Proof. This is nothing more than a disguised version of the spectral theorem. To see this, write

$$A = \sum_1^n \lambda_i x_i x_i'$$

where $x_i \in R^n$, $\lambda_i \in R$, and $\{x_1,\ldots, x_n\}$ is an orthonormal basis for R^n with the usual inner product (here $x_i \Box x_i$ is $x_i x_i'$ since we have the usual inner product on R^n). Let Γ have columns $x_1,\ldots, x_n$ and let D have diagonal elements $\lambda_1,\ldots, \lambda_n$. Then a straightforward computation shows that

$$\sum_1^n \lambda_i x_i x_i' = \Gamma D \Gamma'.$$

The remaining assertions follow immediately from the spectral theorem. □

Our final application of the spectral theorem in this chapter deals with a representation theorem for a linear transformation $A \in \mathcal{L}(V, W)$ where $(V, (\cdot, \cdot))$ and $(W, [\cdot, \cdot])$ are finite dimensional inner product spaces. In this context, eigenvalues and eigenvectors of A make no sense, but something can be salvaged by considering $A'A \in \mathcal{L}(V, V)$. First, $A'A$ is non-negative definite and $\mathfrak{N}(A'A) = \mathfrak{N}(A)$. Let $k = \text{rank}(A) = \text{rank}(A'A)$ and let $\lambda_1 \geqslant \cdots \geqslant \lambda_k > 0$ be the nonzero eigenvalues of $A'A$. There must be exactly k positive eigenvalues of $A'A$ as $\text{rank}(A) = k$. The spectral theorem shows that

$$A'A = \sum_1^k \lambda_i x_i \Box x_i$$

where $\{x_1, \ldots, x_n\}$ is an orthonormal basis for V and $A'Ax_i = \lambda_i x_i$ for $i = 1, \ldots, k$, $A'Ax_i = 0$ for $i = k + 1, \ldots, n$. Therefore, $\mathfrak{N}(A) = \mathfrak{N}(A'A) = (\text{span}\{x_1, \ldots, x_k\})^{\perp}$.

Proposition 1.46. In the notation above, let $w_i = (1/\sqrt{\lambda_i})Ax_i$ for $i = 1, \ldots, k$. Then $\{w_1, \ldots, w_k\}$ is an orthonormal basis for $\mathfrak{R}(A) \subseteq W$ and $A = \sum_1^k \sqrt{\lambda_i}\, w_i \Box x_i$.

Proof. Since $\dim \mathfrak{R}(A) = k$, $\{w_1, \ldots, w_k\}$ is a basis for $\mathfrak{R}(A)$ if $\{w_1, \ldots, w_k\}$ is an orthonormal set. But

$$\begin{aligned}[w_i, w_j] &= (\lambda_i \lambda_j)^{-1/2}[Ax_i, Ax_j] = (\lambda_i \lambda_j)^{-1/2}(x_i, A'Ax_j) \\ &= (\lambda_i \lambda_j)^{-1/2}\lambda_j (x_i, x_j) = \delta_{ij}\end{aligned}$$

and the first assertion holds. To show $A = \sum_1^k \sqrt{\lambda_i}\, w_i \Box x_i$, we verify the two linear transformations agree on the basis $\{x_1, \ldots, x_n\}$. For $1 \leqslant j \leqslant k$, $Ax_j = \sqrt{\lambda_j}\, w_j$ by definition and

$$\left(\sum_1^k \sqrt{\lambda_i}\, w_i \Box x_i\right) x_j = \sum_1^k \sqrt{\lambda_i}\,(x_i, x_j) w_i = \sqrt{\lambda_j}\, w_j.$$

For $k + 1 \leqslant j \leqslant n$, $Ax_j = 0$ since $\mathfrak{N}(A) = (\text{span}\{x_1, \ldots, x_k\})^{\perp}$. Also

$$\left(\sum_1^k \sqrt{\lambda_i}\, w_i \Box x_i\right) x_j = \sum_1^k \sqrt{\lambda_i}\,(x_i, x_j) w_i = 0$$

as $j > k$. □

Some immediate consequences of the above representation are (i) $AA' = \sum_1^k \lambda_i w_i \square w_i$, (ii) $A' = \sum_1^k \sqrt{\lambda_i}\, x_i \square w_i$ and $A'w_i = \sqrt{\lambda_i}\, x_i$ for $i = 1, \ldots, k$. In summary, we have the following.

Theorem 1.3 (Singular Value Decomposition Theorem). Given $A \in \mathcal{L}(V, W)$ of rank k, there exist orthonormal vectors $x_1, \ldots, x_k$ in V and $w_1, \ldots, w_k$ in W and positive numbers $\mu_1, \ldots, \mu_k$ such that

$$A = \sum_1^k \mu_i w_i \square x_i.$$

Also, $\mathcal{R}(A) = \text{span}\{w_1, \ldots, w_k\}$, $\mathcal{N}(A) = (\text{span}\{x_1, \ldots, x_k\})^{\perp}$, $Ax_i = \mu_i w_i$, $i = 1, \ldots, k$, $A' = \sum_1^k \mu_i x_i \square w_i$, $A'A = \sum_1^k \mu_i^2 x_i \square x_i$, $AA' = \sum_1^k \mu_i^2 w_i \square w_i$. The numbers $\mu_1^2, \ldots, \mu_k^2$ are the positive eigenvalues of both AA' and $A'A$.

For matrices, this result takes the following form.

Proposition 1.47. If A is a real $n \times m$ matrix of rank k, then there exist matrices $\Gamma : n \times k$, $D : k \times k$, and $\Psi : k \times m$ that satisfy $\Gamma'\Gamma = I_k$, $\Psi\Psi' = I_k$, D is a diagonal matrix with positive diagonal elements, and

$$A = \Gamma D \Psi.$$

Proof. Take $V = R^m$, $W = R^n$ and apply Theorem 1.3 to get

$$A = \sum_1^k \mu_i w_i x_i'$$

where $x_1, \ldots, x_k$ are orthonormal in R^m, $w_1, \ldots, w_k$ are orthonormal in R^n, and $\mu_i > 0$, $i = 1, \ldots, k$. Let Γ have columns $w_1, \ldots, w_k$, let Ψ have rows $x_1', \ldots, x_k'$, and let D be diagonal with diagonal elements $\mu_1, \ldots, \mu_k$. An easy calculation shows that

$$\sum_1^k \mu_i w_i x_i' = \Gamma D \Psi. \qquad \square$$

In the case that $A \in \mathcal{L}(V, V)$ with rank k, Theorem 1.3 shows that there exist orthonormal sets $\{x_1, \ldots, x_k\}$ and $\{w_1, \ldots, w_k\}$ of V such that

$$A = \sum_1^k \mu_i w_i \square x_i$$

where $\mu_i > 0$, $i = 1, \ldots, k$. Also, $\mathcal{R}(A) = \text{span}\{w_1, \ldots, w_k\}$ and $\mathcal{N}(A) =$

$(\text{span}\{x_1,\ldots,x_k\})^\perp$. Now, consider two subspaces M_1 and M_2 of the inner product space $(V,(\cdot,\cdot))$ and let P_1 and P_2 be the orthogonal projections onto M_1 and M_2. In what follows, the geometrical relationship between the two subspaces (measured in terms of angles, which are defined below) is related to the singular value decomposition of the linear transformation $P_1P_2 \in \mathcal{L}(V,V)$. It is clear that $\mathcal{R}(P_2P_1) \subseteq M_2$ and $\mathcal{N}(P_2P_1) \supseteq M_1^\perp$. Let $k = \text{rank}(P_2P_1)$ so $k \leqslant \dim(M_i)$, $i = 1,2$. Theorem 1.3 implies that

$$P_2P_1 = \sum_1^k \mu_i w_i \,\Box\, x_i$$

where $\mu_i > 0$, $i = 1,\ldots,k$, $\mathcal{R}(P_2P_1) = \text{span}\{w_1,\ldots,w_k\} \subseteq M_2$, and $(\mathcal{N}(P_2P_1))^\perp = \text{span}\{x_1,\ldots,x_k\} \subseteq M_1$. Also, $\{w_1,\ldots,w_k\}$ and $\{x_1,\ldots,x_k\}$ are orthonormal sets. Since $P_2P_1x_j = \mu_j w_j$ and $(P_2P_1)'P_2P_1 = P_1P_2P_2P_1 = P_1P_2P_1 = \Sigma_1^k \mu_i^2 x_i \,\Box\, x_i$, we have

$$\mu_j(x_i, w_j) = (x_i, P_2P_1x_j) = (P_1x_i, P_2P_1x_j) = (x_i, P_1P_2P_1x_j)$$

$$= \left(x_i, \left(\sum_1^k \mu_l^2 x_l \,\Box\, x_l\right)x_j\right) = \mu_j^2(x_i, x_j) = \delta_{ij}\mu_j^2.$$

Therefore, for $i, j = 1,\ldots,k$,

$$(x_i, w_j) = \delta_{ij}\mu_j$$

since $\mu_j > 0$. Furthermore, if $x \in M_1 \cap (\text{span}\{x_1,\ldots,x_k\})^\perp$ and $w \in M_2$, then $(x,w) = (P_1x, P_2w) = (P_2P_1x, w) = 0$ since $P_2P_1x = 0$. Similarly, if $w \in M_2 \cap (\text{span}\{w_1,\ldots,w_k))^\perp$ and $x \in M_1$, then $(x,w) = 0$.

The above discussion yields the following proposition.

Proposition 1.48. Suppose M_1 and M_2 are subspaces of $(V,(\cdot,\cdot))$ and let P_1 and P_2 be the orthogonal projections onto M_1 and M_2. If $k = \text{rank}(P_2P_1)$, then there exist orthonormal sets $\{x_1,\ldots,x_k\} \subseteq M_1$, $\{w_1,\ldots,w_k\} \subseteq M_2$ and positive numbers $\mu_1 \geqslant \cdots \geqslant \mu_k$ such that:

(i) $P_2P_1 = \Sigma_1^k \mu_i w_i \,\Box\, x_i$.
(ii) $P_1P_2P_1 = \Sigma_1^k \mu_i^2 x_i \,\Box\, x_i$.
(iii) $P_2P_1P_2 = \Sigma_1^k \mu_i^2 w_i \,\Box\, w_i$.
(iv) $0 < \mu_j \leqslant 1$ and $(x_i, w_j) = \delta_{ij}\mu_j$ for $i, j = 1,\ldots,k$.
(v) If $x \in M_1 \cap (\text{span}\{x_1,\ldots,x_k\})^\perp$ and $w \in M_2$, then $(x,w) = 0$. If $w \in M_2 \cap (\text{span}\{w_1,\ldots,w_k\})^\perp$ and $x \in M_1$, then $(x,w) = 0$.

Proof. Assertions (i), (ii), (iii), and (v) have been verified as has the relationship $(x_i, w_j) = \delta_{ij}\mu_j$. Since $0 < \mu_j = (x_j, w_j)$, the Cauchy–Schwarz Inequality yields $(x_j, w_j) \leqslant \|x_j\|\,\|w_j\| = 1$. □

In Proposition 1.48, if $k = \text{rank } P_2P_1 = 0$, then M_1 and M_2 are orthogonal to each other and $P_1P_2 = P_2P_1 = 0$. The next result provides the framework in which to relate the numbers $\mu_1 \geqslant \cdots \geqslant \mu_k$ to angles.

Proposition 1.49. In the notation of Proposition 1.48, let $M_{11} = M_1$, $M_{21} = M_2$,

$$M_{1i} = (\text{span}\{x_1, \ldots, x_{i-1}\})^{\perp} \cap M_1,$$

and

$$M_{2i} = (\text{span}\{w_1, \ldots, w_{i-1}\})^{\perp} \cap M_2$$

for $i = 2, \ldots, k + 1$. Also, for $i = 1, \ldots, k$, let

$$D_{1i} = \{x | x \in M_{1i}, \|x\| = 1\}$$

and

$$D_{2i} = \{w | w \in M_{2i}, \|w\| = 1\}.$$

Then

$$\sup_{x \in D_{1i}} \sup_{w \in D_{2i}} (x, w) = (x_i, w_i) = \mu_i$$

for $i = 1, \ldots, k$. Also, $M_{1(k+1)} \perp M_2$ and $M_{2(k+1)} \perp M_1$.

Proof. Since $x_i \in D_{1i}$ and $w_i \in D_{2i}$, the iterated supremum is at least (x_i, w_i) and $(x_i, w_i) = \mu_i$ by (iv) of Proposition 1.48. Thus it suffices to show that for each $x \in D_{1i}$ and $w \in D_{2i}$, we have the inequality $(x, w) \leqslant \mu_i$. However, for $x \in D_{1i}$ and $w \in D_{2i}$,

$$(x, w) = (P_1x, P_2w) = (P_2P_1x, w) \leqslant \|P_2P_1x\|\,\|w\| = \|P_2P_1x\|$$

since $\|w\| = 1$ as $w \in D_{1i}$. Thus

$$(x, w) \leqslant \|P_2P_1x\| = (P_2P_1x, P_2P_1x)^{1/2} = (P_1P_2P_1x, x)^{1/2}$$

$$= \left[\sum_{j=1}^{k} \mu_j^2((x_j \Box x_j)x, x)\right]^{1/2} = \left[\sum_{j=1}^{k} \mu_j^2(x_j, x)^2\right]^{1/2}.$$

Since $x \in D_{1i}$, $(x, x_j) = 0$ for $j = 1, \ldots, i - 1$. Also, the numbers $a_j \equiv$

$(x_j, x)^2$ satisfy $0 \leqslant a_j \leqslant 1$ and $\Sigma_1^k a_j \leqslant 1$ as $\|x\| = 1$. Therefore,

$$(x, w) \leqslant \left[\sum_{j=1}^{k} \mu_j^2 (x_j, x)^2\right]^{1/2} = \left[\sum_{j=i}^{k} a_j \mu_j^2\right]^{1/2} \leqslant \left(\mu_i^2\right)^{1/2} = \mu_i.$$

The last inequality follows from the fact that $\mu_1 \geqslant \cdots \geqslant \mu_k > 0$ and the conditions on the a_j's. Hence,

$$\sup_{x \in D_{1i}} \sup_{w \in D_{2i}} (x, w) = (x_i, w_i) = \mu_i,$$

and the first assertion holds. The second assertion is simply a restatement of (v) of Proposition 1.48. □

Definition 1.28. Let M_1 and M_2 be subspaces of $(V, (\cdot, \cdot))$. Given the numbers $\mu_1 \geqslant \cdots \geqslant \mu_k > 0$, whose existence is guaranteed by Proposition 1.48, define $\theta_i \in [0, \pi/2)$ by

$$\cos\theta_i = \mu_i, \qquad i = 1, \ldots, k.$$

Let $t = \min\{\dim M_1, \dim M_2\}$ and set $\theta_i = \pi/2$ for $i = k+1, \ldots, t$. The numbers $\theta_1 \leqslant \theta_2 \leqslant \cdots \leqslant \theta_t$ are called the *ordered angles* between M_1 and M_2.

The following discussion is intended to provide motivation, explanation, and a geometric interpretation of the above definition. Recall that if y_1 and y_2 are two vectors in $(V, (\cdot, \cdot))$ of length 1, then the cosine of the angle between y_1 and y_2 is defined by $\cos\theta = (y_1, y_2)$ where $0 \leqslant \theta \leqslant \pi$. However, if we want to define the angle between the two lines $\text{span}\{y_1\}$ and $\text{span}\{y_2\}$, then a choice must be made between two angles that are complements of each other. The convention adopted here is to choose the angle in $[0, \pi/2]$. Thus the cosine of the angle between $\text{span}\{y_1\}$ and $\text{span}\{y_2\}$ is just $|(y_1, y_2)|$. To show this agrees with the definition above, we have $M_i = \text{span}\{y_i\}$ and $P_i = y_i \square y_i$ is the orthogonal projection onto M_i, $i = 1, 2$. The rank of $P_2 P_1$ is either zero or one and the rank is zero iff $y_1 \perp y_2$. If $y_1 \perp y_2$, then the angle between M_1 and M_2 is $\pi/2$, which agrees with Definition 1.28. When the rank of $P_2 P_1$ is one, $P_1 P_2 P_1 = (y_1, y_2)^2 y_1 \square y_1$ whose only nonzero eigenvalue is $(y_1, y_2)^2$. Thus $\mu_1^2 = (y_1, y_2)^2$ so $\mu_1 = |(y_1, y_2)| = \cos\theta_1$, and again we have agreement with Definition 1.28.

Now consider the case when $M_1 = \text{span}\{y_1\}$, $\|y_1\| = 1$, and M_2 is an arbitrary subspace of $(V, (\cdot, \cdot))$. Geometrically, it is clear that the angle between M_1 and M_2 is just the angle between M_1 and the orthogonal projection of M_1 onto M_2, say $M_2^* = \text{span}\{P_2 y_1\}$ where P_2 is the orthogonal projection onto M_2. Thus the cosine of the angle between M_1 and M_2 is

$$\cos\theta = \left|\left(y_1, \frac{P_2 y_1}{\|P_2 y_1\|}\right)\right| = \|P_2 y_1\|.$$

If $P_2 y_1 = 0$, then $M_1 \perp M_2$ and $\cos\theta = 0$ so $\theta = \pi/2$ in agreement with Definition 1.28. When $P_2 y_1 \neq 0$, then $P_1 P_2 P_1 = (y_1, P_2 y_1) y_1 \Box y_1$, whose only nonzero eigenvalue is $(y_1, P_2 y_1) = (P_2 y_1, P_2 y_1) = \|P_2 y_1\|^2 = \mu_1^2$. Therefore, $\mu_1 = \|P_2 y_1\|$ and again we have agreement with Definition 1.28.

In the general case when $\dim(M_i) > 1$ for $i = 1, 2$, it is not entirely clear how we should define the angles between M_1 and M_2. However, the following considerations should provide some justification for Definition 1.28. First, if $x \in M_1$ and $w \in M_2$, $\|x\| = \|w\| = 1$. The cosine of the angle between span x and span w is $|(x, w)|$. Thus the largest cosine of any angle (equivalently, the smallest angle in $[0, \pi/2]$) between a one-dimensional subspace of M_1 and a one-dimensional subspace of M_2 is

$$\sup_{x \in D_{11}} \sup_{w \in D_{21}} |(x, w)| = \sup_{x \in D_{11}} \sup_{w \in D_{21}} (x, w).$$

The sets D_{11} and D_{21} are defined in Proposition 1.49. By Proposition 1.49, this iterated supremum is μ_1 and is achieved for $x = x_1 \in D_{11}$ and $w = w_1 \in D_{21}$. Thus the cosine of the angle between $\text{span}\{x_1\}$ and $\text{span}\{w_1\}$ is μ_1. Now, remove $\text{span}\{x_1\}$ from M_1 to get $M_{12} = (\text{span}\{x_1\})^{\perp} \cap M_1$ and remove $\text{span}\{w_1\}$ from M_2 to get $M_{22} = (\text{span}\{w_1\})^{\perp} \cap M_2$. The second largest cosine of any angle between M_1 and M_2 is defined to be the largest cosine of any angle between M_{12} and M_{22} and is given by

$$\sup_{x \in D_{12}} \sup_{w \in D_{22}} (x, w) = (x_2, w_2) = \mu_2.$$

Next $\text{span}\{x_2\}$ is removed from M_{12} and $\text{span}\{w_2\}$ is removed from M_{22}, yielding M_{13} and M_{23}. The third largest cosine of any angle between M_1 and M_2 is defined to be the largest cosine of any angle between M_{13} and M_{23}, and so on. After k steps, we are left with $M_{1(k+1)}$ and $M_{2(k+1)}$, which are orthogonal to each other. Thus the remaining angles are $\pi/2$. The above is precisely the content of Definition 1.28, given the results of Propositions 1.48 and 1.49.

The statistical interpretation of the angles between subspaces is given in a later chapter. In a statistical context, the cosines of these angles are called canonical correlation coefficients and are a measure of the affine dependence between the random vectors.

PROBLEMS

All vector spaces are finite dimensional unless specified otherwise.

1. Let V_{n+1} be the set of all nth degree polynomials (in the real variable t) with real coefficients. With the usual definition of addition and scalar multiplication, prove that V_{n+1} is an $(n+1)$-dimensional real vector space.

2. For $A \in \mathcal{L}(V, W)$, suppose that M is any subspace of V such that $M \oplus \mathfrak{N}(A) = V$.

(i) Show that $\mathfrak{R}(A) = A(M)$ where $A(M) = \{w | w = Ax \text{ for some } x \in M\}$.

(ii) If $x_1, \ldots, x_k$ is any linearly independent set in V such that $\text{span}\{x_1, \ldots, x_k\} \cap \mathfrak{N}(A) = \{0\}$, prove that $Ax_1, \ldots, Ax_k$ is linearly independent.

3. For $A \in \mathcal{L}(V, W)$, fix $w_0 \in W$ and consider the linear equation $Ax = w_0$. If $w_0 \notin \mathfrak{R}(A)$, there is no solution to this equation. If $w_0 \in \mathfrak{R}(A)$, let x_0 be any solution so $Ax_0 = w_0$. Prove that $\mathfrak{N}(A) + x_0$ is the set of all solutions to $Ax = w_0$.

4. For the direct sum space $V_1 \oplus V_2$, suppose $A_{ij} \in \mathcal{L}(V_j, V_i)$ and let

$$T_1 = \begin{pmatrix} A_{11} & A_{12} \\ A_{21} & A_{22} \end{pmatrix}$$

be defined by

$$\begin{pmatrix} A_{11} & A_{12} \\ A_{21} & A_{22} \end{pmatrix} \{v_1, v_2\} = \{A_{11}v_1 + A_{12}v_2, A_{21}v_1 + A_{22}v_2\}$$

for $\{v_1, v_2\} \in V_1 \oplus V_2$.

(i) Prove that T_1 is a linear transformation.

(ii) Conversely, prove that every $T_1 \in \mathcal{L}(V_1 \oplus V_2, V_1 \oplus V_2)$ has such a representation.

(iii) If

$$T = \begin{pmatrix} A_{11} & A_{12} \\ A_{21} & A_{22} \end{pmatrix}$$

and

$$U = \begin{pmatrix} B_{11} & B_{12} \\ B_{21} & B_{22} \end{pmatrix},$$

prove that the representation of TU is

$$\begin{pmatrix} A_{11}B_{11} + A_{12}B_{21} & A_{11}B_{12} + A_{12}B_{22} \\ A_{21}B_{11} + A_{22}B_{21} & A_{21}B_{12} + A_{22}B_{22} \end{pmatrix}.$$

5. Let $x_1,\ldots, x_r, x_{r+1}$ be vectors in V with $x_1,\ldots, x_r$ being linearly independent. For $w_1,\ldots, w_r, w_{r+1}$ in W, give a necessary and sufficient condition for the existence of an $A \in \mathcal{L}(V, W)$ that satisfies $Ax_i = w_i$, $i = 1,\ldots, r + 1$.

6. Suppose $A \in \mathcal{L}(V, V)$ satisfies $A^2 = cA$ where $c \neq 0$. Find a constant k so that $B = kA$ is a projection.

7. Suppose A is an $m \times n$ matrix with columns $a_1,\ldots, a_n$ and B is an $n \times k$ matrix with rows $b'_1,\ldots, b'_n$. Show that $AB = \sum_1^n a_i b'_i$.

8. Let $x_1,\ldots, x_k$ be vectors in R^n, set $M = \text{span}\{x_1,\ldots, x_k\}$, and let A be the $n \times k$ matrix with columns $x_1,\ldots, x_k$ so $A \in \mathcal{L}(R^k, R^n)$.
   (i) Show $M = \mathcal{R}(A)$.
   (ii) Show $\dim(M) = \text{rank}(A'A)$.

9. For linearly independent $x_1,\ldots, x_k$ in $(V, (\cdot, \cdot))$, let $y_1,\ldots, y_k$ be the vectors obtained by applying the Gram–Schmidt (G-S) Process to $x_1,\ldots, x_k$. Show that if $z_i = Ax_i$, $i = 1,\ldots, k$, where $A \in \mathcal{O}(V)$, then the vectors obtained by the G-S Process from $z_1,\ldots, z_k$ are $Ay_1,\ldots, Ay_k$. (In other words, the G-S Process commutes with orthogonal transformations.)

10. In $(V, (\cdot, \cdot))$, let $x_1,\ldots, x_k$ be vectors with $x_1 \neq 0$. Form $y_1^1,\ldots, y_k^1$ by $y_1^1 = x_1/\|x_1\|$ and $y_i^1 = x_i - (x_i, y_1^1)y_1^1$, $i = 2,\ldots, k$:
   (i) Show $\text{span}\{x_1,\ldots, x_r\} = \text{span}\{y_1^1,\ldots, y_r^1\}$ for $r = 1, 2,\ldots, k$.
   (ii) Show $y_1^1 \perp \text{span}\{y_2^1,\ldots, y_k^1\}$ so $\text{span}\{y_1^1,\ldots, y_r^1\} = \text{span}\{y_1^1\} \oplus \text{span}\{y_2^1,\ldots, y_r^1\}$ for $r = 2,\ldots, k$.
   (iii) Now, form $y_2^2,\ldots, y_k^2$ from $y_2^1,\ldots, y_k^1$ as the y^1's were formed from the x's (reordering if necessary to achieve $y_2^1 \neq 0$). Show $\text{span}\{x_1,\ldots, x_k\} = \text{span}\{y_1^1\} \oplus \text{span}\{y_2^2\} \oplus \text{span}\{y_3^2,\ldots, y_k^2\}$.
   (iv) Let $m = \dim(\text{span}\{x_1,\ldots, x_k\})$. Show that after applying the above procedure m times, we get an orthonormal basis $y_1^1, y_2^2,\ldots, y_m^m$ for $\text{span}\{x_1,\ldots, x_k\}$.

(v) If $x_1, \ldots, x_k$ are linearly independent, show that $\operatorname{span}\{x_1, \ldots, x_r\} = \operatorname{span}\{y_1^1, y_2^2, \ldots, y_r^r\}$ for $r = 1, \ldots, k$.

11. Let $x_1, \ldots, x_m$ be a basis for $(V, (\cdot, \cdot))$ and $w_1, \ldots, w_n$ be a basis for $(W, [\cdot, \cdot])$. For $A, B \in \mathcal{L}(V, W)$, show that $[Ax_i, w_j] = [Bx_i, w_j]$ for $i = 1, \ldots, m$ and $j = 1, \ldots, n$ implies that $A = B$.

12. For $x_i \in (V, (\cdot, \cdot))$ and $y_i \in (W, [\cdot, \cdot])$, $i = 1, 2$, suppose that $x_1 \square y_1 = x_2 \square y_2 \neq 0$. Prove that $x_1 = cx_2$ for some scalar $c \neq 0$ and then $y_1 = c^{-1} y_2$.

13. Given two inner products on V, say $(\cdot, \cdot)$ and $[\cdot, \cdot]$, show that there exist positive constants c_1 and c_2 such that $c_1[x, x] \leqslant (x, x) \leqslant c_2[x, x]$, $x \in V$. Using this, show that for any open ball in $(V, (\cdot, \cdot))$, say $B = \{x | (x, x)^{1/2} < \alpha\}$, there exist open balls in $(V, [\cdot, \cdot])$, say $B_i = \{x | [x, x]^{1/2} < \beta_i\}$, $i = 1, 2$, such that $B_1 \subseteq B \subseteq B_2$.

14. In $(V, (\cdot, \cdot))$, prove that $\|x + y\| \leqslant \|x\| + \|y\|$. Using this, prove that $h(x) = \|x\|$ is a convex function.

15. For positive integers I and J, consider the IJ-dimensional real vector space, V, of all real-valued functions defined on $\{1, 2, \ldots, I\} \times \{1, 2, \ldots, J\}$. Denote the value of $y \in V$ at (i, j) by y_{ij}. The inner product on V is taken to be $(y, \tilde{y}) = \Sigma\Sigma y_{ij} \tilde{y}_{ij}$. The symbol $1 \in V$ denotes the vector all of whose coordinates are one.

(i) Define A on V to V by $Ay = \bar{y}_{..} 1$ where $\bar{y}_{..} = (IJ)^{-1} \Sigma\Sigma y_{ij}$. Show that A is the orthogonal projection onto $\operatorname{span}\{1\}$.

(ii) Define linear transformations B_1, B_2, and B_3 on V by

$$(B_1 y)_{ij} = \bar{y}_{i \cdot} - \bar{y}_{..}$$

$$(B_2 y)_{ij} = \bar{y}_{\cdot j} - \bar{y}_{..}$$

$$(B_3 y)_{ij} = y_{ij} - \bar{y}_{i \cdot} - \bar{y}_{\cdot j} + \bar{y}_{..}$$

where

$$\bar{y}_{i \cdot} = J^{-1} \sum_j y_{ij}$$

and

$$\bar{y}_{\cdot j} = I^{-1} \sum_i y_{ij}.$$

Show that B_1, B_2, and B_3 are orthogonal projections and the following holds:

$$AB_k = 0, \qquad k = 1, 2, 3$$

$$B_1B_2 = B_1B_3 = B_2B_3 = 0$$

$$(A + B_1 + B_2 + B_3)y = y, \qquad y \in V.$$

(iii) Show that

$$\|y\|^2 = \|Ay\|^2 + \|B_1y\|^2 + \|B_2y\|^2 + \|B_3y\|^2.$$

16. For $\Gamma \in \mathcal{O}(V)$ and M a subspace of V, suppose that $\Gamma(M) \subseteq M$. Prove that $\Gamma(M^\perp) \subseteq M^\perp$.

17. Given a subspace M of $(V, (\cdot, \cdot))$, show the following are equivalent:
(i) $|(x, y)| \leqslant c\|x\|$ for all $x \in M$.
(ii) $\|P_M y\| \leqslant c$.
Here c is a fixed positive constant and P_M is the orthogonal projection onto M.

18. In $(V, (\cdot, \cdot))$, suppose A and B are positive semidefinite. For $C, D \in \mathcal{L}(V, V)$ prove that $(\text{tr}\, ACBD')^2 \leqslant \text{tr}\, ACBC' \text{tr}\, ADBD'$.

19. Show that $\mathbb{C}^n$ is a $2n$-dimensional real vector space.

20. Let A be an $n \times n$ real matrix. Prove:
(i) If λ_0 is a real eigenvalue of A, then there exists a corresponding real eigenvector.
(ii) If λ_0 is an eigenvalue that is not real, then any corresponding eigenvector cannot be real or pure imaginary.

21. In an n-dimensional space $(V, (\cdot, \cdot))$, suppose P is a rank r orthogonal projection. For $\alpha, \beta \in R$, let $A = \alpha P + \beta(I - P)$. Find eigenvalues, eigenvectors, and the characteristic polynomial of A. Show that A is positive definite iff $\alpha > 0$ and $\beta > 0$. What is A^{-1} when it exists?

22. Suppose A and B are self-adjoint and $A - B \geqslant 0$. Let $\lambda_1 \geqslant \cdots \geqslant \lambda_n$ and $\mu_1 \geqslant \cdots \geqslant \mu_n$ be the eigenvalues of A and B. Show that $\lambda_i \geqslant \mu_i$, $i = 1, \ldots, n$.

23. If $S, T \in \mathcal{L}(V, V)$ and $S > 0$, $T \geqslant 0$, prove that $\langle S, T \rangle = 0$ implies $T = 0$.

24. For $A \in (\mathcal{L}(V, V), \langle \cdot, \cdot \rangle)$, show that $\langle A, I \rangle = \text{tr}\, A$.

25. Suppose A and B in $\mathcal{L}(V, V)$ are self-adjoint and write $A \geqslant B$ to mean $A - B \geqslant 0$.

(i) If $A \geqslant B$, show that $CAC' \geqslant CBC'$ for all $C \in \mathcal{L}(V, V)$.

(ii) Show $I \geqslant A$ iff all the eigenvalues of A are less than or equal to one.

(iii) Assume $A > 0$, $B > 0$, and $A \geqslant B$. Is $A^{1/2} \geqslant B^{1/2}$? Is $A^2 \geqslant B^2$?

26. If P is an orthogonal projection, show that $\operatorname{tr} P$ is the rank of P.

27. Let $x_1, \ldots, x_n$ be an orthonormal basis for $(V, (\cdot, \cdot))$ and consider the vector space $(\mathcal{L}(V, V), \langle \cdot , \cdot \rangle)$. Let M be the subspace of $\mathcal{L}(V, V)$ consisting of all self-adjoint linear transformations and let N be the subspace of all skew symmetric linear transformations. Prove:

(i) $\{x_i \square x_j + x_j \square x_i | i \leqslant j\}$ is an orthogonal basis for M.

(ii) $\{x_i \square x_j - x_j \square x_i | i < j\}$ is an orthogonal basis for N.

(iii) M is orthogonal to N and $M \oplus N = \mathcal{L}(V, V)$.

(iv) The orthogonal projection onto M is $A \to (A + A')/2$, $A \in \mathcal{L}(V, V)$.

28. Consider $\mathcal{L}_{n,n}$ with the usual inner product $\langle A, B \rangle = \operatorname{tr} AB'$, and let $\mathcal{S}_n$ be the subspace of symmetric matrices. Then $(\mathcal{S}_n, \langle \cdot , \cdot \rangle)$ is an inner product space. Show $\dim \mathcal{S}_n = n(n+1)/2$ and for $S, T \in \mathcal{S}_n$, $\langle S, T \rangle = \Sigma_i s_{ii} t_{ii} + 2\Sigma\Sigma_{i<j} s_{ij} t_{ij}$.

29. For $A \in \mathcal{L}(V, W)$, one definition of the norm of A is

$$|||A||| = \sup_{\|v\|=1} \|Av\|$$

where $\| \cdot \|$ is the given norm on W.

(i) Show that $|||A|||$ is the square root of the largest eigenvalue of $A'A$.

(ii) Show that $|||\alpha A||| = |\alpha|\,|||A|||$, $\alpha \in R$ and $|||A + B||| \leqslant |||A||| + |||B|||$.

30. In the inner product spaces $(V, (\cdot, \cdot))$ and $(W, [\cdot, \cdot])$, consider $A \in \mathcal{L}(V, V)$ and $B \in \mathcal{L}(W, W)$, which are both self-adjoint. Write these in spectral form as

$$A = \sum_1^m \lambda_i x_i \square x_i$$

$$B = \sum_1^n \mu_j w_j \square w_j.$$

(Note: The symbol $\square$ has a different meaning in these two equations since the definition of $\square$ depends on the inner product.) Of course, $x_1,\ldots,x_m[w_1,\ldots,w_n]$ is an orthonormal basis for $(V,(\cdot,\cdot))[(W,[\cdot,\cdot])]$. Also, $\{x_i \square w_j | i=1,\ldots,m,\ j=1,\ldots,n\}$ is an orthonormal basis for $(\mathcal{L}(W,V),\langle\cdot,\cdot\rangle)$, and $A\otimes B$ is a linear transformation on $\mathcal{L}(W,V)$ to $\mathcal{L}(W,V)$.

(i) Show that $(A\otimes B)(x_i \square w_j)=\lambda_i\mu_j(x_i \square w_j)$ so $\lambda_i\mu_j$ is an eigenvalue of $A\otimes B$.

(ii) Show that $A\otimes B=\sum\sum\lambda_i\mu_j(x_i \square w_j)\tilde{\square}(x_i \square w_j)$ and this is a spectral decomposition for $A\otimes B$. What are the eigenvalues and corresponding eigenvectors for $A\otimes B$?

(iii) If A and B are positive definite (semidefinite), show that $A\otimes B$ is positive definite (semidefinite).

(iv) Show that $\operatorname{tr} A\otimes B=(\operatorname{tr} A)(\operatorname{tr} B)$ and $\det A\otimes B=(\det A)^n(\det B)^m$.

31. Let $x_1,\ldots,x_p$ be linearly independent vectors in R^n, set $M=\operatorname{span}\{x_1,\ldots,x_p\}$, and let $A: n\times p$ have columns $x_1,\ldots,x_p$. Thus $\mathcal{R}(A)=M$ and $A'A$ is positive definite.

(i) Show that $\psi=A(A'A)^{-1/2}$ is a linear isometry whose columns form an orthonormal basis for M. Here, $(A'A)^{-1/2}$ denotes the inverse of the positive definite square root of $A'A$.

(ii) Show that $\psi\psi'=A(A'A)^{-1}A'$ is the orthogonal projection on to M.

32. Consider two subspaces, M_1 and M_2, of R^n with bases $x_1,\ldots,x_q$ and $y_1,\ldots,y_r$. Let $A(B)$ have columns $x_1,\ldots,x_q$ $(y_1,\ldots,y_r)$. Then $P_1=A(A'A)^{-1}A'$ and $P_2=B(B'B)^{-1}B'$ are the orthogonal projections onto M_1 and M_2, respectively. The cosines of the angles between M_1 and M_2 can be obtained by computing the nonzero eigenvalues of $P_1P_2P_1$. Show that these are the same as the nonzero eigenvalues of

$$(A'A)^{-1}A'B(B'B)^{-1}B'A : q\times q$$

and of

$$(B'B)^{-1}B'A(A'A)^{-1}A'B : r\times r.$$

33. In R^4, set $x_1'=(1,0,0,0)$, $x_2'=(0,1,0,0)$, $y_1'=(1,1,1,1)$, and $y_2'=(1,-1,1,-1)$. Find the cosines of the angles between $M_1=\operatorname{span}\{x_1,x_2\}$ and $M_2=\operatorname{span}\{y_1,y_2\}$.

34. For two subspaces M_1 and M_2 of $(V,(\cdot,\cdot))$, argue that the angles between M_1 and M_2 are the same as the angles between $\Gamma(M_1)$ and $\Gamma(M_2)$ for any $\Gamma\in\mathcal{O}(V)$.

35. This problem has to do with the vector space V of Example 1.9 and V may be infinite dimensional. The results in this problem are not used in the sequel. Write $X_1 \simeq X_2$ if $X_1 = X_2$ a.e. (P_0) for X_1 and X_2 in V. It is easy to verify $\simeq$ is an equivalence relation on V. Let $M = \{X | X \in V, X = 0 \text{ a.e. } (P_0)\}$ so $X_1 \simeq X_2$ iff $X_1 - X_2 \in M$. Let L^2 be the set of equivalence classes in V.

(i) Show that L^2 is a real vector space with the obvious definition of addition and scalar multiplication.

Define $(\cdot, \cdot)$ on L^2 by $(y_1, y_2) = \mathcal{E} X_1 X_2$ where X_i is an element of the equivalence class y_i, $i = 1, 2$.

(ii) Show that $(\cdot, \cdot)$ is well defined and is an inner product on L^2.

Now, let $\mathcal{F}_0$ be a sub σ-algebra of $\mathcal{F}$. For $y \in L^2$, let Py denote the conditional expectation given $\mathcal{F}_0$ of any element in y.

(iii) Show that P is well defined and is a linear transformation on L^2 to L^2.

Let N be the set of equivalence classes of all $\mathcal{F}_0$ measurable functions in V. Clearly, N is a subspace of L^2.

(iv) Show that $P^2 = P$, P is the identity on N, and $\mathcal{R}(P) = N$. Also show that P is self-adjoint—that is $(y_1, Py_2) = (Py_1, y_2)$.

Would you say that P is the orthogonal projection onto N?

NOTES AND REFERENCES

1. The first half of this chapter follows Halmos (1968) very closely. After this, the material was selected primarily for its use in later chapters. The material on outer products and Kronecker products follows the author's tastes more than anything else.
2. The detailed discussion of angles between subspaces resulted from unsuccessful attempts to find a source that meshed with the treatment of canonical correlations given in Chapter 10. A different development can be found in Dempster (1969, Chapter 5).
3. Besides Halmos (1958) and Hoffman and Kunze (1971), I have found the book by Noble and Daniel (1977) useful for standard material on linear algebra.
4. Rao (1973, Chapter 1) gives many useful linear algebra facts not discussed here.

CHAPTER 2

Random Vectors

The basic object of study in this book is the random vector and its induced distribution in an inner product space. Here, utilizing the results outlined in Chapter 1, we introduce random vectors, mean vectors, and covariances. Characteristic functions are discussed and used to give the well known factorization criterion for the independence of random vectors. Two special classes of distributions, the orthogonally invariant distributions and the weakly spherical distributions, are used for illustrative purposes. The vector spaces that occur in this chapter are all finite dimensional.

2.1. RANDOM VECTORS

Before a random vector can be defined, it is necessary to first introduce the Borel sets of a finite dimensional inner product space $(V, (\cdot, \cdot))$. Setting $\|x\| = (x, x)^{1/2}$, the open ball of radius r about x_0 is the set defined by $S_r(x_0) \equiv \{x \mid \|x - x_0\| < r\}$.

Definition 2.1. The Borel σ-algebra of $(V, (\cdot, \cdot))$, denoted by $\mathfrak{B}(V)$, is the smallest σ-algebra that contains all of the open balls.

Since any two inner products on V are related by a positive definite linear transformation, it follows that $\mathfrak{B}(V)$ does not depend on the inner product on V—that is, if we start with two inner products on V and use these inner products to generate a Borel σ-algebra, the two σ-algebras are the same. Thus we simply call $\mathfrak{B}(V)$ the Borel σ-algebra of V without mentioning the inner product.

A probability space is a triple $(\Omega, \mathcal{F}, P_0)$ where Ω is a set, $\mathcal{F}$ is a σ-algebra of subsets of Ω, and P_0 is a probability measure defined on $\mathcal{F}$.

Definition 2.2. A random vector $X \in V$ is a function mapping Ω into V such that $X^{-1}(B) \in \mathcal{F}$ for each Borel set $B \in \mathcal{B}(V)$. Here, $X^{-1}(B)$ is the inverse image of the set B.

Since the space on which a random vector is defined is usually not of interest here, the argument of a random vector X is ordinarily suppressed. Further, it is the induced distribution of X on V that most interests us. To define this, consider a random vector X defined on Ω to V where $(\Omega, \mathcal{F}, P_0)$ is a probability space. For each Borel set $B \in \mathcal{B}(V)$, let $Q(B) = P_0(X^{-1}(B))$. Clearly, Q is a probability measure on $\mathcal{B}(V)$ and Q is called the *induced distribution* of X—that is, Q is induced by X and P_0. The following result shows that any probability measure Q on $\mathcal{B}(V)$ is the induced distribution of some random vector.

Proposition 2.1. Let Q be a probability measure on $\mathcal{B}(V)$ where V is a finite dimensional inner product space. Then there exists a probability space $(\Omega, \mathcal{F}, P_0)$ and a random vector X on Ω to V such that Q is the induced distribution of X.

Proof. Take $\Omega = V$, $\mathcal{F} = \mathcal{B}(V)$, $P_0 = Q$, and let $X(\omega) = \omega$ for $\omega \in V$. Clearly, the induced distribution of X is Q. □

Henceforth, we write things like: "Let X be a random vector in V with distribution Q," to mean that X is a random vector and its induced distribution is Q. Alternatively, the notation $\mathcal{L}(X) = Q$ is also used—this is read: "The distributional law of X is Q."

A function f defined on V to W is called *Borel measurable* if the inverse image of each set $B \in \mathcal{B}(W)$ is in $\mathcal{B}(V)$. Of course, if X is a random vector in V, then $f(X)$ is a random vector in W when f is Borel measurable. In particular, when f is continuous, f is Borel measurable. If $W = R$ and f is Borel measurable on V to R, then $f(X)$ is a real-valued random variable.

Definition 2.3. Suppose X is a random vector in V with distribution Q and f is a real-valued Borel measurable function defined on V. If $\int_V |f(x)| Q(dx) < +\infty$, then we say that $f(X)$ has *finite expectation* and we write $\mathcal{E}f(X)$ for $\int_V f(x) Q(dx)$.

In the above definition and throughout this book, all integrals are Lebesgue integrals, and all functions are assumed Borel measurable.

◆ **Example 2.1.** Take V to be the coordinate space R^n with the usual inner product $(\cdot, \cdot)$ and let dx denote standard Lebesgue measure on R^n. If q is a non-negative function on R^n such that $\int q(x)\, dx = 1$, then q is called a *density function*. It is clear that the measure Q given by $Q(B) = \int_B q(x)\, dx$ is a probability measure on R^n so Q is the distribution of some random vector X on R^n. If $\varepsilon_1, \ldots, \varepsilon_n$ is the standard basis for R^n, then $(\varepsilon_i, X) \equiv X_i$ is the *ith coordinate* of X. Assume that X_i has a finite expectation for $i = 1, \ldots, n$. Then $\mathcal{E}X_i = \int_{R^n}(\varepsilon_i, x)q(x)\, dx \equiv \mu_i$ is called the *mean value* of X_i and the vector $\mu \in R^n$ with coordinates $\mu_1, \ldots, \mu_n$ is the *mean vector* of X. Notice that for any vector $x \in R^n$, $\mathcal{E}(x, X) = \mathcal{E}(\Sigma x_i \varepsilon_i, X) = \Sigma x_i \mathcal{E}(\varepsilon_i, X) = \Sigma x_i \mu_i = (x, \mu)$. Thus the mean vector μ satisfies the equation $\mathcal{E}(x, X) = (x, \mu)$ for all $x \in R^n$ and μ is clearly unique. It is exactly this property of μ that we use to define the mean vector of a random vector in an arbitrary inner product space V. ◆

Suppose X is a random vector in an inner product space $(V, (\cdot, \cdot))$ and assume that for each $x \in V$, the random variable (x, X) has a finite expectation. Let $f(x) = \mathcal{E}(x, X)$, so f is a real-valued function defined on V. Also, $f(\alpha_1 x_1 + \alpha_2 x_2) = \mathcal{E}(\alpha_1 x_1 + \alpha_2 x_2, X) = \mathcal{E}[\alpha_1(x_1, X) + \alpha_2(x_2, X)] = \alpha_1 \mathcal{E}(x_1, X) + \alpha_2 \mathcal{E}(x_2, X) = \alpha_1 f(x_1) + \alpha_2 f(x_2)$. Thus f is a linear function on V. Therefore, there exists a unique vector $\mu \in V$ such that $f(x) = (x, \mu)$ for all $x \in V$. Summarizing, there exists a unique vector $\mu \in V$ that satisfies $\mathcal{E}(x, X) = (x, \mu)$ for all $x \in V$. The vector μ is called the *mean vector* of X and is denoted by $\mathcal{E}X$. This notation leads to the suggestive equation $\mathcal{E}(x, X) = (x, \mathcal{E}X)$, which we know is valid in the coordinate case.

Proposition 2.2. Suppose $X \in (V, (\cdot, \cdot))$ and assume X has a mean vector μ. Let $(W, [\cdot, \cdot])$ be an inner product space and consider $A \in \mathcal{L}(V, W)$ and $w_0 \in W$. Then the random vector $Y = AX + w_0$ has the mean vector $A\mu + w_0$—that is, $\mathcal{E}Y = A\mathcal{E}X + w_0$.

Proof. The proof is a computation. For $w \in W$,

$$\begin{aligned}\mathcal{E}[w, Y] &= \mathcal{E}[w, AX + w_0] = \mathcal{E}[w, AX] + [w, w_0] \\ &= \mathcal{E}(A'w, X) + [w, w_0] = (A'w, \mu) + [w, w_0] \\ &= [w, A\mu] + [w, w_0] = [w, A\mu + w_0].\end{aligned}$$

Thus $A\mu + w_0$ satisfies the defining equation for the mean vector of Y and by the uniqueness of mean vectors, $EY = A\mu + w_0$. □

If X_1 and X_2 are both random vectors in $(V, (\cdot, \cdot))$, which have mean vectors, then it is easy to show that $\mathcal{E}(X_1 + X_2) = \mathcal{E}X_1 + \mathcal{E}X_2$. The following proposition shows that the mean vector μ of a random vector does not depend on the inner product on V.

Proposition 2.3. If X is a random vector in $(V, (\cdot, \cdot))$ with mean vector μ satisfying $\mathcal{E}(x, X) = (x, \mu)$ for all $x \in V$, then μ satisfies $\mathcal{E}f(x, X) = f(x, \mu)$ for every bilinear function f on $V \times V$.

Proof. Every bilinear function f is given by $f(x_1, x_2) = (x_1, Ax_2)$ for some $A \in \mathcal{L}(V, V)$. Thus $\mathcal{E}f(x, X) = \mathcal{E}(x, AX) = (x, A\mu) = f(x, \mu)$ where the second equality follows from Proposition 2.2. □

When the bilinear function f is an inner product on V, the above result establishes that the mean vector is inner product free. At times, a convenient choice of an inner product can simplify the calculation of a mean vector.

The definition and basic properties of the covariance between two real-valued random variables were covered in Example 1.9. Before defining the covariance of a random vector, a review of covariance matrices for coordinate random vectors in R^n is in order.

◆ **Example 2.2.** In the notation of Example 2.1, consider a random vector X in R^n with coordinates $X_i = (\varepsilon_i, X)$ where $\varepsilon_1, \ldots, \varepsilon_n$ is the standard basis for R^n and $(\cdot, \cdot)$ is the standard inner product. Assume that $\mathcal{E}X_i^2 < +\infty$, $i = 1, \ldots, n$. Then $\text{cov}(X_i, X_j) \equiv \sigma_{ij}$ exists for all $i, j = 1, \ldots, n$. Let Σ be the $n \times n$ matrix with elements σ_{ij}. Of course, σ_{ii} is the variance of X_i and σ_{ij} is the covariance between X_i and X_j. The symmetric matrix Σ is called the *covariance matrix* of X. Consider vectors $x, y \in R^n$ with coordinates x_i and y_i, $i = 1, \ldots, n$. Then

$$\begin{aligned}\text{cov}\{(x, X), (y, X)\} &= \text{cov}\left\{\sum_i x_i X_i, \sum_j y_j X_j\right\}\\ &= \sum_i \sum_j x_i y_j \,\text{cov}(X_i, X_j) = \sum_i \sum_j x_i y_j \sigma_{ij}\\ &= (x, \Sigma y).\end{aligned}$$

Hence $\text{cov}\{(x, X), (y, X)\} = (x, \Sigma y)$. It is this property of Σ that is used to define the covariance of a random vector. ◆

With the above example in mind, consider a random vector X in an inner product space $(V, (\cdot, \cdot))$ and assume that $\mathcal{E}(x, X)^2 < \infty$ for all $x \in V$. Thus

(x, X) has a finite variance and the covariance between (x, X) and (y, X) is well defined for each $x, y \in V$.

Proposition 2.4. For $x, y \in V$, define $f(x, y)$ by

$$f(x, y) = \text{cov}\{(x, X), (y, X)\}.$$

Then f is a non-negative definite bilinear function on $V \times V$.

Proof. Clearly, $f(x, y) = f(y, x)$ and $f(x, x) = \text{var}\{(x, X)\} \geqslant 0$, so it remains to show that f is bilinear. Since f is symmetric, it suffices to verify that $f(\alpha_1 x_1 + \alpha_2 x_2, y) = \alpha_1 f(x_1, y) + \alpha_2 f(x_2, y)$. This verification goes as follows:

$$\begin{aligned} f(\alpha_1 x_1 + \alpha_2 x_2, y) &= \text{cov}\{(\alpha_1 x_1 + \alpha_2 x_2, X), (y, X)\} \\ &= \text{cov}\{\alpha_1(x_1, X) + \alpha_2(x_2, X), (y, X)\} \\ &= \alpha_1 \text{cov}\{(x_1, X), (y, X)\} + \alpha_2 \text{cov}\{(x_2, X), (y, X)\} \\ &= \alpha_1 f(x_1, y) + \alpha_2 f(x_2, y). \end{aligned}$$ □

By Proposition 1.26, there exists a unique non-negative definite linear transformation Σ such that $f(x, y) = (x, \Sigma y)$.

Definition 2.4. The unique non-negative definite linear transformation Σ on V to V that satisfies

$$\text{cov}\{(x, X), (y, X)\} = (x, \Sigma y)$$

is called the *covariance* of X and is denoted by $\text{Cov}(X)$.

Implicit in the above definition is the assumption that $\mathcal{E}(x, X)^2 < +\infty$ for all $x \in V$. Whenever we discuss covariances of random vectors, $\mathcal{E}(x, X)^2$ is always assumed finite.

It should be emphasized that the covariance of a random vector in $(V, (\cdot, \cdot))$ depends on the given inner product. The next result shows how the covariance changes as a function of the inner product.

Proposition 2.5. Consider a random vector X in $(V, (\cdot, \cdot))$ and suppose $\text{Cov}(X) = \Sigma$. Let $[\cdot, \cdot]$ be another inner product on V given by $[x, y] = (x, Ay)$ where A is positive definite on $(V, (\cdot, \cdot))$. Then the covariance of X in the inner product space $(V, [\cdot, \cdot])$ is ΣA.

Proof. To verify that ΣA is the covariance for X in $(V, [\cdot, \cdot])$, we must show that $\text{cov}\{[x, X], [y, X]\} = [x, \Sigma Ay]$ for all $x, y \in V$. To do this, use the definition of $[\cdot, \cdot]$ and compute:

$$\text{cov}\{[x, X], [y, X]\} = \text{cov}\{(x, AX), (y, AX)\} = \text{cov}\{(Ax, X), (Ay, X)\}$$
$$= (Ax, \Sigma Ay) = (x, A\Sigma Ay) = [x, \Sigma Ay]. \qquad \square$$

Two immediate consequences of Proposition 2.5 are: (i) if $\text{Cov}(X)$ exists in one inner product, then it exists in all inner products, and (ii) if $\text{Cov}(X) = \Sigma$ in $(V, (\cdot, \cdot))$ and if Σ is positive definite, then the covariance of X in the inner product $[x, y] \equiv (x, \Sigma^{-1}y)$ is the identity linear transformation. The result below often simplifies a computation involving the derivation of a covariance.

Proposition 2.6. Suppose $\text{Cov}(X) = \Sigma$ in $(V, (\cdot, \cdot))$. If Σ_1 is a self-adjoint linear transformation on $(V, (\cdot, \cdot))$ to $(V, (\cdot, \cdot))$ that satisfies

$$\text{(2.1)} \qquad \text{var}\{(x, X)\} = (x, \Sigma_1 x) \qquad \text{for } x \in V,$$

then $\Sigma_1 = \Sigma$.

Proof. Equation (2.1) implies that $(x, \Sigma_1 x) = (x, \Sigma x)$, $x \in V$. Since Σ_1 and Σ are self-adjoint, Proposition 1.16 yields the conclusion $\Sigma_1 = \Sigma$. $\square$

When $\text{Cov}(X) = \Sigma$ is singular, then the random vector X takes values in the translate of a subspace of $(V, (\cdot, \cdot))$. To make this precise, let us consider the following.

Proposition 2.7. Let X be a random vector in $(V, (\cdot, \cdot))$ and suppose $\text{Cov}(X) = \Sigma$ exists. With $\mu = \mathcal{E}X$ and $\mathcal{R}(\Sigma)$ denoting the range of Σ, $P\{X \in \mathcal{R}(\Sigma) + \mu\} = 1$.

Proof. The set $\mathcal{R}(\Sigma) + \mu$ is the set of vectors of the form $x + \mu$ for $x \in \mathcal{R}(\Sigma)$; that is $\mathcal{R}(\Sigma) + \mu$ is the translate, by μ, of the subspace $\mathcal{R}(\Sigma)$. The statement $P\{X \in \mathcal{R}(\Sigma) + \mu\} = 1$ is equivalent to the statement $P\{X - \mu \in \mathcal{R}(\Sigma)\} = 1$. The random vector $Y = X - \mu$ has mean zero and, by Proposition 2.6, $\text{Cov}(Y) = \text{Cov}(X) = \Sigma$ since $\text{var}\{(x, X - \mu)\} = \text{var}\{(x, X)\}$ for $x \in V$. Thus it must be shown that $P\{Y \in \mathcal{R}(\Sigma)\} = 1$. If Σ is nonsingular, then $\mathcal{R}(\Sigma) = V$ and there is nothing to show. Thus assume that the null space of Σ, $\mathcal{N}(\Sigma)$, has dimension $k > 0$ and let $\{x_1, \ldots, x_k\}$ be an orthonormal basis for $\mathcal{N}(\Sigma)$. Since $\mathcal{R}(\Sigma)$ and $\mathcal{N}(\Sigma)$ are perpendicular and $\mathcal{R}(\Sigma) \oplus$

$\mathfrak{N}(\Sigma) = V$, a vector x is not in $\mathfrak{R}(\Sigma)$ iff for some index $i = 1, \ldots, k$, $(x_i, x) \neq 0$. Thus

$$P\{Y \notin \mathfrak{R}(\Sigma)\} = P\{(x_i, Y) \neq 0 \text{ for some } i = 1, \ldots, k\}$$
$$\leqslant \sum_1^k P\{(x_i, Y) \neq 0\}.$$

But (x_i, Y) has mean zero and $\operatorname{var}\{(x_i, Y)\} = (x_i, \Sigma x_i) = 0$ since $x_i \in \mathfrak{N}(\Sigma)$. Thus (x_i, Y) is zero with probability one, so $P\{(x_i, Y) \neq 0\} = 0$. Therefore $P\{Y \notin \mathfrak{R}(\Sigma)\} = 0$. □

Proposition 2.2 describes how the mean vector changes under linear transformations. The next result shows what happens to the covariance under linear transformations.

Proposition 2.8. Suppose X is a random vector in $(V, (\cdot, \cdot))$ with $\operatorname{Cov}(X) = \Sigma$. If $A \in \mathcal{L}(V, W)$ where $(W, [\cdot, \cdot])$ is an inner product space, then

$$\operatorname{Cov}(AX + w_0) = A\Sigma A'$$

for all $w_0 \in W$.

Proof. By Proposition 2.6, it suffices to show that for each $w \in W$,

$$\operatorname{var}[w, AX + w_0] = [w, A\Sigma A'w].$$

However,

$$\operatorname{var}[w, AX + w_0] = \operatorname{var}([w, AX] + [w, w_0]) = \operatorname{var}[w, AX]$$
$$= \operatorname{var}(A'w, X) = (A'w, \Sigma A'w) = [w, A\Sigma A'w].$$

Thus $\operatorname{Cov}(AX + w_0) = A\Sigma A'$. □

2.2. INDEPENDENCE OF RANDOM VECTORS

With the basic properties of mean vectors and covariances established, the next topic of discussion is characteristic functions and independence of random vectors. Let X be a random vector in $(V, (\cdot, \cdot))$ with distribution Q.

Definition 2.5. The complex valued function on V defined by

$$\phi(v) \equiv \mathcal{E} e^{i(v, X)} = \int_V e^{i(v, x)} Q(dx)$$

is the *characteristic function* of X.

In the above definition, $e^{it} = \cos t + i \sin t$ where $i = \sqrt{-1}$ and $t \in R$. Since e^{it} is a bounded continuous function of t, characteristic functions are well defined for all distributions Q on $(V, (\cdot, \cdot))$. Forthcoming applications of characteristic functions include the derivation of distributions of certain functions of random vectors and a characterization of the independence of two or more random vectors.

One basic property of characteristic functions is their uniqueness, that is, if Q_1 and Q_2 are probability distributions on $(V, (\cdot, \cdot))$ with characteristic functions ϕ_1 and ϕ_2, and if $\phi_1(x) = \phi_2(x)$ for all $x \in V$, then $Q_1 = Q_2$. A proof of this is based on the multidimensional Fourier inversion formula, which can be found in Cramér (1946). A consequence of this uniqueness is that, if X_1 and X_2 are random vectors in $(V, (\cdot, \cdot))$ such that $\mathcal{L}((x, X_1)) = \mathcal{L}((x, X_2))$ for all $x \in V$, then $\mathcal{L}(X_1) = \mathcal{L}(X_2)$. This follows by observing that $\mathcal{L}((x, X_1)) = \mathcal{L}((x, X_2))$ for all x implies the characteristic functions of X_1 and X_2 are the same and hence their distributions are the same.

To define independence, consider a probability space $(\Omega, \mathcal{F}, P_0)$ and let $X \in (V, (\cdot, \cdot))$ and $Y \in (W, [\cdot, \cdot])$ be two random vectors defined on Ω.

Definition 2.6. The random vectors X and Y are *independent* if for any Borel sets $B_1 \in \mathcal{B}(V)$ and $B_2 \in \mathcal{B}(W)$,

$$P_0\{X^{-1}(B_1) \cap Y^{-1}(B_2)\} = P_0\{X^{-1}(B_1)\}P_0\{Y^{-1}(B_2)\}.$$

In order to describe what independence means in terms of the induced distributions of $X \in (V, (\cdot, \cdot))$ and $Y \in (W, [\cdot, \cdot])$, it is necessary to define what is meant by the joint induced distribution of X and Y. The natural vector space in which to have X and Y take values is the direct sum $V \oplus W$ defined in Chapter 1. For $\{v_i, w_i\} \in V \oplus W$, $i = 1, 2$, define the inner product $(\cdot, \cdot)_1$ by

$$(\{v_1, w_1\}, \{v_2, w_2\})_1 = (v_1, v_2) + [w_1, w_2].$$

That $(\cdot, \cdot)_1$ is an inner product on $V \oplus W$ is routine to check. Thus $\{X, Y\}$ takes values in the inner product space $V \oplus W$. However, it must be shown that $\{X, Y\}$ is a Borel measurable function. Briefly, this argument goes as follows. The space $V \oplus W$ is a Cartesian product space—that is, $V \oplus W$ consists of all pairs $\{v, w\}$ with $v \in V$ and $w \in W$. Thus one way to get a σ-algebra on $V \oplus W$ is to form the product σ-algebra $\mathcal{B}(V) \times \mathcal{B}(W)$, which is the smallest σ-algebra containing all the product Borel sets $B_1 \times B_2 \subseteq V \oplus W$ where $B_1 \in \mathcal{B}(V)$ and $B_2 \in \mathcal{B}(W)$. It is not hard to verify that inverse images, under $\{X, Y\}$, of sets in $\mathcal{B}(V) \times \mathcal{B}(W)$ are in the σ-algebra $\mathcal{F}$. But the product σ-algebra $\mathcal{B}(V) \times \mathcal{B}(W)$ is just the σ-algebra $\mathcal{B}(V \oplus W)$ defined earlier. Thus $\{X, Y\} \in V \oplus W$ is a random vector and hence has an induced distribution Q defined on $\mathcal{B}(V \oplus W)$. In addition, let Q_1 be the induced distribution of X on $\mathcal{B}(V)$ and let Q_2 be the induced distribution of Y on $\mathcal{B}(W)$. It is clear that $Q_1(B_1) = Q(B_1 \times W)$ for $B_1 \in \mathcal{B}(V)$ and $Q_2(B_2) = Q(V \times B_2)$ for $B_2 \in \mathcal{B}(W)$. Also, the characteristic function of $\{X, Y\} \in V \oplus W$ is

$$\phi(\{v, w\}) = \mathcal{E} \exp\left[i(\{v, w\}, \{X, Y\})_1\right] = \mathcal{E} \exp(i(v, X) + i[w, Y])$$

and the marginal characteristic functions of X and Y are

$$\phi_1(v) = \mathcal{E} e^{i(v, X)}$$

and

$$\phi_2(w) = \mathcal{E} e^{i[w, Y]}.$$

Proposition 2.9. Given random vectors $X \in (V, (\cdot, \cdot))$ and $Y \in (W, [\cdot, \cdot])$, the following are equivalent:

(i) X and Y are independent.
(ii) $Q(B_1 \times B_2) = Q_1(B_1)Q_2(B_2)$ for all $B_1 \in \mathcal{B}(V)$ and $B_2 \in \mathcal{B}(W)$.
(iii) $\phi(\{v, w\}) = \phi_1(v)\phi_2(w)$ for all $v \in V$ and $w \in W$.

Proof. By definition,

$$Q(B_1 \times B_2) = P_0\{\{X, Y\} \in B_1 \times B_2\} = P_0\{X \in B_1, Y \in B_2\}.$$

The equivalence of (i) and (ii) follows immediately from the above equation. To show (ii) implies (iii), first note that, if f_1 and f_2 are integrable complex valued functions on V and W, then when (ii) holds,

$$\int_{V \oplus W} f_1(v) f_2(w) Q(dv, dw) = \int_V \int_W f_1(v) f_2(w) Q_1(dv) Q_2(dw)$$

$$= \int_V f_1(v) Q_1(dv) \int_W f_2(w) Q_2(dw)$$

by Fubini's Theorem (see Chung, 1968). Taking $f_1(v) = e^{i(v_1, v)}$ for v_1, $v \in V$, and $f_2(w) = e^{i[w_1, w]}$ for w_1, $w \in W$, we have

$$\phi(\{v_1, w_1\}) = \int \exp(i(v_1, v) + i[w_1, w]) Q(dv, dw)$$

$$= \int_V \exp[i(v_1, v)] Q_1(dv) \int_W \exp(i[w_1, w]) Q_2(dw)$$

$$= \phi_1(v_1)\phi_2(w_1).$$

Thus (ii) implies (iii). For (iii) implies (ii), note that the product measure $Q_1 \times Q_2$ has characteristic function $\phi_1\phi_2$. The uniqueness of characteristic functions then implies that $Q = Q_1 \times Q_2$. □

Of course, all of the discussion above extends to the case of more than two random vectors. For completeness, we briefly describe the situation. Given a probability space $(\Omega, \mathcal{F}, P_0)$ and random vectors $X_j \in (V_j, (\cdot, \cdot)_j)$, $j = 1, \ldots, k$, let Q_j be the induced distribution of X_j and let ϕ_j be the characteristic function of X_j. The random vectors $X_1, \ldots, X_k$ are *independent* if for all $B_j \in \mathcal{B}(V_j)$,

$$P_0\{X_j \in B_j, j = 1, \ldots, k\} = \prod_{j=1}^{k} P_0\{X_j \in B_j\}.$$

To construct one random vector from $X_1, \ldots, X_k$, consider the direct sum $V_1 \oplus \cdots \oplus V_k$ with the inner product $(\cdot, \cdot) = \Sigma_1^k(\cdot, \cdot)_j$. In other words, if $\{v_1, \ldots, v_k\}$ and $\{w_1, \ldots, w_k\}$ are elements of $V_1 \oplus \cdots \oplus V_k$, then the inner product between these vectors is $\Sigma_1^k(v_j, w_j)_j$. An argument analogous to that given earlier shows that $\{X_1, \ldots, X_k\}$ is a random vector in $V_1 \oplus \cdots \oplus V_k$ and the Borel σ-algebra of $V_1 \oplus \cdots \oplus V_k$ is just the product σ-algebra $\mathcal{B}(V_1) \times \cdots \times \mathcal{B}(V_k)$. If Q denotes the induced distribution of $\{X_1, \ldots, X_k\}$, then the independence of $X_1, \ldots, X_k$ is equivalent to the assertion that

$$Q(B_1 \times \cdots \times B_k) = \prod_{j=1}^{k} Q_j(B_j)$$

for all $B_j \in \mathcal{B}(V_j)$, $j = 1, \ldots, k$, and this is equivalent to

$$\mathcal{E} \exp\left[i \sum_{1}^{k} (v_j, X_j)_j\right] = \prod_{j=1}^{k} \phi_j(v_j).$$

Of course, when $X_1, \ldots, X_k$ are independent and f_j is an integrable real valued function on V_j, $j = 1, \ldots, k$, then

$$\mathcal{E} \prod_{j=1}^{k} f_j(X_j) = \prod_{j=1}^{k} \mathcal{E} f_j(X_j).$$

This equality follows from the fact that

$$Q(B_1 \times \cdots \times B_k) = \prod_{j=1}^{k} Q_j(B_j)$$

and Fubini's Theorem.

◆ **Example 2.3.** Consider the coordinate space R^p with the usual inner product and let Q_0 be a fixed distribution on R^p. Suppose $X_1,\ldots, X_n$ are independent with each $X_i \in R^p$, $i = 1,\ldots, n$, and $\mathcal{L}(X_i) = Q_0$. That is, there is a probability space $(\Omega, \mathcal{F}, P_0)$, each X_i is a random vector on Ω with values in R^p, and for Borel sets,

$$P_0\{X_i \in B_i, i = 1,\ldots, n\} = \prod_1^n Q_0(B_i).$$

Thus $\{X_1,\ldots, X_n\}$ is a random vector in the direct sum $R^p \oplus \cdots \oplus R^p$ with n terms in the sum. However, there are a variety of ways to think about the above direct sum. One possibility is to form the coordinate random vector

$$Y = \begin{pmatrix} X_1 \\ X_2 \\ \vdots \\ X_n \end{pmatrix} \in R^{np}$$

and simply consider Y as a random vector in R^{np} with the usual inner product. A disadvantage of this representation is that the independence of $X_1,\ldots, X_n$ becomes slightly camouflaged by the notation. An alternative is to form the random matrix

$$X = \begin{pmatrix} X_1' \\ X_2' \\ \vdots \\ X_n' \end{pmatrix} \in \mathcal{L}_{p,n}.$$

Thus X has rows X_i', $i = 1,\ldots, n$, which are independent and each has distribution Q_0. The inner product on $\mathcal{L}_{p,n}$ is just that inherited from the standard inner products on R^n and R^p. Therefore X is a random vector in the inner product space $(\mathcal{L}_{p,n}, \langle \cdot , \cdot \rangle)$. In the sequel, we ordinarily represent $X_1,\ldots, X_n$ by the random vector $X \in \mathcal{L}_{p,n}$. The advantages of this representation are far from clear at this point, but the reader should be convinced by the end of this book that such a choice is not unreasonable. The derivation of the mean and covariance of $X \in \mathcal{L}_{p,n}$ given in the next section should provide some evidence that the above representation is useful. ◆

2.3. SPECIAL COVARIANCE STRUCTURES

In this section, we derive the covariances of some special random vectors. The orthogonally invariant probability distributions on a vector space are shown to have covariances that are a constant times the identity transformation. In addition, the covariance of the random vector given in Example 2.3 is shown to be a Kronecker product. The final example provides an expression for the covariance of an outer product of a random vector with itself.

Suppose $(V, (\cdot, \cdot))$ is an inner product space and recall that $\mathcal{O}(V)$ is the group of orthogonal transformations on V to V.

Definition 2.7. A random vector X in $(V, (\cdot, \cdot))$ with distribution Q has an *orthogonally invariant distribution* if $\mathcal{L}(X) \equiv \mathcal{L}(\Gamma X)$ for all $\Gamma \in \mathcal{O}(V)$, or equivalently if $Q(B) = Q(\Gamma B)$ for all Borel sets B and $\Gamma \in \mathcal{O}(V)$.

Many properties of orthogonally invariant distributions follow from the following proposition.

Proposition 2.10. Let $x_0 \in V$ with $\|x_0\| = 1$. If $\mathcal{L}(X) = \mathcal{L}(\Gamma X)$ for $\Gamma \in \mathcal{O}(V)$, then for $x \in V$, $\mathcal{L}((x, X)) = \mathcal{L}(\|x\|(x_0, X))$.

Proof. The assertion is that the distribution of the real-valued random variable (x, X) is the same as the distribution of $\|x\|(x_0, X)$. Thus knowing the distribution of (x, X) for one particular nonzero $x \in V$ gives us the distribution of (x, X) for all $x \in V$. If $x = 0$, the assertion of the proposition is trivial. For $x \neq 0$, choose $\Gamma \in \mathcal{O}(V)$ such that $\Gamma x_0 = x/\|x\|$. This is possible since x_0 and $x/\|x\|$ both have norm 1. Thus

$$\mathcal{L}((x, X)) = \mathcal{L}\left(\|x\|\left(\frac{x}{\|x\|}, X\right)\right) = \mathcal{L}\left(\|x\|(\Gamma x_0, X)\right) = \mathcal{L}\left(\|x\|(x_0, \Gamma' X)\right)$$
$$= \mathcal{L}\left(\|x\|(x_0, X)\right)$$

where the last equality follows from the assumption that $\mathcal{L}(X) = \mathcal{L}(\Gamma X)$ for all $\Gamma \in \mathcal{O}(V)$ and the fact that $\Gamma \in \mathcal{O}(V)$ implies $\Gamma' \in \mathcal{O}(V)$. □

Proposition 2.11. Let $x_0 \in V$ with $\|x_0\| = 1$. Suppose the distribution of X is orthogonally invariant. Then:

(i) $\phi(x) \equiv \mathcal{E} e^{i(x, X)} = \phi(\|x\| x_0)$.
(ii) If $\mathcal{E}X$ exists, then $\mathcal{E}X = 0$.
(iii) If $\operatorname{Cov}(X)$ exists, then $\operatorname{Cov}(X) = \sigma^2 I$ where $\sigma^2 = \operatorname{var}\{(x_0, X)\}$, and I is the identity linear transformation.

Proof. Assertion (i) follows from Proposition 2.10 and

$$\mathcal{E}e^{i(x,X)} = \mathcal{E}e^{i\|x\|(x_0;X)} = \mathcal{E}e^{i(\|x\|x_0,X)} = \phi(\|x\|x_0).$$

For (ii), let $\mu = \mathcal{E}X$. Since $\mathcal{L}(X) = \mathcal{L}(\Gamma X)$, $\mu = \mathcal{E}X = \mathcal{E}\Gamma X = \Gamma\mathcal{E}X = \Gamma\mu$ for all $\Gamma \in \mathcal{O}(V)$. The only vector μ that satisfies $\mu = \Gamma\mu$ for all $\Gamma \in \mathcal{O}(V)$ is $\mu = 0$. To prove (iii), we must show that $\sigma^2 I$ satisfies the defining equation for Cov(X). But by Proposition 2.10,

$$\text{var}\{(x,X)\} = \text{var}\{\|x\|(x_0,X)\} = \|x\|^2\,\text{var}\{x_0,X\} = \sigma^2(x,x) = (x,\sigma^2 Ix)$$

so Cov$(X) = \sigma^2 I$ by Proposition 2.6. □

Assertion (i) of Proposition 2.11 shows that the characteristic function ϕ of an orthogonally invariant distribution satisfies $\phi(\Gamma x) = \phi(x)$ for all $x \in V$ and $\Gamma \in \mathcal{O}(V)$. Any function f defined on V and taking values in some set is called *orthogonally invariant* if $f(x) = f(\Gamma x)$ for all $\Gamma \in \mathcal{O}(V)$. A characterization of orthogonal invariant functions is given by the following proposition.

Proposition 2.12. A function f defined on $(V,(\cdot,\cdot))$ is orthogonally invariant iff $f(x) = f(\|x\|x_0)$ where $x_0 \in V$, $\|x_0\| = 1$.

Proof. If $f(x) = f(\|x\|x_0)$, then $f(\Gamma x) = f(\|\Gamma x\|x_0) = f(\|x\|x_0) = f(x)$ so f is orthogonally invariant. Conversely, suppose f is orthogonally invariant and $x_0 \in V$ with $\|x_0\| = 1$. For $x = 0$, $f(0) = f(\|x\|x_0)$ since $\|x\| = 0$. If $x \neq 0$, let $\Gamma \in \mathcal{O}(V)$ be such that $\Gamma x_0 = x/\|x\|$. Then $f(x) = f(\Gamma\|x\|x_0) = f(\|x\|x_0)$. □

If X has an orthogonally invariant distribution in $(V,(\cdot,\cdot))$ and h is a function on R to R, then

$$f(x) \equiv \mathcal{E}h((x,X))$$

clearly satisfies $f(\Gamma x) = f(x)$ for $\Gamma \in \mathcal{O}(V)$. Thus $f(x) = f(\|x\|x_0) = \mathcal{E}h(\|x\|(x_0,X))$, so to calculate $f(x)$, one only needs to calculate $f(\alpha x_0)$ for $\alpha \in (0,\infty)$. We have more to say about orthogonally invariant distributions in later chapters.

A random vector $X \in V(\cdot,\cdot)$ is called *orthogonally invariant* about x_0 if $X - x_0$ has an orthogonally invariant distribution. It is not difficult to show, using characteristic functions, that if X is orthogonally invariant about both x_0 and x_1, then $x_0 = x_1$. Further, if X is orthogonally invariant about x_0 and if $\mathcal{E}X$ exists, then $\mathcal{E}(X - x_0) = 0$ by Proposition 2.11. Thus $x_0 = \mathcal{E}X$ when $\mathcal{E}X$ exists.

It has been shown that if X has an orthogonally invariant distribution and if $\text{Cov}(X)$ exists, then $\text{Cov}(X) = \sigma^2 I$ for some $\sigma^2 \geqslant 0$. Of course there are distributions other than orthogonally invariant distributions for which the covariance is a constant times the identity. Such distributions arise in the chapter on linear models.

Definition 2.8. If $X \in (V, (\cdot, \cdot))$ and

$$\text{Cov}(X) = \sigma^2 I \quad \text{for some } \sigma^2 > 0,$$

X has a *weakly spherical* distribution.

The justification for the above definition is provided by Proposition 2.13.

Proposition 2.13. Suppose X is a random vector in $(V, (\cdot, \cdot))$ and $\text{Cov}(X)$ exists. The following are equivalent:

(i) $\text{Cov}(X) = \sigma^2 I$ for some $\sigma^2 \geqslant 0$.
(ii) $\text{Cov}(X) = \text{Cov}(\Gamma X)$ for all $\Gamma \in \mathcal{O}(V)$.

Proof. That (i) implies (ii) follows from Proposition 2.8. To show (ii) implies (i), let $\Sigma = \text{Cov}(X)$. From (ii) and Proposition 2.8, the non-negative definite linear transformation Σ must satisfy $\Sigma = \Gamma\Sigma\Gamma'$ for all $\Gamma \in \mathcal{O}(V)$. Thus for all $x \in V$, $\|x\| = 1$,

$$(x, \Sigma x) = (x, \Gamma\Sigma\Gamma' x) = (\Gamma' x, \Sigma\Gamma' x).$$

But $\Gamma' x$ can be any vector in V with length one since Γ' can be any element of $\mathcal{O}(V)$. Thus for all x, y, $\|x\| = \|y\| = 1$,

$$(x, \Sigma x) = (y, \Sigma y).$$

From the spectral theorem, write $\Sigma = \Sigma_1^n \lambda_i x_i \Box x_i$ and choose $x = x_j$ and $y = x_k$. Then we have

$$\lambda_j = (x_j, \Sigma x_j) = (x_k, \Sigma x_k) = \lambda_k$$

for all j, k. Setting $\sigma^2 = \lambda_1$,

$$\Sigma = \Sigma_1^n \sigma^2 x_i \Box x_i = \sigma^2 \Sigma_1^n x_i \Box x_i = \sigma^2 I.$$

That $\sigma^2 \geqslant 0$ follows from the positive semidefiniteness of Σ. $\square$

Orthogonally invariant distributions are sometimes called *spherical distributions*. The term weakly spherical results from weakening the assumption that the entire distribution is orthogonally invariant to the assumption that just the covariance structure is orthogonally invariant (condition (ii) of Proposition 2.13). A slight generalization of Proposition 2.13, given in its algebraic context, is needed for use later in this chapter.

Proposition 2.14. Suppose f is a bilinear function on $V \times V$ where $(V, (\cdot, \cdot))$ is an inner product space. If $f[\Gamma x_1, \Gamma x_2] = f[x_1, x_2]$ for all $x_1, x_2 \in V$ and $\Gamma \in \mathcal{O}(V)$, then $f[x_1, x_2] = c(x_1, x_2)$ where c is some real constant. If A is a linear transformation on V to V that satisfies $\Gamma' A \Gamma = A$ for all $\Gamma \in \mathcal{O}(V)$, then $A = cI$ for some real c.

Proof. Every bilinear function on $V \times V$ has the form (x_1, Ax_2) for some linear transformation A on V to V. The assertion that $f[\Gamma x_1, \Gamma x_2] = f[x_1, x_2]$ is clearly equivalent to the assertion that $\Gamma' A \Gamma = A$ for all $\Gamma \in \mathcal{O}(V)$. Thus it suffices to verify the assertion concerning the linear transformation A. Suppose $\Gamma' A \Gamma = A$ for all $\Gamma \in \mathcal{O}(V)$. Then for $x_1, x_2 \in V$,

$$(x_1, Ax_2) = (x_1, \Gamma' A \Gamma x_2) = (\Gamma x_1, A \Gamma x_2).$$

By Proposition 1.20, there exists a Γ such that

$$\Gamma \frac{x_1}{\|x_1\|} = \frac{x_2}{\|x_2\|}, \qquad \Gamma \frac{x_2}{\|x_2\|} = \frac{x_1}{\|x_1\|}$$

when x_1 and x_2 are not zero. Thus for x_1 and x_2 not zero,

$$(x_1, Ax_2) = (\Gamma x_1, A\Gamma x_2) = (x_2, Ax_1) = (Ax_1, x_2).$$

However, this relationship clearly holds if either x_1 or x_2 is zero. Thus for all $x_1, x_2 \in V$, $(x_1, Ax_2) = (Ax_1, x_2)$, so A must be self-adjoint. Now, using the spectral theorem, we can argue as in the proof of Proposition 2.13 to conclude that $A = cI$ for some real number c. □

◆ **Example 2.4.** Consider coordinate space R^n with the usual inner product. Let f be a function on $[0, \infty)$ to $[0, \infty)$ so that

$$\int_{R^n} f(\|x\|^2)\, dx = 1.$$

Thus $f(\|x\|^2)$ is a density on R^n. If the coordinate random vector

$X \in R^n$ has $f(\|x\|^2)$ as its density, then for $\Gamma \in \mathcal{O}_n$ (the group of $n \times n$ orthogonal matrices), the density of ΓX is again $f(\|x\|^2)$. This follows since $\|\Gamma x\| = \|x\|$ and the Jacobian of the linear transformation determined by Γ is equal to one. Hence the distribution determined by the density is $\mathcal{O}_n$ invariant. One particular choice for f is $f(u) = (2\pi)^{-n/2}e^{-1/2u}$ and the density for X is then

$$f(\|x\|^2) = (2\pi)^{-n/2}\exp\left[-\tfrac{1}{2}\Sigma_1^n x_i^2\right] = \prod_{i=1}^{n}(2\pi)^{-1/2}\exp\left[-\tfrac{1}{2}x_i^2\right].$$

Each of the factors in the above product is a density on R (corresponding to a normal distribution with mean zero and variance one). Therefore, the coordinates of X are independent and each has the same distribution. An example of a distribution on R^n that is weakly spherical, but not spherical, is provided by the density (with respect to Lebesgue measure)

$$p(x) = 2^{-n}\exp\left[-\Sigma_1^n|x_i|\right]$$

where $x \in R^n$, $x' = (x_1, x_2, \ldots, x_n)$. More generally, if the random variables $X_1, \ldots, X_n$ are independent with the same distribution on R, and $\sigma^2 = \text{var}(X_1)$, then the random vector X with coordinates $X_1, \ldots, X_n$ is easily shown to satisfy $\text{Cov}(X) = \sigma^2 I_n$ where I_n is the $n \times n$ identity matrix. ◆

The next topic in this section concerns the covariance between two random vectors. Suppose $X_i \in (V_i, (\cdot, \cdot)_i)$ for $i = 1, 2$ where X_1 and X_2 are defined on the same probability space. Then the random vector $\{X_1, X_2\}$ takes values in the direct sum $V_1 \oplus V_2$. Let $[\cdot, \cdot]$ denote the usual inner product on $V_1 \oplus V_2$ inherited from $(\cdot, \cdot)_i$, $i = 1, 2$. Assume that $\Sigma_{ii} = \text{Cov}(X_i)$, $i = 1, 2$, both exist. Then, let

$$f(x_1, x_2) = \text{cov}\{(x_1, X_1)_1, (x_2, X_2)_2\}$$

and note that the Cauchy–Schwarz Inequality (Example 1.9) shows that

$$|f(x_1, x_2)|^2 \leqslant (x_1, \Sigma_{11}x_1)_1(x_2, \Sigma_{22}x_2)_2.$$

Further, it is routine to check that $f(\cdot, \cdot)$ is a bilinear function on $V_1 \times V_2$ so there exists a linear transformation $\Sigma_{12} \in \mathcal{L}(V_2, V_1)$ such that

$$f(x_1, x_2) = (x_1, \Sigma_{12}x_2)_1.$$

The next proposition relates Σ_{11}, Σ_{12}, and Σ_{22} to the covariance of $\{X_1, X_2\}$ in the vector space $(V_1 \oplus V_2, [\cdot, \cdot])$.

Proposition 2.15. Let $\Sigma = \text{Cov}\{X_1, X_2\}$. Define a linear transformation A on $V_1 \oplus V_2$ to $V_1 \oplus V_2$ by

$$A\{x_1, x_2\} = \{\Sigma_{11}x_1 + \Sigma_{12}x_2, \Sigma'_{12}x_1 + \Sigma_{22}x_2\}$$

where Σ'_{12} is the adjoint of Σ_{12}. Then $A = \Sigma$.

Proof. It is routine to check that

$$[A\{x_1, x_2\}, \{x_3, x_4\}] = [\{x_1, x_2\}, A\{x_3, x_4\}]$$

so A is self-adjoint. To show $A = \Sigma$, it is sufficient to verify

$$[\{x_1, x_2\}, A\{x_1, x_2\}] = [\{x_1, x_2\}, \Sigma\{x_1, x_2\}]$$

by Proposition 1.16. However,

$$\begin{aligned}
[\{x_1, x_2\}, \Sigma\{x_1, x_2\}] &= \text{var}[\{x_1, x_2\}, \{X_1, X_2\}] \\
&= \text{var}\{(x_1, X_1)_1 + (x_2, X_2)_2\} \\
&= \text{var}(x_1, X_1)_1 + \text{var}(x_2, X_2)_2 \\
&\quad + 2\,\text{cov}\{(x_1, X_1)_1, (x_2, X_2)_2\} \\
&= (x_1, \Sigma_{11}x_1)_1 + (x_2, \Sigma_{22}x_2)_2 + 2(x_1, \Sigma_{12}x_2)_1 \\
&= (x_1, \Sigma_{11}x_1)_1 + (x_2, \Sigma_{22}x_2)_2 \\
&\quad + (x_1, \Sigma_{12}x_2)_1 + (\Sigma'_{12}x_1, x_2)_2 \\
&= [\{x_1, x_2\}, \{\Sigma_{11}x_1 + \Sigma_{12}x_2, \Sigma'_{12}x_1 + \Sigma_{22}x_2\}] \\
&= [\{x_1, x_2\}, A\{x_1, x_2\}]. \qquad \square
\end{aligned}$$

It is customary to write the linear transformation A in partitioned form as

$$\begin{pmatrix} \Sigma_{11} & \Sigma_{12} \\ \Sigma'_{12} & \Sigma_{22} \end{pmatrix}\{x_1, x_2\} = \{\Sigma_{11}x_1 + \Sigma_{12}x_2, \Sigma'_{12}x_1 + \Sigma_{22}x_2\}.$$

With this notation,

$$\operatorname{Cov}\{X_1, X_2\} = \begin{pmatrix} \Sigma_{11} & \Sigma_{12} \\ \Sigma'_{12} & \Sigma_{22} \end{pmatrix}.$$

Definition 2.9. The random vectors X_1 and X_2 are uncorrelated if $\Sigma_{12} = 0$.

In the above definition, it is assumed that $\operatorname{Cov}(X_i)$ exists for $i = 1, 2$. It is clear that X_1 and X_2 are uncorrelated iff

$$\operatorname{cov}\{(x_1, X_1)_1, (x_2, X_2)_2\} = 0 \qquad \text{for all } x_i \in V_i,\ i = 1, 2.$$

Also, if X_1 and X_2 are uncorrelated in the two given inner products, then they are uncorrelated in all inner products on V_1 and V_2. This follows from the fact that any two inner products are related by a positive definite linear transformation.

Given $X_i \in (V_i, (\cdot,\cdot)_i)$ for $i = 1, 2$, suppose

$$\operatorname{Cov}\{X_1, X_2\} = \begin{pmatrix} \Sigma_{11} & \Sigma_{12} \\ \Sigma'_{12} & \Sigma_{22} \end{pmatrix}.$$

We want to show that there is a linear transformation $B \in \mathcal{L}(V_2, V_1)$ such that $X_1 + BX_2$ and X_2 are uncorrelated random vectors. However, before this can be established, some preliminary technical results are needed.

Consider an inner product space $(V, (\cdot,\cdot))$ and suppose $A \in \mathcal{L}(V, V)$ is self-adjoint of rank k. Then, by the spectral theorem, $A = \Sigma_1^k \lambda_i x_i \,\square\, x_i$ where $\lambda_i \neq 0$, $i = 1, \ldots, k$, and $\{x_1, \ldots, x_k\}$ is an orthonormal set that is a basis for $\mathcal{R}(A)$. The linear transformation

$$A^- \equiv \Sigma_1^k \frac{1}{\lambda_i} x_i \,\square\, x_i$$

is called the *generalized inverse* of A. If A is nonsingular, then it is clear that A^- is the inverse of A. Also, A^- is self-adjoint and $AA^- = A^-A = \Sigma_1^k x_i \,\square\, x_i$, which is just the orthogonal projection onto $\mathcal{R}(A)$. A routine computation shows that $A^-AA^- = A^-$ and $AA^-A = A$.

In the notation established previously (see Proposition 2.15), suppose $\{X_1, X_2\} \in V_1 \oplus V_2$ has a covariance

$$\Sigma = \operatorname{Cov}\{X_1, X_2\} = \begin{pmatrix} \Sigma_{11} & \Sigma_{12} \\ \Sigma'_{12} & \Sigma_{22} \end{pmatrix}.$$

Proposition 2.16. For the covariance above, $\mathcal{N}(\Sigma_{22}) \subseteq \mathcal{N}(\Sigma_{12})$ and $\Sigma_{12} = \Sigma_{12}\Sigma_{22}^-\Sigma_{22}$.

Proof. For $x_2 \in \mathfrak{N}(\Sigma_{22})$, it must be shown that $\Sigma_{12}x_2 = 0$. Consider $x_1 \in V_1$ and $\alpha \in R$. Then $\Sigma_{22}(\alpha x_2) = 0$ and since Σ is positive semidefinite,

$$0 \leqslant [\{x_1, \alpha x_2\}, \Sigma\{x_1, \alpha x_2\}] = [\{x_1, \alpha x_2\}, \{\Sigma_{11}x_1 + \alpha\Sigma_{12}x_2, \Sigma'_{12}x_1\}]$$

$$= (x_1, \Sigma_{11}x_1)_1 + \alpha(x_1, \Sigma_{12}x_2)_1 + \alpha(x_2, \Sigma'_{12}x_1)_2$$

$$= (x_1, \Sigma_{11}x_1) + 2\alpha(x_1, \Sigma_{12}x_2)_1.$$

As this inequality holds for all $\alpha \in R$, for each $x_1 \in V$, $(x_1, \Sigma_{12}x_2)_1 = 0$. Hence $\Sigma_{12}x_2 = 0$ and the first claim is proved. To verify that $\Sigma_{12} = \Sigma_{12}\Sigma_{22}^{-}\Sigma_{22}$, it suffices to establish the identity $\Sigma_{12}(I - \Sigma_{22}^{-}\Sigma_{22}) = 0$. However, $I - \Sigma_{22}^{-}\Sigma_{22}$ is the orthogonal projection onto $\mathfrak{N}(\Sigma_{22})$. Since $\mathfrak{N}(\Sigma_{22}) \subseteq \mathfrak{N}(\Sigma_{12})$, it follows that $\Sigma_{12}(I - \Sigma_{22}^{-}\Sigma_{22}) = 0$. □

We are now in a position to show that $X_1 - \Sigma_{12}\Sigma_{22}^{-}X_2$ and X_2 are uncorrelated.

Proposition 2.17. Suppose $\{X_1, X_2\} \in V_1 \oplus V_2$ has a covariance

$$\Sigma = \operatorname{Cov}\{X_1, X_2\} = \begin{pmatrix} \Sigma_{11} & \Sigma_{12} \\ \Sigma'_{12} & \Sigma_{22} \end{pmatrix}.$$

Then $X_1 - \Sigma_{12}\Sigma_{22}^{-}X_2$ and X_2 are uncorrelated, and $\operatorname{Cov}(X_1 - \Sigma_{12}\Sigma_{22}^{-}X_2) = \Sigma_{11} - \Sigma_{12}\Sigma_{22}^{-}\Sigma_{21}$ where $\Sigma_{21} \equiv \Sigma'_{12}$.

Proof. For $x_i \in V_i$, $i = 1, 2$, it must be verified that

$$\operatorname{cov}\{(x_1, X_1 - \Sigma_{12}\Sigma_{22}^{-}X_2)_1, (x_2, X_2)_2\} = 0.$$

This calculation goes as follows:

$$\operatorname{cov}\{(x_1, X_1 - \Sigma_{12}\Sigma_{22}^{-}X_2)_1, (x_2, X_2)_2\}$$

$$= \operatorname{cov}\{(x_1, X_1)_1, (x_2, X_2)_2\}$$

$$- \operatorname{cov}\{(\Sigma_{22}^{-}\Sigma'_{12}x_1, X_2)_2, (x_2, X_2)_2\}$$

$$= (x_1, \Sigma_{12}x_2)_1 - (\Sigma_{22}^{-}\Sigma'_{12}x_1, \Sigma_{22}x_2)_2$$

$$= (x_1, \Sigma_{12}x_2)_1 - (x_1, \Sigma_{12}\Sigma_{22}^{-}\Sigma_{22}x_2)_1$$

$$= (x_1, (\Sigma_{12} - \Sigma_{12}\Sigma_{22}^{-}\Sigma_{22})x_2)_1 = 0.$$

The last equality follows from Proposition 2.15 since $\Sigma_{12} = \Sigma_{12}\Sigma_{22}^{-}\Sigma_{22}$. To verify the second assertion, we need to establish the identity

$$\operatorname{var}(x_1, X_1 - \Sigma_{12}\Sigma_{22}^{-}X_2)_1 = (x_1, (\Sigma_{11} - \Sigma_{12}\Sigma_{22}^{-}\Sigma_{21})x_1)_1.$$

But

$$\begin{aligned}\operatorname{var}(x_1, X_1 - \Sigma_{12}\Sigma_{22}^{-}X_2)_1 &= \operatorname{var}(x_1, X_1)_1 + \operatorname{var}(x_1, \Sigma_{12}\Sigma_{22}^{-}X_2)_1 \\ &\quad -2\operatorname{cov}\{(x_1, X_1)_1, (x_1, \Sigma_{12}\Sigma_{22}^{-}X_2)_1\} \\ &= (x_1, \Sigma_{11}x_1)_1 + (x_1, \Sigma_{12}\Sigma_{22}^{-}\Sigma_{22}\Sigma_{22}^{-}\Sigma_{12}'x_1)_1 \\ &\quad -2(x_1, \Sigma_{12}\Sigma_{22}^{-}\Sigma_{12}'x_1)_1 \\ &= (x_1, (\Sigma_{11} - \Sigma_{12}\Sigma_{22}^{-}\Sigma_{12}')x_1)_1.\end{aligned}$$

In the above, the identity $\Sigma_{22}^{-}\Sigma_{22}\Sigma_{22}^{-} = \Sigma_{22}^{-}$ has been used. □

We now return to the situation considered in Example 2.4. Consider independent coordinate random vectors $X_1, \ldots, X_n$ with each $X_i \in R^p$, and suppose that $\mathcal{E}X_i = \mu \in R^p$, and $\operatorname{Cov}(X_i) = \Sigma$ for $i = 1, \ldots, n$. Form the random matrix $X \in \mathcal{L}_{p,n}$ with rows $X_1', \ldots, X_n'$. Our purpose is to describe the mean vector and covariance of X in terms of Σ and μ. The inner product on $\mathcal{L}_{p,n}$, $\langle \cdot, \cdot \rangle$ is that inherited from the standard inner products on the coordinate spaces R^p and R^n. Recall that, for matrices $A, B \in \mathcal{L}_{p,n}$,

$$\langle A, B \rangle = \operatorname{tr} AB' = \operatorname{tr} B'A = \operatorname{tr} A'B = \operatorname{tr} BA'.$$

Let e denote the vector in R^n whose coordinates are all equal to 1.

Proposition 2.18. In the above notation,

(i) $\mathcal{E}X = e\mu'$.
(ii) $\operatorname{Cov}(X) = I_n \otimes \Sigma$.

Here I_n is the $n \times n$ identity matrix and $\otimes$ denotes the Kronecker product.

Proof. The matrix $e\mu'$ has each row equal to μ' and, since each row of X has mean μ', the first assertion is fairly obvious. To verify (i) formally, it must be shown that, for $A \in \mathcal{L}_{p,n}$,

$$\mathcal{E}\langle A, X \rangle = \langle A, e\mu' \rangle.$$

Let $a_1', \ldots, a_n'$, $a_i \in R^p$, be the rows of A. Then

$$\mathcal{E}\langle A, X\rangle = \mathcal{E}\operatorname{tr} AX' = \mathcal{E}\Sigma_1^n a_i'X_i = \Sigma_1^n a_i'\mathcal{E}X_i = \Sigma_1^n a_i'\mu = \operatorname{tr} A\mu e' = \langle A, e\mu'\rangle.$$

Thus (i) holds. To verify (ii) it suffices to establish the identity

$$\operatorname{var}\langle A, X\rangle = \langle A, (I \otimes \Sigma)A\rangle$$

for $A \in \mathcal{L}_{p,n}$. In the notation above,

$$\operatorname{var}\langle A, X\rangle = \operatorname{var}(\Sigma_1^n a_i'X_i) = \Sigma_1^n \operatorname{var}(a_i'X_i) + \sum_{i \neq j} \operatorname{cov}\{a_i'X_i, a_j'X_j\} = \Sigma_1^n a_i'\Sigma a_i$$

$$= \operatorname{tr} A'A\Sigma = \operatorname{tr} A\Sigma A' = \operatorname{tr} A(A\Sigma)' = \langle A, (I_n \otimes \Sigma)A\rangle.$$

The third equality follows from $\operatorname{var}(a_i'X) = a_i'\Sigma a_i$ and, for $i \neq j$, $a_i'X_i$ and $a_j'X_j$ are uncorrelated. □

The assumption of the independence of $X_1, \ldots, X_n$ was not used to its full extent in the proof of Proposition 2.18. In fact the above proof shows that, if $X_1, \ldots, X_n$ are random variables in R^p with $\mathcal{E}X_i = \mu$, $i = 1, \ldots, n$, then $\mathcal{E}X = e\mu'$. Further, if $X_1, \ldots, X_n$ in R^p are uncorrelated with $\operatorname{Cov}(X_i) = \Sigma$, $i = 1, \ldots, n$, then $\operatorname{Cov}(X) = I_n \otimes \Sigma$. One application of this formula for $\operatorname{Cov}(X)$ describes how $\operatorname{Cov}(X)$ transforms under Kronecker products. For example, if $A \in \mathcal{L}_{n,n}$ and $B \in \mathcal{L}_{p,p}$, then $(A \otimes B)X = AXB'$ is a random vector in $\mathcal{L}_{p,n}$. Proposition 2.8 shows that

$$\operatorname{Cov}((A \otimes B)X) = (A \otimes B)\operatorname{Cov}(X)(A \otimes B)'.$$

In particular, if $\operatorname{Cov}(X) = I_n \otimes \Sigma$, then

$$\operatorname{Cov}((A \otimes B)X) = (A \otimes B)(I_n \otimes \Sigma)(A \otimes B)' = (AA') \otimes (B\Sigma B').$$

Since $A \otimes B = (A \otimes I_p)(I_n \otimes B)$, the interpretation of the above covariance formula reduces to an interpretation for $A \otimes I_p$ and $I_n \otimes B$. First, $(I_n \otimes B)X$ is a random matrix with rows $X_i'B' = (BX_i)'$, $i = 1, \ldots, n$. If $\operatorname{Cov}(X_i) = \Sigma$, then $\operatorname{Cov}(BX_i) = B\Sigma B'$. Thus it is clear from Proposition 2.18 that $\operatorname{Cov}((I_n \times B)X) = I_n \otimes (B\Sigma B')$. Second, $(A \otimes I_p)$ applied to X is the same as applying the linear transformation A to each column of X. When $\operatorname{Cov}(X) = I_n \otimes \Sigma$, the rows of X are uncorrelated and, if A is an $n \times n$ orthogonal matrix, then

$$\operatorname{Cov}\big((A \otimes I_p)X\big) = I_n \otimes \Sigma = \operatorname{Cov}(X).$$

Thus the absence of correlation between the rows is preserved by an orthogonal transformation of the columns of X.

A converse to the observation that $\text{Cov}((A \otimes I_p)X) = I_n \otimes \Sigma$ for all $A \in \mathcal{O}(n)$ is valid for random linear transformations. To be more precise, we have the following proposition.

Proposition 2.19. Suppose $(V_i, (\cdot, \cdot)_i)$, $i = 1, 2$, are inner product spaces and X is a random vector in $(\mathcal{L}(V_1, V_2), \langle \cdot , \cdot \rangle)$. The following are equivalent:

(i) $\text{Cov}(X) = I_2 \otimes \Sigma$.
(ii) $\text{Cov}((\Gamma \otimes I_1)X) = \text{Cov}(X)$ for all $\Gamma \in \mathcal{O}(V_2)$.

Here, I_i is identity linear transformation on V_i, $i = 1, 2$, and Σ is a non-negative definite linear transformation on V_1 to V_1.

Proof. Let $\Psi = \text{Cov}(X)$ so Ψ is a positive semidefinite linear transformation on $\mathcal{L}(V_1, V_2)$ to $\mathcal{L}(V_1, V_2)$ and Ψ is characterized by the equation

$$\text{cov}\{\langle A, X\rangle, \langle B, X\rangle\} = \langle A, \Psi B\rangle$$

for all $A, B \in \mathcal{L}(V_1, V_2)$. If (i) holds, then we have

$$\begin{aligned}\text{Cov}((\Gamma \otimes I_1)X) &= (\Gamma \otimes I_1)\text{Cov}(X)(\Gamma \otimes I_1)' \\ &= (\Gamma \otimes I_1)(I_2 \otimes \Sigma)(\Gamma' \otimes I_1) = (\Gamma I_2 \Gamma') \otimes (I_1 \Sigma I_1) \\ &= I_2 \otimes \Sigma = \text{Cov}(X),\end{aligned}$$

so (ii) holds.

Now, assume (ii) holds. Since outer products form a basis for $\mathcal{L}(V_1, V_2)$, it is sufficient to show there exists a positive semidefinite Σ on V_1 to V_1 such that, for $x_1, x_2 \in V_1$ and $y_1, y_2 \in V_2$,

$$\langle y_1 \square x_1, \Psi(y_2 \square x_2)\rangle = \langle y_1 \square x_1, (I_2 \otimes \Sigma)(y_2 \square x_2)\rangle.$$

Define H by

$$H(x_1, x_2, y_1, y_2) \equiv \text{cov}\{\langle y_1 \square x_1, X\rangle, \langle y_2 \square x_2, X\rangle\}$$

for $x_1, x_2 \in V_1$ and $y_1, y_2 \in V_2$. From assumption (ii), we know that Ψ

satisfies $\Psi = (\Gamma \otimes I_1)\Psi(\Gamma \otimes I_1)'$ for all $\Gamma \in \mathcal{O}(V_2)$. Thus

$$\begin{aligned} H(x_1, x_2, y_1, y_2) &= \langle y_1 \square x_1, \Psi(y_2 \square x_2)\rangle \\ &= \langle y_1 \square x_1, (\Gamma \otimes I_1)\Psi(\Gamma \otimes I_1)'(y_2 \square x_2)\rangle \\ &= \langle (\Gamma \otimes I_1)'(y_1 \square x_1), \Psi(\Gamma \otimes I_1)'(y_2 \square x_2)\rangle \\ &= \langle (\Gamma' y_1) \square x_1, \Psi(\Gamma' y_2) \square x_2\rangle = H(x_1, x_2, \Gamma' y_1, \Gamma' y_2) \end{aligned}$$

for all $\Gamma \in \mathcal{O}(V_2)$. It is clear that H is a linear function of each of its four arguments when the other three are held fixed. Therefore, for x_1 and x_2 fixed, G is a bilinear function on $V_2 \times V_2$ and this bilinear function satisfies the assumption of Proposition 2.14. Thus there is a constant, which depends on x_1 and x_2, say $c[x_1, x_2]$, and

$$H(x_1, x_2, y_1, y_2) = c[x_1, x_2](y_1, y_2)_2.$$

However, for $y_1 = y_2 \neq 0$, H, as a function of x_1 and x_2, is bilinear and non-negative definite on $V_1 \times V_1$. In other words, $c[x_1, x_2]$ is a non-negative definite bilinear function on $V_1 \times V_1$, so

$$c[x_1, x_2] = (x_1, \Sigma x_2)_1$$

for some non-negative definite Σ. Thus

$$H(x_1, x_2, y_1, y_2) = (x_1, \Sigma x_2)_1 (y_1, y_2)_2 = \langle y_1 \square x_1, (I_2 \otimes \Sigma)(y_2 \square x_2)\rangle,$$

so $\Psi = I_2 \otimes \Sigma$. □

The next topic of consideration in the section concerns the calculation of means and covariances for outer products of random vectors. These results are used throughout the sequel to simplify proofs and provide convenient formulas. Suppose X_i is a random vector in $(V_i, (\cdot, \cdot)_i)$ for $i = 1, 2$ and let $\mu_i = \mathcal{E} X_i$, and $\Sigma_{ii} = \text{Cov}(X_i)$ for $i = 1, 2$. Thus $\{X_1, X_2\}$ takes values in $V_1 \oplus V_2$ and

$$\text{Cov}\{X_1, X_2\} = \begin{pmatrix} \Sigma_{11} & \Sigma_{12} \\ \Sigma'_{12} & \Sigma_{22} \end{pmatrix}$$

where Σ_{12} is characterized by

$$\text{cov}\{(x_1, X_1)_1, (x_2, X_2)_2\} = (x_1, \Sigma_{12} x_2)_1$$

for $x_i \in V_i$, $i = 1, 2$. Of course, $\text{Cov}\{X_1, X_2\}$ is expressed relative to the natural inner product on $V_1 \oplus V_2$ inherited from $(V_1, (\cdot, \cdot)_1)$ and $(V_2, (\cdot, \cdot)_2)$.

Proposition 2.20. For $X_i \in (V_i, (\cdot, \cdot))$, $i = 1, 2$, as above,

$$\mathcal{E} X_1 \square X_2 = \Sigma_{12} + \mu_1 \square \mu_2.$$

Proof. The random vector $X_1 \square X_2$ takes values in the inner product space $(\mathcal{L}(V_2, V_1), \langle \cdot , \cdot \rangle)$. To verify the above formula, it must be shown that

$$\mathcal{E}\langle A, X_1 \square X_2 \rangle = \langle A, \Sigma_{12} \rangle + \langle A, \mu_1 \square \mu_2 \rangle$$

for $A \in \mathcal{L}(V_2, V_1)$. However, it is sufficient to verify this equation for $A = x_1 \square x_2$ since both sides of the equation are linear in A and every A is a linear combination of elements in $\mathcal{L}(V_2, V_1)$ of the form $x_1 \square x_2$, $x_i \in V_i$, $i = 1, 2$. For $x_1 \square x_2 \in \mathcal{L}(V_2, V_1)$,

$$\begin{aligned} \mathcal{E}\langle x_1 \square x_2, X_1 \square X_2 \rangle &= \mathcal{E}(x_1, X_1)_1 (x_2, X_2)_2 \\ &= \text{cov}\{(x_1, X_1)_1, (x_2, X_2)_2\} + \mathcal{E}(x_1, X_1)_1 \mathcal{E}(x_2, X_2)_2 \\ &= (x_1, \Sigma_{12} x_2)_1 + (x_1, \mu_1)_1 (x_2, \mu_2)_2 \\ &= \langle x_1 \square x_2, \Sigma_{12} \rangle + \langle x_1 \square x_2, \mu_1 \square \mu_2 \rangle. \end{aligned}$$

□

A couple of interesting applications of Proposition 2.20 are given in the following proposition.

Proposition 2.21. For X_1, X_2 in $(V, (\cdot, \cdot))$, let $\mu_i = \mathcal{E} X_i$, $\Sigma_{ii} = \text{Cov}(X_i)$ for $i = 1, 2$. Also, let Σ_{12} be the unique linear transformation satisfying

$$\text{cov}\{(x_1, X_1), (x_2, X_2)\} = (x_1, \Sigma_{12} x_2)$$

for all $x_1, x_2 \in V$. Then:

(i) $\mathcal{E} X_1 \square X_1 = \Sigma_{11} + \mu_1 \square \mu_1$.
(ii) $\mathcal{E}(X_1, X_2) = \langle I, \Sigma_{12} \rangle + (\mu_1, \mu_2)$.
(iii) $\mathcal{E}(X_1, X_1) = \langle I, \Sigma_{11} \rangle + (\mu_1, \mu_1)$.

Here $I \in \mathcal{L}(V, V)$ is the identity linear transformation and $\langle \cdot , \cdot \rangle$ is the inner product on $\mathcal{L}(V, V)$ inherited from $(V, (\cdot, \cdot))$.

Proof. For (i), take $X_1 = X_2$ and $(V_1, (\cdot, \cdot)_1) = (V_2, (\cdot, \cdot)_2) = (V, (\cdot, \cdot))$ in Proposition 2.20. To verify (ii), first note that

$$\mathcal{E} X_1 \square X_2 = \Sigma_{12} + \mu_1 \square \mu_2$$

by the previous proposition. Thus for $I \in \mathcal{L}(V, V)$,

$$\mathcal{E}\langle I, X_1 \square X_2 \rangle = \langle I, \Sigma_{12} \rangle + \langle I, \mu_1 \square \mu_2 \rangle.$$

However, $\langle I, X_1 \square X_2 \rangle = (X_1, X_2)$ and $\langle I, \mu_1 \square \mu_2 \rangle = (\mu_1, \mu_2)$ so (ii) holds. Assertion (iii) follows from (ii) by taking $X_1 = X_2$. □

One application of the preceding result concerns the affine prediction of one random vector by another random vector. By an affine function on a vector space V to W, we mean a function f given by $f(v) = Av + w_0$ where $A \in \mathcal{L}(V, W)$ and w_0 is a fixed vector in W. The term linear transformation is reserved for those affine functions that map zero into zero. In the notation of Proposition 2.21, consider $X_i \in (V_i, (\cdot, \cdot)_i$ for $i = 1, 2$, let $\mu_i = \mathcal{E} X_i$, $i = 1, 2$, and suppose

$$\Sigma \equiv \operatorname{Cov}\{X_1, X_2\} = \begin{pmatrix} \Sigma_{11} & \Sigma_{12} \\ \Sigma'_{12} & \Sigma_{22} \end{pmatrix}$$

exists. An affine predictor of X_2 based on X_1 is any function of the form $AX_1 + x_0$ where $A \in \mathcal{L}(V_1, V_2)$ and x_0 is a fixed vector in V_2. If we assume that μ_1, μ_2, and Σ are known, then A and x_0 are allowed to depend on these known quantities. The statistical interpretation is that we observe X_1, but not X_2, and X_2 is to be predicted by $AX_1 + x_0$. One intuitively reasonable criterion for selecting A and x_0 is to ask that the choice of A and x_0 minimize

$$\mathcal{E}\|X_2 - (AX_1 + x_0)\|_2^2.$$

Here, the expectation is over the joint distribution of X_1 and X_2 and $\|\cdot\|_2$ is the norm in the vector space $(V_2, (\cdot, \cdot)_2)$. The quantity $\mathcal{E}\|X_2 - (AX_1 + x_0)\|_2^2$ is the average distance of $X_2 - (AX_1 + x_0)$ from 0. Since $AX_1 + x_0$ is supposed to predict X_2, it is reasonable that A and x_0 be chosen to minimize this average distance. A solution to this minimization problem is given in Proposition 2.22.

Proposition 2.22. For X_1 and X_2 as above,

$$\mathcal{E}\|X_2 - (AX_1 + x_0)\|_2^2 \geqslant \langle I_2, \Sigma_{22} - \Sigma'_{12}\Sigma_{11}^{-}\Sigma_{12} \rangle$$

with equality for $A = \Sigma'_{12}\Sigma_{11}^{-}$ and $x_0 = \mu_2 - \Sigma'_{12}\Sigma_{11}^{-}\mu_1$.

Proof. The proof is a calculation. It essentially consists of completing the square and applying (ii) of Proposition 2.21. Let $Y_i = X_i - \mu_i$ for $i = 1, 2$. Then

$$\begin{aligned}\mathcal{E}\|X_2 - (AX_1 + x_0)\|_2^2 &= \mathcal{E}\|Y_2 - AY_1 + \mu_2 - A\mu_1 - x_0\|_2^2 = \mathcal{E}\|Y_2 - AY_1\|_2^2 \\ &\quad + 2\mathcal{E}(Y_2 - AY_1, \mu_2 - A\mu_1 - x_0)_2 + \|\mu_2 - A\mu_1 - x_0\|_2^2 \\ &= \mathcal{E}\|Y_2 - AY_1\|_2^2 + \|\mu_2 - A\mu_1 - x_0\|_2^2.\end{aligned}$$

The last equality holds since $\mathcal{E}(Y_2 - AY_1) = 0$. Thus for each $A \in \mathcal{L}(V_1, V_2)$,

$$\mathcal{E}\|X_2 - (AX_1 + x_0)\|_2^2 \geqslant \mathcal{E}\|Y_2 - AY_1\|_2^2$$

with equality for $x_0 = \mu_2 - A\mu_1$. For notational convenience let $\Sigma_{21} = \Sigma'_{12}$. Then

$$\begin{aligned}\mathcal{E}\|Y_2 - AY_1\|_2^2 &= \mathcal{E}\|Y_2 - \Sigma_{21}\Sigma_{11}^- Y_1 + (\Sigma_{21}\Sigma_{11}^- - A)Y_1\|_2^2 \\ &= \mathcal{E}\|Y_2 - \Sigma_{21}\Sigma_{11}^- Y_1\|_2^2 + \mathcal{E}\|(\Sigma_{21}\Sigma_{11}^- - A)Y_1\|_2^2 \\ &\quad + 2\mathcal{E}(Y_2 - \Sigma_{21}\Sigma_{11}^- Y_1, (\Sigma_{21}\Sigma_{11}^- - A)Y_1)_2 \\ &= \mathcal{E}\|Y_2 - \Sigma_{21}\Sigma_{11}^- Y_1\|_2^2 + \mathcal{E}\|(\Sigma_{21}\Sigma_{11}^- - A)Y_1\|_2^2 \\ &\geqslant \mathcal{E}\|Y_2 - \Sigma_{21}\Sigma_{11}^- Y_1\|_2^2.\end{aligned}$$

The last equality holds since $\mathcal{E}(Y_2 - \Sigma_{21}\Sigma_{11}^- Y_1) = 0$ and $Y_2 - \Sigma_{21}\Sigma_{11}^- Y_1$ is uncorrelated with Y_1 (Proposition 2.17) and hence is uncorrelated with $(\Sigma_{21}\Sigma_{11}^- - A)Y_1$. By (ii) of Proposition 2.21, we see that $\mathcal{E}(Y_2 - \Sigma_{21}\Sigma_{11}^- Y_1, (\Sigma_{21}\Sigma_{11}^- - A)Y_1)_2 = 0$. Therefore, for each $A \in \mathcal{L}(V_1, V_2)$,

$$\mathcal{E}\|Y_2 - AY_1\|_2^2 \geqslant \mathcal{E}\|Y_2 - \Sigma_{21}\Sigma_{11}^- Y_1\|_2^2$$

with equality for $A = \Sigma_{21}\Sigma_{11}^-$. However, $\mathrm{Cov}(Y_2 - \Sigma_{21}\Sigma_{11}^- Y_1) = \Sigma_{22} - \Sigma_{21}\Sigma_{11}^-\Sigma_{12}$ and $\mathcal{E}(Y_2 - \Sigma_{21}\Sigma_{11}^- Y_1) = 0$ so (iii) of Proposition 2.21 shows that

$$\mathcal{E}\|Y_2 - \Sigma_{21}\Sigma_{11}^- Y_1\|_2^2 = \langle I_2, \Sigma_{22} - \Sigma_{21}\Sigma_{11}^-\Sigma_{12}\rangle.$$

Therefore,

$$\mathcal{E}\|X_2 - (AX_1 + x_0)\|_2^2 \geqslant \langle I_2, \Sigma_{22} - \Sigma_{21}\Sigma_{11}^-\Sigma_{12}\rangle$$

with equality for $A = \Sigma_{21}\Sigma_{11}^-$ and $x_0 = \mu_2 - \Sigma_{21}\Sigma_{11}^-\mu_1$. □

The last topic in this section concerns the covariance of $X \square X$ when X is a random vector in $(V, (\cdot, \cdot))$. The random vector $X \square X$ is an element of the vector space $(\mathcal{L}(V, V), \langle \cdot , \cdot \rangle)$. However, $X \square X$ is a self-adjoint linear transformation so $X \square X$ is also a random vector in $(M_s, \langle \cdot , \cdot \rangle)$ where M_s is the linear subspace of self-adjoint transformations in $\mathcal{L}(V, V)$. In what follows, we regard $X \square X$ as a random vector in $(M_s, \langle \cdot , \cdot \rangle)$. Thus the covariance of $X \square X$ is a positive semidefinite linear transformation on $(M_s, \langle \cdot , \cdot \rangle)$. In general, this covariance is quite complicated and we make some simplifying assumptions concerning the distribution of X.

Proposition 2.23. Suppose X has an orthogonally invariant distribution in $(V, (\cdot, \cdot))$ where $\mathcal{E}\|X\|^4 < +\infty$. Let v_1 and v_2 be fixed vectors in V with $\|v_i\| = 1$, $i = 1, 2$, and $(v_1, v_2) = 0$. Set $c_1 = \text{var}\{(v_1, X)^2\}$ and $c_2 = \text{cov}\{(v_1, X)^2, (v_2, X)^2\}$. Then

$$\text{Cov}(X \square X) = (c_1 - c_2) I \otimes I + c_2 T_1,$$

where T_1 is the linear transformation on M_s given by $T_1(A) = \langle I, A \rangle I$. In other words, for $A, B \in M_s$,

$$\begin{aligned}\text{cov}\{\langle A, X \square X \rangle, \langle B, X \square X \rangle\} &= \langle A, ((c_1 - c_2) I \otimes I + c_2 T_1) B \rangle \\ &= (c_1 - c_2)\langle A, B \rangle + c_2 \langle I, A \rangle \langle I, B \rangle.\end{aligned}$$

Proof. Since $(c_1 - c_2) I \otimes I + c_2 T_1$ is self-adjoint on $(M_s, \langle \cdot , \cdot \rangle)$, Proposition 2.6 shows that it suffices to verify the equation

$$\text{var}\langle A, X \square X \rangle = (c_1 - c_2)\langle A, A \rangle + c_2 \langle I, A \rangle^2$$

for $A \in M_s$ in order to prove that

$$\text{Cov}(X \square X) = (c_1 - c_2) I \otimes I + c_2 T_1.$$

First note that, for $x \in V$,

$$\text{var}\langle x \square x, X \square X \rangle = \text{var}(x, X)^2 = \|x\|^4 \text{var}\left(\frac{x}{\|x\|}, X\right)^2 = \|x\|^4 \text{var}(v_1, X)^2.$$

This last equality follows from Proposition 2.10 as the distribution of X is orthogonally invariant. Also, for $x_1, x_2 \in V$ with $(x_1, x_2) = 0$,

$$\operatorname{cov}\{(x_1, X)^2, (x_2, X)^2\} = \|x_1\|^2\|x_2\|^2 \operatorname{cov}\left\{\left(\frac{x_1}{\|x_1\|}, X\right)^2, \left(\frac{x_2}{\|x_2\|}, X\right)^2\right\}$$

$$= \|x_1\|^2\|x_2\|^2 \operatorname{cov}\{(v_1, X)^2, (v_2, X)^2\}.$$

Again, the last equality follows since $\mathcal{L}(X) = \mathcal{L}(\Psi X)$ for $\Psi \in \mathcal{O}(V)$ so

$$\operatorname{cov}\left\{\left(\frac{x_1}{\|x_1\|}, X\right)^2, \left(\frac{x_2}{\|x_2\|}, X\right)^2\right\} = \operatorname{cov}\left\{\left(\Psi\frac{x_1}{\|x_1\|}, X\right)^2, \left(\Psi\frac{x_2}{\|x_2\|}, X\right)^2\right\}$$

and Ψ can be chosen so that

$$\Psi\frac{x_i}{\|x_i\|} = v_i, \qquad i = 1, 2.$$

For $A \in M_s$, apply the spectral theorem and write $A = \Sigma_1^n a_i x_i \square x_i$ where $x_1, \dots, x_n$ is an orthonormal basis for $(V, (\cdot, \cdot))$. Then

$$\begin{aligned}
\operatorname{var}\langle A, X\square X\rangle &= \operatorname{var}\langle \Sigma a_i x_i \square x_i, X\square X\rangle \\
&= \Sigma a_i^2 \operatorname{var}\langle x_i \square x_i, X\square X\rangle \\
&\quad + \sum_{i \neq j}\sum a_i a_j \operatorname{cov}\{\langle x_i \square x_i, X\square X\rangle, \langle x_j \square x_j, X\square X\rangle\} \\
&= \Sigma a_i^2 \operatorname{var}(x_i, X)^2 + \sum_{i \neq j}\sum a_i a_j \operatorname{cov}\{(x_i, X)^2, (x_j, X)^2\} \\
&= c_1 \Sigma a_i^2 + c_2 \sum_{i \neq j}\sum a_i a_j = (c_1 - c_2)\sum_i a_i^2 + c_2 \sum_i \sum_j a_i a_j \\
&= (c_1 - c_2)\langle A, A\rangle + c_2\langle I, A\rangle^2. \qquad \square
\end{aligned}$$

When X has an orthogonally invariant normal distribution, then the constant $c_2 = 0$ so $\operatorname{Cov}(X\square X) = c_1 I \otimes I$. The following result provides a slight generalization of Proposition 2.23.

Proposition 2.24. Let X, v_1, and v_2 be as in Proposition 2.23. For $C \in \mathcal{L}(V, V)$, let $\Sigma = CC'$ and suppose Y is a random vector in $(V, (\cdot, \cdot))$ with

$\mathcal{L}(Y) = \mathcal{L}(CX)$. Then

$$\operatorname{Cov}(Y \square Y) = (c_1 - c_2)\Sigma \otimes \Sigma + c_2 T_2$$

where $T_2(A) = \langle A, \Sigma \rangle \Sigma$ for $A \in M_s$.

Proof. We apply Proposition 2.8 and the calculational rules for Kronecker products. Since $(CX)\square(CX) = (C \otimes C)(X \square X)$,

$$\begin{aligned}
\operatorname{Cov}(Y \square Y) &= \operatorname{Cov}((CX \square CX)) = \operatorname{Cov}((C \otimes C)(X \square X)) \\
&= (C \otimes C)\operatorname{Cov}(X \square X)(C \otimes C)' \\
&= (C \otimes C)((c_1 - c_2)I \otimes I + c_2 T_1)(C' \otimes C') \\
&= (c_1 - c_2)(C \otimes C)(I \otimes I)(C' \otimes C') \\
&\quad + c_2(C \otimes C)T_1(C' \otimes C') \\
&= (c_1 - c_2)\Sigma \otimes \Sigma + c_2(C \otimes C)T_1(C' \otimes C').
\end{aligned}$$

It remains to show that $(C \otimes C)T_1(C' \otimes C') = T_2$. For $A \in M_s$,

$$\begin{aligned}
(C \otimes C)T_1(C' \otimes C')(A) &= C \otimes C(\langle I, (C' \otimes C')A \rangle I) \\
&= \langle (C \otimes C)I, A \rangle (C \otimes C)(I) = \langle CC', A \rangle CC' \\
&= \langle \Sigma, A \rangle \Sigma = T_2(A).
\end{aligned}$$

□

PROBLEMS

1. If $x_1, \dots, x_n$ is a basis for $(V, (\cdot, \cdot))$ and if (x_i, X) has finite expectation for $i = 1, \dots, n$, show that (x, X) has finite expectation for all $x \in V$. Also, show that if $(x_i, X)^2$ has finite expectation for $i = 1, \dots, n$, then $\operatorname{Cov}(X)$ exists.

2. Verify the claim that if $X_1(X_2)$ with values in $V_1(V_2)$ are uncorrelated for one pair of inner products on V_1 and V_2, then they are uncorrelated no matter what the inner products are on V_1 and V_2.

3. Suppose $X_i \in V_i$, $i = 1, 2$ are uncorrelated. If f_i is a linear function on V_i, $i = 1, 2$, show that

   (2.2) $$\operatorname{cov}\{f_1(X_1), f_2(X_2)\} = 0.$$

   Conversely, if (2.2) holds for all linear functions f_1 and f_2, then X_1 and X_2 are uncorrrelated (assuming the relevant expectations exist).

4. For $X \in R^n$, partition X as

$$X = \begin{pmatrix} \dot{X} \\ \ddot{X} \end{pmatrix}$$

with $\dot{X} \in R^r$ and suppose X has an orthogonally invariant distribution. Show that $\dot{X}$ has an orthogonally invariant distribution on R^r. Argue that the conditional distribution of $\dot{X}$ given $\ddot{X}$ has an orthogonally invariant distribution.

5. Suppose $X_1, \ldots, X_k$ in $(V, (\cdot, \cdot))$ are pairwise uncorrelated. Prove that $\text{Cov}(\Sigma_1^k X_i) = \Sigma_1^k \text{Cov}(X_i)$.

6. In R^k, let $\varepsilon_1, \ldots, \varepsilon_k$ denote the standard basis vectors. Define a random vector U in R^k by specifying that U takes on the value ε_i with probability p_i where $0 \leqslant p_i \leqslant 1$ and $\Sigma_1^k p_i = 1$. (U represents one of k mutually exclusive and exhaustive events that can occur). Let $p \in R^k$ have coordinates $p_1, \ldots, p_k$. Show that $\mathcal{E}U = p$, $\text{Cov}(U) = D_p - pp'$ where D_p is a diagonal matrix with diagonal entries $p_1, \ldots, p_k$. When $0 < p_i < 1$, show that $\text{Cov}(U)$ has rank $k - 1$ and identify the null space of $\text{Cov}(U)$. Now, let $X_1, \ldots, X_n$ be i.i.d. each with the distribution of U. The random vector $Y = \Sigma_1^n X_i$ has a multinomial distribution (prove this) with parameters k (the number of cells), the vector of probabilities p, and the number of trials n. Show that $\mathcal{E}Y = np$, $\text{Cov}(Y) = n(D_p - pp')$.

7. Fix a vector x in R^n and let π denote a permutation of $1, 2, \ldots, n$ (there are $n!$ such permutations). Define the permuted vector πx to be the vector whose ith coordinate is $x(\pi^{-1}(i))$ where $x(j)$ denotes the jth coordinate of x. (This choice is justified in Chapter 7.) Let X be a random vector such that $P_r\{X = \pi x\} = 1/n!$ for each possible permutation π. Find $\mathcal{E}X$ and $\text{Cov}(X)$.

8. Consider a random vector $X \in R^n$ and suppose $\mathcal{L}(X) = \mathcal{L}(DX)$ for each diagonal matrix D with diagonal elements $d_{ii} = \pm 1$, $i = 1, \ldots, n$. If $\mathcal{E}\|X\|^2 < +\infty$, show that $\mathcal{E}X = 0$ and $\text{Cov}(X)$ is a diagonal matrix (the coordinates of X are uncorrelated).

9. Given $X \in (V, (\cdot, \cdot))$ with $\text{Cov}(X) = \Sigma$, let A_i be a linear transformation on $(V, (\cdot, \cdot))$ to $(W_i, [\cdot, \cdot]_i)$, $i = 1, 2$. Form $Y = \{A_1 X, A_2 X\}$ with values in the direct sum $W_1 \oplus W_2$. Show

$$\text{Cov}(Y) = \begin{pmatrix} A_1 \Sigma A_1' & A_1 \Sigma A_2' \\ A_2 \Sigma A_1' & A_2 \Sigma A_2' \end{pmatrix}$$

in $W_1 \oplus W_2$ with its usual inner product.

10. For X in $(V, \cdot, \cdot))$ with $\mu = \mathcal{E}X$ and $\Sigma = \text{Cov}(X)$, show that $\mathcal{E}(X, AX) = \langle A, \Sigma \rangle + (\mu, A\mu)$ for any $A \in \mathcal{L}(V, V)$.

11. In $(\mathcal{L}_{p,n}, \langle \cdot, \cdot \rangle)$, suppose the $n \times p$ random matrix X has the covariance $I_n \otimes \Sigma$ for some $p \times p$ positive semidefinite Σ. Show that the rows of X are uncorrelated. If $\mu = \mathcal{E}X$ and A is an $n \times n$ matrix, show that $\mathcal{E}X'AX = (\text{tr}\, A)\Sigma + \mu' A \mu$.

12. The usual inner product on the space of $p \times p$ symmetric matrices, denoted by $\mathcal{S}_p$, is $\langle \cdot, \cdot \rangle$, given by $\langle A, B \rangle = \text{tr}\, AB'$. (This is the natural inner product inherited from $(\mathcal{L}_{p,p}, \langle \cdot, \cdot \rangle)$ by regarding $\mathcal{S}_p$ as a subspace of $\mathcal{L}_{p,p}$.) Let S be a random matrix with values in $\mathcal{S}_p$ and suppose that $\mathcal{L}(\Gamma S \Gamma') = \mathcal{L}(S)$ for all $\Gamma \in \mathcal{O}_p$. (For example, if $X \in R^p$ has an orthogonally invariant distribution and $S = XX'$, then $\mathcal{L}(\Gamma S \Gamma') = \mathcal{L}(S)$.) Show that $\mathcal{E}S = cI_p$ where c is constant.

13. Given a random vector X in $(\mathcal{L}(V, W), \langle \cdot, \cdot \rangle)$, suppose that $\mathcal{L}(X) = \mathcal{L}((\Gamma \otimes \psi)X)$ for all $\Gamma \in \mathcal{O}(W)$ and $\psi \in \mathcal{O}(V)$.

(i) If X has a covariance, show $\mathcal{E}X = 0$ and $\text{Cov}(X) = cI_W \otimes I_V$ where $c \geqslant 0$.

(ii) If $Y \in \mathcal{L}(V, W)$ has a density (with respect to Lebesgue measure) given by $f(y) = p(\langle y, y \rangle)$, $y \in \mathcal{L}(V, W)$, show that $\mathcal{L}(Y) = \mathcal{L}((\Gamma \otimes \psi)Y)$ for $\Gamma \in \mathcal{O}(W)$ and $\psi \in \mathcal{O}(V)$.

14. Let $X_1, \ldots, X_n$ be uncorrelated random vectors in R^p with $\text{Cov}(X_i) = \Sigma$, $i = 1, \ldots, n$. Form the $n \times p$ random matrix X with rows $X_1', \ldots, X_n'$ and values in $(\mathcal{L}_{p,n}, \langle \cdot, \cdot \rangle)$. Thus $\text{Cov}(X) = I_n \otimes \Sigma$.

(i) Form $\tilde{X}$ in the coordinate space R^{np} with the coordinate inner product where

$$\tilde{X} = \begin{pmatrix} X_1 \\ \vdots \\ X_n \end{pmatrix}.$$

In the space R^{np} show that

$$\text{Cov}(\tilde{X}) = \begin{pmatrix} \Sigma & 0 & \cdots & 0 \\ 0 & \Sigma & \cdots & 0 \\ \vdots & & \ddots & \vdots \\ 0 & 0 & \cdots & \Sigma \end{pmatrix}$$

where each block is $p \times p$.

(ii) Now, form $\tilde{\tilde{X}}$ in the space R^{np} where

$$\tilde{\tilde{X}} = \begin{pmatrix} Z_1 \\ \vdots \\ Z_p \end{pmatrix}; \qquad Z_i \in R^n$$

and Z_i has coordinates $X_{1i}, \ldots, X_{ni}$ for $i = 1, \ldots, p$. Show that

$$\operatorname{Cov}(\tilde{\tilde{X}}) = \begin{pmatrix} \sigma_{11}I_n & \sigma_{12}I_n & \cdots & \sigma_{1p}I_n \\ \sigma_{21}I_n & \sigma_{22}I_n & \cdots & \sigma_{2p}I_n \\ \vdots & & \ddots & \vdots \\ \sigma_{p1}I_n & \sigma_{p2}I_n & \cdots & \sigma_{pp}I_n \end{pmatrix}$$

where each block is $n \times n$, $\Sigma = \{\sigma_{ij}\}$.

15. The unit sphere in R^n is the set $\{x | x \in R^n, \|x\| = 1\} = \mathcal{X}$. A random vector X with values in $\mathcal{X}$ has a *uniform* distribution on $\mathcal{X}$ if $\mathcal{L}(X) = \mathcal{L}(\Gamma X)$ for all $\Gamma \in \mathcal{O}_n$. (There is one and only one uniform distribution on $\mathcal{X}$—this is discussed in detail in Chapters 6 and 7.)

(i) Show that $\mathcal{E}X = 0$ and $\operatorname{Cov}(X) = (1/n)I_n$.

(ii) Let X_1 be the first coordinate of X and let $\dot{X} \in R^{n-1}$ be the remaining $n - 1$ coordinates. What is the best affine predictor of X_1 based on $\dot{X}$? How would you predict X_1 on the basis of $\dot{X}$?

16. Show that the linear transformation T_2 in Proposition 2.24 is $\Sigma \,\square\, \Sigma$ where $\square$ denotes the outer product of the vector space $(M_s, \langle \cdot, \cdot \rangle)$. Here, $\langle \cdot, \cdot \rangle$ is the natural inner product on $\mathcal{L}(V, V)$.

17. Suppose $X \in R^2$ has coordinates X_1 and X_2 that are independent with a standard normal distribution. Let $S = XX'$ and denote the elements of S by s_{11}, s_{22}, and $s_{12} = s_{21}$.

(i) What is the covariance matrix of

$$\begin{pmatrix} s_{11} \\ s_{12} \\ s_{22} \end{pmatrix} \in R^3?$$

(ii) Regard S as a random vector in $(\mathcal{S}_2, \langle \cdot, \cdot \rangle)$ (see Problem 12). What is $\operatorname{Cov}(S)$ in the space $(\mathcal{S}_2, \langle \cdot, \cdot \rangle)$?

(iii) How do you reconcile your answers to (i) and (ii)?

NOTES AND REFERENCES

1. In the first two sections of this chapter, we have simply translated well known coordinate space results into their inner product space versions. The coordinate space results can be found in Billingsley (1979). The inner product space versions were used by Kruskal (1961) in his work on missing and extra values in analysis of variance problems.

2. In the third section, topics with multivariate flavor emerge. The reader may find it helpful to formulate coordinate versions of each proposition. If nothing else, this exercise will soon explain my acquired preference for vector space, as opposed to coordinate, methods and notation.

3. Proposition 2.14 is a special case of Schur's Lemma—a basic result in group representation theory. The book by Serre (1977) is an excellent place to begin a study of group representations.

CHAPTER 3

The Normal Distribution on a Vector Space

The univariate normal distribution occupies a central position in the statistical theory of analyzing random samples consisting of one-dimensional observations. This situation is even more pronounced in multivariate analysis due to the paucity of analytically tractable multivariate distributions—one notable exception being the multivariate normal distribution. Ordinarily, the nonsingular multivariate normal distribution is defined on R^n by specifying the density function of the distribution with respect to Lebesgue measure. For our purposes, this procedure poses some problems. First, it is desirable to have a definition that does not require the covariance to be nonsingular. In addition, we have not, as yet, constructed what will be called Lebesgue measure on a finite dimensional inner product space. The definition of the multivariate normal distribution we have chosen circumvents the above technical difficulties by specifying the distribution of each linear function of the random vector. Of course, this necessitates a proof that such normal distributions exist.

After defining the normal distribution in a finite dimensional vector space and establishing some basic properties of the normal distribution, we derive the distribution of a quadratic form in a normal random vector. Conditions for the independence of two quadratic forms are then presented followed by a discussion of conditional distributions for normal random vectors. The chapter ends with a derivation of Lebesgue measure on a finite dimensional vector space and of the density function of a nonsingular normal distribution on a vector space.

3.1. THE NORMAL DISTRIBUTION

Recall that a random variable $Z_0 \in R$ has a normal distribution with mean zero and variance one if the density function of Z_0 is

$$p(z) = (2\pi)^{-1/2}\exp\left[-\tfrac{1}{2}z^2\right], \qquad z \in R$$

with respect to Lebesgue measure. We write $\mathcal{L}(Z_0) = N(0, 1)$ when Z_0 has density p. More generally, a random variable $Z \in R$ has a normal distribution with mean $\mu \in R$ and variance $\sigma^2 \geqslant 0$ if $\mathcal{L}(Z) = \mathcal{L}(\sigma Z_0 + \mu)$ where $\mathcal{L}(Z_0) = N(0, 1)$. In this case, we write $\mathcal{L}(Z) = N(\mu, \sigma^2)$. When $\sigma^2 = 0$, the distribution $N(\mu, \sigma^2)$ is to be interpreted as the distribution degenerate at μ. If $\mathcal{L}(Z) = N(\mu, \sigma^2)$, then the characteristic function of Z is easily shown to be

$$\phi(t) = \exp\left[i\mu t - \tfrac{1}{2}\sigma^2 t^2\right], \qquad t \in R.$$

The phrase "Z has a normal distribution" means that for some μ and some $\sigma \geqslant 0$, $\mathcal{L}(Z) = N(\mu, \sigma^2)$. If $Z_1, \ldots, Z_k$ are independent with $\mathcal{L}(Z_j) = N(\mu_j, \sigma_j^2)$, then $\mathcal{L}(\Sigma\alpha_j Z_j) = N(\Sigma\alpha_j\mu_j, \Sigma\alpha_j^2\sigma_j^2)$. To see this, consider the characteristic function

$$\mathcal{E}\exp\left[it\Sigma\alpha_j Z_j\right] = \mathcal{E}\prod_{j=1}^{k}\exp\left[it\alpha_j Z_j\right] = \prod_{j=1}^{k}\mathcal{E}\exp\left[it\alpha_j Z_j\right]$$

$$= \prod_{j=1}^{k}\exp\left[it\alpha_j\mu_j - \tfrac{1}{2}t^2\alpha_j^2\sigma_j^2\right]$$

$$= \exp\left[it\left(\Sigma\alpha_j\mu_j\right) - \tfrac{1}{2}t^2\left(\Sigma\alpha_j^2\sigma_j^2\right)\right].$$

Thus the characteristic function of $\Sigma\alpha_j Z_j$ is that of a normal distribution with mean $\Sigma\alpha_j\mu_j$ and variance $\Sigma\alpha_j^2\sigma_j^2$. In summary, linear combinations of independent normal random variables are normal.

We are now in a position to define the normal distribution on a finite dimensional inner product space $(V, (\cdot, \cdot))$.

Definition 3.1. A random vector $X \in V$ has a normal distribution if, for each $x \in V$, the random variable (x, X) has a normal distribution on R.

To show that a normal distribution exists on $(V, (\cdot, \cdot))$, let $\{x_1, \ldots, x_n\}$ be an orthonormal basis for $(V, (\cdot, \cdot))$. Also, let $Z_1, \ldots, Z_n$ be independent

$N(0, 1)$ random variables. Then $X \equiv \Sigma Z_i x_i$ is a random vector and $(x, X) = \Sigma(x, x_i)Z_i$, which is a linear combination of independent normals. Thus (x, X) has a normal distribution for each $x \in V$. Since $\mathcal{E}(x, X) = \Sigma(x_i, x)\mathcal{E}Z_i = 0$, the mean vector of X is $0 \in V$. Also,

$$\operatorname{var}(x, X) = \operatorname{var}(\Sigma(x, x_i)Z_i) = \Sigma(x, x_i)^2 \operatorname{var}(Z_i) = \Sigma(x, x_i)^2 = (x, x).$$

Therefore, $\operatorname{Cov}(X) = I \in \mathcal{L}(V, V)$. The particular normal distribution we have constructed on $(V, (\cdot, \cdot))$ has mean zero and covariance equal to the identity linear transformation.

Now, we want to describe all the normal distributions on $(V, (\cdot, \cdot))$. The first result in this direction shows that linear transformations of normal random vectors are again normal random vectors.

Proposition 3.1. Suppose X has a normal distribution on $(V, (\cdot, \cdot))$ and let $A \in \mathcal{L}(V, W)$, $w_0 \in W$. Then $AX + w_0$ has a normal distribution on $(W, [\cdot, \cdot])$.

Proof. It must be shown that, for each $w \in W$, $[w, AX + w_0]$ has a normal distribution on R. But $[w, AX + w_0] = [w, AX] + [w, w_0] = (A'w, X) + [w, w_0]$. By assumption, $(A'w, X)$ is normal. Since $[w, w_0]$ is a constant, $(A'w, X) + [w, w_0]$ is normal. □

If X has a normal distribution on $(V, (\cdot, \cdot))$ with mean zero and covariance I, consider $A \in \mathcal{L}(V, V)$ and $\mu \in V$. Then $AX + \mu$ has a normal distribution on $(V, (\cdot, \cdot))$ and we know $\mathcal{E}(AX + \mu) = A(\mathcal{E}X) + \mu = \mu$ and $\operatorname{Cov}(AX + \mu) = A \operatorname{Cov}(X) A' = AA'$. However, every positive semidefinite linear transformation Σ can be expressed as AA' (take A to be the positive semidefinite square root of Σ). Thus given $\mu \in V$ and a positive semidefinite Σ, there is a random vector that has a normal distribution in V with mean vector μ and covariance Σ. If X has such a distribution, we write $\mathcal{L}(X) = N(\mu, \Sigma)$. To show that all the normal distributions on V have been described, suppose $X \in V$ has a normal distribution. Since (x, X) is normal on R, $\operatorname{var}(x, X)$ exists for each $x \in V$. Thus $\mu = \mathcal{E}X$ and $\Sigma = \operatorname{Cov}(X)$ both exist and $\mathcal{L}(X) = N(\mu, \Sigma)$. Also, $\mathcal{L}((x, X)) = N((x, \mu), (x, \Sigma x))$ for $x \in V$. Hence the characteristic function of (x, X) is

$$\phi(t) = \mathcal{E}\exp[it(x, X)] = \exp\left[it(x, \mu) - \tfrac{1}{2}t^2(x, \Sigma x)\right].$$

Setting $t = 1$, we obtain the characteristic function of X:

$$\xi(x) = \mathcal{E}\exp[i(x, X)] = \exp\left[i(x, \mu) - \tfrac{1}{2}(x, \Sigma x)\right].$$

Summarizing this discussion yields the following.

Proposition 3.2. Given $\mu \in V$ and a positive semidefinite $\Sigma \in \mathcal{L}(V, V)$, there exists a random vector $X \in V$ with distribution $N(\mu, \Sigma)$ and characteristic function

$$\xi(x) = \exp\left[i(x, \mu) - \tfrac{1}{2}(x, \Sigma x)\right].$$

Conversely, if X has a normal distribution on V, then with $\mu = \mathcal{E}X$ and $\Sigma = \text{Cov}(X)$, $\mathcal{L}(X) = N(\mu, \Sigma)$ and the characteristic function of X is given by ξ.

Consider random vectors X_i with values in $(V_i, (\cdot, \cdot)_i)$ for $i = 1, 2$. Then $\{X_1, X_2\}$ is a random vector in the direct sum $V_1 \oplus V_2$. The inner product on $V_1 \oplus V_2$ is $[\cdot, \cdot]$ where

$$\left[\{v_1, v_2\}, \{v_3, v_4\}\right] \equiv (v_1, v_3)_1 + (v_2, v_4)_2,$$

$v_1, v_3 \in V_1$ and $v_2, v_4 \in V_2$. If $\text{Cov}(X_i) = \Sigma_{ii}$, $i = 1, 2$, exists, then $\mathcal{E}\{X_1, X_2\} = \{\mu_1, \mu_2\}$ where $\mu_i = \mathcal{E}X_i$, $i = 1, 2$. Also,

$$\Sigma \equiv \text{Cov}\{X_1, X_2\} = \begin{pmatrix} \Sigma_{11} & \Sigma_{12} \\ \Sigma_{21} & \Sigma_{22} \end{pmatrix} \in \mathcal{L}(V_1 \oplus V_2, V_1 \oplus V_2)$$

as defined in Chapter 2 and $\Sigma_{21} \equiv \Sigma_{12}'$.

Proposition 3.3. If $\{X_1, X_2\}$ has a normal distribution on $V_1 \oplus V_2$, then X_1 and X_2 are independent iff $\Sigma_{12} = 0$.

Proof. If X_1 and X_2 are independent, then clearly $\Sigma_{12} = 0$. Conversely, if $\Sigma_{12} = 0$, the characteristic function of $\{X_1, X_2\}$ is

$$\begin{aligned}
\mathcal{E}\exp\{i[\{v_1, v_2\}, \{X_1, X_2\}]\} &= \exp\{i[\{v_1, v_2\}, \{\mu_1, \mu_2\}] \\
&\quad - \tfrac{1}{2}[\{v_1, v_2\}, \Sigma\{v_1, v_2\}]\} \\
&= \exp\{i(v_1, \mu_1)_1 + i(v_2, \mu_2)_2 \\
&\quad - \tfrac{1}{2}(v_1, \Sigma_{11}v_1)_1 - \tfrac{1}{2}(v_2, \Sigma_{22}v_2)_2\} \\
&= \exp\{i(v_1, \mu_1)_1 - \tfrac{1}{2}(v_1, \Sigma_{11}v_1)_1\} \\
&\quad \times \exp\{i(v_2, \mu_2)_2 - \tfrac{1}{2}(v_2, \Sigma_{22}v_2)_2\}
\end{aligned}$$

since $\Sigma_{12} = \Sigma'_{21} = 0$. However, for $v_1 \in V_1$, $(v_1, X_1)_1 = [\{v_1, 0\}, \{X_1, X_2\}]$, which has a normal distribution for all $v_1 \in V_1$. Thus $\mathcal{L}(X_1) = N(\mu_1, \Sigma_1)$ on V_1 and similarly $\mathcal{L}(X_2) = N(\mu_2, \Sigma_2)$ on V_2. The characteristic function of $\{X_1, X_2\}$ is just the product of the characteristic functions of X_1 and X_2. Thus independence follows and the proof is complete. □

The result of Proposition 3.3 is often paraphrased as "for normal random vectors, X_1 and X_2 are independent iff they are uncorrelated." A useful consequence of Proposition 3.3 is shown in Proposition 3.4.

Proposition 3.4. Suppose $\mathcal{L}(X) = N(\mu, \Sigma)$ on $(V, (\cdot, \cdot))$, and consider $A \in \mathcal{L}(V, W_1)$, $B \in \mathcal{L}(V, W_2)$ where $(W_1, [\cdot, \cdot]_1)$ and $(W_2, [\cdot, \cdot]_2)$ are inner product spaces. AX and BX are independent iff $A\Sigma B' = 0$.

Proof. We apply the previous proposition to $X_1 = AX$ and $X_2 = BX$. That $\{X_1, X_2\}$ has a normal distribution on $W_1 \oplus W_2$ follows from

$$[w_1, X_1]_1 + [w_2, X_2]_2 = (A'w_1, X) + (B'w_2, X) = (A'w_1 + B'w_2, X)$$

and the normality of (x, X) for all $x \in V$. However,

$$\begin{aligned}\operatorname{cov}\{[w_1, X_1]_1, [w_2, X_2]_2\} &= \operatorname{cov}\{(A'w_1, X), (B'w_2, X)\} \\ &= (A'w_1, \Sigma B'w_2) \\ &= [w_1, A\Sigma B'w_2]_1.\end{aligned}$$

Thus $X_1 = AX$ and $X_2 = BX$ are uncorrelated iff $A\Sigma B' = 0$. Since $\{X_1, X_2\}$ has a normal distribution, the condition $A\Sigma B' = 0$ is equivalent to the independence of X_1 and X_2. □

One special case of Proposition 3.4 is worthy of mention. If $\mathcal{L}(X) = N(\mu, I)$ on $(V, (\cdot, \cdot))$ and P is an orthogonal projection in $\mathcal{L}(V, V)$, then PX and $(I - P)X$ are independent since $P(I - P) = 0$. Also, it should be mentioned that the result of Proposition 3.3 extends to the case of k random vectors—that is, if $\{X_1, X_2, \ldots, X_k\}$ has a normal distribution on the direct sum space $V_1 \oplus V_2 \oplus \cdots \oplus V_k$, then $X_1, X_2, \ldots, X_k$ are independent iff X_i and X_j are uncorrelated for all $i \neq j$. The proof of this is essentially the same as that given for the case of $k = 2$ and is left to the reader.

A particularly useful result for the multivariate normal distribution is the following.

Proposition 3.5. Suppose $\mathcal{L}(X) = N(\mu, \Sigma)$ on the n-dimensional vector space $(V, (\cdot, \cdot))$. Write $\Sigma = \Sigma_1^n \lambda_i x_i \square x_i$ in spectral form, and let $X_i = (x_i, X)$, $i = 1, \ldots, n$. Then $X_1, \ldots, X_n$ are independent random variables that have a normal distribution on R with $\mathcal{E} X_i = (x_i, \mu)$ and $\text{var}(X_i) = \lambda_i$, $i = 1, \ldots, n$. In particular, if $\Sigma = I$, then for any orthonormal basis $\{x_1, \ldots, x_n\}$ for V, the random variables $X_i = (x_i, X)$ are independent and normal with $\mathcal{E} X_i = (x_i, \mu)$ and $\text{var}(X_i) = 1$.

Proof. For any scalars $\alpha_1, \ldots, \alpha_n$ in R, $\Sigma_1^n \alpha_i X_i = \Sigma_1^n \alpha_i (x_i, X) = (\Sigma_1^n \alpha_i x_i, X)$, which has a normal distribution. Thus the random vector $\tilde{X} \in R^n$ with coordinates $X_1, \ldots, X_n$ has a normal distribution in the coordinate vector space R^n. Thus $X_1, \ldots, X_n$ are independent iff they are uncorrelated. However,

$$\text{cov}\{X_j, X_k\} = \text{cov}\{(x_j, X), (x_k, X)\} = (x_j, \Sigma x_k)$$

$$= (x_j, (\Sigma_1^n \lambda_i x_i \square x_i) x_k) = \lambda_j \delta_{jk}.$$

Thus independence follows. It is clear that each X_i is normal with $\mathcal{E} X_i = (x_i, \mu)$ and $\text{var}(X_i) = \lambda_i$, $i = 1, \ldots, n$. When $\Sigma = I$, then $\Sigma_1^n x_i \square x_i = I$ for any orthonormal basis $x_1, \ldots, x_n$. This completes the proof. □

The following is a technical discussion having to do with representations of the normal distribution that are useful when establishing properties of the normal distribution. It seems preferable to dispose of the issues here rather than repeat the same argument in a variety of contexts later. Suppose $X \in (V, (\cdot, \cdot))$ has a normal distribution, say $\mathcal{L}(X) = N(\mu, \Sigma)$, and let Q be the probability distribution of X on $(V, (\cdot, \cdot))$. If we are interested in the distribution of some function of X, say $f(X) \in (W, [\cdot, \cdot])$, then the underlying space on which X is defined is irrelevant since the distribution Q determines the distribution of $f(X)$—that is, if $B \in \mathcal{B}(W)$, then

$$P\{f(X) \in B\} = P\{X \in f^{-1}(B)\} = Q(f^{-1}(B)).$$

Therefore, if Y is another random vector in $(V, (\cdot, \cdot))$ with $\mathcal{L}(X) = \mathcal{L}(Y)$, then $f(X)$ and $f(Y)$ have the same distribution. At times, it is convenient to represent $\mathcal{L}(X)$ by $\mathcal{L}(CZ + \mu)$ where $\mathcal{L}(Z) = N(0, \text{I})$ and $CC' = \Sigma$. Thus

$\mathcal{L}(X) = \mathcal{L}(CZ + \mu)$ so $f(X)$ and $f(CZ + \mu)$ have the same distribution. A slightly more subtle point arises when we discuss the independence of two functions of X, say $f_1(X)$ and $f_2(X)$, taking values in $(W_1, [\cdot, \cdot]_1)$ and $(W_2, [\cdot, \cdot]_2)$. To show that independence of $f_1(X)$ and $f_2(X)$ depends only on Q, consider $B_i \in \mathcal{B}(W_i)$ for $i = 1, 2$. Then independence is equivalent to

$$P\{f_1(X) \in B_1, f_2(X) \in B_2\} = P\{f_1(X) \in B_1\}P\{f_2(X) \in B_2\}.$$

But both of these probabilities can be calculated from Q:

$$P\{f_1(X) \in B_1, f_2(X) \in B_2\} = P\{X \in f_1^{-1}(B_1) \cap f_2^{-1}(B_2)\}$$

$$= Q\big(f_1^{-1}(B_1) \cap f_2^{-1}(B_2)\big)$$

and

$$P\{f_i(X) \in B_i\} = Q\big(f_i^{-1}(B_i)\big), \qquad i = 1, 2.$$

Again, if $\mathcal{L}(Y) = \mathcal{L}(X)$, then $f_1(X)$ and $f_2(X)$ are independent iff $f_1(Y)$ and $f_2(Y)$ are independent. More generally, if we are trying to prove something about the random vector X, $\mathcal{L}(X) = N(\mu, \Sigma)$, and if what we are trying to prove depends only on the distribution Q_1 of X, then we can represent X by any other random vector Y as long as $\mathcal{L}(Y) = \mathcal{L}(X)$. In particular, we can take $Y = CZ + \mu$ where $\mathcal{L}(Z) = N(0, I)$ and $CC' = \Sigma$. This representation of X is often used in what follows.

3.2. QUADRATIC FORMS

The problem in this section is to derive, or at least describe, the distribution of (X, AX) where $X \in (V, (\cdot, \cdot))$, A is self-adjoint in $\mathcal{L}(V, V)$ and $\mathcal{L}(X) = N(\mu, \Sigma)$. First, consider the special case of $\Sigma = I$, and by the spectral theorem, write $A = \Sigma_1^n \lambda_i x_i \Box x_i$. Thus

$$(X, AX) = \big(X, (\Sigma_1^n \lambda_i x_i \Box x_i) X\big) = \Sigma_1^n \lambda_i (x_i, X)^2.$$

But $X_i \equiv (x_i, X)$, $i = 1, \dots, n$, are independent since $\Sigma = I$ (Proposition 3.5) and $\mathcal{L}(X_i) = N((x_i, \mu), 1)$. Thus our first task is to derive the distribution of X_i^2 when $\mathcal{L}(X_i) = N((x_i, \mu), 1)$.

Recall that a random variable Z has a chi-square distribution with m degrees of freedom, written $\mathcal{L}(Z) = \chi_m^2$, if Z has a density on $(0, \infty)$ given

by

$$p_m(z) = \frac{z^{(m/2)-1}}{\Gamma(m/2)2^{m/2}} \exp\left[-\tfrac{1}{2}z\right], \qquad z > 0.$$

Here m is a positive integer and $\Gamma(\cdot)$ is the gamma function. The characteristic function of a χ_m^2 random variable is easily shown to be

$$\mathcal{E}e^{it\chi_m^2} = (1 - 2it)^{-m/2}, \qquad t \in R^1.$$

Thus, if $\mathcal{L}(Z_1) = \chi_m^2$, $\mathcal{L}(Z_2) = \chi_n^2$, and Z_1 and Z_2 are independent, then

$$\begin{aligned}\mathcal{E}\exp[it(Z_1 + Z_2)] &= \mathcal{E}\exp[itZ_1]\mathcal{E}\exp[itZ_2] \\ &= (1 - 2it)^{-m/2}(1 - 2it)^{-n/2} = (1 - 2it)^{-(m+n)/2}.\end{aligned}$$

Therefore, $\mathcal{L}(Z_1 + Z_2) = \chi_{m+n}^2$. This argument clearly extends to more than two factors. In particular, if $\mathcal{L}(Z) = \chi_m^2$, then, for independent random variables $Z_1, \ldots, Z_m$ with $\mathcal{L}(Z_i) = \chi_1^2$, $\mathcal{L}(\Sigma_1^m Z_i) = \mathcal{L}(Z)$. It is not difficult to show that if $\mathcal{L}(X) = N(0,1)$ on R, then $\mathcal{L}(X^2) = \chi_1^2$. However, if $\mathcal{L}(X) = N(\alpha, 1)$ on R, the distribution of X^2 is a bit harder to derive. To this end, we make the following definition.

Definition 3.2. Let p_m, $m = 1, 2, \ldots$, be the density of a χ_m^2 random variable and, for $\lambda \geqslant 0$, let

$$q_j = \exp\left[-\frac{\lambda}{2}\right]\frac{1}{j!}\left(\frac{\lambda}{2}\right)^j.$$

For $\lambda = 0$, $q_0 = 1$ and $q_j = 0$ for $j > 0$. A random variable with density

$$h(z) = \sum_{j=0}^{\infty} q_j p_{m+2j}(z), \qquad z > 0$$

is said to have a *noncentral chi-square distribution* with m degrees of freedom and noncentrality parameter λ. If Z has such a distribution, we write $\mathcal{L}(Z) = \chi_m^2(\lambda)$.

When $\lambda = 0$, it is clear that $\mathcal{L}(\chi_m^2(0)) = \chi_m^2$. The weights q_j, $j = 0, 1, \ldots$, are Poisson probabilities with parameter $\lambda/2$ (the reason for the 2 becomes clear in a bit). The characteristic function of a $\chi_m^2(\lambda)$ random variable is calculated as follows:

$$
\begin{aligned}
\mathcal{E}\exp\left[it\chi_m^2(\lambda)\right] &= \sum_{j=0}^{\infty} q_j \int_0^{\infty} \exp(itx)\, p_{m+2j}(x)\, dx \\
&= \sum_{j=0}^{\infty} q_j (1-2it)^{-(m/2+j)} \\
&= (1-2it)^{-m/2} \sum_{j=0}^{\infty} q_j (1-2it)^{-j} \\
&= (1-2it)^{-m/2} \exp(-\lambda/2) \sum_{j=0}^{\infty} \left(\frac{\lambda}{2}\right)^j \frac{(1-2it)^{-j}}{j!} \\
&= (1-2it)^{-m/2} \exp\left[-\frac{\lambda}{2} + \frac{\lambda}{2}\frac{1}{1-2it}\right] \\
&= (1-2it)^{-m/2} \exp \frac{\lambda}{2}\left[\frac{2it}{1-2it}\right].
\end{aligned}
$$

From this expression for the characteristic function, it follows that if $\mathcal{L}(Z_i) = \chi_{m_i}^2(\lambda_i)$, $i = 1, 2$, with Z_1 and Z_2 independent, then $\mathcal{L}(Z_1 + Z_2) = \chi_{m_1+m_2}^2(\lambda_1 + \lambda_2)$. This result clearly extends to the sum of k independent noncentral chi-square variables. The reason for introducing the noncentral chi-square distribution is provided in the next result.

Proposition 3.6. Suppose $\mathcal{L}(X) = N(\alpha, 1)$ on R. Then $\mathcal{L}(X^2) = \chi_1^2(\alpha^2)$.

Proof. The proof consists of calculating the characteristic function of X^2. A justification of the change of variable in the calculation below can be given using contour integration. The characteristic function of X^2 is

$$
\begin{aligned}
\mathcal{E}\exp(itX^2) &= \int_{-\infty}^{\infty} \frac{1}{\sqrt{2\pi}} \exp\left[itx^2 - \tfrac{1}{2}(x-\alpha)^2\right] dx \\
&= \int_{-\infty}^{\infty} \frac{1}{\sqrt{2\pi}} \exp\left[-\tfrac{1}{2}(1-2it)x^2 + \alpha x - \tfrac{1}{2}\alpha^2\right] dx \\
&= \frac{(1-2it)^{-1/2}}{\sqrt{2\pi}} \int_{-\infty}^{\infty} \exp\Big[-\tfrac{1}{2}w^2 + \alpha(1-2it)^{-1/2} \\
&\qquad \times w - \tfrac{1}{2}\alpha^2\Big]\, dw
\end{aligned}
$$

$$= \frac{(1-2it)^{-1/2}}{\sqrt{2\pi}} \int_{-\infty}^{\infty} \exp\left[-\tfrac{1}{2}\left(w - \alpha(1-2it)^{-1/2}\right)^2 + \frac{\alpha^2}{2}\left(\frac{2it}{1-2it}\right)\right] dw$$

$$= (1-2it)^{-1/2} \exp \frac{\alpha^2}{2}\left(\frac{2it}{1-2it}\right).$$

By the uniqueness of characteristic functions, $\mathcal{L}(X^2) = \chi_1^2(\alpha^2)$. □

Proposition 3.7. Suppose the random vector X in $(V, (\cdot, \cdot))$ has a $N(\mu, I)$ distribution. If $A \in \mathcal{L}(V, V)$ is an orthogonal projection of rank k, then $\mathcal{L}((X, AX)) = \chi_k^2((\mu, A\mu))$.

Proof. Let $\{x_1, \ldots, x_k\}$ be an orthonormal basis for the range of A. Thus $A = \Sigma_1^k x_i \square x_i$ and

$$(X, AX) = \Sigma_1^k (x_i, X)^2.$$

But the random variables $(x_i, X)^2$, $i = 1, \ldots, k$, are independent (Proposition 3.5) and, by Proposition 3.6, $\mathcal{L}(X_i^2) = \chi_1^2((x_i, \mu)^2)$. From the additive property of independent noncentral chi-square variables,

$$\mathcal{L}\left(\Sigma_1^k (x_i, X)^2\right) = \chi_k^2\left(\Sigma_1^k (x_i, \mu)^2\right).$$

Noting that $(\mu, A\mu) = \Sigma_1^k (x_i, \mu)^2$, the proof is complete. □

When $\mathcal{L}(X) = N(\mu, \Sigma)$, the distribution of the quadratic form (X, AX), with A self-adjoint, is reasonably complicated, but there is something that can be said. Let B be the positive semidefinite square root of Σ and assume that $\mu \in \mathcal{R}(\Sigma)$. Thus $\mu \in \mathcal{R}(B)$ since $\mathcal{R}(B) = \mathcal{R}(\Sigma)$. Therefore, for some vector $\tau \in V$, $\mu = B\tau$. Thus $\mathcal{L}(X) = \mathcal{L}(BY)$ where $\mathcal{L}(Y) = N(\tau, I)$ and it suffices to describe the distribution of $(BY, ABY) = (Y, BABY)$. Since A and B are self-adjoint, BAB is self-adjoint. Write BAB in spectral form:

$$BAB = \Sigma_1^n \lambda_i x_i \square x_i$$

where $\{x_1, \ldots, x_n\}$ is an orthonormal basis for $(V, (\cdot, \cdot))$. Then

$$(Y, BABY) = \Sigma_1^n \lambda_i (x_i, Y)^2$$

and the random variables (x_i, Y), $i = 1, \ldots, n$, are independent with $\mathcal{L}((x_i, Y)^2) = \chi_1^2((x_i, \tau)^2)$. It follows that the quadratic form $(Y, BABY)$ has the same distribution as a linear combination of independent noncentral chi-square random variables. Symbolically,

$$\mathcal{L}((Y, BABY)) = \mathcal{L}\left(\Sigma_1^n \lambda_i \chi_{1,i}^2\left((x_i, \tau)^2\right)\right).$$

In general not much more can be said about this distribution without some assumptions concerning the eigenvalues $\lambda_1, \ldots, \lambda_n$. However, when BAB is an orthogonal projection of rank k, then Proposition 3.7 is applicable and

$$\mathcal{L}((Y, BABY)) = \chi_k^2((\tau, BAB\tau)) = \chi_k^2((B\tau, AB\tau)) = \chi_k^2((\mu, A\mu)).$$

In summary, we have the following.

Proposition 3.8. Suppose $\mathcal{L}(X) = N(\mu, \Sigma)$ where $\mu \in \mathcal{R}(\Sigma)$, and let B be the positive semidefinite square root of Σ. If A is self-adjoint and BAB is a rank k orthogonal projection, then

$$\mathcal{L}((X, AX)) = \chi_k^2((\mu, A\mu)).$$

We can use a slightly different set of assumptions and reach the same conclusion as Proposition 3.8, as follows.

Proposition 3.9. Suppose $\mathcal{L}(X) = N(\mu, \Sigma)$ and let B be the positive semidefinite square root of Σ. Write $\mu = \mu_1 + \mu_2$ where $\mu_1 \in \mathcal{R}(\Sigma)$ and $\mu_2 \in \mathcal{N}(\Sigma)$. If A is a self-adjoint such that $A\mu_2 = 0$ and BAB is a rank k orthogonal projection, then

$$\mathcal{L}((X, AX)) = \chi_k^2((\mu, A\mu)).$$

Proof. Since $A\mu_2 = 0$, $(X, AX) = (X - \mu_2, A(X - \mu_2))$. Let $Y = X - \mu_2$ so $\mathcal{L}(Y) = N(\mu_1, \Sigma)$ and $\mathcal{L}((X, AX)) = \mathcal{L}((Y, AY))$. Since $\mu_1 \in \mathcal{R}(\Sigma)$, Proposition 3.8 shows that

$$\mathcal{L}((Y, AY)) = \chi_k^2((\mu_1, A\mu_1)).$$

However, $(\mu, A\mu) = (\mu_1, A\mu_1)$ as $A\mu_2 = 0$. □

3.3. INDEPENDENCE OF QUADRATIC FORMS

Thus far, necessary and sufficient conditions for the independence of different linear transformations of a normal random vector have been given and the distribution of a quadratic form in a normal random vector has been described. In this section, we give sufficient conditions for the independence of different quadratic forms in normal random vectors.

Suppose $X \in (V, (\cdot, \cdot))$ has an $N(\mu, \Sigma)$ distribution and consider two self-adjoint linear transformations, A_i, $i = 1, 2$, on V to V. To discuss the independence of $(X, A_1 X)$ and $(X, A_2 X)$, it is convenient to first reduce the discussion to the case when $\mu = 0$ and $\Sigma = I$. Let B be the positive semidefinite square root of Σ so if $\mathcal{L}(Y) = N(0, I)$, then $\mathcal{L}(X) = \mathcal{L}(BY + \mu)$. Thus it suffices to discuss the independence of $(BY + \mu, A_1(BY + \mu))$ and $(BY + \mu, A_2(BY + \mu))$ when $\mathcal{L}(Y) = N(0, I)$. However,

$$(BY + \mu, A_i(BY + \mu)) = (Y, BA_iBY) + 2(BA_i\mu, Y) + (\mu, A_i\mu)$$

for $i = 1, 2$. Let $C_i = BA_iB$, $i = 1, 2$, and let $x_i = 2BA_i\mu$. Then we want to know conditions under which $(Y, C_1 Y) + (x_1, Y)$ and $(Y, C_2 Y) + (x_2, Y)$ are independent when $\mathcal{L}(Y) = N(0, I)$. Clearly, the constants $(\mu, A_i\mu)$, $i = 1, 2$, do not affect the independence of the two quadratic forms. It is this problem, in reduced form, that is treated now. Before stating the principal result, the following technical proposition is needed.

Proposition 3.10. For self-adjoint linear transformations A_1 and A_2 on $(V, (\cdot, \cdot))$ to $(V, (\cdot, \cdot))$, the following are equivalent:

(i) $A_1 A_2 = 0$.
(ii) $\mathcal{R}(A_1) \perp \mathcal{R}(A_2)$.

Proof. If $A_1 A_2 = 0$, then $A_1 A_2 x = 0$ for all $x \in V$ so $\mathcal{R}(A_2) \subseteq \mathcal{N}(A_1)$. Since $\mathcal{N}(A_1) \perp \mathcal{R}(A_1)$, $\mathcal{R}(A_2) \perp \mathcal{R}(A_1)$. Conversely, if $\mathcal{R}(A_1) \perp \mathcal{R}(A_2)$, then $\mathcal{R}(A_2) \subseteq \mathcal{R}(A_1)^{\perp} = \mathcal{N}(A_1)$ and this implies that $A_1 A_2 x = 0$ for all $x \in V$. Therefore, $A_1 A_2 = 0$. □

Proposition 3.11. Let $Y \in (V, (\cdot, \cdot))$ have a $N(0, I)$ distribution and suppose $Z_i = (Y, A_i Y) + (x_i, Y)$ where A_i is self-adjoint and $x_i \in V$, $i = 1, 2$. If $A_1 A_2 = 0$, $A_1 x_2 = 0$, $A_2 x_1 = 0$, and $(x_1, x_2) = 0$, then Z_1 and Z_2 are independent random variables.

Proof. The idea of the proof is to show that Z_1 and Z_2 are functions of two different independent random vectors. To this end, let P_i be the orthogonal projection onto $\mathcal{R}(A_i)$ for $i = 1, 2$. It is clear that $P_i A_i P_i = A_i$ for $i = 1, 2$. Thus $Z_i = (P_i Y, A_i P_i Y) + (x_i, Y)$ for $i = 1, 2$. The random vector $\{P_1 Y, (x_1, Y)\}$ takes values in the direct sum $V \oplus R$ and Z_1 is a function of this vector. Also, $\{P_2Y, (x_2, Y)\}$ takes values in $V \oplus R$ and Z_2 is a function of this vector. The remainder of the proof is devoted to showing that $\{P_1Y, (x_1, Y)\}$ and $\{P_2Y, (x_2, Y)\}$ are independent random vectors. This is done by verifying that the random vectors are jointly normal and that they are uncorrelated. Let $[\cdot, \cdot]$ denote the induced inner product on the direct sum $V \oplus R$. The inner product of the vector $\{\{y_1, \alpha_1\}, \{y_2, \alpha_2\}\}$ in $(V \oplus R) \oplus (V \oplus R)$ with $\{\{P_1Y, (x_1, Y)\}, \{P_2Y, (x_2, Y)\}\}$ is

$$(y_1, P_1Y) + \alpha_1(x_1, Y) + (y_2, P_2Y) + \alpha_2(x_2, Y)$$

$$= (P_1y_1 + \alpha_1 x + P_2y_2 + \alpha_2 x_2, Y),$$

which has a normal distribution since Y is normal. Thus $\{\{P_1Y, (x_1, Y)\}, \{P_2Y, (x_2, Y)\}\}$ has a normal distribution. The independence of these two vectors follows from the calculation below, which shows the vectors are uncorrelated. For $\{y_1, \alpha_1\} \in V \oplus R$ and $\{y_2, \alpha_2\} \in V \oplus R$,

$$\operatorname{cov}\{[\{y_1, \alpha_1\}, \{P_1Y, (x_1, Y)\}], [\{y_2, \alpha_2\}, \{P_2Y, (x_2, Y)\}]\}$$

$$= \operatorname{cov}\{(y_1, P_1Y) + \alpha_1(x_1, Y), (y_2, P_2Y) + \alpha_2(x_2, Y)\}$$

$$= \operatorname{cov}\{(P_1y_1, Y), (P_2y_2, Y)\} + \alpha_1 \operatorname{cov}\{(x_1, Y), (P_2y_2, Y)\}$$

$$+ \alpha_2 \operatorname{cov}\{(P_1y_1, Y), (x_2, Y)\} + \alpha_1\alpha_2 \operatorname{cov}\{(x_1, Y), (x_2, Y)\}$$

$$= (P_1y_1, P_2y_2) + \alpha_1(x_1, P_2y_2) + \alpha_2(x_2, P_1y_1) + \alpha_1\alpha_2(x_1, x_2)$$

$$= (y_1, P_1P_2y_2) + \alpha_1(P_2x_1, y_2) + \alpha_2(P_1x_2, y_1) + \alpha_1\alpha_2(x_1, x_2).$$

However, $P_1P_2 = 0$ since $\mathfrak{R}(A_1) \perp \mathfrak{R}(A_2)$. Also, $P_2x_1 = 0$ as $x_1 \in \mathfrak{N}(A_2)$ and, similarly, $P_1x_2 = 0$. Further, $(x_1, x_2) = 0$ by assumption. Thus the above covariance is zero so Z_1 and Z_2 are independent. □

A useful consequence of Proposition 3.11 is Proposition 3.12.

Proposition 3.12. Suppose $\mathcal{L}(X) = N(\mu, \Sigma)$ on $(V, (\cdot, \cdot))$ and let C_i, $i = 1, 2$, be self-adjoint linear transformations. If $C_1\Sigma C_2 = 0$, then (X, C_1X) and (X, C_2X) are independent.

Proof. Let B denote the positive semidefinite square root of Σ, and suppose $\mathcal{L}(Y) = N(0, I)$. It suffices to show that $Z_1 \equiv (BY + \mu, C_1(BY +$

$\mu))$ is independent of $Z_2 \equiv (BY + \mu, C_2(BY + \mu))$ since $\mathcal{L}(X) = \mathcal{L}(BY + \mu)$. But

$$Z_i = (Y, BC_iBY) + 2(BC_i\mu, Y) + (\mu, C_i\mu)$$

for $i = 1, 2$. Proposition 3.11 can now be applied with $A_i = BC_iB$ and $x_i = 2BC_i\mu$ for $i = 1, 2$. Since $\Sigma = BB$, $A_1A_2 = BC_1BBC_2B = BC_1\Sigma C_2B = 0$ as $C_1\Sigma C_2 = 0$ by assumption. Also, $A_1x_2 = 2BC_1BBC_2\mu = 2BC_1\Sigma C_2\mu = 0$. Similarly, $A_2x_1 = 0$ and $(x_1, x_2) = 4(BC_1\mu, BC_2\mu) = 4(\mu, C_1\Sigma C_2\mu) = 0$. Thus $(Y, BC_1BY) + 2(BC_1\mu, Y)$ and $(Y, BC_2BY) + 2(BC_2\mu, Y)$ are independent. Hence Z_1 and Z_2 are independent. □

The results of this section are general enough to handle most situations that arise when dealing with quadratic forms. However, in some cases we need a sufficient condition for the independence of k quadratic forms. An examination of the proof of Proposition 3.11 shows that when $\mathcal{L}(Y) = N(0, I)$, the quadratic forms $Z_i = (Y, A_iY) + (x_i, Y)$, $i = 1, \ldots, k$, are mutually independent if, for each $i \neq j$, $A_iA_j = 0$, $A_ix_j = 0$, $A_jx_i = 0$, and $(x_i, x_j) = 0$. The details of this verification are left to the reader.

3.4. CONDITIONAL DISTRIBUTIONS

The basic result of this section gives the conditional distribution of one normal random vector given another normal random vector. It is this result that underlies many of the important distributional and independence properties of the normal and related distributions that are established in later chapters.

Consider random vectors $X_i \in (V_i, (\cdot, \cdot)_i)$, $i = 1, 2$, and assume that the random vector $\{X_1, X_2\}$ in the direct sum $V_1 \oplus V_2$ has a normal distribution with mean vector $\{\mu_1, \mu_2\} \in V_1 \oplus V_2$ and covariance given by

$$\operatorname{Cov}(X) = \begin{pmatrix} \Sigma_{11} & \Sigma_{12} \\ \Sigma_{12}' & \Sigma_{22} \end{pmatrix}.$$

Thus $\mathcal{L}(X_i) = N(\mu_i, \Sigma_{ii})$ on $(V_i, (\cdot, \cdot)_i)$ for $i = 1, 2$. The conditional distribution of X_1 given $X_2 = x_2 \in V_2$ is described in the next result.

Proposition 3.13. Let $\mathcal{L}(X_1 | X_2 = x_2)$ denote the conditional distribution of X_1 given $X_2 = x_2$. Then, under the above normality assumptions,

$$\mathcal{L}(X_1 | X_2 = x_2) = N\left(\mu_1 + \Sigma_{12}\Sigma_{22}^{-}(x_2 - \mu_2), \Sigma_{11} - \Sigma_{12}\Sigma_{22}^{-}\Sigma_{12}'\right).$$

Here, Σ_{22}^{-} denotes the generalized inverse of Σ_{22}.

Proof. The proof consists of calculating the conditional characteristic function of X_1 given $X_2 = x_2$. To do this, first note that $X_1 - \Sigma_{12}\Sigma_{22}^{-}X_2$ and X_2 are jointly normal on $V_1 \oplus V_2$ and are uncorrelated by Proposition 2.17. Thus $X_1 - \Sigma_{12}\Sigma_{22}^{-}X_2$ and X_2 are independent. Therefore, for $x \in V_1$,

$$\phi(x) \equiv \mathcal{E}\left(\exp\left[i(x, X_1)_1\right] | X_2 = x_2\right)$$

$$= \mathcal{E}\left(\exp\left[i(x, X_1)_1 - i(x, \Sigma_{12}\Sigma_{22}^{-}X_2)_1 + i(x, \Sigma_{12}\Sigma_{22}^{-}X_2)_1\right] | X_2 = x_2\right)$$

$$= \exp\left[i(x, \Sigma_{12}\Sigma_{22}^{-}x_2)_1\right]\mathcal{E}\left(\exp\left[i(x, X_1 - \Sigma_{12}\Sigma_{22}^{-}X_2)_1\right] | X_2 = x_2\right)$$

$$= \exp\left[i(x, \Sigma_{12}\Sigma_{22}^{-}x_2)_1\right]\mathcal{E}\exp\left[i(x, X_1 - \Sigma_{12}\Sigma_{22}^{-}X_2)_1\right]$$

where the last equality follows from the independence of X_2 and $X_1 - \Sigma_{12}\Sigma_{22}^{-}X_2$. However, it is clear that

$$\mathcal{L}(X_1 - \Sigma_{12}\Sigma_{22}^{-}X_2) = N(\mu_1 - \Sigma_{12}\Sigma_{22}^{-}\mu_2, \Sigma_{11} - \Sigma_{12}\Sigma_{22}^{-}\Sigma_{12}')$$

as $X_1 - \Sigma_{12}\Sigma_{22}^{-}X_2$ is normal on V_1 and has the given mean vector and covariance (Proposition 2.17). Thus

$$\phi(x) = \exp\left[i(x, \Sigma_{12}\Sigma_{22}^{-}x_2)_1\right]\exp\left[i(x, \mu_1 - \Sigma_{12}\Sigma_{22}^{-}\mu_2)_1\right]$$

$$\times\exp\left[-\tfrac{1}{2}(x, (\Sigma_{11} - \Sigma_{12}\Sigma_{22}^{-}\Sigma_{12}')_1 x)\right]$$

$$= \exp\left[i(x, \mu_1 + \Sigma_{12}\Sigma_{22}^{-}(x_2 - \mu_2))_1 - \tfrac{1}{2}(x, (\Sigma_{11} - \Sigma_{12}\Sigma_{22}^{-}\Sigma_{12}')x)_1\right].$$

The uniqueness of characteristic functions yields the desired conclusion. □

For normal random vectors, $X_i \in (V_i, (\cdot, \cdot)_i)$, $i = 1, 2$, Proposition 3.13 shows that the conditional mean of X_1 given $X_2 = x_2$ is an affine function of x_2 (affine means a linear transformation, plus a constant vector so zero does not necessarily get mapped into zero). In other words,

$$\mathcal{E}(X_1 | X_2 = x_2) = \mu_1 + \Sigma_{12}\Sigma_{22}^{-}(x_2 - \mu_2).$$

Further, the conditional covariance of X_1 does not depend on the value of X_2. Also, this conditional covariance is the same as the unconditional covariance of the normal random vector $X_1 - \Sigma_{12}\Sigma_{22}^{-}X_2$. Of course, the specification of the conditional mean vector and covariance specifies the conditional distribution of X_1 given $X_2 = x_2$ as this conditional distribution is normal.

◆ **Example 3.1.** Let $W_1, \ldots, W_n$ be independent coordinate random vectors in R^p where R^p has the usual inner product. Assume that $\mathcal{L}(W_i) = N(\mu, \Sigma)$ so $\mu \in R^p$ is the coordinate mean vector of each W_i and Σ is the $p \times p$ covariance matrix of each W_i. Form the random matrix $X \in \mathcal{L}_{p,n}$ with rows W_i', $i = 1, \ldots, n$. We know that

$$\mathcal{E}X = e\mu'$$

and

$$\operatorname{Cov}(X) = I_n \otimes \Sigma$$

where $e \in R^n$ is the vector of ones. To show X has a normal distribution on the inner product space $(\mathcal{L}_{p,n}, \langle \cdot, \cdot \rangle)$, it must be verified that for each $A \in \mathcal{L}_{p,n}$, $\langle A, X \rangle$ has a normal distribution. To do this, let the rows of A be $a_1', \ldots, a_n'$, $a_i \in R^p$. Then

$$\langle A, X \rangle = \operatorname{tr} AX' = \sum_1^n a_i' W_i.$$

However, $a_i'W_i$ has a normal distribution on R since $\mathcal{L}(W_i) = N(\mu, \Sigma)$ on R^p. Also, since $W_1, \ldots, W_n$ are independent, $a_1'W_1, \ldots, a_n'W_n$ are independent. Since a linear combination of independent normal random variables is normal, $\langle A, X \rangle$ has a normal distribution for each $A \in \mathcal{L}_{p,n}$. Thus

$$\mathcal{L}(X) = N(e\mu', I_n \otimes \Sigma)$$

on the inner product space $(\mathcal{L}_{p,n}, \langle \cdot, \cdot \rangle)$. We now want to describe the conditional distribution of the first q columns of X given the last r columns of X where $q + r = p$. After some relabeling and a bit of manipulation, this conditional distribution follows from Proposition 3.13. Partition each W_i into Y_i and Z_i where $Y_i \in R^q$ consists of the first q coordinates of W_i and $Z_i \in R^r$ consists of the last r coordinates of W_i. Let $X_1 \in \mathcal{L}_{q,n}$ have rows $Y_1', \ldots, Y_n'$ and let $X_2 \in \mathcal{L}_{r,n}$ have rows $Z_1', \ldots, Z_n'$. Also, partition μ into $\mu_1 \in R^q$ and $\mu_2 \in R^r$ so $\mathcal{E}Y_i = \mu_1$ and $\mathcal{E}Z_i = \mu_2$, $i = 1, \ldots, n$. Further, partition the covariance matrix Σ of each W_i so that

$$\operatorname{Cov}\{Y_i, Z_i\} = \begin{pmatrix} \Sigma_{11} & \Sigma_{12} \\ \Sigma_{21} & \Sigma_{22} \end{pmatrix}$$

where $\Sigma_{21} = \Sigma_{12}'$. From the independence of $W_1, \ldots, W_n$, it follows that

$$\mathcal{L}(X_1) = N(e\mu_1', I_n \otimes \Sigma_{11}),$$

$$\mathcal{L}(X_2) = N(e\mu_2', I_n \otimes \Sigma_{22})$$

and $\{X_1, X_2\}$ has a normal distribution on $\mathcal{L}_{q,n} \oplus \mathcal{L}_{r,n}$ with mean vector $\{e\mu_1', e\mu_2'\}$ and

$$\operatorname{Cov}\{X_1, X_2\} = \begin{pmatrix} I_n \otimes \Sigma_{11} & I_n \otimes \Sigma_{12} \\ I_n \otimes \Sigma_{21} & I_n \otimes \Sigma_{22} \end{pmatrix}.$$

Now, Proposition 3.13 is directly applicable to $\{X_1, X_2\}$ where we make the parameter correspondence

$$\mu_i \leftrightarrow e\mu_i', \qquad i = 1, 2$$

and

$$\Sigma_{ij} \leftrightarrow I_n \otimes \Sigma_{ij}.$$

Therefore, the conditional distribution of X_1 given $X_2 = x_2 \in \mathcal{L}_{r,n}$ is normal with mean vector

$$\mathcal{E}(X_1 | X_2 = x_2) = e\mu_1' + (I_n \otimes \Sigma_{12})(I_n \otimes \Sigma_{22})^- (x_2 - e\mu_2')$$

and

$$\operatorname{Cov}(X_1 | X_2 = x_2)$$

$$= I_n \otimes \Sigma_{11} - (I_n \otimes \Sigma_{12})(I_n \otimes \Sigma_{22})^- (I_n \otimes \Sigma_{21}).$$

However, it is not difficult to show that $(I_n \otimes \Sigma_{22})^- = I_n \otimes \Sigma_{22}^-$. Using the manipulation rules for Kronecker products, we have

$$\mathcal{E}(X_1 | X_2 = x_2) = e\mu_1' + (x_2 - e\mu_2')\Sigma_{22}^-\Sigma_{21}$$

and

$$\operatorname{Cov}(X_1 | X_2 = x_2) = I_n \otimes (\Sigma_{11} - \Sigma_{12}\Sigma_{22}^-\Sigma_{21}).$$

This result is used in a variety of contexts in later chapters. ◆

3.5. THE DENSITY OF THE NORMAL DISTRIBUTION

The problem considered here is how to define the density function of a nonsingular normal distribution on an inner product space $(V, (\cdot, \cdot))$. By nonsingular, we mean that the covariance of the distribution is nonsingular. To motivate the technical considerations given below, the density function of a nonsingular normal distribution is first given for the standard coordinate space R^n with the usual inner product.

Consider a random vector X in R^n with coordinates $X_1, \dots, X_n$ and assume that $X_1, \dots, X_n$ are independent with $\mathcal{L}(X_i) = N(0, 1)$. The symbol dx denotes Lebesgue measure on R^n. Since $X_1, \dots, X_n$ are independent, the joint density of $X_1, \dots, X_n$ in R^n is just the product of the marginal densities, that is, X has a density with respect to dx given by

$$p(x) = \prod_{i=1}^{n} \frac{1}{\sqrt{2\pi}} \exp\left[-\tfrac{1}{2}x_i^2\right] = \frac{1}{(2\pi)^{n/2}} \exp\left[-\tfrac{1}{2}\Sigma_1^n x_i^2\right]$$

where $x \in R^n$ has coordinates $x_1, \dots, x_n$. Thus

$$p(x) = (2\pi)^{-n/2} \exp\left[-\tfrac{1}{2}x'x\right]$$

and $x'x$ is just the inner product of x with x in R^n. To derive the density of an arbitrary nonsingular normal distribution in R^n, let A be an $n \times n$ nonsingular matrix and set $Y = AX + \mu$ where $\mu \in R^n$. Since $\mathcal{L}(X) = N(0, I_n)$, $\mathcal{L}(Y) = N(\mu, \Sigma)$ where $\Sigma = AA'$ is positive definite. Thus $X = A^{-1}(Y - \mu)$ and the Jacobian of the nonsingular linear transformation on R^n to R^n sending x into $A^{-1}(x - \mu)$ is $|\det(A^{-1})|$ where $|\cdot|$ denotes absolute value. Therefore, the density function of Y with respect to dy is

$$\begin{aligned} p_1(y) &= |\det(A^{-1})| p(A^{-1}(y - \mu)) = (\det \Sigma)^{-1/2}(2\pi)^{-n/2} \\ &\quad \times \exp\left[-\tfrac{1}{2}(y - \mu)'A'^{-1}A^{-1}(y - \mu)\right] \\ &= (\det \Sigma)^{-1/2}(2\pi)^{-n/2} \exp\left[-\tfrac{1}{2}(y - \mu)'\Sigma^{-1}(y - \mu)\right]. \end{aligned}$$

Thus we have the density function with respect to dy of any nonsingular normal distribution on R^n. Of course, this expression makes no sense when Σ is singular.

Now, suppose Y is a random vector in an n-dimensional vector space $(V, (\cdot, \cdot))$ and $\mathcal{L}(Y) = N(\mu, \Sigma)$ where Σ is positive definite. The expression

$$p_2(y) = (2\pi)^{-n/2}(\det \Sigma)^{-1/2} \exp\left[-\tfrac{1}{2}((y - \mu), \Sigma^{-1}(y - \mu))\right],$$

for $y \in V$, certainly makes sense and it is tempting to call this the density function of $Y \in (V, (\cdot, \cdot))$. The problem is: What is the measure on $(V, (\cdot, \cdot))$ with respect to which p_2 is a density? In other words, what is the analog of Lebesgue measure on $(V, (\cdot, \cdot))$? To answer the question, we now show that there is a natural measure on $(V, (\cdot, \cdot))$, which is constructed from Lebesgue measure on R^n, and p_2 is the density function of Y with respect to this measure.

The details of the construction of "Lebesgue measure" on an n- dimensional inner product space $(V, (\cdot, \cdot))$ follow. First, we review some basic topological notions for $(V, (\cdot, \cdot))$. Recall that $S_r(x_0) \equiv \{x \| \|x - x_0\| < r\}$ is called the open ball of radius r with center x_0. A set $B \subseteq V$ is called *open* if, for each $x_0 \in B$, there is an $r > 0$ such that $S_r(x_0) \subseteq B$. Since all inner products on V are related by positive definite linear transformations, the definition of open does not depend on the given inner product. A set is *closed* iff its complement is open and a set if *bounded* iff it is contained in $S_r(0)$ for some $r > 0$. Just as in R^n, a set is compact iff it is closed and bounded (see Rudin, 1953, for the definition and characterization of compact sets in R^n). As with openness, the definitions and characterizations of closedness, boundedness, and compactness do not depend on the particular inner product on V. Let l denote standard Lebesgue measure on R^n. To move l over to the space V, let $x_1, \ldots, x_n$ be a fixed orthonormal basis in $(V, (\cdot, \cdot))$ and define the linear transformation T on R^n to V by

$$T(a) = \Sigma_1^n a_i x_i$$

where $a \in R^n$ has coordinates $a_1, \ldots, a_n$. Clearly, T is one-to-one, onto, and maps open, closed, bounded, and compact sets of R^n into open, closed, bounded, and compact sets of V. Also, T^{-1} on V to R^n maps $x \in V$ into the vector with coordinates (x_i, x), $i = 1, \ldots, n$. Now, define the measure ν_0 on Borel sets $B \in \mathcal{B}(V)$ by

$$\nu_0(B) = l(T^{-1}(B)).$$

Notice that $\nu_0(B + x) = l(T^{-1}(B + x)) = l(T^{-1}(B) + T^{-1}x) = l(T^{-1}(B)) = \nu_0(B)$ since Lebesgue measure is invariant under translations. Also, $\nu_0(B) < +\infty$ if B is a compact set. This leads to the following definition.

Definition 3.3. A nonzero measure ν defined on the Borel sets $\mathcal{B}(V)$ of $(V, (\cdot, \cdot))$ is *invariant* if:

(i) $\nu(B + x) = \nu(B)$ for $x \in V$ and $B \in \mathcal{B}(V)$.

(ii) $\nu(B) < +\infty$ for all compact sets B.

The measure ν_0 defined above is invariant and it is shown that, if ν is any invariant measure on $\mathcal{B}(V)$, then $\nu = c\nu_0$ for some constant $c > 0$. Condition (ii) of Definition 3.3 relates the topology of V to the measure ν. The measure that counts the number of points in a set satisfies (i) but not (ii) of Definition 3.3 and this measure is not equal to a positive constant times ν_0.

Before characterizing the measure ν_0, it is now shown that ν_0 is a dominating measure for the density function of a nonsingular normal distribution on $(V, (\cdot, \cdot))$.

Proposition 3.14. Suppose $\mathcal{L}(Y) = N(\mu, \Sigma)$ on the inner product space $(V, (\cdot, \cdot))$ where Σ is nonsingular. The density function of Y with respect to the measure ν_0 is given by

$$p(y) = (2\pi)^{-n/2}(\det \Sigma)^{-1/2} \exp\left[-\tfrac{1}{2}(y - \mu, \Sigma^{-1}(y - \mu))\right]$$

for $y \in V$.

Proof. It must be shown that, for each Borel set B,

$$P\{Y \in B\} = \int I_B(y) p(y) \nu_0(dy),$$

where I_B is the indicator function of the set B. From the definition of the measure ν_0, it follows that (see Lehmann, 1959, p. 38)

$$\int I_B(y) p(y) \nu_0(dy) = \int I_B(T(a)) p(T(a)) l(da).$$

Let $X = T^{-1}(Y) \in R^n$ so X is a random vector with coordinates (x_i, Y), $i = 1, \dots, n$. Thus X has a normal distribution in R^n with mean vector $T^{-1}(\mu)$ and covariance matrix $[\Sigma]$ where $[\Sigma]$ is the matrix of Σ in the given orthonormal basis $x_1, \dots, x_n$. Therefore,

$$\begin{aligned} P\{Y \in B\} &= P\{T^{-1}(Y) \in T^{-1}(B)\} = P\{X \in T^{-1}(B)\} \\ &= \int I_{T^{-1}(B)}(a)(2\pi)^{-n/2}(\det[\Sigma])^{-1/2} \\ &\quad \times \exp\left[-\tfrac{1}{2}(a - T^{-1}(\mu))'[\Sigma]^{-1}(a - T^{-1}(\mu))\right] l(da) \\ &= \int I_B(T(a)) p(T(a)) l(da). \end{aligned}$$

The last equality follows since $I_{T^{-1}(B)}(a) = I_B(T(a))$ and

$$\begin{aligned} p(T(a)) &= (2\pi)^{-n/2}(\det \Sigma)^{-1/2} \\ &\quad\times \exp\left[-\tfrac{1}{2}(T(a)-\mu, \Sigma^{-1}(T(a)-\mu))\right] \\ &= (2\pi)^{-n/2}(\det[\Sigma])^{-1/2} \\ &\quad\times \exp\left[-\tfrac{1}{2}(a - T^{-1}(\mu))'[\Sigma]^{-1}(a - T^{-1}(\mu))\right]. \end{aligned}$$

Thus

$$\begin{aligned} P\{Y \in B\} &= \int I_B(T(a))p(T(a))l(da) \\ &= \int I_B(y)p(y)\nu_0(dy). \end{aligned}$$

□

We now want to show that the measure ν_0, constructed from Lebesgue measure on R^n, is the unique translation invariant measure that satisfies

$$\int p(y)\nu_0(dy) = 1.$$

Let $\mathcal{K}^+$ be the collection of all bounded non-negative Borel measurable functions defined on V that satisfy the following: given $f \in \mathcal{K}^+$, there is a compact set B such that $f(v) = 0$ if $v \notin B$. If ν is any invariant measure on V and $f \in \mathcal{K}^+$, then $\int f(v)\nu(dv) < +\infty$ since f is bounded and the ν-measure of every compact set is finite. It is clear that, if ν_1 and ν_2 are invariant measures such that

$$\int f(v)\nu_1(dv) = \int f(v)\nu_2(dv) \qquad \text{for all } f \in \mathcal{K}^+,$$

then $\nu_1 = \nu_2$. From the definition of an invariant measure, we also have

$$\int f(v+x)\nu(dv) = \int f(v)\nu(dv)$$

for all $f \in \mathcal{K}^+$ and $x \in V$. Furthermore, the definition of ν_0 shows that

$$\int f(x)\nu_0(dx) = \int f(T(a))l(da) = \int f(T(-a))l(da)$$

$$= \int f(-T(a))l(da) = \int f(-x)\nu_0(dx)$$

for all $f \in \mathcal{K}^+$. Here, we have used the linearity of T and the invariance of Lebesgue measure under multiplication of the argument of integration by a minus one.

Proposition 3.15. If ν is an invariant measure on $\mathcal{B}(V)$, then there exists a positive constant c such that $\nu = c\nu_0$.

Proof. For $f, g \in \mathcal{K}^+$, we have

$$\int f(x)\nu(dx)\int g(y)\nu_0(dy) = \int\int f(x-y)g(y)\nu(dx)\nu_0(dy)$$

$$= \int\int f(-(y-x))g(y-x+x)\nu_0(dy)\nu(dx)$$

$$= \int\int f(-w)g(w+x)\nu_0(dw)\nu(dx)$$

$$= \int\int f(-w)g(w+x)\nu(dx)\nu_0(dw)$$

$$= \int f(-w)\nu_0(dw)\int g(x)\nu(dx)$$

$$= \int f(w)\nu_0(dw)\int g(x)\nu(dx).$$

Therefore,

$$\int f(x)\nu(dx)\int g(y)\nu_0(dy) = \int f(w)\nu_0(dw)\int g(y)\nu(dy)$$

for all $f, g \in \mathcal{K}^+$. Fix $f \in \mathcal{K}^+$ such that $\int f(w)\nu_0(dw) = 1$ and set $c = \int f(x)\nu(dx)$. Then

$$\int g(y)\nu(dy) = c\int g(y)\nu_0(dy)$$

for all $g \in \mathcal{K}^+$. The constant c cannot be zero as the measure ν is not zero. Thus $c > 0$ and $\nu = c\nu_0$. □

The measure ν_0 is called the Lebesgue measure on V and is henceforth denoted by dv or dx, as is the Lebesgue measure on R^n. It is possible to show that ν_0 does not depend on the particular orthonormal basis used to define it by using a Jacobian argument in R^n. However, the argument given above contains more information than this. In fact, some minor technical modifications of the proof of Proposition 3.15 yield the uniqueness (up to a positive constant) of invariant measures on locally compact topological groups. This topic is discussed in detail in Chapter 6.

An application of Proposition 3.14 to the situation treated in Example 3.1 follows.

◆ **Example 3.2.** For independent coordinate random vectors $W_i \in R^p$, $i = 1,\ldots,n$, with $\mathcal{L}(W_i) = N(\mu, \Sigma)$, form the random matrix $X \in \mathcal{L}_{p,n}$ with rows W_i', $i = 1,\ldots,n$. As shown in Example 3.1,

$$\mathcal{L}(X) = N(e\mu', I_n \otimes \Sigma)$$

on the inner product space $(\mathcal{L}_{p,n}, \langle\cdot,\cdot\rangle)$, where $e \in R^n$ is the vector of ones. Let dX denote Lebesgue measure on the vector space $\mathcal{L}_{p,n}$. If Σ is nonsingular, then $I_n \otimes \Sigma$ is nonsingular and $(I_n \otimes \Sigma)^{-1} = I_n \otimes \Sigma^{-1}$. Thus when Σ is nonsingular, the density of X with respect to dX is

$$(3.1)\quad p(X) = \sqrt{2\pi})^{-np}(\det(I_n \otimes \Sigma))^{-1/2} \times \exp\left[-\tfrac{1}{2}\langle X - e\mu', (I_n \otimes \Sigma^{-1})(X - e\mu')\rangle\right].$$

It is shown in Chapter 5 that $\det(I_n \otimes \Sigma) = (\det \Sigma)^n$. Since the inner product $\langle\cdot,\cdot\rangle$ is given by the trace, the density p can be written

$$p(X) = (\sqrt{2\pi})^{-np}(\det \Sigma)^{-n/2} \times \exp\left[-\tfrac{1}{2}\operatorname{tr}(X - e\mu')'(X - e\mu')\Sigma^{-1}\right].$$

However, this form of the density is somewhat less revealing, from a statistical point of view, than (3.1). In order to make this statement more precise and to motivate some future statistical considerations, we now think of $\mu \in R^p$ and Σ as unknown parameters. Thus, we can write (3.1) as

$$(3.2)\quad p(X|\mu,\Sigma) = (\sqrt{2\pi})^{-np}(\det\Sigma)^{-n/2} \times \exp\left[-\tfrac{1}{2}\langle X - e\mu', (I_n \otimes \Sigma^{-1})(X - e\mu')\rangle\right]$$

where μ ranges over R^p and Σ ranges over all $p \times p$ positive definite matrices. Thus we have a parametric family of densities for the distribution of the random vector X. As a first step in analyzing this parametric family, let

$$M = \{x \in \mathcal{L}_{p,n} | x = e\mu', \mu \in R^p\}.$$

It is clear that M is a p-dimensional linear subspace of $\mathcal{L}_{p,n}$ and M is simply the space of possible values for the mean vector of X. Let $P_e = (1/n)ee'$ so P_e is the orthogonal projection onto span$\{e\} \subseteq R^n$. Thus $P_e \otimes I_p$ is an orthogonal projection and it is easily verified that the range of $P_e \otimes I_p$ is M. Therefore, the orthogonal projection onto M is $P_e \otimes I_p$. Let $Q_e = I_n - P_e$ so $Q_e \otimes I_p$ is the orthogonal projection onto $M^\perp$ and $(Q_e \otimes I_p)(P_e \otimes I_p) = 0$. We now decompose X into the part of X in M and the part of X in $M^\perp$—that is, write $X = (P_e \otimes I_p)X + (Q_e \otimes I_p)X$. Substituting this into the exponential part of (3.2) and using the relation $(P_e \otimes I_p)(I_n \otimes \Sigma)(Q_e \otimes I_p) = 0$, we have

$$\begin{aligned}
\langle X - e\mu', (I_n \otimes \Sigma^{-1})(X - e\mu')\rangle
&= \langle P_e(X - e\mu'), (I_n \otimes \Sigma^{-1})P_e(X - e\mu')\rangle \\
&\quad + \langle Q_eX, (I_n \otimes \Sigma^{-1})Q_eX\rangle \\
&= \langle P_eX - e\mu', (I_n \otimes \Sigma^{-1})(P_eX - e\mu')\rangle + \operatorname{tr} Q_eX\Sigma^{-1}(Q_eX)' \\
&= \langle P_eX - e\mu', (I_n \otimes \Sigma^{-1})(P_eX - e\mu')\rangle + \operatorname{tr} X'Q_eX\Sigma^{-1}.
\end{aligned}$$

Thus the density $p(X|\mu,\Sigma)$ is a function of the pair P_eX and $X'Q_eX$ so P_eX and $X'Q_eX$ is a sufficient statistic for the parametric family (3.2). Proposition 3.4 shows that $(P_e \otimes I_p)X$ and $(Q_e \otimes I_p)X$ are independent since $(P_e \otimes I_p)(I_n \otimes \Sigma)(Q_e \otimes I_p) = (P_eQ_e) \otimes \Sigma = 0$ as $P_eQ_e = 0$. Therefore, P_eX and $X'Q_eX$ are independent since $P_eX = (P_e \otimes I_p)X$ and $X'Q_eX = ((Q_e \otimes I_p)X)'((Q_e \otimes I_p)X)$. To interpret the sufficient statistic in terms of the original random vectors $W_1, \ldots, W_n$, first note that

$$P_e X = \frac{1}{n} ee'X = e\overline{W}'$$

where $\overline{W} = (1/n)\Sigma W_i$ is the sample mean. Also,

$$\begin{aligned} X'Q_e X &= (Q_e X)'(Q_e X) = ((I_n - P_e)X)'((I - P_e)X) \\ &= (X - e\overline{W}')'(X - e\overline{W}') = \Sigma_1^n (W_i - \overline{W})(W_i - \overline{W})'. \end{aligned}$$

The quantity $(1/n)X'Q_e X$ is often called the sample covariance matrix. Since $e\overline{W}'$ and $\overline{W}$ are one-to-one functions of each other, we have that the sample mean and sample covariance matrix form a sufficient statistic and they are independent. It is clear that

$$\mathcal{L}(\overline{W}) = N\left(\mu, \frac{1}{n}\Sigma\right).$$

The distribution of $X'Q_e X$, commonly called the Wishart distribution, is derived later. The procedure of decomposing X into the projection onto the mean space (the subspace M) and the projection onto the orthogonal complement of the mean space is fundamental in multivariate analysis as in univariate statistical analysis. In fact, this procedure is at the heart of analyzing linear models—a topic to be considered in the next chapter. ◆

PROBLEMS

1. Suppose $X_1, \ldots, X_n$ are independent with values in $(V, (\cdot, \cdot))$ and $\mathcal{L}(X_i) = N(\mu_i, A_i)$, $i = 1, \ldots, n$. Show that $\mathcal{L}(\Sigma X_i) = N(\Sigma \mu_i, \Sigma A_i)$.

2. Let X and Y be random vectors in R^n with a joint normal distribution given by

$$\mathcal{L}\begin{pmatrix} X \\ Y \end{pmatrix} = N\left(0, \begin{pmatrix} I_n & \rho I_n \\ \rho I_n & I_n \end{pmatrix}\right)$$

where ρ is a scalar. Show that $|\rho| \leqslant 1$ and the covariance is positive definite iff $|\rho| < 1$. Let $Q(Y) = I_n - (Y'Y)^{-1}YY'$. Prove that $W = X'Q(Y)X$ has the distribution of $(1 - \rho^2)\chi^2_{n-1}$ (the constant $1 - \rho^2$ times a chi-squared random variable with $n - 1$ degrees of freedom).

3. When $X \in R^n$ and $\mathcal{L}(X) = N(0, \Sigma)$ with Σ nonsingular, then $\mathcal{L}(X) = \mathcal{L}(CZ)$ where $\mathcal{L}(Z) = N(0, I_n)$ and $CC' = \Sigma$. Hence, $\mathcal{L}(C^{-1}X) = \mathcal{L}(Z)$ so C^{-1} transforms X into a vector of i.i.d. $N(0, 1)$ random variables. There are many C^{-1}'s that do this. The problem at hand concerns the construction of one such C^{-1}. Given any $p \times p$ positive definite matrix A, $p \geqslant 2$, partition A as

$$A = \begin{pmatrix} a_{11} & A_{12} \\ A_{21} & A_{22} \end{pmatrix}$$

where $a_{11} \in R^1$, $A_{21} = A'_{12} \in R^{p-1}$. Define $T_p(A)$ by

$$T_p(A) = \begin{pmatrix} a_{11}^{-1/2} & 0 \\ -\dfrac{A_{21}}{a_{11}} & I_{p-1} \end{pmatrix}.$$

(i) Partition $\Sigma : n \times n$ as A is partitioned and set $X^{(1)} = T_n(\Sigma)X$. Show that

$$\text{Cov}(X^{(1)}) = \begin{pmatrix} 1 & 0 \\ 0 & \Sigma^{(1)} \end{pmatrix}$$

where $\Sigma^{(1)} = \Sigma_{22} - \Sigma_{21}\Sigma_{12}/\sigma_{11}$.

(ii) For $k = 1, 2, \ldots, n-2$, define $X^{(k+1)}$ by

$$X^{(k+1)} = \begin{pmatrix} I_k & 0 \\ 0 & T_{n-k}(\Sigma^{(k)}) \end{pmatrix} X^{(k)}.$$

Prove that

$$\text{Cov}(X^{(k+1)}) = \begin{pmatrix} I_{k+1} & 0 \\ 0 & \Sigma^{(k+1)} \end{pmatrix}$$

for some positive definite $\Sigma^{(k+1)}$.

(iii) For $k = 0, \ldots, n-2$, let

$$T^{(k)} = \begin{pmatrix} I_k & 0 \\ 0 & T_{n-k}(\Sigma^{(k)}) \end{pmatrix},$$

where $T^{(0)} = T_n(\Sigma)$. With $T = T^{(n-2)} \ldots T^{(0)}$, show that $X^{(n-1)} = TX$ and $\text{Cov}(X^{(n-1)}) = I_n$. Also, show that T is lower triangular and $\Sigma^{-1} = T'T$.

4. Suppose $X \in R^2$ has coordinates X_1 and X_2, and has a density

$$p(x) = \begin{cases} \frac{1}{\pi} \exp\left[-\frac{1}{2}(x_1^2 + x_2^2)\right] & \text{if } x_1 x_2 > 0 \\ 0 & \text{otherwise} \end{cases}$$

so p is zero in the second and fourth quadrants. Show X_1 and X_2 are both normal but X is not normal.

5. Let $X_1, \dots, X_n$ be i.i.d. $N(\mu, \sigma^2)$ random variables. Show that $U = \Sigma\Sigma(X_i - X_j)^2$ and $W = \Sigma X_i$ are independent. What is the distribution of U?

6. For $X \in (V, (\cdot, \cdot))$ with $\mathcal{L}(X) = N(0, I)$, suppose (X, AX) and (X, BX) are independent. If A and B are both positive semidefinite, prove that $AB = 0$. Hint: Show that $\operatorname{tr} AB = 0$ by using $\operatorname{cov}\{(X, AX), (X, BX)\} = 0$. Then use the positive semidefiniteness and $\operatorname{tr} AB = 0$ to conclude that $AB = 0$.

7. The method used to define the normal distribution on $(V, (\cdot, \cdot))$ consisted of three steps: (i) first, an $N(0, 1)$ distribution was defined on R^1; (ii) next, if $\mathcal{L}(Z) = N(0, 1)$, then W is $N(\mu, \sigma^2)$ if $\mathcal{L}(W) = \mathcal{L}(\sigma Z + \mu)$; and (iii) X with values in $(V, (\cdot, \cdot))$ is normal if (x, X) is normal on R^1 for each $x \in V$. It is natural to ask if this procedure can be used to define other types of distributions on $(V, (\cdot, \cdot))$. Here is an attempt for the Cauchy distribution. For $X \in R^1$, say Z is standard Cauchy (which we write as $\mathcal{L}(Z) = C(0, 1)$) if the density of Z is

$$p(z) = \frac{1}{\pi} \frac{1}{1 + z^2}, \qquad z \in R^1.$$

Say W has a Cauchy distribution on R^1 if $\mathcal{L}(W) = \mathcal{L}(\sigma Z + \mu)$ for some $\mu \in R^1$ and $\sigma \geqslant 0$—in this case write $\mathcal{L}(W) = C(\mu, \sigma)$. Finally, say $X \in (V, (\cdot, \cdot))$ is Cauchy if (x, X) is Cauchy on R^1.

(i) Let $W_1, \dots, W_n$ be independent $C(\mu_j, \sigma_j)$, $j = 1, \dots, n$. Show that $\mathcal{L}(\Sigma a_j W_j) = C(\Sigma a_j \mu_j, \Sigma |a_j| \sigma_j)$. Hint: The characteristic function of a $C(0, 1)$ distribution is $\exp[-|t|]$, $t \in R^1$.

(ii) Let $Z_1, \dots, Z_n$ be i.i.d. $C(0, 1)$ and let $x_1, \dots, x_n$ be any basis for $(V, (\cdot, \cdot))$. Show $X = \Sigma Z_j x_j$ has a Cauchy distribution on $(V, (\cdot, \cdot))$.

8. Consider a density on R^1 given by

$$f(u) = \int_0^\infty t^{-1} \phi(u/t) G(dt)$$

where ϕ is the density of an $N(0, 1)$ distribution and G is a distribution function with $G(0) = 0$. The distribution defined by f is called a *scale mixture of normals*.

(i) Let Z_0 be $N(0, 1)$ and let R be independent of Z_0 with $\mathcal{L}(R) = G$. Show that $U = RZ_0$ has f as its density function.

If $\mathcal{L}(Y) = \mathcal{L}(cU)$ for some $c > 0$, we can say that Y has a *type-f distribution*.

(ii) In $(V, (\cdot, \cdot))$, suppose $\mathcal{L}(Z) = N(0, I)$ and form $X = RZ$ where R and Z are independent and $\mathcal{L}(R) = G$. For each $x \in V$, show (x, X) has a type-f distribution.

Remark. The distribution of X in $(V, (\cdot, \cdot))$ provides a possible vector space generalization of a type-f distribution on R^1.

9. In the notation of Example 3.1, assume that $\mu = 0$ so $\mathcal{L}(X) = N(0, I_n \otimes \Sigma)$ on $(\mathcal{L}_{p,n}, \langle \cdot , \cdot \rangle)$. Also,

$$\mathcal{L}(X_1 | X_2 = x_2) = N\left(x_2 \Sigma_{22}^{-1} \Sigma_{21}, I_n \otimes \Sigma_{11 \cdot 2}\right)$$

where $\Sigma_{11 \cdot 2} = \Sigma_{11} - \Sigma_{12} \Sigma_{22}^{-1} \Sigma_{21}$. Show that the conditional distribution of $X_2' X_1$ given X_2 is the same as the conditional distribution of $X_2' X_1$ given $X_2' X_2$.

10. The map T of Section 3.5 has been defined on R^n to $(V, (\cdot, \cdot))$ by $Ta = \sum_1^n a_i x_i$ where $x_1, \ldots, x_n$ is an orthonormal basis for $(V, (\cdot, \cdot))$. Also, we have defined ν_0 by $\nu_0(B) = l(T^{-1}(B))$ for $B \in \mathcal{B}(V)$. Consider another orthonormal basis $y_1, \ldots, y_n$ for $(V(\cdot, \cdot))$ and define T_1 by $T_1 a = \sum_1^n a_i y_i$, $a \in R^n$. Define ν_1 by $\nu_1(B) = l(T_1^{-1}(B))$ for $B \in \mathcal{B}(V)$. Prove that $\nu_0 = \nu_1$.

11. The measure ν_0 in Problem 10 depends on the inner product $(\cdot, \cdot)$ on V. Suppose $[\cdot, \cdot]$ is another inner product given by $[x, y] = (x, Ay)$ where $A > 0$. Let ν_1 be the measure constructed on $(V, [\cdot, \cdot])$ in the same manner that ν_0 was constructed on $(V, (\cdot, \cdot))$. Show that $\nu_1 = c\nu_0$ where $c = (\det(A))^{1/2}$.

12. Consider the space $\mathcal{S}_p$ of $p \times p$ symmetric matrices with the inner product given by $\langle S_1, S_2 \rangle = \operatorname{tr} S_1 S_2$. Show that the density function of an $N(0, I)$ distribution on $(\mathcal{S}_p, \langle \cdot , \cdot \rangle)$ with respect to the measure ν_0 is

$$p(S) = (2\pi)^{-p(p+1)/4} \exp\left[-\tfrac{1}{2}\left(\textstyle\sum_1^p s_{ii}^2 + 2 \sum\sum_{i<j} s_{ij}^2 \right) \right]$$

where $S = \{s_{ij}\}$, $i, j = 1, \ldots, p$. Explain your answer (what is ν_0)?

13. Consider $X_1, \ldots, X_n$, which are i.i.d. $N(\mu, \Sigma)$ on R^p. Let $X \in \mathcal{L}_{p,n}$ have rows $X_1', \ldots, X_n'$ so $\mathcal{L}(X) = N(e\mu', I_n \otimes \Sigma)$. Assume that Σ has the form

$$\Sigma = \sigma^2 \begin{pmatrix} 1 & \rho & \cdots & \rho \\ \rho & 1 & \cdots & \rho \\ \vdots & & \ddots & \vdots \\ \rho & \rho & \cdots & 1 \end{pmatrix}$$

where $\sigma^2 > 0$ and $-1/(p-1) < \rho < 1$ so Σ is positive definite. Such a covariance matrix is said to have intraclass covariance structure.

(i) On R^p, let $A = (1/p)e_1 e_1'$ where $e_1 \in R^p$ is the vector of ones. Show that a positive definite covariance matrix has intraclass covariance structure iff $\Sigma = \alpha A + \beta(I - A)$ for some positive scalars α and β. In this case $\Sigma^{-1} = \alpha^{-1}A + \beta^{-1}(I - A)$.

(ii) Using the notation and methods of Example 3.2, show that when (μ, σ^2, ρ) are unknown parameters, then $(\overline{X}, \operatorname{tr} AX'Q_eX, \operatorname{tr}(I - A)X'Q_eX)$ is a sufficient statistic.

NOTES AND REFERENCES

1. A coordinate treatment of the normal distribution similar to the treatment given here can be found in Muirhead (1982).

2. Examples 3.1 and 3.2 indicate some of the advantages of vector space techniques over coordinate techniques. For comparison, the reader may find it instructive to formulate coordinate versions of these examples.

3. The converse of Proposition 3.11 is true. The only proof I know involves characteristic functions. For a discussion of this, see Srivastava and Khatri (1979, p. 64).

CHAPTER 4

Linear Statistical Models

The purpose of this chapter is to develop a theory of linear unbiased estimation that is sufficiently general to be applicable to the linear models arising in multivariate analysis. Our starting point is the classical regression model where the Gauss–Markov Theorem is formulated in vector space language. The approach taken here is to first isolate the essential aspects of a regression model and then use the vector space machinery developed thus far to derive the Gauss–Markov estimator of a mean vector.

After presenting a useful necessary and sufficient condition for the equality of the Gauss–Markov and least-squares estimators of a mean vector, we then discuss the existence of Gauss–Markov estimators for what might be called generalized linear models. This discussion leads to a version of the Gauss–Markov Theorem that is directly applicable to the general linear model of multivariate analysis.

4.1. THE CLASSICAL LINEAR MODEL

The linear regression model arises from the following considerations. Suppose we observe a random variable $Y_i \in R$ and associated with Y_i are known numbers $z_{i1}, \ldots, z_{ik}$, $i = 1, \ldots, n$. The numbers $z_{i1}, \ldots, z_{ik}$ might be indicator variables denoting the presence or absence of a treatment as in the case of an analysis of variance situation or they might be the numerical levels of some physical parameters that affect the observed value of Y_i. It is assumed that the mean value of Y_i is $\mathcal{E}Y_i = \Sigma_1^k z_{ij}\beta_j$ where the β_j are unknown parameters. It is also assumed that $\operatorname{var}(Y_i) = \sigma^2 > 0$ and $\operatorname{cov}(Y_i, Y_j) = 0$ if $i \neq j$. Let $Y \in R^n$ be the random vector with coordinates $Y_1, \ldots, Y_n$, let $Z = \{z_{ij}\}$ be the $n \times k$ matrix of z_{ij}'s, and let $\beta \in R^k$ be the vector with coordinates $\beta_1, \ldots, \beta_k$. In vector form, the assumptions we have made concerning Y are that $\mathcal{E}Y = Z\beta$ and $\text{Cov}(Y) = \sigma^2 I_n$. In summary, we observe the vector Y whose mean is $Z\beta$ where Z is a known $n \times k$ matrix, $\beta \in R^k$ is a vector of unknown parameters, and $\text{Cov}(Y) = \sigma^2 I_n$ where σ^2 is an unknown parameter. The two essential features of this parametric model are: (i) the mean vector of Y is an unknown element of a known subspace of R^n—namely, $\mathcal{E}Y$ is an element of the range of the known linear transformation determined by Z that maps R^k to R^n; (ii) $\text{Cov}(Y) = \sigma^2 I_n$—that is, the distribution of Y is weakly spherical. For a discussion of the classical statistical problems related to the above model, the reader is referred to Scheffé (1959).

Now, consider a finite dimensional inner product space $(V, (\cdot, \cdot))$. With the above regression model in mind, we define a weakly spherical linear model for a random vector with values in $(V, (\cdot, \cdot))$.

Definition 4.1. Let M be a subspace of V and let ε_0 be a random vector in V with a distribution that satisfies $\mathcal{E}\varepsilon_0 = 0$ and $\text{Cov}(\varepsilon_0) = I$. For each $\mu \in M$ and $\sigma > 0$, let $Q_{\mu, \sigma}$ denote the distribution of $\mu + \sigma\varepsilon_0$. The family $\{Q_{\mu, \sigma} | \mu \in M, \sigma > 0\}$ is a *weakly spherical linear model* for $Y \in V$ if the distribution of Y is in $\{Q_{\mu, \sigma} | \mu \in M, \sigma > 0\}$.

This definition is just a very formal statement of the assumption that the mean vector of Y is an element of the subspace of M and the distribution of Y is weakly spherical so $\text{Cov}(Y) = \sigma^2 I$ for some $\sigma^2 > 0$. In an abuse of notation, we often write $Y = \mu + \varepsilon$ for $\mu \in M$ where ε is a random vector with $\mathcal{E}\varepsilon = 0$ and $\text{Cov}(\varepsilon) = \sigma^2 I$. This is to indicate the assumption that we have a weakly spherical linear parametric model for the distribution of Y. The unobserved random vector ε is often called the error vector. The subspace M is called the regression subspace (or manifold) and the subspace $M^\perp$ is called the error subspace. Further, the parameter $\mu \in M$ is assumed unknown as is the parameter σ^2. It is clear that the regression model used to motivate Definition 4.1 is a weakly spherical linear model for the observed random vector and the subspace M is just the range of Z.

Given a linear model $Y = \mu + \varepsilon$, $\mu \in M$, $\mathcal{E}\varepsilon = 0$, $\text{Cov}(\varepsilon) = \sigma^2 I$, we now want to discuss the problem of estimating μ. The classical Gauss–Markov approach to estimating μ is to first restrict attention to linear transformations of Y that are unbiased estimators and then, within this class of estimators, find the estimator with minimum expected norm-squared deviation from μ. To make all of this precise, we proceed as follows. By a linear estimator of μ, we mean an estimator of the form AY where $A \in \mathcal{L}(V, V)$. (We could consider affine estimators $AY + v_0$, $v_0 \in V$, but the unbiasedness restriction would imply $v_0 = 0$.) A linear estimator AY of μ is unbiased iff, when $\mu \in M$ is the mean of Y, we have $\mathcal{E}(AY) = \mu$. This is equivalent to the condition that $A\mu = \mu$ for all $\mu \in M$ since $\mathcal{E}AY = A\mathcal{E}Y = A\mu$. Thus AY is an unbiased estimator of μ iff $A\mu = \mu$ for all $\mu \in M$. Let

$$\mathcal{A} = \{A | A \in \mathcal{L}(V, V), A\mu = \mu \text{ for } \mu \in M\}.$$

The linear unbiased estimators of μ are those estimators of the form AY with $A \in \mathcal{A}$. We now want to choose the one estimator (i.e., $A \in \mathcal{A}$) that minimizes the expected norm-squared deviation of the estimator from μ. In other words, the problem is to find an element $A \in \mathcal{A}$ that minimizes $\mathcal{E}\|AY - \mu\|^2$. The justification for choosing such an A is that $\|AY - \mu\|^2$ is the squared distance between AY and μ so $\mathcal{E}\|AY - \mu\|^2$ is the average squared distance between AY and μ. Since we would like AY to be close to μ, such a criterion for choosing $A \in \mathcal{A}$ seems reasonable. The first result in this chapter, the Gauss–Markov Theorem, shows that the orthogonal projection onto M, say P, is the unique element in $\mathcal{A}$ that minimizes $\mathcal{E}\|AY - \mu\|^2$.

Theorem 4.1 (Gauss–Markov Theorem). For each $A \in \mathcal{A}$, $\mu \in M$, and $\sigma^2 > 0$,

$$\mathcal{E}\|AY - \mu\|^2 \geqslant \mathcal{E}\|PY - \mu\|^2$$

where P is the orthogonal projection onto M. There is equality in this inequality iff $A = P$.

Proof. Write $A = P + C$ so $C = A - P$. Since $A\mu = \mu$ for $\mu \in M$, $C\mu = 0$ for $\mu \in M$ and this implies that $CP = 0$. Therefore, $C(Y - \mu)$ and $P(Y - \mu)$ are uncorrelated random vectors, so $\mathcal{E}(C(Y - \mu), P(Y - \mu)) = 0$ (see Proposition 2.21). Now,

$$\begin{aligned}\mathcal{E}\|AY - \mu\|^2 &= \mathcal{E}\|A(Y - \mu)\|^2 = \mathcal{E}\|P(Y - \mu) + C(Y - \mu)\|^2 \\ &= \mathcal{E}\|P(Y - \mu)\|^2 + \mathcal{E}\|C(Y - \mu)\|^2 \\ &\geqslant \mathcal{E}\|P(Y - \mu)\|^2 = \mathcal{E}\|PY - \mu\|^2.\end{aligned}$$

The third equality results from the fact that the cross product term is zero. This establishes the desired inequality. It is clear that there is equality in this inequality iff $\mathcal{E}\|C(Y - \mu)\|^2 = 0$. However, $C(Y - \mu)$ has mean zero and covariance $\sigma^2 CC'$ so

$$\mathcal{E}\|C(Y - \mu)\|^2 = \sigma^2 \langle I, CC' \rangle$$

by Proposition 2.21. Since $\sigma^2 > 0$, there is equality iff $\langle I, CC' \rangle = 0$. But $\langle I, CC' \rangle = \langle C, C \rangle$ and this is zero iff $C = A - P = 0$. □

The estimator PY of $\mu \in M$ is called the Gauss–Markov estimator of the mean vector and the notation $\hat{\mu} \equiv PY$ is used here. A moment's reflection shows that the validity of Theorem 4.1 has nothing to do with the parameter σ^2, be it known or unknown, as long as $\sigma^2 > 0$. The estimator $\hat{\mu} = PY$ is also called the least-squares estimator of μ for the following reason. Given the observation vector Y, we ask for that vector in M that is closest, in the given norm, to Y—that is, we want to minimize, over $x \in M$, the expression $\|Y - x\|^2$. But $Y = PY + QY$ where $Q = (I - P)$ so, for $x \in M$,

$$\|Y - x\|^2 = \|PY - x + QY\|^2 = \|PY - x\|^2 + \|QY\|^2.$$

The second equality is a consequence of $Qx = 0$ and $QP = 0$. Thus

$$\|Y - x\|^2 \geqslant \|QY\|^2$$

with equality iff $x = PY$. In other words, the point in M that is closest to Y is $\hat{\mu} = PY$. When the vector space V is R^n with the usual inner product, then $\|Y - x\|^2$ is just a sum of squares and $\hat{\mu} = PY \in M$ minimizes this sum of squares—hence the term least-squares estimator.

◆ **Example 4.1.** Consider the regression model used to motivate Definition 4.1. Here, $Y \in R^n$ has a mean vector $Z\beta$ when $\beta \in R^k$ and Z is an $n \times k$ known matrix with $k \leqslant n$. Also, it is assumed that $\text{Cov}(Y) = \sigma^2 I_n$, $\sigma^2 > 0$. Therefore, we have a weakly spherical linear model for Y and $\mu \equiv Z\beta$ is the mean vector of Y. The regression manifold M is just the range of Z. To compute the Gauss–Markov estimator of μ, the orthogonal projection onto M, relative to the usual inner product on R^n, must be found. To find this projection explicitly in terms of Z, it is now assumed that the rank of Z is k. The claim is that $P \equiv Z(Z'Z)^{-1}Z'$ is the orthogonal projection onto M. Clearly, $P^2 = P$ and P is self-adjoint so P is the orthogonal projection onto its range. However, Z' maps R^n onto R^k since the rank of Z' is k. Thus $(Z'Z)^{-1}Z'$ maps R^n onto R^k. Therefore, the range of $Z(Z'Z)^{-1}Z'$ is $Z(R^k)$, which is just M, so P is the orthogonal projection onto M. Hence $\hat{\mu} = Z(Z'Z)^{-1}Z'Y$ is the Gauss–Markov and least-squares estimator of μ. Since $\mu = Z\beta$, $Z'\mu = Z'Z\beta$ and thus $\beta = (Z'Z)^{-1}Z'\mu$. There is the obvious temptation to call

$$\hat{\beta} \equiv (Z'Z)^{-1}Z'\hat{\mu} = (Z'Z)^{-1}Z'Z(Z'Z)^{-1}Z'Y = (Z'Z)^{-1}Z'Y$$

the Gauss–Markov and least-squares estimator of the parameter β.

Certainly, calling $\hat{\beta}$ the least-squares estimator of β is justified since

$$\|Y - Z\gamma\|^2 \geqslant \|Y - Z\hat{\beta}\|^2$$

for all $\gamma \in R^k$, as $Z\hat{\beta} = \hat{\mu}$ and $Z\gamma \in M$. Thus $\hat{\beta}$ minimizes the sum of squares $\|Y - Z\gamma\|^2$ as a function of γ. However, it is not clear why $\hat{\beta}$ should be called the Gauss–Markov estimator of β. The discussion below rectifies this situation. ◆

Again, consider the linear model in $(V, (\cdot, \cdot))$, $Y = \mu + \varepsilon$, where $\mu \in M$, $\mathcal{E}\varepsilon = 0$, and $\text{Cov}(\varepsilon) = \sigma^2 I$. As usual, M is a linear subspace of V and ε is a random vector in V. Let $(W, [\cdot, \cdot])$ be an inner product space. Motivated by the considerations in Example 4.1, consider the problem of estimating $B\mu$, $B \in \mathcal{L}(V, W)$, by a linear unbiased estimator AY where $A \in \mathcal{L}(V, W)$. That AY is an unbiased estimator of $B\mu$ for each $\mu \in M$ is clearly equivalent to $A\mu = B\mu$ for $\mu \in M$ since $\mathcal{E}AY = A\mu$. Let

$$\mathcal{Q}_1 = \{A | A \in \mathcal{L}(V, W), A\mu = B\mu \text{ for } \mu \in M\},$$

so AY is an unbiased estimator of $B\mu$, $\mu \in M$ iff $A \in \mathcal{Q}_1$. The following result, which is a generalization of Theorem 4.1, shows that $B\hat{\mu}$ is the Gauss–Markov estimator for $B\mu$ in the sense that, for all $A \in \mathcal{Q}_1$,

$$\mathcal{E}\|AY - B\mu\|_1^2 \geqslant \mathcal{E}\|BPY - B\mu\|_1^2.$$

Here $\|\cdot\|_1$ is the norm on the space $(W, [\cdot, \cdot])$.

Proposition 4.1. For each $A \in \mathcal{Q}_1$,

$$\mathcal{E}\|AY - B\mu\|_1^2 \geqslant \mathcal{E}\|BPY - B\mu\|_1^2$$

where P is the orthogonal projection onto M. There is equality in this inequality iff $A = BP$.

Proof. The proof is very similar to the proof of Theorem 4.1. Define $C \in \mathcal{L}(V, W)$ by $C = A - BP$ and note that $C\mu = A\mu - BP\mu = B\mu - B\mu = 0$ since $A \in \mathcal{Q}_1$ and $P\mu = \mu$ for $\mu \in M$. Thus $CP = 0$, and this implies that $BP(Y - \mu)$ and $C(Y - \mu)$ are uncorrelated random vectors. Since these random vectors have zero means,

$$\mathcal{E}[BP(Y - \mu), C(Y - \mu)] = 0.$$

For $A \in \mathcal{A}_1$,

$$\mathcal{E}\|AY - B\mu\|_1^2 = \mathcal{E}\|BP(Y-\mu) + C(Y-\mu)\|_1^2$$
$$= \mathcal{E}\|BP(Y-\mu)\|_1^2 + \mathcal{E}\|C(Y-\mu)\|_1^2$$
$$\geqslant \mathcal{E}\|BP(Y-\mu)\|_1^2 = \mathcal{E}\|BPY - B\mu\|_1^2.$$

This establishes the desired inequality. There is equality in this inequality iff $\mathcal{E}\|C(Y-\mu)\|_1^2 = 0$. The argument used in Theorem 4.1 applies here so there is equality iff $C = A - BP = 0$. □

Proposition 4.1 makes precise the statement that the Gauss–Markov estimator of a linear transformation of μ is just the linear transformation applied to the Gauss–Markov estimator of μ. In other words, the Gauss–Markov estimator of $B\mu$ is $B\hat{\mu}$ where $B \in \mathcal{L}(V, W)$. There is one particular case of this that is especially interesting. When $W = R$, the real line, then a linear transformation on V to W is just a linear functional on V. By Proposition 1.10, every linear functional on V has the form (x_0, x) for some $x_0 \in V$. Thus the Gauss–Markov estimator of (x_0, μ) is just $(x_0, \hat{\mu}) = (x_0, PY) = (Px_0, Y)$. Further, a linear estimator of (x_0, μ), say (z, Y), is an unbiased estimator of (x_0, μ) iff $(z, \mu) = (x_0, \mu)$ for all $\mu \in M$. For any such vector z, Proposition 4.1 shows that

$$\operatorname{var}(z, Y) \geqslant \operatorname{var}(Px_0, Y).$$

Thus the minimum of var(z, Y), over the class of all z's such that (z, Y) is an unbiased estimator of (x_0, μ), is achieved uniquely for $z = Px_0$. In particular, if $x_0 \in M$, $z = x_0$ achieves the minimum variance.

In the definition of a linear model, $Y = \mu + \varepsilon$, no distributional assumptions concerning ε were made, other than the first and second moment assumptions $\mathcal{E}\varepsilon = 0$ and $\operatorname{Cov}(\varepsilon) = \sigma^2 I$. One of the attractive features of Proposition 4.1 is its validity under these relatively weak assumptions. However, very little can be said concerning the distribution of $\hat{\mu} = PY$ other than $\mathcal{E}\hat{\mu} = \mu$ and $\operatorname{Cov}(\hat{\mu}) = \sigma^2 P$. In the following example, some of the implications of assuming that ε has a normal distribution are discussed.

◆ **Example 4.2.** Consider the situation treated in Example 4.1. A coordinate random vector $Y \in R^n$ has a mean vector $\mu = Z\beta$ where Z is an $n \times k$ known matrix of rank k $(k \leqslant n)$ and $\beta \in R^k$ is a vector of unknown parameters. It is also assumed that $\operatorname{Cov}(Y) = \sigma^2 I_n$. The Gauss–Markov estimator of μ is $\hat{\mu} = Z(Z'Z)^{-1}Z'Y$. Since

$\beta = (Z'Z)^{-1}Z'\mu$, Proposition 4.1 shows that the Gauss–Markov estimator of β is $\hat{\beta} = (Z'Z)^{-1}Z'\hat{\mu} = (Z'Z)^{-1}Z'Y$. Now, add the assumption that Y has a normal distribution—that is, $\mathcal{L}(Y) = N(\mu, \sigma^2 I_n)$ where $\mu \in M$ and M is the range of Z. For this particular parametric model, we want to find a minimal sufficient statistic and the maximum likelihood estimators of the unknown parameters. The density function of Y, with respect to Lebesgue measure, is

$$p(y|\mu, \sigma^2) = (2\pi\sigma^2)^{-n/2}\exp\left[-\frac{1}{2\sigma^2}\|y - \mu\|^2\right]$$

where $y \in R^n$, $\mu \in M$, and $\sigma^2 > 0$. Let P denote the orthogonal projection onto M, so $Q \equiv I - P$ is the orthogonal projection onto $M^{\perp}$. Since $\|y - \mu\|^2 = \|Py - \mu\|^2 + \|Qy\|^2$, the density of y can be written

$$p(y|\mu, \sigma^2) = (2\pi\sigma^2)^{-n/2}\exp\left[-\frac{1}{2\sigma^2}\|Py - \mu\|^2 - \frac{1}{2\sigma^2}\|Qy\|^2\right].$$

This shows that the pair $\{Py, \|Qy\|^2\}$ is a sufficient statistic as the density is a function of the pair $\{Py, \|Qy\|^2\}$. The normality assumption implies that PY and QY are independent random vectors as they are uncorrelated (see Proposition 3.4). Thus PY and $\|QY\|^2$ are independent. That the pair $\{Py, \|Qy\|^2\}$ is minimal sufficient and complete follows from results about exponential families (see Lehmann 1959, Chapter 2). To find the maximum likelihood estimators of $\mu \in M$ and σ^2, the density $p(y|\mu, \sigma^2)$ must be maximized over all values of $\mu \in M$ and σ^2. For each fixed $\sigma^2 > 0$,

$$\begin{aligned} p(y|\mu, \sigma^2) &= (2\pi\sigma^2)^{-n/2}\exp\left[-\frac{1}{2\sigma^2}\|Py - \mu\|^2 - \frac{1}{2\sigma^2}\|Qy\|^2\right] \\ &\leqslant (2\pi\sigma^2)^{-n/2}\exp\left[-\frac{1}{2\sigma^2}\|Qy\|^2\right] \end{aligned}$$

with equality iff $\mu = Py$. Therefore, the Gauss–Markov estimator $\hat{\mu} = PY$ is the maximum likelihood estimator for μ. Of course, this also shows that $\hat{\beta} = (Z'Z)^{-1}Z'Y$ is the maximum likelihood estimator of β. To find the maximum likelihood estimator of σ^2, it remains to maximize

$$p(y|Py, \sigma^2) = (2\pi\sigma^2)^{-n/2}\exp\left[-\frac{1}{2\sigma^2}\|Qy\|^2\right].$$

An easy differentiation argument shows that $p(y|Py, \sigma^2)$ is maximized for σ^2 equal to $\|Qy\|^2/n$. Thus $\tilde{\sigma}^2 \equiv \|Qy\|^2/n$ is the maximum likelihood estimator of σ^2. From our previous observation, $\hat{\mu} = PY$ and $\tilde{\sigma}^2$ are independent. Since $\mathcal{L}(Y) = N(\mu, \sigma^2 I)$,

$$\mathcal{L}(\hat{\mu}) = \mathcal{L}(PY) = N(\mu, \sigma^2 P)$$

and

$$\mathcal{L}(\hat{\beta}) = \mathcal{L}\big((Z'Z)^{-1}Z'Y\big) = N\big(\beta, \sigma^2(Z'Z)^{-1}\big).$$

Also,

$$\mathcal{L}(QY) = N(0, \sigma^2 Q)$$

since $Q\mu = 0$ and $Q^2 = Q = Q'$. Hence from Proposition 3.7,

$$\mathcal{L}\left(\frac{\|QY\|^2}{\sigma^2}\right) = \chi^2_{n-k}$$

since Q is a rank $n - k$ orthogonal projection. Therefore,

$$\mathcal{E}\tilde{\sigma}^2 = \frac{n-k}{n}\sigma^2.$$

It is common practice to replace the estimator $\tilde{\sigma}^2$ by the unbiased estimator

$$\hat{\sigma}^2 \equiv \frac{\|QY\|^2}{n-k}.$$

It is clear that $\hat{\sigma}^2$ is distributed as the constant $\sigma^2/(n-k)$ times a χ^2_{n-k} random variable. ◆

The final result of this section shows that the unbiased estimator of σ^2, derived in the example above, is in fact unbiased without the normality assumption. Let $Y = \mu + \varepsilon$ be a random vector in V where $\mu \in M \subseteq V$, $\mathcal{E}\varepsilon = 0$, and $\text{Cov}(\varepsilon) = \sigma^2 I$. Given this linear model for Y, let P be the orthogonal projection onto M and set $Q = I - P$.

Proposition 4.2. Let $n = \dim V$, $k = \dim M$, and assume that $k < n$. Then the estimator

$$\hat{\sigma}^2 \equiv \frac{\|QY\|^2}{n-k}$$

is an unbiased estimator of σ^2.

Proof. The random vector QY has mean zero and $\text{Cov}(QY) = \sigma^2 Q$. By Proposition 2.21,

$$\mathcal{E}\|QY\|^2 = \langle I, \sigma^2 Q\rangle = \sigma^2\langle I, Q\rangle = \sigma^2(n-k).$$

The last equality follows from the observation that for any self-adjoint operator S, $\langle I, S\rangle$ is just the sum of the eigenvalues of S. Specializing this to the projection Q yields $\langle I, Q\rangle = n - k$. □

4.2. MORE ABOUT THE GAUSS–MARKOV THEOREM

The purpose of this section is to investigate to what extent Theorem 4.1 depends on the weak sphericity assumption. In this regard, Proposition 4.1 provides some information. If we take $W = V$ and $B = I$, then Proposition 4.1 implies that

$$\mathcal{E}\|AY - \mu\|_1^2 \geqslant \mathcal{E}\|PY - \mu\|_1^2$$

where $\|\cdot\|_1$ is the norm obtained from an inner product $[\cdot,\cdot]$. Thus the orthogonal projection P minimizes $\mathcal{E}\|AY - \mu\|_1^2$ over $A \in \mathcal{A}$ no matter what inner product is used to measure deviations of AY from μ. The key to the proof of Theorem 4.1 is the relationship

$$\mathcal{E}[P(Y-\mu), (A-P)(Y-\mu)] = 0.$$

This follows from the fact that the random vectors $P(Y-\mu)$ and $(A - P)(Y-\mu)$ are uncorrelated and

$$\mathcal{E}P(Y-\mu) = \mathcal{E}(A-P)(Y-\mu) = 0 \qquad \text{for } A \in \mathcal{A}.$$

This observation is central to the presentation below. The following alternative development of linear estimation theory provides the needed generality to apply the theory to multivariate linear models.

Consider a random vector Y with values in an inner product space $(V, (\cdot,\cdot))$ and assume that the mean vector of Y, say $\mu = \mathcal{E}Y$, lies in a known regression manifold $M \subseteq V$. For the moment, we suppose that $\text{Cov}(Y) = \Sigma$ where Σ is fixed and known (Σ is not necessarily nonsingular). As in the previous section, a linear estimator of μ, say AY, is unbiased iff

$$A \in \mathcal{A} \equiv \{A | A\mu = \mu, \mu \in M\}.$$

Given any inner product $[\cdot,\cdot]$ on V, the problem is to choose $A \in \mathcal{A}$ to minimize

$$\Psi(A) = \mathcal{E}\|AY - \mu\|_1^2 = \mathcal{E}[AY - \mu, AY - \mu]$$

where the expectation is computed under the assumption that $\mathcal{E}Y = \mu$ and $\text{Cov}(Y) = \Sigma$. Because of Proposition 4.1, it is reasonable to expect that the minimum of $\Psi(A)$ occurs at a point $P_0 \in \mathcal{Q}$ where P_0 is a projection onto M along some subspace N such that $M \cap N = \{0\}$ and $M + N = V$. Of course, N is the null space of P_0 and the pair M, N determines P_0. To find the appropriate subspace N, write $\Psi(A)$ as

$$\begin{aligned}\Psi(A) &= \mathcal{E}\|AY - \mu\|_1^2 \\ &= \mathcal{E}\|P_0(Y-\mu) + (A - P_0)(Y-\mu)\|_1^2 \\ &= \mathcal{E}\|P_0(Y-\mu)\|_1^2 + \mathcal{E}\|(A-P_0)(Y-\mu)\|_1^2 \\ &\quad + 2\mathcal{E}[P_0(Y-\mu), (A-P_0)(Y-\mu)].\end{aligned}$$

When the third term in the final expression for $\Psi(A)$ is zero, then P_0 minimizes $\Psi(A)$. If $P_0(Y - \mu)$ and $(A - P_0)(Y - \mu)$ are uncorrelated, the third term will be zero (shown below), so the proper choice of P_0, and hence N, will be to make $P_0(Y - \mu)$ and $(A - P_0)(Y - \mu)$ uncorrelated. Setting $C = A - P_0$, it follows that $\mathfrak{N}(C) \supseteq M$. The absence of correlation between $P_0(Y - \mu)$ and $C(Y - \mu)$ is equivalent to the condition

$$P_0 \Sigma C' = 0.$$

Here, C' is the adjoint of C relative to the initial inner product $(\cdot, \cdot)$ on V. Since $\mathfrak{N}(C) \supseteq M$, we have

$$\mathfrak{R}(C') = (\mathfrak{N}(C))^{\perp} \subseteq M^{\perp}$$

and

$$\mathfrak{R}(\Sigma C') \subseteq \Sigma(M^{\perp}).$$

The symbol $\perp$ refers to the inner product $(\cdot, \cdot)$. Therefore, if the null space of P_0, namely N, is chosen so that $N \supseteq \Sigma(M^{\perp})$, then $P_0 \Sigma C' = 0$ and P_0 minimizes $\Psi(A)$. Now, it remains to clean up the technical details of the above argument. Obviously, the subspace $\Sigma(M^{\perp})$ is going to play a role in what follows.

First, a couple of preliminary results.

Proposition 4.3. Suppose $\Sigma = \text{Cov}(Y)$ in $(V, (\cdot, \cdot))$ and M is a linear subspace of V. Then:

(i) $\Sigma(M^\perp) \cap M = \{0\}$.
(ii) The subspace $\Sigma(M^\perp)$ does not depend on the inner product on V.

Proof. To prove (i), recall that the null space of Σ is

$$\{x|(x, \Sigma x) = 0\}$$

since Σ is positive semidefinite. If $u \in \Sigma(M^\perp) \cap M$, then $u = \Sigma u_1$ for some $u_1 \in M^\perp$. Since $\Sigma u_1 \in M$, $(u_1, \Sigma u_1) = 0$ so $u = \Sigma u_1 = 0$. Thus (i) holds. For (ii), let $[\cdot, \cdot]$ be any other inner product on V. Then

$$[x, y] = (x, A_0 y)$$

for some positive definite linear transformation A_0. The covariance transformation of Y with respect to the inner product $[\cdot, \cdot]$ is ΣA_0 (see Proposition 2.5). Further, the orthogonal complement of M relative to the inner product $[\cdot, \cdot]$ is

$$\{y|[x, y] = 0 \text{ for all } x \in M\} = \{y|(x, A_0 y) = 0 \text{ for all } x \in M\}$$

$$= \{A_0^{-1}u|(x, u) = 0 \text{ for all } x \in M\} = A_0^{-1}(M^\perp).$$

Thus $\Sigma(M^\perp) = (\Sigma A_0)(A_0^{-1}(M^\perp))$. Therefore, the image of the orthogonal complement of M under the covariance transformation of Y is the same no matter what inner product is used on V. □

Proposition 4.4. Suppose X_1 and X_2 are random vectors with values in $(V, (\cdot, \cdot))$. If X_1 and X_2 are uncorrelated and $\mathcal{E}X_2 = 0$, then

$$\mathcal{E}f[X_1, X_2] = 0$$

for every bilinear function f defined on $V \times V$.

Proof. Since X_1 and X_2 are uncorrelated and X_2 has mean zero, for $x_1, x_2 \in V$, we have

$$0 = \text{cov}\{(x_1, X_1), (x_2, X_2)\} = \mathcal{E}(x_1, X_1)(x_2, X_2) - \mathcal{E}(x_1, X_1)\mathcal{E}(x_2, X_2)$$

$$= \mathcal{E}(x_1, X_1)(x_2, X_2).$$

However, every bilinear form f on $(V,(\cdot,\cdot))$ is given by

$$f[u_1, u_2] = (u_1, Bu_2)$$

where $B \in \mathcal{L}(V, V)$. Also, every B can be written as

$$B = \sum_i \sum_j b_{ij} y_i \square y_j$$

where $y_1, \ldots, y_n$ is a basis for V. Therefore,

$$\mathcal{E} f[X_1, X_2] = \mathcal{E} \sum \sum b_{ij} (X_1, y_i \square y_j X_2) = \sum \sum b_{ij} \mathcal{E} (y_i, X_1)(y_j, X_2) = 0.$$

□

We are now in a position to generalize Theorem 4.1. To review the assumptions, Y is a random vector in $(V,(\cdot,\cdot))$ with $\mathcal{E}Y = \mu \in M$ and $\text{Cov}(Y) = \Sigma$. Here, M is a known subspace of V and Σ is the covariance of Y relative to the given inner product $(\cdot,\cdot)$. Let $[\cdot,\cdot]$ be another product on V and set

$$\Psi(A) = \mathcal{E}\|AY - \mu\|_1^2$$

for $A \in \mathcal{A}$, where $\|\cdot\|_1$ is the norm defined by $[\cdot,\cdot]$.

Theorem 4.2. Let N be any subspace of V that is complementary to M and contains the subspace $\Sigma(M^\perp)$. Here $M^\perp$ is the orthogonal complement of M relative to $(\cdot,\cdot)$. Let P_0 be the projection onto M along N. Then

$$\Psi(A) \geqslant \Psi(P_0) \quad \text{for } A \in \mathcal{A}. \tag{4.1}$$

If Σ is nonsingular, define a new inner product $(\cdot,\cdot)_\Sigma$ by

$$(x, y)_\Sigma \equiv (x, \Sigma^{-1} y).$$

Then P_0 is the unique element of $\mathcal{A}$ that minimizes $\Psi(A)$. Further, P_0 is the orthogonal projection, relative to the inner product $(\cdot,\cdot)_\Sigma$, onto M.

Proof. The existence of a subspace $N \supseteq \Sigma(M^\perp)$, which is complementary to M, is guaranteed by Proposition 4.3. Let $C \in \mathcal{L}(V, V)$ be such that $M \subseteq \mathfrak{N}(C)$. Therefore,

$$\mathfrak{R}(C') = (\mathfrak{N}(C))^\perp \subseteq M^\perp$$

so

$$\mathfrak{R}(\Sigma C') \subseteq \Sigma(M^{\perp}).$$

This implies that

$$P_0 \Sigma C' = 0$$

since $\mathfrak{N}(P_0) = N \supseteq \Sigma(M^{\perp})$. However, the condition $P_0\Sigma C' = 0$ is equivalent to the condition that $P_0(Y - \mu)$ and $C(Y - \mu)$ are uncorrelated.

With these preliminaries out of the way, consider $A \in \mathcal{Q}$ and let $C = A - P_0$ so $\mathfrak{N}(C) \supseteq M$. Thus

$$\begin{aligned}\Psi(A) &= \mathcal{E}\|A(Y-\mu)\|_1^2 = \mathcal{E}\|P_0(Y-\mu) + C(Y-\mu)\|_1^2 \\ &= \mathcal{E}\|P_0(Y-\mu)\|_1^2 + \mathcal{E}\|C(Y-\mu)\|_1^2 + 2\mathcal{E}\left[P_0(Y-\mu), C(Y-\mu)\right] \\ &= \mathcal{E}\|P_0(Y-\mu)\|_1^2 + \mathcal{E}\|C(Y-\mu)\|_1^2.\end{aligned}$$

The last equality follows by applying Proposition 4.4 to $P_0(Y - \mu)$ and $C(Y - \mu)$. Therefore,

$$\Psi(A) = \Psi(P_0) + \mathcal{E}\|C(Y-\mu)\|_1^2$$

so P_0 minimizes Ψ over $A \in \mathcal{Q}$.

Now, assume that Σ is nonsingular. Then the subspace N is uniquely defined ($N = \Sigma(M^{\perp})$) since $\dim(\Sigma(M^{\perp})) = \dim(M^{\perp})$ and $M + \Sigma(M^{\perp}) = V$. Therefore, P_0 is uniquely defined as its range and null space have been specified. To show that P_0 uniquely minimizes Ψ, for $A \in \mathcal{Q}$, we have

$$\Psi(A) = \Psi(P_0) + \mathcal{E}\|C(Y-\mu)\|_1^2$$

where $C = A - P_0$. Thus $\Psi(A) > \Psi(P_0)$ with equality iff

$$\mathcal{E}\|C(Y-\mu)\|_1^2 = 0.$$

This expectation can be zero iff $C(Y - \mu) = 0$ (a.e.) and this happens iff the covariance transformation of $C(Y - \mu)$ is zero in some (and hence every) inner product. But in the inner product $(\cdot, \cdot)$,

$$\operatorname{Cov}(C(Y-\mu)) = C\Sigma C'$$

and this is zero iff $C = 0$ as Σ is nonsingular. Therefore, P_0 is the unique minimizer of Ψ. For the last assertion, let N_1 be the orthogonal complement of M relative to the inner product $(\cdot, \cdot)_\Sigma$. Then,

$$N_1 = \{y|(x, y)_\Sigma = 0 \text{ for all } x \in M\} = \{y|(x, \Sigma^{-1}y) = 0 \text{ for all } x \in M\}$$

$$= \{\Sigma y|(x, y) = 0 \text{ for all } x \in M\} = \Sigma(M^\perp).$$

Since $\mathfrak{N}(P_0) = \Sigma(M^\perp)$, it follows that P_0 is the orthogonal projection onto M relative to $(\cdot, \cdot)_\Sigma$. □

In all of the applications of Theorem 4.2 in this book, the covariance of Y is nonsingular. Thus the projection P_0 is unique and $\hat{\mu} = P_0 Y$ is called the Gauss–Markov estimator of $\mu \in M$. In the context of Theorem 4.2, if $\text{Cov}(Y) = \sigma^2\Sigma$ where Σ is known and nonsingular and $\sigma^2 > 0$ is unknown, then $P_0 Y$ is still the Gauss–Markov estimator for $\mu \in M$ since $(\sigma^2\Sigma)(M^\perp) = \Sigma(M^\perp)$ for each $\sigma^2 > 0$. That is, the presence of an unknown scale parameter σ^2 does not affect the projection P_0. Thus P_0 still minimizes Ψ for each fixed $\sigma^2 > 0$.

Consider a random vector Y taking values in $(V, (\cdot, \cdot))$ with $\mathcal{E}Y = \mu \in M$ and

$$\text{Cov}(Y) = \sigma^2\Sigma_1, \qquad \sigma^2 > 0.$$

Here, Σ_1 is assumed known and positive definite while $\sigma^2 > 0$ is unknown. Theorem 4.2 implies that the Gauss–Markov estimator of μ is $\hat{\mu} = P_0 Y$ where P_0 is the projection onto M along $\Sigma_1(M^\perp)$. Recall that the least-squares estimator of μ is PY where P is the orthogonal projection onto M in the given inner product, that is, P is the projection onto M along $M^\perp$.

Proposition 4.5. The Gauss–Markov and least-squares estimators of μ are the same iff $\Sigma_1(M) \subseteq M$.

Proof. Since P_0 and P are both projections onto M, $P_0 Y = PY$ iff both P_0 and P have the same null spaces—that is, the Gauss–Markov and least-squares estimators are the same iff

$$\Sigma_1(M^\perp) = M^\perp.$$

Since Σ_1 is nonsingular and self-adjoint, this condition is equivalent to the condition $\Sigma_1(M) \subseteq M$. □

The above result shows that if $\Sigma_1(M) \subseteq M$, we are free to compute either P or P_0 to find $\hat{\mu}$. The implications of this observation become clearer in the next section.

4.3. GENERALIZED LINEAR MODELS

First, consider the linear model introduced in Section 4.2. The random vector Y in $(V, (\cdot, \cdot))$ has a mean vector $\mu \in M$ where M is a subspace of V and $\text{Cov}(Y) = \sigma^2\Sigma_1$. Here, Σ_1 is a fixed positive definite linear transformation and $\sigma^2 > 0$. The essential features of this linear model are: (i) the mean vector of Y is assumed to be an element of a known subspace M and (ii) the covariance of Y is an element of the set $\{\sigma^2\Sigma_1 | \sigma^2 > 0\}$. The assumption concerning the mean vector of Y is not especially restrictive since no special assumptions have been made about the subspace M. However, the covariance structure of Y is quite restricted. The set $\{\sigma^2\Sigma_1 | \sigma^2 > 0\}$ is an open half line from $0 \in \mathcal{L}(V, V)$ through the point $\Sigma_1 \in \mathcal{L}(V, V)$ so the set of the possible covariances for Y is a one-dimensional set. It is this assumption concerning the covariance of Y that we want to modify so that linear models become general enough to include certain models in multivariate analysis. In particular, we would like to discuss Example 3.2 within the framework of linear models.

Now, let M be a fixed subspace of $(V, (\cdot, \cdot))$ and let γ be an arbitrary set of positive definite linear transformations on V to V. We say that $\{M, \gamma\}$ is the *parameter set* of a linear model for Y if $\mathcal{E}Y = \mu \in M$ and $\text{Cov}(Y) \in \gamma$. For a general parameter set $\{M, \gamma\}$, not much can be said about a linear model for Y. In order to restrict the class of parameter sets under consideration, we now turn to the question of existence of Gauss–Markov estimators (to be defined below) for μ. As in Section 4.1, let

$$\mathcal{A} = \{A | A \in \mathcal{L}(V, V), A\mu = \mu \text{ for } \mu \in M\}.$$

Thus a linear transformation of Y is an unbiased estimator of $\mu \in M$ iff it has the form AY for $A \in \mathcal{A}$. The following definition is motivated by Theorem 4.2.

Definition 4.2. Let $\{M, \gamma\}$ be the parameter set of a linear model for Y. For $A_0 \in \mathcal{A}$, A_0Y is a Gauss–Markov estimator of μ iff

$$\mathcal{E}_\Sigma\|AY - \mu\|^2 \geqslant \mathcal{E}_\Sigma\|A_0Y - \mu\|^2$$

for all $A \in \mathcal{A}$ and $\Sigma \in \gamma$. The subscript Σ on the expectation means that the expectation is computed when $\text{Cov}(Y) = \Sigma$.

When $\gamma = \{\sigma^2 I | \sigma^2 > 0\}$, Theorem 4.1 establishes the existence and uniqueness of a Gauss–Markov estimator for μ. More generally, when $\gamma = \{\sigma^2\Sigma_1 | \sigma^2 > 0\}$, Theorem 4.2 shows that the Gauss–Markov estimator for μ is $P_1 Y$ where P_1 is the orthogonal projection onto M relative to the inner product $(\cdot, \cdot)_1$ given by

$$(x, y)_1 \equiv (x, \Sigma_1^{-1} y), \qquad x, y \in V.$$

The problem of the existence of a Gauss–Markov estimator for general γ is taken up in the next paragraph.

Suppose that $\{M, \gamma\}$ is the parameter set for a linear model for Y. Consider a fixed element $\Sigma_1 \in \gamma$, and let $(\cdot, \cdot)_1$ be the inner product on V defined by

$$(x, y)_1 \equiv (x, \Sigma_1^{-1} y), \qquad x, y \in V.$$

As asserted in Theorem 4.2, the unique element in $\mathcal{Q}$ that minimizes $\mathcal{E}_{\Sigma_1}\|AY - \mu\|^2$ is P_1—the orthogonal projection onto M relative to $(\cdot, \cdot)_1$. Thus if a Gauss–Markov estimator $A_0 Y$ exists according to Definition 4.2, A_0 must be P_1. However, exactly the same argument applies for $\Sigma_2 \in \gamma$, so A_0 must be P_2—the orthogonal projection onto M relative to the inner product defined by Σ_2. These two projections are the same iff $\Sigma_1(M^\perp) = \Sigma_2(M^\perp)$—see Theorem 4.2. Since Σ_1 and Σ_2 were arbitrary elements of γ, the conclusion is that a Gauss–Markov estimator can exist iff $\Sigma_1(M^\perp) = \Sigma_2(M^\perp)$ for all $\Sigma_1, \Sigma_2 \in \gamma$. Summarizing this leads to the following.

Proposition 4.6. Suppose that $\{M, \gamma\}$ is the parameter set of a linear model for Y in $(V, (\cdot, \cdot))$. Let Σ_1 be a fixed element of γ. A Gauss–Markov estimator of μ exists iff

$$\Sigma(M^\perp) = \Sigma_1(M^\perp) \quad \text{for all } \Sigma \in \gamma.$$

When a Gauss–Markov estimator of μ exists, it is $\hat{\mu} = PY$ where P is the orthogonal projection onto M relative to any inner product $[\cdot, \cdot]$ given by $[x, y] = (x, \Sigma^{-1} y)$ for some $\Sigma \in \gamma$.

Proof. It has been argued that a Gauss–Markov estimator for μ can exist iff $\Sigma_1(M^\perp) = \Sigma_2(M^\perp)$ for all $\Sigma_1, \Sigma_2 \in \gamma$. This is clearly equivalent to $\Sigma(M^\perp) = \Sigma_1(M^\perp)$ for all $\Sigma \in M$. The second assertion follows from the observation that when $\Sigma(M^\perp) = \Sigma_1(M^\perp)$, then all the projections onto M, relative to the inner products determined by elements of γ, are the same. That $\hat{\mu} = PY$ is a consequence of Theorem 4.2. □

An interesting special case of Proposition 4.6 occurs when $I \in \gamma$. In this case, choose $\Sigma_1 = I$ so a Gauss–Markov estimator exists iff $\Sigma(M^\perp) = M^\perp$ for all $\Sigma \in \gamma$. This is clearly equivalent to $\Sigma(M) = M$ for all $\Sigma \in \gamma$, which is equivalent to the condition

$$\Sigma(M) \subseteq M \quad \text{for all } \Sigma \in \gamma$$

since each $\Sigma \in \gamma$ is nonsingular. It is this condition that is verified in the examples that follow.

◆ **Example 4.3.** As motivation for the discussion of the general multivariate linear model, we first consider the multivariate version of the k-sample situation. Suppose X_{ij}'s, $j = 1,\ldots, n_i$ and $i = 1,\ldots, k$, are random vectors in R^p. It is assumed that $\mathcal{E}X_{ij} = \mu_i$, $\text{Cov}(X_{ij}) = \Sigma$, and different random vectors are uncorrelated. Form the random matrix X whose first n_1 rows are $X'_{1j}, j = 1,\ldots, n_1$, the next n_2 rows of X are X'_{2j}, $j = 1,\ldots, n_2$, and so on. Then X is a random vector in $(\mathcal{L}_{p,n}, \langle \cdot , \cdot \rangle)$ where $n = \Sigma_1^k n_i$. It was argued in the discussion following Proposition 2.18 that

$$\text{Cov}(X) = I_n \otimes \Sigma$$

relative to the inner product $\langle \cdot , \cdot \rangle$ on $\mathcal{L}_{p,n}$. The mean of X, say $\mu = \mathcal{E}X$, is an $n \times p$ matrix whose first n_1 rows are all μ'_1, whose next n_2 rows are all μ'_2, and so on. Let B be the $k \times p$ matrix with rows $\mu'_1,\ldots, \mu'_k$. Thus the mean of X can be written $\mu = ZB$ where Z is an $n \times k$ matrix with the following structure: the first column of Z consists of n_1 ones followed by $n - n_1$ zeroes, the second column of Z consists of n_1 zeroes followed by n_2 ones followed by $n - n_1 - n_2$ zeroes, and so on. Define the linear subspace M of $\mathcal{L}_{p,n}$ by

$$M = \{\mu | \mu = ZB, B \in \mathcal{L}_{p,k}\}$$

so M is the range of $Z \otimes I_p$ as a linear transformation on $\mathcal{L}_{p,k}$ to $\mathcal{L}_{p,n}$. Further, set

$$\gamma = \{I_n \otimes \Sigma | \Sigma \in \mathcal{L}_{p,p}, \Sigma \text{ positive definite}\}$$

and note that γ is a set of positive definite linear transformations on $\mathcal{L}_{p,n}$ to $\mathcal{L}_{p,n}$. Therefore, $\mathcal{E}X \in M$ and $\text{Cov}(X) \in \gamma$, and $\{M, \gamma\}$ is a parameter set for a linear model for X. Since $I_n \otimes I_p$ is the identity linear transformation on $\mathcal{L}_{p,n}$ and $I_n \otimes I_p \in \gamma$, to show that a Gauss–Markov estimator for $\mu \in M$ exists, it is sufficient to verify that, if $x \in M$, then $(I_n \otimes \Sigma)x \in M$. For $x \in M$, $x = ZB$ for some $B \in \mathcal{L}_{p,k}$. Therefore,

$$(I_n \otimes \Sigma)(ZB) = ZB\Sigma = (Z \otimes I_p)(B\Sigma),$$

which is an element of M. Thus M is invariant under each element of γ so a Gauss–Markov estimator for μ exists. Since the identity is an element of γ, the Gauss–Markov estimator is just the orthogonal projection of X on M relative to the given inner product $\langle \cdot , \cdot \rangle$. To find this projection, we argue as in Example 4.1. The regression subspace M is the range of $Z \otimes I_p$ and, clearly, Z has rank k. Let

$$\begin{aligned} P &= (Z \otimes I_p)\left[(Z \otimes I_p)'(Z \otimes I_p)\right]^{-1}(Z \otimes I_p)' \\ &= (Z \otimes I_p)\left[(Z'Z) \otimes I_p\right]^{-1}(Z' \otimes I_p) = Z(Z'Z)^{-1}Z' \otimes I_p, \end{aligned}$$

which is an orthogonal projection; see Proposition 1.28. To verify that P is the orthogonal projection onto M, it suffices to show that the range of P is M. For any $x \in \mathcal{L}_{p,n}$,

$$Px = \left(Z(Z'Z)^{-1}Z' \otimes I_p\right)x = (Z \otimes I_p)\left[(Z'Z)^{-1}Z'x\right],$$

which is an element of M since $(Z'Z)^{-1}Z'x \in \mathcal{L}_{p,k}$. However, if $x \in M$, then $x = ZB$ and $Px = P(ZB) = ZB$—that is, P is the identity on M. Hence, the range of P is M and the Gauss–Markov estimator of μ is

$$\hat{\mu} = PX = Z(Z'Z)^{-1}Z'X.$$

Since $\mu = ZB$,

$$B = (Z'Z)^{-1}Z'\mu = \left((Z'Z)^{-1}Z' \otimes I_p\right)\mu$$

and, by Proposition 4.1,

$$\hat{B} = \left((Z'Z)^{-1}Z' \otimes I_p\right)\hat{\mu} = (Z'Z)^{-1}Z'X$$

is the Gauss–Markov estimator of the matrix B. Further, $\mathcal{E}(\hat{B}) = B$

and

$$\begin{aligned}\operatorname{Cov}(\hat{B}) &= \operatorname{Cov}\left[\left((Z'Z)^{-1}Z' \otimes I_p\right)X\right] \\ &= \left((Z'Z)^{-1}Z' \otimes I_p\right)(I_n \otimes \Sigma)\left(Z(Z'Z)^{-1} \otimes I_p\right) \\ &= (Z'Z)^{-1} \otimes \Sigma.\end{aligned}$$

For the particular matrix Z, $Z'Z$ is a $k \times k$ diagonal matrix with diagonal entries $n_1, \ldots, n_k$ so $(Z'Z)^{-1}$ is diagonal with diagonal elements $n_1^{-1}, \ldots, n_k^{-1}$. A bit of calculation shows that the matrix $\hat{B} = (Z'Z)^{-1}Z'X$ has rows $\overline{X}_1', \ldots, \overline{X}_k'$ where

$$\overline{X}_i = \frac{1}{n_i} \sum_{j=1}^{n_i} X_{ij}$$

is the sample mean in the ith sample. Thus the Gauss–Markov estimator of the ith mean μ_i is $\overline{X}_i$, $i = 1, \ldots, k$. ◆

It is fairly clear that the explicit form of the matrix Z in the previous example did not play a role in proving that a Gauss–Markov estimator for the mean vector exists. This observation leads quite naturally to what is usually called the general linear model of multivariate analysis. After introducing this model in the next example, we then discuss the implications of adding the assumption of normality.

◆ **Example 4.4 (Multivariate General Linear Model).** As in Example 4.3, consider a random matrix X in $(\mathcal{L}_{p,n}, \langle \cdot, \cdot \rangle)$ and assume that (i) $\mathcal{E}X = ZB$ where Z is a known $n \times k$ matrix of rank k and B is a $k \times p$ matrix of parameters, (ii) $\operatorname{Cov}(X) = I_n \otimes \Sigma$ where Σ is a $p \times p$ positive definite matrix—that is, the rows of X are uncorrelated and each row of X has covariance matrix Σ. It is clear we have simply abstracted the essential features of the linear model in Example 4.3 into assumptions for the linear model of this example. The similarity between the current example and Example 4.1 should also be noted. Each component of the observation vector in Example 4.1 has become a vector, the parameter vector has become a matrix, and the rows of the observation matrix are still uncorrelated. Of course, the rows of the observation vector in Example 4.1 are just scalars. For the example at hand, it is clear that

$$M \equiv \{\mu | \mu = ZB,\, B \in \mathcal{L}_{p,k}\}$$

is a subspace of $\mathcal{L}_{p,n}$ and is the range of $Z \otimes I_p$. Setting

$$\gamma = \{I_n \otimes \Sigma | \Sigma \text{ is a } p \times p \text{ positive definite matrix}\},$$

$\{M, \gamma\}$ is the parameter set of a linear model for X. More specifically, the linear model for X is that $\mathcal{E}X = \mu \in M$ and $\text{Cov}(X) \in \gamma$. Just as in Example 4.3, M is invariant under each element of γ so a Gauss–Markov estimator of $\mu = \mathcal{E}X$ exists and is PX where

$$P \equiv Z(Z'Z)^{-1}Z' \otimes I_p$$

is the orthogonal projection onto M relative to $\langle \cdot, \cdot \rangle$. Mimicking the argument given in Example 4.3 yields

$$\hat{B} = (Z'Z)^{-1}Z'X = \left((Z'Z)^{-1}Z' \otimes I_p\right)X$$

and

$$\text{Cov}(\hat{B}) = (Z'Z)^{-1} \otimes \Sigma.$$

In addition to the linear model assumptions for X, we now assume that $\mathcal{L}(X) = N(ZB, I_n \otimes \Sigma)$ so X has a normal distribution in $(\mathcal{L}_{p,n}, \langle \cdot, \cdot \rangle)$. As in Example 4.2, a discussion of sufficient statistics and maximum likelihood estimators follows. The density function of X with respect to Lebesgue measure is

$$p(x|\mu, \Sigma) = (2\pi)^{-np/2}|\Sigma|^{-n/2}$$
$$\times \exp\left[-\tfrac{1}{2}\langle x - \mu, \left(I_n \otimes \Sigma^{-1}\right)(x - \mu)\rangle\right],$$

as discussed in Chapter 3. Let $P_0 = Z(Z'Z)^{-1}Z'$ and $Q_0 = I - P_0$ so $P = P_0 \otimes I_p$ is the orthogonal projection onto M and $Q \equiv Q_0 \otimes I_p$ is the orthogonal projection onto $M^{\perp}$. Note that both P and Q commute with $I_n \otimes \Sigma$ for any Σ. Since $\mu \in M$, we have

$$\langle x - \mu, \left(I_n \otimes \Sigma^{-1}\right)(x - \mu)\rangle$$
$$= \langle P(x - \mu) + Qx, \left(I_n \otimes \Sigma^{-1}\right)(P(x - \mu) + Qx)\rangle$$
$$= \langle P(x - \mu), \left(I_n \otimes \Sigma^{-1}\right)P(x - \mu)\rangle + \langle Qx, \left(I_n \otimes \Sigma^{-1}\right)Qx\rangle$$

because $\langle Qx, (I_n \otimes \Sigma^{-1})P(x - \mu)\rangle = \langle x, Q(I_n \otimes \Sigma^{-1})P(x - \mu)\rangle$

$= 0$ since $Q(I_n \otimes \Sigma^{-1})P = QP(I_n \otimes \Sigma^{-1}) = 0$. However,

$$\begin{aligned}\langle Qx, (I_n \otimes \Sigma^{-1})Qx\rangle &= \langle x, Q(I_n \otimes \Sigma^{-1})Qx\rangle = \langle x, Q(I_n \otimes \Sigma^{-1})x\rangle \\ &= \langle x, (Q_0 \otimes \Sigma^{-1})x\rangle = \langle x, Q_0 x\Sigma^{-1}\rangle \\ &= \operatorname{tr}(x\Sigma^{-1}x'Q_0) = \operatorname{tr}(x'Q_0 x\Sigma^{-1}).\end{aligned}$$

Thus

$$\begin{aligned}\langle x - \mu, (I_n \otimes \Sigma^{-1})(x - \mu)\rangle \\ = \langle Px - \mu, (I_n \otimes \Sigma^{-1})(Px - \mu)\rangle + \operatorname{tr}(x'Q_0 x\Sigma^{-1}).\end{aligned}$$

Therefore, the density $p(x|\mu, \Sigma)$ is a function of the pair $\{Px, x'Q_0x\}$ so the pair $\{Px, x'Q_0x\}$ is sufficient. That this pair is minimal sufficient and complete for the parametric family $\{p(\cdot|\mu, \Sigma); \mu \in M$, Σ positive definite$\}$ follows from exponential family theory. Since $P(I_n \otimes \Sigma)Q = PQ(I_n \otimes \Sigma) = 0$, the random vectors PX and QX are independent. Also, $X'Q_0X = (QX)'(QX)$ so the random vectors PX and $X'Q_0X$ are independent. In other words, $\{PX, X'Q_0X\}$ is a sufficient statistic and PX and $X'Q_0X$ are independent. To derive the maximum likelihood estimator of $\mu \in M$, fix Σ. Then

$$\begin{aligned}p(x|\mu, \Sigma) &= (2\pi)^{-np/2}|\Sigma|^{-n/2} \\ &\quad\times \exp\left[-\tfrac{1}{2}\langle Px - \mu, (I_n \otimes \Sigma^{-1})(Px - \mu)\rangle - \tfrac{1}{2}\operatorname{tr} x'Q_0x\Sigma^{-1}\right] \\ &\leqslant (2\pi)^{-np/2}|\Sigma|^{-n/2}\exp\left[-\tfrac{1}{2}\operatorname{tr} x'Q_0x\Sigma^{-1}\right]\end{aligned}$$

with equality iff $\mu = Px$. Thus the maximum likelihood estimator of μ is $\hat{\mu} = PX$, which is also the Gauss–Markov and least-squares estimator of μ. It follows immediately that

$$\hat{B} = (Z'Z)Z'X$$

is the maximum likelihood estimator of B, and

$$\mathcal{L}(\hat{B}) = N\left(B, (Z'Z)^{-1} \otimes \Sigma\right).$$

To find the maximum likelihood estimator of Σ, the function

$$p(x|\hat{\mu}, \Sigma) = (2\pi)^{-np/2}|\Sigma|^{-n/2}\exp\left[-\tfrac{1}{2}\operatorname{tr} x'Q_0x\Sigma^{-1}\right]$$

must be maximized over all $p \times p$ positive definite matrices Σ. When $x'Q_0x$ is positive definite, this maximum occurs uniquely at

$$\hat{\Sigma} \equiv \frac{1}{n}x'Q_0x$$

so the maximum likelihood estimator of Σ is stochastically independent of $\hat{\mu}$. A proof that $\hat{\Sigma}$ is the maximum likelihood estimator of Σ and a derivation of the distribution of $\hat{\Sigma}$ is deferred until later. ◆

The principal result of this chapter, Proposition 4.6, gives necessary and sufficient conditions on the parameter set $\{M, \gamma\}$ of a linear model in order that the Gauss–Markov estimator of $\mu \in M$ exists. Many of the classical parametric models in multivariate analysis are in fact linear models with a parameter set $\{M, \gamma\}$ so that there is a Gauss–Markov estimator for $\mu \in M$. For such models, the additional assumption of normality implies that $\hat{\mu}$ is also the maximum likelihood estimator of μ, and the estimation of μ is relatively easy if we are satisfied with the maximum likelihood estimator. For the time being, let us agree that the problem of estimating μ has been solved in these models. However, very little has been said about the estimation of the covariance other than in Example 4.4. To be specific, assume $\mathcal{L}(X) = N(\mu, \Sigma)$ where $\mu \in M \subseteq (V, (\cdot, \cdot))$ and $\{M, \gamma\}$ is the parameter set of this linear model for x. Assume that $I \in \gamma$ and $\hat{\mu} = PX$ is the Gauss–Markov estimator for μ so $\Sigma M = M$ for all $\Sigma \in \gamma$. Here, P is the orthogonal projection onto M in the given inner product on V. It follows immediately from Proposition 4.6 that $\hat{\mu} = PX$ is also the maximum likelihood estimator of $\mu \in M$. Substituting $\hat{\mu}$ into the density of X yields

$$p(x|\hat{\mu}, \Sigma) = (2\pi)^{-n/2}|\Sigma|^{-1/2}\exp\left[-\tfrac{1}{2}(Qx, \Sigma^{-1}Qx)\right]$$

where $n = \dim V$ and $Q = I - P$ is the orthogonal projection onto $M^{\perp}$. Thus to find the maximum likelihood estimator of $\Sigma \in \gamma$, we must compute

$$\sup_{\Sigma \in \gamma} p(x|\hat{\mu}, \Sigma) \equiv p(x|\hat{\mu}, \hat{\Sigma});$$

assuming that the supremum is attained at a point $\hat{\Sigma} \in \gamma$. Although many examples of explicit sets γ are known where $\hat{\Sigma}$ is not too difficult to find, general conditions on γ that yield an explicit $\hat{\Sigma}$ are not available. This overview of the maximum likelihood estimation problem in linear models where Gauss–Markov estimators exists has been given to provide the reader with a general framework in which to view many of the estimation and testing problems to be discussed in later chapters.

PROBLEMS

1. Let Z be an $n \times k$ matrix (not necessarily of full rank) so Z defines a linear transformation on R^k to R^n. Let M be the range of Z and let $z_1,\ldots, z_k$ be the columns of Z.

(i) Show that $M = \text{span}\{z_1,\ldots, z_k\}$.

(ii) Show that $Z(Z'Z)^- Z'$ is the orthogonal projection onto M where $(Z'Z)^-$ is the generalized inverse of $Z'Z$.

2. Suppose $X_1,\ldots, X_n$ are i.i.d. from a density $p(x|\beta) = f(x - \beta)$ where f is a symmetric density on R^1 and $\int x^2 f(x)\,dx = 1$. Here, β is an unknown translation parameter. Let $X \in R^n$ have coordinates $X_1,\ldots, X_n$.

(i) Show that $\mathcal{L}(X) = \mathcal{L}(\beta e + \varepsilon)$ where $\varepsilon_1,\ldots, \varepsilon_n$ are i.i.d. with density f. Show that $\mathcal{E}X = \beta e$ and $\text{Cov}(X) = I_n$.

(ii) Based on (i), find the Gauss–Markov estimator of β.

(iii) Let U be the vector of order statistics for X ($U_1 < U_2 < \cdots < U_n$) so $\mathcal{L}(U) = \mathcal{L}(\beta e + \nu)$ where ν is the vector of order statistics of ε. Show that $\mathcal{E}(U) = \beta e + a_0$ where $a_0 = \mathcal{E}\nu$ is a known vector (f is assumed known), and $\text{Cov}(U) = \Sigma_0 \equiv \text{Cov}(\nu)$ where Σ_0 is also known. Thus $\mathcal{L}(U - a_0) = \mathcal{L}(\beta e + (\nu - a_0))$ where $\mathcal{E}(\nu - a_0) = 0$ and $\text{Cov}(\nu - a_0) = \Sigma_0$. Based on this linear model, find the Gauss–Markov estimator for β.

(iv) How do these two estimators of β compare?

3. Consider the linear model $Y = \mu + \varepsilon$ where $\mu \in M$, $\mathcal{E}\varepsilon = 0$, and $\text{Cov}(\varepsilon) = \sigma^2 I_n$. At times, a submodel of this model is of interest. In particular, assume $\mu \in \omega$ where ω is a linear subspace of M.

(i) Let $M - \omega = \{x | x \in M, x \perp \omega\}$. Show that $M - \omega = M \cap \omega^\perp$.

(ii) Show that $P_M - P_\omega$ is the orthogonal projection onto $M - \omega$ and verify that $\|(P_M - P_\omega)x\|^2 = \|P_M x\|^2 - \|P_\omega x\|^2$.

4. For this problem, we use the notation of Problem 1.15. Consider subspaces of R^{IJ} given by

$$M_0 = \{y | y_{ij} = y_{..} \quad \text{for all } i, j\}$$

$$M_1 = \{y | y_{ij} = y_{ik} \quad \text{for all } j, k;\ i = 1, \dots, I\}$$

$$M_2 = \{y | y_{ij} = y_{kj} \quad \text{for all } i, k;\ j = 1, \dots, J\}$$

(i) Show that $\mathfrak{R}(A) = M_0$, $\mathfrak{R}(B_1) = M_1 - M_0$, and $\mathfrak{R}(B_2) = M_2 - M_0$.

Let M_3 be the range of B_3.

(ii) Show that $R^{IJ} = M_0 \oplus (M_1 - M_0) \oplus (M_2 - M_0) \oplus M_3$.

(iii) Show that a vector μ is in $M = M_0 \oplus (M_1 - M_0) \oplus (M_2 - M_0)$ iff μ can be written as $\mu_{ij} = \alpha + \beta_i + \gamma_j$, $i = 1, \dots, I$, $j = 1, \dots, J$, where α, β_i, and γ_j are scalars that satisfy $\Sigma \beta_i = \Sigma \gamma_j = 0$.

5. (The $\mathfrak{F}$-test.) Most of the classical hypothesis testing problems in regression analysis or ANOVA can be described as follows. A linear model $Y = \mu + \varepsilon$, $\mu \in M$, $\mathcal{E}\varepsilon = 0$, and $\text{Cov}(\varepsilon) = \sigma^2 I$ is given in $(V, (\cdot, \cdot))$. A subspace ω of M $(\omega \neq M)$ is given and the problem is to test $H_0 : \mu \in \omega$ versus $H_1 : \mu \notin \omega$, $\mu \in M$. Assume that $\mathcal{L}(Y) = N(\mu, \sigma^2 I)$ in $(V, (\cdot, \cdot))$.

(i) Show that the likelihood ratio test of H_0 versus H_1 rejects for large values of $F = \|P_{M-\omega} Y\|^2 / \|Q_M Y\|^2$ where $Q_M = I - P_M$.

(ii) Under H_0, show that F is distributed as the ratio of two independent chi-squared variables.

6. In the notation of Problem 4, consider $Y \in R^{IJ}$ with $\mathcal{E}Y = \mu \in M$ (M is given in (iii) of Problem 4). Under the assumption of normality, use the results of Problem 5 to show that the $\mathfrak{F}$-test for testing $H_0 : \beta_1 = \beta_2 = \cdots = \beta_I$ rejects for large values of

$$\frac{J \Sigma_i (\bar{y}_{i.} - \bar{y}_{..})^2}{\Sigma_i \Sigma_j (y_{ij} - \bar{y}_{i.} - \bar{y}_{.j} + \bar{y}_{..})^2}.$$

Identify ω for this problem.

7. (The normal equations.) Suppose the elements of the regression subspace $M \subseteq R^n$ are given by $\mu = X\beta$ where X is $n \times k$ and $\beta \in R^k$. Given an observation vector y, the problem is to find $\hat{\mu} = P_M y$. The

equations (in β)

$$X'y = X'X\beta, \qquad \beta \in R^k \tag{4.2}$$

are often called the normal equations.

(i) Show that (4.2) always has a solution $b \in R^k$.

(ii) If b is any solution to (4.2), show that $Xb = P_M y$.

8. For $Y \in R^n$, assume $\mu = \mathcal{E}Y \in M$ and $\text{Cov}(Y) \in \gamma$ where $\gamma = \{\Sigma | \Sigma = \alpha P_e + \beta Q_e, \alpha > 0, \beta > 0\}$. As usual, e is the vector of ones, P_e is the orthogonal projection onto span$\{e\}$, and $Q_e = I - P_e$.

(i) If $e \in M$ or $e \in M^\perp$, show that the Gauss–Markov and least-squares estimators for μ are the same for each α and β.

(ii) If $e \notin M$ and $e \notin M^\perp$, show that there are values of α and β so that the least-squares and Gauss–Markov estimators of μ differ.

(iii) If $\mathcal{L}(Y) = N(\mu, \Sigma)$ with $\Sigma \in \gamma$ and $M \subseteq (\text{span}\{e\})^\perp$ ($M \neq (\text{span}\{e\})^\perp$), find the maximum likelihood estimates for μ, α, and β. What happens when $M = \text{span}\{e\}$?

9. In the linear model $Y = X\beta + \varepsilon$ on R^n with $X: n \times k$ of full rank, $\mathcal{E}\varepsilon = 0$, and $\text{Cov}(\varepsilon) = \sigma^2\Sigma_1$ (Σ_1 is positive definite and known), show that $\hat{\mu} = X(X'\Sigma_1^{-1}X)^{-1}X'\Sigma_1^{-1}Y$ and $\hat{\beta} = (X'\Sigma_1^{-1}X)^{-1}X'\Sigma_1^{-1}Y$.

10. (Invariance in the simple linear model.) In $(V, (\cdot, \cdot))$, suppose that $\{M, \gamma\}$ is the parameter set for a linear model for Y where $\gamma = \{\Sigma | \Sigma = \sigma^2 I, \sigma > 0\}$. Thus $\mathcal{E}Y = \mu \in M$ and $\text{Cov}(Y) \in \gamma$. This problem has to do with the invariance of this linear model under affine transformations:

(i) If $\Gamma \in \mathcal{O}(V)$ satisfies $\Gamma(M) \subseteq M$, show that $\Gamma'(M) \subseteq M$.

Let $\mathcal{O}_M(V)$ be those $\Gamma \in \mathcal{O}(V)$ that satisfy $\Gamma(M) \subseteq M$.

(ii) For $x_0 \in M$, $c > 0$, and $\Gamma \in \mathcal{O}_M(V)$, define the function (c, Γ, x_0) on V to V by $(c, \Gamma, x_0)y = c\Gamma y + x_0$. Show that this function is one-to-one and onto and find the inverse of this function. Show that this function maps M onto M.

(iii) Let $\tilde{Y} = (c, \Gamma, x_0)Y$. Show that $\mathcal{E}\tilde{Y} \in M$ and $\text{Cov}(\tilde{Y}) \in \gamma$. Thus (M, γ) is the parameter set for $\tilde{Y}$ and we say that the linear model for Y is invariant under the transformation (c, Γ, x_0).

Since $\mathcal{E}Y = \mu$, it follows that $\mathcal{E}\tilde{Y} = (c, \Gamma, x_0)\mu$ for $\mu \in M$. If $t(Y)$ (t maps V into M) is *any* point estimator for μ, then it seems plausible to use $t(\tilde{Y})$ as a point estimator for $(c, \Gamma, x_0)\mu = c\Gamma\mu + x_0$. Solving for μ, it then seems plausible to use $c^{-1}\Gamma'(t(\tilde{Y}) - x_0)$ as a point estimator for μ. Equating these estimators of μ leads to $t(Y) = c^{-1}\Gamma'(t(c\Gamma Y +$

$x_0) - x_0)$ or

(4.3) $$t(c\Gamma Y + x_0) = c\Gamma t(Y) + x_0.$$

An estimator that satisfies (4.3) for all $c > 0$, $\Gamma \in \mathcal{O}_M(V)$, and $x_0 \in M$ is called *equivariant*.

(iv) Show that $t_0(Y) = P_M Y$ is equivariant.

(v) Show that if t maps V into M and satisfies the equation $t(\Gamma Y + x_0) = \Gamma t(Y) + x_0$ for all $\Gamma \in \mathcal{O}_M(V)$ and $x_0 \in M$, then $t(Y) = P_M Y$.

11. Consider $U \in R^n$ and $V \in R^n$ and assume $\mathcal{L}(U) = N(Z_1\beta_1, \sigma_{11}I_n)$ and $\mathcal{L}(V) = N(Z_2\beta_2, \sigma_{22}I_n)$. Here, Z_i is $n \times k$ of rank k and $\beta_i \in R^k$ is an unknown vector of parameters, $i = 1, 2$. Also, $\sigma_{ii} > 0$ is unknown, $i = 1, 2$. Now, let $X = (UV) : n \times 2$ so $\mu = \mathcal{E}X$ has first column $Z_1\beta_1$ and second column $Z_2\beta_2$.

(i) When U and V are independent, then $\text{Cov}(X) = I_n \otimes A$ where

$$A = \begin{pmatrix} \sigma_{11} & 0 \\ 0 & \sigma_{22} \end{pmatrix}.$$

In this case, show that the Gauss–Markov and least-squares estimates for μ are the same. Further, show that the Gauss–Markov estimates for β_1 and β_2 are the same as what we obtain by treating the two regression problems separately.

(ii) Now, suppose $\text{Cov}(X) = I_n \otimes \Sigma$ where

$$\Sigma = \begin{pmatrix} \sigma_{11} & \sigma_{12} \\ \sigma_{12} & \sigma_{22} \end{pmatrix}$$

is positive definite and unknown. For general Z_1 and Z_2, show that the regression subspace of X is not invariant under all $I_n \otimes \Sigma$ so the Gauss–Markov and least-squares estimators are not the same in general. However, if $Z_1 = Z_2$, show that the results given in Example 4.4 apply directly.

(iii) If the column space of Z_1 equals the column space of Z_2, show that the Gauss–Markov and least-squares estimators of μ are the same for each $I_n \otimes \Sigma$.

NOTES AND REFERENCES

1. Scheffé (1959) contains a coordinate account of what might be called univariate linear model theory. The material in the first section here follows Kruskal (1961) most closely.

2. The result of Proposition 4.5 is due to Kruskal (1968).

3. Proposition 4.3 suggests that a theory of best linear unbiased estimation can be developed in vector spaces without inner products (i.e., dual spaces are not identified with the vector space via the inner product). For a version of such a theory, see Eaton (1978).

4. The arguments used in Section 4.3 were used in Eaton (1970) to help answer the following question. Given $X \in \mathcal{L}_{p,n}$ with $\text{Cov}(X) = I_n \otimes \Sigma$ where Σ is unknown but positive definite, for what subspaces M does there exist a Gauss–Markov estimator for $\mu \in M$? In other words, with γ as in Example 4.4, for what M's can the parameter set $\{M, \gamma\}$ admit a Gauss–Markov estimator? The answer to this question is that M must have the form of the subspaces considered in Example 4.4. Further details and other examples can be found in Eaton (1970).

CHAPTER 5

Matrix Factorizations and Jacobians

This chapter contains a collection of results concerning the factorization of matrices and the Jacobians of certain transformations on Euclidean spaces. The factorizations and Jacobians established here do have some intrinsic interest. Rather than interrupt the flow of later material to present these results, we have chosen to collect them together for easy reference. The reader is asked to mentally file the results and await their application in future chapters.

5.1. MATRIX FACTORIZATIONS

We begin by fixing some notation. As usual, R^n denotes n-dimensional coordinate space and $\mathcal{L}_{m,n}$ is the space of $n \times m$ real matrices. The linear space of $n \times n$ symmetric real matrices, a subspace of $\mathcal{L}_{n,n}$, is denoted by $\mathcal{S}_n$. If $S \in \mathcal{S}_n$, we write $S > 0$ to mean S is positive definite and $S \geqslant 0$ means that S is positive semidefinite.

Recall that $\mathcal{F}_{p,n}$ is the set of all $n \times p$ linear isometries of R^p into R^n, that is, $\Psi \in \mathcal{F}_{p,n}$ iff $\Psi'\Psi = I_p$. Also, if $T \in \mathcal{L}_{n,n}$, then $T = \{t_{ij}\}$ is *lower triangular* if $t_{ij} = 0$ for $i < j$. The set of all $n \times n$ lower triangular matrices with $t_{ii} > 0$, $i = 1, \ldots, n$, is denoted by G_T^+. The dependence of G_T^+ on the dimension n is usually clear from context. A matrix $U \in \mathcal{L}_{n,n}$ is upper triangular if U' is lower triangular and G_U^+ denotes the set of all $n \times n$ upper triangular matrices with positive diagonal elements.

Our first result shows that G_T^+ and G_U^+ are closed under matrix multiplication and matrix inverse. In other words, G_T^+ and G_U^+ are groups of matrices with the group operation being matrix multiplication.

Proposition 5.1. If $T = \{t_{ij}\} \in G_T^+$, then $T^{-1} \in G_T^+$ and the ith diagonal element of T^{-1} is $1/t_{ii}$, $i = 1, \ldots, n$. If T_1 and $T_2 \in G_T^+$, then $T_1 T_2 \in G_T^+$.

Proof. To prove the first assertion, we proceed by induction on n. Assume the result is true for integers $1, 2, \ldots, n-1$. When T is $n \times n$, partition T as

$$T = \begin{pmatrix} T_{11} & 0 \\ T_{21} & t_{nn} \end{pmatrix}$$

where T_{11} is $(n-1) \times (n-1)$, T_{21} is $1 \times (n-1)$, and t_{nn} is the (n, n) diagonal element of T. In order to be T^{-1}, the matrix

$$A \equiv \begin{pmatrix} A_{11} & 0 \\ A_{21} & a_{nn} \end{pmatrix}$$

must satisfy the equation $TA = I_n$. Thus

$$\begin{pmatrix} T_{11} & 0 \\ T_{21} & t_{nn} \end{pmatrix}\begin{pmatrix} A_{11} & 0 \\ A_{21} & a_{nn} \end{pmatrix} = \begin{pmatrix} T_{11}A_{11} & 0 \\ T_{21}A_{11} + t_{nn}A_{21} & t_{nn}a_{nn} \end{pmatrix} = \begin{pmatrix} I_{n-1} & 0 \\ 0 & 1 \end{pmatrix}$$

so $A_{11} = T_{11}^{-1}$, $a_{nn} = 1/t_{nn}$, and

$$A_{21} = -\frac{T_{21}T_{11}^{-1}}{t_{nn}}.$$

The induction hypothesis implies that T_{11}^{-1} is lower triangular with diagonal elements $1/t_{ii}$, $i = 1, \ldots, n-1$. Thus the first assertion holds. The second assertion follows easily from the definition of matrix multiplication. □

Arguing in exactly the same way, G_U^+ is closed under matrix inverse and matrix multiplication. The first factorization result in this chapter is next.

Proposition 5.2. Suppose $A \in \mathcal{L}_{p,n}$ where $p \leqslant n$ and A has rank p. Then $A = \Psi U$ where $\Psi \in \mathfrak{F}_{p,n}$ and $U \in G_U^+$ is $p \times p$. Further, Ψ and U are unique.

Proof. The idea of the proof is to apply the Gram–Schmidt orthogonalization procedure to the columns of the matrix A. Let $a_1, \ldots, a_p$ be the columns of A so $a_i \in R^n$, $i = 1,\ldots, p$. Since A is of rank p, the vectors $a_1,\ldots, a_p$ are linearly independent. Let $\{b_1,\ldots, b_p\}$ be the orthonormal set of vectors obtained by applying the Gram–Schmidt process to $a_1,\ldots, a_p$ in the order $1,2,\ldots, p$. Thus the matrix Ψ with columns $b_1,\ldots, b_p$ is an element of $\mathcal{F}_{p,n}$ as $\Psi'\Psi = I_p$. Since $\text{span}\{a_1,\ldots, a_i\} = \text{span}\{b_1,\ldots, b_i\}$ for $i = 1,\ldots, p$, $b_j'a_i = 0$ if $j > i$, and an examination of the Gram–Schmidt Process shows that $b_i'a_i > 0$ for $i = 1,\ldots, p$. Thus the matrix $U \equiv \Psi'A$ is an element of G_U^+, and

$$\Psi U = \Psi\Psi'A.$$

But $\Psi\Psi'$ is the orthogonal projection onto $\text{span}\{b_1,\ldots, b_p\} = \text{span}\{a_1,\ldots, a_p\}$ so $\Psi\Psi'A = A$, as $\Psi\Psi'$ is the identity transformation on its range. This establishes the first assertion. For the uniqueness of Ψ and U, assume that $A = \Psi_1 U_1$ for $\Psi_1 \in \mathcal{F}_{p,n}$ and $U_1 \in G_U^+$. Then $\Psi_1 U_1 = \Psi U$, which implies that $\Psi'\Psi_1 = UU_1^{-1}$. Since A is of rank p, U_1 must have rank p so $\mathcal{R}(A) = \mathcal{R}(\Psi_1) = \mathcal{R}(\Psi)$. Therefore, $\Psi_1\Psi_1'\Psi = \Psi$ since $\Psi_1\Psi_1'$ is the orthogonal projection onto its range. Thus $\Psi'\Psi_1\Psi_1'\Psi = I_p$—that is, $\Psi'\Psi_1$ is a $p \times p$ orthogonal matrix. Therefore, $UU_1^{-1} = \Psi'\Psi_1$ is an orthogonal matrix and $UU_1^{-1} \in G_U^+$. However, a bit of reflection shows that the only matrix that is both orthogonal and an element of G_U^+ is I_p. Thus $U = U_1$ so $\Psi = \Psi_1$ as U has rank p. □

The main statistical application of Proposition 5.2 is the decomposition of the random matrix Y discussed in Example 2.3. This decomposition is used to give a derivation of the Wishart density function and, under certain assumptions on the distribution of $Y = \Psi U$, it can be proved that Ψ and U are independent. The above decomposition also has some numerical applications. For example, the proof of Proposition 5.2 shows that if $A = \Psi U$, then the orthogonal projection onto the range of A is $\Psi\Psi' = A(A'A)^{-1}A'$. Hence this projection can be computed without computing $(A'A)^{-1}$. Also, if $p = n$ and $A = \Psi U$, then $A^{-1} = U^{-1}\Psi'$. Thus to compute A^{-1}, we need only to compute U^{-1} and this computation can be done iteratively, as the proof of Proposition 5.1 shows.

Our next decomposition result establishes a one-to-one correspondence between positive definite matrices and elements of G_T^+. First, a property of positive definite matrices is needed.

Proposition 5.3. For $S \in \mathcal{S}_p$ and $S > 0$, partition S as

$$S = \begin{pmatrix} S_{11} & S_{12} \\ S_{21} & S_{22} \end{pmatrix}$$

where S_{11} and S_{22} are both square matrices. Then S_{11}, S_{22}, $S_{11} - S_{12}S_{22}^{-1}S_{21}$, and $S_{22} - S_{21}S_{11}^{-1}S_{12}$ are all positive definite.

Proof. For $x \in R^p$, partition x into y and z to be comformable with the partition of S. Then, for $x \neq 0$,

$$0 < x'Sx = y'S_{11}y + 2z'S_{21}y + z'S_{22}z.$$

For $y \neq 0$ and $z = 0$, $x \neq 0$ so $y'S_{11}y > 0$, which shows that $S_{11} > 0$. Similarly, $S_{22} > 0$. For $y \neq 0$ and $z = -S_{22}^{-1}S_{21}y$,

$$0 < x'Sx = y'\left(S_{11} - S_{12}S_{22}^{-1}S_{21}\right)y,$$

which shows that $S_{11} - S_{12}S_{22}^{-1}S_{21} > 0$. Similarly, $S_{22} - S_{21}S_{11}^{-1}S_{12} > 0$. □

Proposition 5.4. If $S > 0$, then $S = TT'$ for a unique element $T \in G_T^+$.

Proof. First, we establish the existence of T and then prove it is unique. The proof is by induction on dimension. If $S \in \mathbb{S}_p$ with $S > 0$, partition S as

$$S = \begin{pmatrix} S_{11} & S_{12} \\ S_{21} & S_{22} \end{pmatrix}$$

where S_{11} is $(p-1) \times (p-1)$ and $S_{22} \in (0, \infty)$. By the induction hypothesis, $S_{11} = T_{11}T'_{11}$ for $T_{11} \in G_T^+$. Consider the equation

$$\begin{pmatrix} S_{11} & S_{12} \\ S_{21} & S_{22} \end{pmatrix} = \begin{pmatrix} T_{11} & 0 \\ T_{21} & T_{22} \end{pmatrix} \begin{pmatrix} T_{11} & 0 \\ T_{21} & T_{22} \end{pmatrix}',$$

which is to be solved for $T_{21} : 1 \times (p-1)$ and $T_{22} \in (0, \infty)$. This leads to the two equations $T_{21}T'_{11} = S_{21}$ and $T_{21}T'_{21} + T_{22}^2 = S_{22}$. Thus $T_{21} = S_{21}(T'_{11})^{-1}$, so

$$\begin{aligned} S_{22} &= T_{22}^2 + S_{21}\left(T'_{11}\right)^{-1}\left(S_{21}\left(T'_{11}\right)^{-1}\right)' \\ &= T_{22}^2 + S_{21}\left(T_{11}T'_{11}\right)^{-1}S_{12} = T_{22}^2 + S_{21}S_{11}^{-1}S_{12}. \end{aligned}$$

Therefore, $T_{22}^2 = S_{22} - S_{21}S_{11}^{-1}S_{12}$, which is positive by Proposition 5.3. Hence, $T_{22} = (S_{22} - S_{21}S_{11}^{-1}S_{12})^{1/2}$ is the solution for $T_{22} > 0$. This shows that $S = TT'$ for some $T \in G_T^+$. For uniqueness, if $S = TT' = T_1T_1'$, then $T_1^{-1}TT'(T_1')^{-1} = I_p$ so $T_1^{-1}T$ is an orthogonal matrix. But $T_1^{-1}T \in G_T^+$ and the only matrix that is both orthogonal and in G_T^+ is I_p. Hence, $T_1^{-1}T = I_p$ and uniqueness follows. □

Let $\mathbb{S}_p^+$ denote the set of $p \times p$ positive definite matrices. Proposition 5.4 shows that the function $F: G_T^+ \to \mathbb{S}_p^+$ defined by $F(T) = TT'$ is both one-to-one and onto. Of course, the existence of $F^{-1}: \mathbb{S}_p^+ \to G_T^+$ is also part of the content of Proposition 5.4. For $T_1 \in G_T^+$, the uniqueness part of Proposition 5.4 yields $F^{-1}(T_1ST_1') = T_1F^{-1}(S)$. This relationship is used later in this chapter. It is clear that the above result holds for G_T^+ replaced by G_U^+. In other words, every $S \in \mathbb{S}_p^+$ has a unique decomposition $S = UU'$ for $U \in G_U^+$.

Proposition 5.5. Suppose $A \in \mathcal{L}_{p,n}$ where $p \leqslant n$ and A has rank p. Then $A = \Psi S$ where $\Psi \in \mathfrak{F}_{p,n}$ and S is positive definite. Furthermore, Ψ and S are unique.

Proof. Since A has rank p, $A'A$ has rank p and is positive definite. Let S be the positive definite square root of $A'A$, so $A'A = SS$. From Proposition 1.31, there exists a linear isometry $\Psi \in \mathfrak{F}_{p,n}$ such that $A = \Psi S$. To establish the uniqueness of Ψ and S, suppose that $A = \Psi S = \Psi_1 S_1$ where $\Psi, \Psi_1 \in \mathfrak{F}_{p,n}$, and S and S_1 are both positive definite. Then $\mathcal{R}(A) = \mathcal{R}(\Psi) = \mathcal{R}(\Psi_1)$. As in the proof of Proposition 5.2, this implies that $\Psi'\Psi_1\Psi_1'\Psi = I_p$ since $\Psi_1\Psi_1'$ is the orthogonal projection onto $\mathcal{R}(\Psi_1) = \mathcal{R}(\Psi)$. Therefore, $SS_1^{-1} = \Psi'\Psi_1$ is a $p \times p$ orthogonal matrix so the eigenvalues of SS_1^{-1} are all on the unit circle in the complex plane. But the eigenvalues of SS_1^{-1} are the same as the eigenvalues of $S^{1/2}S_1^{-1}S^{1/2}$ (see Proposition 1.39) where $S^{1/2}$ is the positive definite square root of S. Since $S^{1/2}S_1^{-1}S^{1/2}$ is positive definite, the eigenvalues of $S^{1/2}S_1^{-1}S^{1/2}$ are all positive. Therefore, the eigenvalues of $S^{1/2}S_1^{-1}S^{1/2}$ must all be equal to one, as this is the only point of intersection of $(0, \infty)$ with the unit circle in the complex plane. Since the only $p \times p$ matrix with all eigenvalues equal to one is the identity, $S^{1/2}S_1^{-1}S^{1/2} = I_p$ so $S = S_1$. Since S is nonsingular, $\Psi = \Psi_1$. □

The factorizations established this far were concerned with writing one matrix as the product of two other matrices with special properties. The results below are concerned with factorizations for two or more matrices. Statistical applications of these factorizations occur in later chapters.

Proposition 5.6. Suppose A is a $p \times p$ positive definite matrix and B is a $p \times p$ symmetric matrix. There exists a nonsingular $p \times p$ matrix C and a

$p \times p$ diagonal matrix D such that $A = CC'$ and $B = CDC'$. The diagonal elements of D are the eigenvalues of $A^{-1}B$.

Proof. Let $A^{1/2}$ be the positive definite square root of A and $A^{-1/2} = (A^{1/2})^{-1}$. By the spectral theorem for matrices, there exists a $p \times p$ orthogonal matrix Γ such that $\Gamma' A^{-1/2} B A^{-1/2} \Gamma \equiv D$ is diagonal (see Proposition 1.45), and the eigenvalues of $A^{-1/2}BA^{-1/2}$ are the diagonal elements of D. Let $C = A^{1/2}\Gamma$. Then $CC' = A^{1/2}\Gamma\Gamma' A^{1/2} = A$ and $CDC' = B$. Since the eigenvalues of $A^{-1/2}BA^{-1/2}$ are the same as the eigenvalues of $A^{-1}B$, the proof is complete. □

Proposition 5.7. Suppose S is a $p \times p$ positive definite matrix and partition S as

$$S = \begin{pmatrix} S_{11} & S_{12} \\ S'_{12} & S_{22} \end{pmatrix}$$

where S_{11} is $p_1 \times p_1$ and S_{22} is $p_2 \times p_2$ with $p_1 \leqslant p_2$. Then there exist nonsingular matrices A_{ii} of dimension $p_i \times p_i$, $i = 1, 2$, such that $A_{ii} S_{ii} A'_{ii} = I_{p_i}$, $i = 1, 2$, and $A_{11} S_{12} A'_{22} = (D0)$ where D is a $p_1 \times p_1$ diagonal matrix and 0 is a $p_1 \times (p_2 - p_1)$ matrix of zeroes. The diagonal elements of D^2 are the eigenvalues of $S_{11}^{-1} S_{12} S_{22}^{-1} S_{21}$ where $S_{21} = S'_{12}$, and these eigenvalues are all in the interval [0, 1].

Proof. Since S is positive definite, S_{11} and S_{22} are positive definite. Let $S_{11}^{1/2}$ and $S_{22}^{1/2}$ be the positive definite square roots of S_{11} and S_{22}. Using Proposition 1.46, write the matrix $S_{11}^{-1/2} S_{12} S_{22}^{-1/2}$ in the form

$$S_{11}^{-1/2} S_{12} S_{22}^{-1/2} = \Gamma D \Psi$$

where Γ is a $p_1 \times p_1$ orthogonal matrix, D is a $p_1 \times p_1$ diagonal matrix, and Ψ is a $p_1 \times p_2$ linear isometry. The p_1 rows of Ψ form an orthonormal set in R^{p_2} and $p_2 - p_1$ orthonormal vectors can be adjoined to Ψ to obtain a $p_2 \times p_2$ orthogonal matrix Ψ_1 whose first p_1 rows are the rows of Ψ. It is clear that

$$D\Psi = (D0)\Psi_1$$

where 0 is a $p_1 \times (p_2 - p_1)$ matrix of zeroes. Set $A_{11} = \Psi' S_{11}^{-1/2}$ and $A_{22} = \Psi_1 S_{22}^{-1/2}$ so $A_{ii} S_{ii} A'_{ii} = I_{p_i}$ for $i = 1, 2$. Obviously, $A_{11} S_{12} A'_{22} = (D0)$. Since $S_{11}^{-1/2} S_{12} S_{22}^{-1/2} = \Gamma D \Psi$,

$$S_{11}^{-1/2} S_{12} S_{22}^{-1} S_{21} S_{11}^{-1/2} = \Gamma D^2 \Gamma'$$

so the eigenvalues of $S_{11}^{-1/2}S_{12}S_{22}^{-1}S_{21}S_{11}^{-1/2}$ are the diagonal elements of D^2. Since the eigenvalues of $S_{11}^{-1/2}S_{12}S_{22}^{-1}S_{21}S_{11}^{-1/2}$ are the same as the eigenvalues of $S_{11}^{-1}S_{12}S_{22}^{-1}S_{21}$, it remains to show that these eigenvalues are in [0, 1]. By Proposition 5.3, $S_{11} - S_{12}S_{22}^{-1}S_{21}$ is positive definite so $I_{p_1} - S_{11}^{-1/2}S_{12}S_{22}^{-1}S_{21}S_{11}^{-1/2}$ is positive definite. Thus for $x \in R^{p_1}$,

$$0 \leqslant x'S_{11}^{-1/2}S_{12}S_{22}^{-1}S_{21}S_{11}^{-1/2}x \leqslant x'x,$$

which implies that (see Proposition 1.44) the eigenvalues of $S_{11}^{-1/2}S_{12}S_{22}^{-1}S_{21}S_{11}^{-1/2}$ are in the interval [0, 1]. □

It is shown later that the eigenvalues of $S_{11}^{-1}S_{12}S_{22}^{-1}S_{21}$ are related to the angles between two subspaces of R^p. However, it is also shown that these eigenvalues have a direct statistical interpretation in terms of correlation coefficients, and this establishes the connection between canonical correlation coefficients and angles between subspaces. The final decomposition result in this section provides a useful result for evaluating integrals over the space of $p \times p$ positive definite matrices.

Proposition 5.8. Let $\mathbb{S}_p^+$ denote the space of $p \times p$ positive definite matrices. For $S \in \mathbb{S}_p^+$, partition S as

$$S = \begin{pmatrix} S_{11} & S_{12} \\ S_{21} & S_{22} \end{pmatrix}$$

where S_{ii} is $p_i \times p_i$, $i = 1, 2$, S_{12} is $p_1 \times p_2$, and $S_{21} = S'_{12}$. The function f defined on $\mathbb{S}_p^+$ to $\mathbb{S}_{p_1}^+ \times \mathbb{S}_{p_2}^+ \times \mathcal{L}_{p_2, p_1}$ by

$$f(S) = \left(S_{11} - S_{12}S_{22}^{-1}S_{21}, S_{22}, S_{12}S_{22}^{-1}\right)$$

is a one-to-one onto function. The function h on $\mathbb{S}_{p_1}^+ \times \mathbb{S}_{p_2}^+ \times \mathcal{L}_{p_2, p_1}$ to $\mathbb{S}_p^+$ given by

$$h(A_{11}, A_{22}, A_{12}) = \begin{pmatrix} A_{11} + A_{12}A_{22}A'_{12} & A_{12}A_{22} \\ A_{22}A'_{12} & A_{22} \end{pmatrix}$$

is the inverse of f.

Proof. It is routine to verify that $f \circ h$ is the identity function on $\mathbb{S}_{p_1}^+ \times \mathbb{S}_{p_2}^+ \times \mathcal{L}_{p_2, p_1}$ and $h \circ f$ is the identity function on $\mathbb{S}_p^+$. This implies the assertions of the proposition. □

5.2. JACOBIANS

Jacobians provide the basic technical tool for describing how multivariate integrals over open subsets of R^n transform under a change of variable. To describe the situation more precisely, let B_0 and B_1 be fixed open subsets of R^n and let g be a one-to-one onto mapping from B_0 to B_1. Recall that the differential of g, assuming the differential exists, is a function D_g defined on B_0 that takes values in $\mathcal{L}_{n,n}$ and satisfies

$$\lim_{\delta \to 0} \frac{\|g(x+\delta) - g(x) - D_g(x)\delta\|}{\|\delta\|} = 0$$

for each $x \in B_0$. Here δ is a vector in R^n chosen small enough so that $x + \delta \in B_0$. Also, $D_g(x)\delta$ is the matrix $D_g(x)$ applied to the vector δ, and $\|\cdot\|$ denotes the standard norm on R^n. Let $g_1, \dots, g_n$ denote the coordinate functions of the vector valued function g. It is well known that the matrix $D_g(x)$ is given by

$$D_g(x) = \left\{ \frac{\partial g_i}{\partial x_j}(x) \right\}, \qquad x \in B_0.$$

In other words, the (i, j) element of the matrix $D_g(x)$ is the partial derivative of g_i with respect to x_j evaluated at $x \in B_0$. The Jacobian of g is defined by

$$J_g(x) = |\det D_g(x)|, \qquad x \in B_0$$

so the Jacobian is the absolute value of the determinant of D_g. A formal statement of the change of variables theorem goes as follows. Consider any real valued Borel measurable function f defined on the open set B_1 such that

$$\int_{B_1} |f(y)|\, dy < +\infty$$

where dy means Lebesgue measure. Introduce the change of variables $y = g(x)$, $x \in B_0$ in the integral $\int_{B_1} f(y)\, dy$. Then the change of variables theorem asserts that

$$(5.1) \qquad \int_{B_1} f(y)\, dy = \int_{B_0} f(g(x)) J_g(x)\, dx.$$

An alternative way to express (5.1) is by the formal expression

$$d(g(x)) = J_g(x)\,dx, \qquad x \in B_0. \tag{5.2}$$

To give a precise meaning to (5.2), proceed as follows. For each Borel measurable function h defined on B_0 such that $\int_{B_0}|h(x)|J_g(x)\,dx < +\infty$, define

$$I_1(h) \equiv \int_{B_0} h(x)J_g(x)\,dx,$$

and define

$$I_2(h) \equiv \int_{B_0} h(x)d(g(x)) \equiv \int_{g(B_0)} h(g^{-1}(x))\,dx.$$

Then (5.2) means that $I_1(h) = I_2(h)$ for all h such that $I_1(|h|) < +\infty$. To show that (5.1) and the equality of I_1 and I_2 are equivalent, simply set $f = h \circ g^{-1}$ so $f \circ g = h$. Thus $I_1(h) = I_2(h)$ iff

$$\int_{B_0} f(g(x))J_g(x)\,dx = \int_{B_1} f(x)\,dx$$

since $B_1 = g(B_0)$.

One property of Jacobians that is often useful in simplifying computations is the following. Let B_0, B_1, and B_2 be open subsets of R^n, suppose g_1 is a one-to-one onto map from B_0 to B_1, and suppose D_{g_1} exists. Also, suppose g_2 is a one-to-one onto map from B_1 to B_2 and assume that D_{g_2} exists. Then, $g_2 \circ g_1$ is a one-to-one onto map from B_0 to B_2 and it is not difficult to show that

$$D_{g_2 \circ g_1}(x) = D_{g_2}(g_1(x))D_{g_1}(x), \qquad x \in B_0.$$

Of course, the right-hand side of this equality means the matrix product of $D_{g_2}(g_1(x))$ and $D_{g_1}(x)$. From this equality, it follows that

$$J_{g_2 \circ g_1}(x) = J_{g_2}(g_1(x))J_{g_1}(x), \qquad x \in B_0.$$

In particular, if $B_2 = B_0$ and $g_2 = g_1^{-1}$, then $g_2 \circ g_1 = g_1^{-1} \circ g_1$ is the identity function on B_0 so its Jacobian is one. Thus

$$1 = J_{g_2 \circ g_1}(x) = J_{g_2}(g_1(x))J_{g_1}(x), \qquad x \in B_0$$

and

$$J_{g_1^{-1}}(y) = \frac{1}{J_{g_1}(g_1^{-1}(y))}, \qquad y \in B_1.$$

We now turn to the problem of explicitly computing some Jacobians that are needed later. The first few results present Jacobians for linear transformations.

Proposition 5.9. Let A be an $n \times n$ nonsingular matrix and define g on R^n to R^n by $g(x) = A(x)$. Then $J_g(x) = |\det(A)|$ for $x \in R^n$.

Proof. We must compute the differential matrix of g. It is clear that the ith coordinate function of f is g_i where

$$g_i(x) = \sum_{k=1}^{n} a_{ik} x_k.$$

Here $A = \{a_{ij}\}$ and x has coordinates $x_1, \ldots, x_n$. Thus

$$\frac{\partial g_i}{\partial x_j}(x) = a_{ij}$$

so $D_g(x) = \{a_{ij}\}$. Thus $J_g(x) = |\det(A)|$. □

Proposition 5.10. Let A be an $n \times n$ nonsingular matrix and let B be a $p \times p$ nonsingular matrix. Define g on the np-dimensional coordinate space $\mathcal{L}_{p,n}$ to $\mathcal{L}_{p,n}$ by

$$g(X) = AXB' = (A \otimes B)X.$$

Then $J_g(X) = |\det A|^p |\det B|^n$.

Proof. First note that $A \otimes B = (I_n \otimes B)(A \otimes I_p)$. Setting $g_1(X) = (A \otimes I_p)X$ and $g_2(X) = (I_n \otimes B)X$, it is sufficient to verify that

$$J_{g_1}(X) = |\det A|^p$$

and

$$J_{g_2}(X) = |\det B|^n.$$

Let $x_1,\ldots, x_p$ be the columns of the $n \times p$ matrix X so $x_i \in R^n$. Form the np-dimensional vector

$$[X] = \begin{pmatrix} x_1 \\ x_2 \\ \vdots \\ x_p \end{pmatrix}.$$

Since $(A \otimes I_p)X$ has columns $Ax_1,\ldots, Ax_n$, the matrix of $A \otimes I_p$ as a linear transformation on $[X]$ is

$$\begin{pmatrix} A & & & \\ & A & & \\ & & \ddots & \\ & & & A \end{pmatrix} : (np) \times (np)$$

where the elements not indicated are zero. Clearly, the determinant of this matrix is $(\det A)^p$ since A occurs p times on the diagonal. Since the determinant of a linear transformation is independent of a matrix representation, we have that

$$\det(A \otimes I_p) = (\det A)^p.$$

Applying Proposition 5.9, it follows that

$$J_{g_1}(X) = |\det A|^p.$$

Using the rows instead of the columns, we find that

$$\det(I_n \otimes B) = (\det B)^n,$$

so

$$J_{g_2}(X) = |\det B|^n. \qquad \square$$

Proposition 5.11. Let A be a $p \times p$ nonsingular matrix and define the function g on the linear space $\mathcal{S}_p$ of $p \times p$ real symmetric matrices by

$$g(S) = ASA' = (A \otimes A)S.$$

Then $J_g(S) = |\det A|^{p+1}$.

Proof. The result of the previous proposition shows that $\det(A \otimes A) = (\det A)^{2p}$ when $A \otimes A$ is regarded as a linear transformation on $\mathcal{L}_{p,p}$. However, this result is not applicable to the current case since we are considering the restriction of $A \otimes A$ to the subspace $\mathbb{S}_p$ of $\mathcal{L}_{p,p}$.

To establish the present result, write $A = \Gamma_1 D \Gamma_2$ where Γ_1 and Γ_2 are $p \times p$ orthogonal matrices and D is a diagonal matrix with positive diagonal elements (see Proposition 1.47). Then,

$$ASA' = (A \otimes A)S = (\Gamma_1 \otimes \Gamma_1)(D \otimes D)(\Gamma_2 \otimes \Gamma_2)S$$

so the linear transformation $A \otimes A$ has been decomposed into the composition of three linear transformations, two of which are determined by orthogonal matrices.

We now claim that if Γ is a $p \times p$ orthogonal matrix and g_1 is defined on $\mathbb{S}_p$ by

$$g_1(S) = \Gamma S \Gamma' = (\Gamma \otimes \Gamma)S,$$

then $J_{g_1} = 1$. To see this, let $\langle \cdot , \cdot \rangle$ be the natural inner product on $\mathcal{L}_{p,p}$ restricted to $\mathbb{S}_p$, that is, let

$$\langle S_1, S_2 \rangle = \operatorname{tr} S_1 S_2 .$$

Then

$$\begin{aligned}\langle (\Gamma \otimes \Gamma)S_1, (\Gamma \otimes \Gamma)S_2 \rangle &= \operatorname{tr} \Gamma S_1 \Gamma' \Gamma S_2 \Gamma' = \operatorname{tr} \Gamma S_1 S_2 \Gamma' \\ &= \operatorname{tr} \Gamma' \Gamma S_1 S_2 = \operatorname{tr} S_1 S_2 = \langle S_1, S_2 \rangle .\end{aligned}$$

Therefore, $\Gamma \otimes \Gamma$ is an orthogonal transformation on the inner product space $(\mathbb{S}_p, \langle \cdot , \cdot \rangle)$, so the determinant of this linear transformation on $\mathbb{S}_p$ is ± 1. Thus g_1 is a linear transformation that is also orthogonal so $J_{g_1} = 1$ and the claim is established.

The next claim is that if D is a $p \times p$ diagonal matrix with positive diagonal elements and g_2 is defined on $\mathbb{S}_p$ by

$$g_2(S) = DSD,$$

then $J_{g_2} = (\det D)^{p+1}$. In the $[p(p+1)/2]$-dimensional space $\mathbb{S}_p$, let s_{ij}, $1 \leqslant j \leqslant i \leqslant p$, denote the coordinates of S. Then it is routine to show that the (i, j) coordinate function of g_2 is $g_{2,ij}(S) = \lambda_i \lambda_j s_{ij}$ where $\lambda_1, \ldots, \lambda_p$ are the diagonal elements of D. Thus the matrix of the linear transformation g_2 is a $[p(p+1)/2] \times [p(p+1)/2]$ diagonal matrix with diagonal entries

$\lambda_i\lambda_j$ for $1 \leqslant j \leqslant i \leqslant p$. Hence the determinant of this matrix is the product of the $\lambda_i\lambda_j$ for $1 \leqslant j \leqslant i \leqslant p$. A bit of calculation shows this determinant is $(\Pi\lambda_i)^{p+1}$. Since $\det D = \Pi\lambda_i$, the second claim is established.

To complete the proof, note that

$$g(S) = ASA' = (\Gamma_1 \otimes \Gamma_1)(D \otimes D)(\Gamma_2 \otimes \Gamma_2)S = h_1(h_2(h_3(S)))$$

where $h_1(S) = (\Gamma_1 \otimes \Gamma_1)S$, $h_2(S) = (D \otimes D)S$, and $h_3(S) = (\Gamma_2 \otimes \Gamma_2)S$. A direct argument shows that

$$\begin{aligned} J_{h_1 \circ h_2 \circ h_3}(S) &= J_{h_1 \circ h_2}(h_3(S)) J_{h_3}(S) \\ &= J_{h_1}(h_2(h_3(S))) J_{h_2}(h_3(S)) J_{h_3}(S). \end{aligned}$$

But $J_{h_1} = 1 = J_{h_3}$ and $J_{h_2} = (\det D)^{p+1}$. Since $A = \Gamma_1 D \Gamma_2$, $|\det A| = \det D$, which entails $J_g = |\det A|^{p+1}$. □

Proposition 5.12. Let M be the linear space of $p \times p$ skew-symmetric matrices and define g on M to M by

$$g(S) = ASA'$$

where A is a $p \times p$ nonsingular matrix. Then $J_g(S) = |\det A|^{p-1}$.

Proof. The proof is similar to that of Proposition 5.11 and is left to the reader. □

Proposition 5.13. Let G_T^+ be the set of $p \times p$ lower triangular matrices with positive diagonal elements and let A be a fixed element of G_T^+. The function g defined on G_T^+ to G_T^+ by

$$g(T) = AT, \qquad T \in G_T^+$$

has a Jacobian given by $J_g(T) = \prod_1^p a_{ii}^i$ where $a_{11}, \ldots, a_{pp}$ are the diagonal elements of A.

Proof. The set G_T^+ is an open subset of $[\frac{1}{2}p(p+1)]$-dimensional coordinate space and g is a one-to-one onto function by Proposition 5.1. For $T \in G_T^+$, form the vector $[T]$ with coordinates $t_{11}, t_{21}, t_{22}, t_{31}, \ldots, t_{pp}$ and write the coordinate functions of g in the same order. Then the matrix of partial derivatives is lower triangular with diagonal elements

$a_{11}, a_{22}, a_{22}, a_{33}, \ldots, a_{pp}$ where a_{ii} occurs i times on the diagonal. Thus the determinant of this matrix of partial derivatives is $\prod_1^p a_{ii}^i$ so $J_g = \prod_1^p a_{ii}^i$. □

Proposition 5.14. In the notation of Proposition 5.13, define g on G_T^+ to G_T^+ by

$$g(T) = TB, \qquad T \in G_T^+$$

where B is a fixed element of G_T^+. Then $J_g(T) = \prod_1^p b_{ii}^{p-i+1}$ where $b_{11}, \ldots, b_{pp}$ are the diagonal elements of B.

Proof. The proof is similar to that of Proposition 5.13 and is omitted. □

Proposition 5.15. Let G_U^+ be the set of all $p \times p$ upper triangular matrices with positive diagonal elements. For fixed elements A and B of G_U^+, define g by

$$g(U) = AUB, \qquad U \in G_U^+.$$

Then,

$$J_g(U) = \prod_1^p a_{ii}^{p-i+1} \prod_1^p b_{ii}^i$$

where $a_{11}, \ldots, a_{pp}$ and $b_{11}, \ldots, b_{pp}$ are diagonal elements of A and B.

Proof. The proof is similar to that given for lower triangular matrices and is left to the reader. □

Thus far, only Jacobians of linear transformations have been computed explicitly, and, of course, these Jacobians have been constant functions. In the next proposition, the Jacobian of the nonlinear transformation described in Proposition 5.8 is computed.

Proposition 5.16. Let p_1 and p_2 be positive integers and set $p = p_1 + p_2$. Using the notation of Proposition 5.8, define h on $\mathcal{S}_{p_1}^+ \times \mathcal{S}_{p_2}^+ \times \mathcal{L}_{p_2, p_1}$ to $\mathcal{S}_p^+$ by

$$h(A_{11}, A_{22}, A_{12}) = \begin{pmatrix} A_{11} + A_{12}A_{22}A_{12}' & A_{12}A_{22} \\ A_{22}A_{12}' & A_{22} \end{pmatrix}.$$

Then $J_h(A_{11}, A_{22}, A_{12}) = (\det A_{22})^{p_1}$.

Proof. For notational convenience, set $S = h(A_{11}, A_{22}, A_{12})$ and partition S as

$$S = \begin{pmatrix} S_{11} & S_{12} \\ S_{12} & S_{22} \end{pmatrix}$$

where S_{ij} is $p_i \times p_j$, $i, j = 1, 2$. The partial derivatives of the elements of S, as functions of the elements of A_{11}, A_{12} and A_{22}, need to be computed. Since $S_{11} = A_{11} + A_{12}A_{22}A'_{12}$, the matrix of partial derivatives of the $p_1(p_1+1)/2$ elements of S_{11} with respect to the $p_1(p_1+1)/2$ elements of A_{11} is just the $[p_1(p_1+1)/2]$-dimensional identity matrix. Since $S_{12} = A_{12}A_{22}$, the matrix of partial derivatives of the p_1p_2 elements of S_{12} with respect to the elements of A_{11} is the $p_1p_2 \times p_1p_2$ zero matrix. Also, since $S_{22} = A_{22}$, the partial derivative of elements of S_{22} with respect to the elements of A_{11} or A_{12} are all zero and the matrix of partial derivatives of the $p_2(p_2+1)/2$ elements of S_{22} with respect to the $p_2(p_2+1)/2$ elements of A_{22} is the identity matrix. Thus the matrix of partial derivatives has the form

$$\begin{array}{c} \\ S_{11} \\ S_{12} \\ S_{22} \end{array} \begin{array}{c} \begin{array}{ccc} A_{11} & A_{12} & A_{22} \end{array} \\ \begin{pmatrix} I_1 & - & - \\ 0 & B & - \\ 0 & 0 & I_2 \end{pmatrix} \end{array}$$

so the determinant of this matrix is just the determinant of the $p_1p_2 \times p_1p_2$ matrix B, which must be found. However, B is the matrix of partial derivatives of the elements of S_{12} with respect to the elements of A_{12} where $S_{12} = A_{12}A_{22}$. Hence the determinant of B is just the Jacobian of the transformation $g(A_{12}) = A_{12}A_{22}$ with A_{22} fixed. This Jacobian is $(\det A_{22})^{p_1}$ by Proposition 5.10. □

As an application of Proposition 5.16, a special integral over the space $\mathbb{S}_p^+$ is now evaluated.

◆ **Example 5.1.** Let dS denote Lebesgue measure on the set $\mathbb{S}_p^+$. The integral below arises in our discussion of the Wishart distribution. For a positive integer p and a real number $r > p - 1$, let

$$c(r, p) = \int_{\mathbb{S}_p^+} |S|^{(r-p-1)/2} \exp\left[-\tfrac{1}{2}\operatorname{tr} S\right] dS.$$

In this example, the constant $c(r, p)$ is calculated. When $p = 1$,

$\mathbb{S}_p^+ = (0, \infty)$ so for $r > 0$,

$$c(r, 1) = \int_0^\infty s^{(r/2)-1} \exp\left[-\frac{s}{2}\right] ds = 2^{r/2}\Gamma\left(\frac{r}{2}\right)$$

where $\Gamma(r/2)$ is the gamma function evaluated at $r/2$. The first claim is that

$$c(r, p+1) = (2\pi)^{p/2} c(r-1, p) c(r, 1),$$

for $r > p$ and $p \geqslant 1$. To verify this claim, consider $S \in \mathbb{S}_{p+1}^+$ and partition S as

$$S = \begin{pmatrix} S_{11} & S_{12} \\ S_{12}' & S_{22} \end{pmatrix}$$

where $S_{11} \in \mathbb{S}_p^+$, $S_{22} \in (0, \infty)$, and S_{12} is $p \times 1$. Introduce the change of variables

$$\begin{pmatrix} S_{11} & S_{12} \\ S_{12}' & S_{22} \end{pmatrix} = \begin{pmatrix} A_{11} + A_{12}A_{22}A_{12}' & A_{12}A_{22} \\ A_{22}A_{12}' & A_{22} \end{pmatrix}$$

where $A_{11} \in \mathbb{S}_p^+$, $A_{22} \in (0, \infty)$, and $A_{12} \in R^p$. By Proposition 5.16, the Jacobian of this transformation is A_{22}^p. Since $\det S = \det(S_{11} - S_{12}S_{22}^{-1}S_{12}')\det S_{22} = (\det A_{11})A_{22}$, we have

$$c(r, p+1) = \int_{\mathbb{S}_{p+1}^+} |S|^{(r-p-2)/2} \exp\left[-\tfrac{1}{2}\operatorname{tr} S\right] dS$$

$$= \int_0^\infty \int_{R^p} \int_{\mathbb{S}_p^+} |A_{11}|^{(r-p-2)/2} A_{22}^{(r-p-2)/2}$$

$$\times \exp\left[-\tfrac{1}{2}\operatorname{tr} A_{11} - \tfrac{1}{2}A_{22}A_{12}'A_{12} - \tfrac{1}{2}A_{22}\right]$$

$$\times A_{22}^p \, dA_{11} \, dA_{12} \, dA_{22}.$$

Integrating with respect to A_{12} yields

$$\int_{R^p} \exp\left[-\tfrac{1}{2}A_{22}A_{12}'A_{12}\right] dA_{12} = (2\pi)^{p/2} A_{22}^{-p/2}.$$

Substituting this into the second integral expression for $c(r, p+1)$

and then integrating on A_{22} shows that

$$c(r, p+1) = (2\pi)^{p/2} c(r,1) \int_{\mathbb{S}_p^+} |A_{11}|^{(r-p-2)/2} \exp\left[-\tfrac{1}{2} \operatorname{tr} A_{11}\right] dA_{11}$$

$$= (2\pi)^{p/2} c(r,1) c(r-1, p).$$

This establishes the first claim. Now, it is an easy matter to solve for $c(r, p)$. A bit of manipulation shows that with

$$c(r, p) = \pi^{p(p-1)/4} 2^{rp/2} \prod_{j=1}^{p} \Gamma\left(\frac{r-j+1}{2}\right),$$

for $p = 1, 2, \ldots,$ and $r > p - 1$, the equation

$$c(r, p+1) = (2\pi)^{p/2} c(r,1) c(r-1, p)$$

is satisfied. Further,

$$c(r,1) = 2^{r/2} \Gamma\left(\frac{r}{2}\right).$$

Uniqueness of the solution to the above equation is clear. In summary,

$$\int_{\mathbb{S}_p^+} |S|^{(r-p-1)/2} \exp\left[-\tfrac{1}{2} \operatorname{tr} S\right] dS = \pi^{p(p-1)/4} 2^{rp/2} \prod_{j=1}^{p} \Gamma\left(\frac{r-j+1}{2}\right)$$

and this is valid for $p = 1, 2, \ldots$ and $r > p - 1$. The restriction that r be greater than $p - 1$ is necessary so that $\Gamma[(r - p + 1)/2]$ be well defined. It is not difficult to show that the above integral is $+\infty$ if $r \leqslant p - 1$. Now, set $\omega(r, p) = 1/c(r, p)$ so

$$f(S) \equiv \omega(r, p) |S|^{(r-p-1)/2} \exp\left[-\tfrac{1}{2} \operatorname{tr} S\right]$$

is a density function on $\mathbb{S}_p^+$. When r is an integer, $r \geqslant p$, f turns out to be the density of the Wishart distribution. ◆

Proposition 5.4 shows that there is a one-to-one correspondence between elements of $\mathbb{S}_p^+$ and elements of G_T^+. More precisely, the function g defined on G_T^+ by

$$g(T) = TT', \qquad T \in G_T^+$$

is one-to-one and onto $\mathbb{S}_p^+$. It is clear that g has a differential since each coordinate function of g is a polynomial in the elements of T. One way to find the Jacobian of g is to simply compute the matrix of partial derivatives and then find its determinant. As motivation for some considerations in the next chapter, a different derivation of the Jacobian of g is given here. The first observation is as follows.

Proposition 5.17. Let dS denote Lebesgue measure on $\mathbb{S}_p^+$ and consider the measure μ on $\mathbb{S}_p^+$ given by $\mu(dS) = dS/|S|^{(p+1)/2}$. For each Borel measurable function f on $\mathbb{S}_p^+$, which is integrable with respect to μ, and for each nonsingular matrix A,

$$\int_{\mathbb{S}_p^+} f(S)\mu(dS) = \int_{\mathbb{S}_p^+} f(ASA')\mu(dS).$$

Proof. Set $B = ASA'$. By Proposition 5.11, the Jacobian of this transformation on $\mathbb{S}_p^+$ to $\mathbb{S}_p^+$ is $|\det A|^{p+1}$. Thus

$$\int_{\mathbb{S}_p^+} f(ASA')\mu(dS) = \int_{\mathbb{S}_p^+} f(ASA')\frac{dS}{|S|^{(p+1)/2}}$$

$$= \int_{\mathbb{S}_p^+} \frac{f(ASA')|\det A|^{p+1}}{|ASA'|^{(p+1)/2}}\,dS$$

$$= \int_{\mathbb{S}_p^+} \frac{f(B)}{|B|^{(p+1)/2}}\,dB = \int_{\mathbb{S}_p^+} f(S)\mu(dS). \qquad \square$$

The result of Proposition 5.17 is often paraphrased by saying that the measure μ is invariant under each of the transformations g_A defined on $\mathbb{S}_p^+$ by $g_A(S) = ASA'$. The following calculation gives a heuristic proof of this result:

$$\mu(dg_A(S)) = \frac{d(g_A(S))}{|ASA'|^{(p+1)/2}} = \frac{J_{g_A}(S)\,dS}{|ASA'|^{(p+1)/2}}$$

$$= \frac{|\det A|^{p+1}}{|AA'|^{(p+1)/2}}\frac{dS}{|S|^{(p+1)/2}} = \frac{dS}{|S|^{(p+1)/2}} = \mu(dS).$$

In fact, a similar calculation suggests that μ is the only invariant measure in $\mathbb{S}_p^+$ (up to multiplication of μ by a positive constant). Consider a measure ν

of the form $\nu(dS) = h(S)\,dS$ where h is a positive Borel measurable function and dS is Lebesgue measure. In order that ν be invariant, we must have

$$h(S)\,dS = \nu(dS) = \nu(dg_A(S)) = h(g_A(S))d(g_A(S))$$
$$= h(g_A(S))|\det A|^{p+1}\,dS$$

so h should satisfy the equation

$$h(S) = h(ASA')|AA'|^{(p+1)/2},$$

since $g_A(S) = ASA'$ and $|\det A|^{p+1} = |AA'|^{(p+1)/2}$. Set $S = I_p$, $B = AA'$, and $c = h(I_p)$. Then

$$h(B) = \frac{c}{|B|^{(p+1)/2}}, \qquad B \in \mathbb{S}_p^+$$

so

$$\nu(dS) = c\mu(dS)$$

where c is a positive constant. Making this argument rigorous is one of the topics treated in the next chapter.

The calculation of the Jacobian of g on G_T^+ to $\mathbb{S}_p^+$ is next.

Proposition 5.18. For $g(T) = TT'$, $T \in G_T^+$,

$$J_g(T) = 2^p \prod_{i=1}^{p} t_{ii}^{p-i+1}$$

where $t_{11}, \ldots, t_{pp}$ are the diagonal elements of T.

Proof. The Jacobian J_g is the unique continuous function defined on G_T^+ that satisfies the equation

$$\int_{\mathbb{S}_p^+} f(S) \frac{dS}{|S|^{(p+1)/2}} = \int_{G_T^+} \frac{f(g(T))J_g(T)}{|g(T)|^{(p+1)/2}}\,dT$$

for all Borel measurable functions f for which the integral over $\mathbb{S}_p^+$ exists. But the left-hand side of this equation is invariant under the replacement of $f(S)$ by $f(ASA')$ for any nonsingular $p \times p$ matrix. Thus the right-hand side must have the same property. In particular, for $A \in G_T^+$, we have

$$\int_{G_T^+} \frac{f(TT')}{|TT'|^{(p+1)/2}} J_g(T)\, dT = \int_{G_T^+} \frac{f(ATT'A')}{|TT'|^{(p+1)/2}} J_g(T)\, dT.$$

In this second integral, we make the change of variable $T = A^{-1}B$ for $A \in G_T^+$ fixed and $B \in G_T^+$. By Proposition 5.12, the Jacobian of this transformation is $1/\prod_1^p a_{ii}^i$ where $a_{11}, \ldots, a_{pp}$ are the diagonal elements of A. Thus

$$\int_{G_T^+} \frac{f(TT')}{|TT'|^{(p+1)/2}} J_g(T)\, dT = \int_{G_T^+} \frac{f(BB')}{|BB'|^{(p+1)/2}} \frac{J_g(A^{-1}B)}{|A^{-1}|^{p+1}} \frac{1}{\prod_1^p a_{ii}^i}\, dB.$$

Since this must hold for all Borel measurable f and since J_g is a continuous function, it follows that for all $T \in G_T^+$ and $A \in G_T^+$,

$$J_g(T) = J_g(A^{-1}T) \frac{|A|^{p+1}}{\prod_1^p a_{ii}^i}.$$

Setting $A = T$ and noting that $|T| = \prod_1^p t_{ii}$, we have

$$J_g(T) = J_g(I_p) \prod_1^p t_{ii}^{p-i+1}.$$

Thus $J_g(T)$ is a constant k times $\prod_1^p t_{ii}^{p-i+1}$. Hence

$$\int_{\mathbb{S}_p^+} f(S) \frac{dS}{|S|^{(p+1)/2}} = \int_{G_T^+} k \frac{f(TT')}{|T|^{p+1}} \prod_{i=1}^p t_{ii}^{p-i+1}\, dT = \int_{G_T^+} kf(TT') \prod_{i=1}^p t_{ii}^{-i}\, dT.$$

To evaluate the constant k, pick

$$f(S) = |S|^{r/2} \exp[-\tfrac{1}{2} \operatorname{tr} S], \qquad r > p - 1.$$

But

$$\int_{\mathbb{S}_p^+} |S|^{r/2} \exp[-\tfrac{1}{2} \operatorname{tr} S] \frac{dS}{|S|^{(p+1)/2}} = c(r, p)$$

where $c(r, p)$ is defined in Example 5.1. However,

$$k\int_{G_T^+} |TT'|^{r/2}\exp[-\tfrac{1}{2}\operatorname{tr} TT']\prod_1^p t_{ii}^{-i}\,dT$$

$$= k\int_{G_T^+}\prod_1^p t_{ii}^{r-i}\exp\left[-\tfrac{1}{2}\sum_{j\leqslant i} t_{ij}^2\right]dT = k2^{-p}c(r, p)$$

so $k = 2^p$. The evaluation of the last integral is carried out by noting that t_{ii} ranges from 0 to ∞ and t_{ij} for $j < i$ ranges from $-\infty$ to ∞. Thus the integral is a product of $p(p+1)/2$ integrals on R, each of which is easy to evaluate. □

A by-product of this proof is that

$$h(T) = \frac{\prod_1^p t_{ii}^{r-i}}{2^p c(r, p)}\exp\left[-\tfrac{1}{2}\sum_{j\leqslant i} t_{ij}^2\right]$$

is a density function on G_T^+. Since the density h factors into a product of densities, the elements of T, t_{ij} for $j \leqslant i$, are independent. Clearly,

$$\mathcal{L}(t_{ij}) = N(0, 1) \quad \text{for } j < i$$

and

$$\mathcal{L}(t_{ii}^2) = \chi_{n-i+1}^2$$

when r is the integer $n \geqslant p$.

Proposition 5.19. Define g on G_U^+ to $\mathcal{S}_p^+$ by $g(U) = UU'$. Then $J_g(U)$ is given by

$$J_g(U) = 2^p\prod_{i=1}^p u_{ii}^i$$

where $u_{11}, \ldots, u_{pp}$ are the diagonal elements of U.

Proof. The proof is essentially the same as the proof of Proposition 5.18 and is left to the reader. □

The technique used to prove Proposition 5.18 is an important one. Given g on G_T^+ to $\mathcal{S}_p^+$, the idea of the proof was to write down the equation the

Jacobian satisfies, namely,

$$\int_{\mathbb{S}_p^+} \frac{f(S)}{|S|^{(p+1)/2}}\, dS = \int_{G_T^+} \frac{f(g(T))}{|T|^{p+1}} J_g(T)\, dT$$

for all integrable f. Since this equation must hold for all integrable f, J_g is uniquely defined (up to sets of Lebesgue measure zero) by this equation. It is clear that any property satisfied by the left-hand integral must also be satisfied by the right-hand integral and this was used to characterize J_g. In particular, it was noted that the left-hand integral remained the same if $f(S)$ was replaced by $f(ASA')$ for an nonsingular A. For $A \in G_T^+$, this led to the equation

$$J_g(T) = J_g(A^{-1}T)\frac{|A|^{p+1}}{\prod_1^p a_{ii}^i},$$

which determined J_g. It should be noted that only Jacobians of the linear transformations discussed in Propositions 5.11 and 5.13 were used to determine the Jacobian of the nonlinear transformation g. Arguments similar to this are used throughout Chapter 6 to derive invariant integrals (measures) on matrix groups and spaces that are acted upon by matrix groups.

PROBLEMS

1. Given $A \in \mathcal{L}_{p,n}$ with rank$(A) = p$, show that $A = \Psi T$ where $\Psi \in \mathcal{F}_{p,n}$ and $T \in G_T^+$. Prove that Ψ and T are unique.

2. Define the function F on $\mathbb{S}_p^+$ to G_T^+ as follows. For each $S \in \mathbb{S}_p^+$, $F(S)$ is the unique element in G_T^+ such that $S = F(S)(F(S))'$. Show that $F(TST') = TF(S)$ for $T \in G_T^+$ and $S \in \mathbb{S}_p^+$.

3. Given $S \in \mathbb{S}_p^+$, show there exists a unique $U \in G_U^+$ such that $S = UU'$.

4. For $S \in \mathbb{S}_p^+$, partition S as

$$S = \begin{pmatrix} S_{11} & S_{12} \\ S_{21} & S_{22} \end{pmatrix}$$

where S_{ij} is $p_i \times p_j$, $i, j = 1, 2$. Assume for definiteness that $p_1 \leqslant p_2$.

Show that S can be written as

$$\begin{pmatrix} S_{11} & S_{12} \\ S_{21} & S_{22} \end{pmatrix} = \begin{pmatrix} A_1 & 0 \\ 0 & A_2 \end{pmatrix} \begin{pmatrix} I_{p_1} & (D0) \\ (D0)' & I_{p_2} \end{pmatrix} \begin{pmatrix} A_1 & 0 \\ 0 & A_2 \end{pmatrix}'$$

where A_i is $p_i \times p_i$ and nonsingular, D is $p_1 \times p_1$ and diagonal with diagonal elements in $[0, 1)$.

5. Let $\mathcal{L}^0_{p,n}$ be those elements in $\mathcal{L}_{p,n}$ that have rank p. Define F on $\mathfrak{F}_{p,n} \times G_U^+$ to $\mathcal{L}^0_{p,n}$ by $F(\Psi, U) = \Psi U$.

(i) Show that F is one-to-one onto, and describe the inverse of F.

(ii) For $\Gamma \in \mathcal{O}_n$ and $T \in G_T^+$, define $\Gamma \otimes \mathrm{T}$ on $\mathcal{L}^0_{p,n}$ to $\mathcal{L}^0_{p,n}$ by $(\Gamma \otimes \mathrm{T})A = \Gamma A \mathrm{T}'$. Show that $(\Gamma \otimes T)F(\Psi, U) = F(\Gamma\Psi, UT')$. Also, show that $F^{-1}((\Gamma \otimes T)A) = (\Gamma\Psi, UT')$ where $F^{-1}(A) = (\Psi, U)$.

6. Let B_0 and B_1 be open sets in R^n and fix $x_0 \in B_0$. Suppose g maps B_0 into B_1 and $g(x) = g(x_0) + A(x - x_0) + R(x - x_0)$ where A is an $n \times n$ matrix and $R(\cdot)$ is a function that satisfies

$$\lim_{u \to 0} \frac{\|R(u)\|}{\|u\|} = 0.$$

Prove that $A = D_g(x_0)$ so $J_g(x_0) = |\det(A)|$.

7. Let V be the linear coordinate space of all $p \times p$ lower triangular real matrices so V is of dimension $p(p+1)/2$. Let $\mathcal{S}_p$ be the linear coordinate space of all $p \times p$ real symmetric matrices so $\mathcal{S}_p$ is also of dimension $p(p+1)/2$.

(i) Show that G_T^+ is an open subset of V.

(ii) Define g on G_T^+ to $\mathcal{S}_p$ by $g(T) = TT'$. For fixed $T_0 \in G_T^+$, show that $g(T) = g(T_0) + L(T - T_0) + (T - T_0)(T - T_0)'$ where L is defined on V to $\mathcal{S}_p$ by $L(x) = xT_0' + T_0 x'$, $x \in V$. Also show that $R(T - T_0) = (T - T_0)(T - T_0)'$ satisfies

$$\lim_{x \to 0} \frac{\|R(x)\|}{\|x\|} = 0.$$

(iii) Prove by induction that $\det L = 2^p \prod_1^p t_{ii}^{p-i+1}$ where $t_{11}, \dots, t_{pp}$ are the diagonal elements of T_0.

(iv) Using (iii) and Problem 6, show that $J_g(T) = 2^p \prod_1^p t_{ii}^{p-i+1}$. (This is just Proposition 5.18).

8. When S is a positive definite matrix, partition S and S^{-1} as

$$S = \begin{pmatrix} S_{11} & S_{12} \\ S_{21} & S_{22} \end{pmatrix}, \qquad S^{-1} = \begin{pmatrix} S^{11} & S^{12} \\ S^{21} & S^{22} \end{pmatrix}.$$

Show that

$$S^{11} = \left(S_{11} - S_{12}S_{22}^{-1}S_{21}\right)^{-1}$$

$$S^{12} = -S^{11}S_{12}S_{22}^{-1}$$

$$S^{22} = \left(S_{22} - S_{21}S_{11}^{-1}S_{12}\right)^{-1}$$

$$S^{21} = -S^{22}S_{21}S_{11}^{-1}$$

and verify the identity

$$S_{22}^{-1}S_{21}S^{11} = S^{22}S_{21}S_{11}^{-1}.$$

9. In coordinate space R^p, partition x as $x = \binom{y}{z}$, and for $\Sigma > 0$, partition $\Sigma : p \times p$ conformably as

$$\Sigma = \begin{pmatrix} \Sigma_{11} & \Sigma_{12} \\ \Sigma_{21} & \Sigma_{22} \end{pmatrix}.$$

Define the inner product $(\cdot, \cdot)$ on R^p by $(u, v) = u'\Sigma^{-1}v$.

(i) Show that the matrix

$$P = \begin{pmatrix} I & -\Sigma_{12}\Sigma_{22}^{-1} \\ 0 & 0 \end{pmatrix}$$

defines an orthogonal projection in the inner product $(\cdot, \cdot)$. What is $\mathfrak{R}(P)$?

(ii) Show that the identity

$$\binom{y}{z}'\Sigma^{-1}\binom{y}{z} = \left(y - \Sigma_{12}\Sigma_{22}^{-1}z\right)'\Sigma^{11}\left(y - \Sigma_{12}\Sigma_{22}^{-1}z\right) + z'\Sigma_{22}^{-1}z$$

is the same as the identity

$$\|x\|^2 = \|Px\|^2 + \|(I - P)x\|^2$$

where $(x, x) = \|x\|^2$ and $x = \binom{y}{z}$.

(iii) For a random vector

$$X = \begin{pmatrix} Y \\ Z \end{pmatrix} \in R^p$$

with $\mathcal{L}(X) = N(0, \Sigma)$, $\Sigma > 0$, use part (ii) to give a direct proof via densities that the conditional distribution of Y given Z is $N(\Sigma_{12}\Sigma_{22}^{-1}Z, \Sigma_{11} - \Sigma_{12}\Sigma_{22}^{-1}\Sigma_{21})$.

10. Verify the equation

$$\int_{G_T^+} \prod_1^p t_{ii}^{r-i} \exp\left[-\tfrac{1}{2} \sum_{j \leqslant i}^{p} t_{ij}^2\right] dT = 2^{-p} c(r, p)$$

where $c(r, p)$ is given in Example 5.1. Here, r is real, $r > p - 1$.

NOTES AND REFERENCES

1. Other matrix factorizations of interest in statistical problems can be found in Anderson (1958), Rao (1973), and Muirhead (1982). Many matrix factorizations can be viewed as results that give a maximal invariant under the action of a group—a topic discussed in detail in Chapter 7.

2. Only the most elementary facts concerning the transformation of measures under a change of variable have been given in the second section. The Jacobians of other transformations that occur naturally in statistical problems can be found in Deemer and Olkin (1951), Anderson (1958), James (1954), Farrell (1976), and Muirhead (1982). Some of these transformations involve functions defined on manifolds (rather than open subsets of R^n) and the corresponding Jacobian calculations require a knowledge of differential forms on manifolds. Otherwise, the manipulations just look like magic that somehow yields answers we do not know how to check. Unfortunately, the amount of mathematics behind these calculations is substantial. The mastery of this material is no mean feat. Farrell (1976) provides one treatment of the calculus of differential forms. James (1954) and Muirhead (1982) contain some background material and references.

3. I have found Lang (1969, Part Six, Global Analysis) to be a very readable introduction to differential forms and manifolds.

CHAPTER 6

Topological Groups and Invariant Measures

The language of vector spaces has been used in the previous chapters to describe a variety of properties of random vectors and their distributions. Apart from the discussion in Chapter 4, not much has been said concerning the structure of parametric probability models for distributions of random vectors. Groups of transformations acting on spaces provide a very useful framework in which to generate and describe many parametric statistical models. Furthermore, the derivation of induced distributions of a variety of functions of random vectors is often simplified and clarified using the existence and uniqueness of invariant measures on locally compact topological groups. The ideas and techniques presented in this chapter permeate the remainder of this book.

Most of the groups occurring in multivariate analysis are groups of nonsingular linear transformations or related groups of affine transformations. Examples of matrix groups are given in Section 6.1 to illustrate the definition of a group. Also, examples of quotient spaces that arise naturally in multivariate analysis are discussed.

In Section 6.2, locally compact topological groups are defined. The existence and uniqueness theorem concerning invariant measures (integrals) on these groups is stated and the matrix groups introduced in Section 6.1 are used as examples. Continuous homomorphisms and their relation to relatively invariant measures are described with matrix groups again serving as examples. Some of the material in this section and the next is modeled after Nachbin (1965). Rather than repeat the proofs given in Nachbin (1965), we have chosen to illustrate the theory with numerous examples.

Section 6.3 is concerned with the existence and uniqueness of relatively invariant measures on spaces that are acted on transitively by groups of transformations. In fact, this situation is probably more relevant to statistical problems than that discussed in Section 6.2. Of course, the examples are selected with statistical applications in mind.

6.1. GROUPS

We begin with the definition of a group and then give examples of matrix groups.

Definition 6.1. A group $(G, \circ)$ is a set G together with a binary operation $\circ$ such that the following properties hold for all elements in G:

(i) $(g_1 \circ g_2) \circ g_3 = g_1 \circ (g_2 \circ g_3)$.
(ii) There is a unique element of G, denoted by e, such that $g \circ e = e \circ g = g$ for all $g \in G$. The element e is the *identity* in G.
(iii) For each $g \in G$, there is a unique element in G, denoted by g^{-1}, such that $g \circ g^{-1} = g^{-1} \circ g = e$. The element g^{-1} is the *inverse* of g.

Henceforth, the binary operation is ordinarily deleted and we write $g_1 g_2$ for $g_1 \circ g_2$. Also, parentheses are usually not used in expressions involving more than two group elements as these expressions are unambiguously defined in (i). A group G is called *commutative* if $g_1 g_2 = g_2 g_1$ for all $g_1, g_2 \in G$. It is clear that a vector space V is a commutative group where the group operation is addition, the identity element is $0 \in V$, and the inverse of x is $-x$.

♦ **Example 6.1.** If $(V, (\cdot, \cdot))$ is a finite dimensional inner product space, it has been shown that the set of all orthogonal transformations $\mathcal{O}(V)$ is a group. The group operation is the composition of linear transformations, the identity element is the identity linear transformation, and if $\Gamma \in \mathcal{O}(V)$, the inverse of Γ is Γ'. When V is the coordinate space R^n, $\mathcal{O}(V)$ is denoted by $\mathcal{O}_n$, which is just the group of $n \times n$ orthogonal matrices. ♦

♦ **Example 6.2.** Consider the coordinate space R^p and let G_T^+ be the set of all $p \times p$ lower triangular matrices with positive diagonal elements. The group operation in G_T^+ is taken to be matrix multiplication. It has been verified in Chapter 5 that G_T^+ is a group, the identity in G_T^+ is the $p \times p$ identity matrix, and if $T \in G_T^+$, T^{-1} is just the matrix inverse of T. Similarly, the set of $p \times p$ upper triangular matrices with positive diagonal elements G_U^+ is a group with the group operation of matrix multiplication. ◆

◆ **Example 6.3.** Let V be an n-dimensional vector space and let $Gl(V)$ be the set of all nonsingular linear transformations of V onto V. The group operation in $Gl(V)$ is defined to be composition of linear transformations. With this operation, it is easy to verify that $Gl(V)$ is a group, the identity in $Gl(V)$ is the identity linear transformation, and if $g \in Gl(V)$, g^{-1} is the inverse linear transformation of g. The group $Gl(V)$ is often called the *general linear group* of V. When V is the coordinate space R^n, $Gl(V)$ is denoted by Gl_n. Clearly, Gl_n is just the set of $n \times n$ nonsingular matrices and the group operation is matrix multiplication. ◆

It should be noted that $\mathcal{O}(V)$ is a subset of $Gl(V)$ and the group operation in $\mathcal{O}(V)$ is that of $Gl(V)$. Further, G_T^+ and G_U^+ are subsets of Gl_n with the inherited group operations. This observation leads to the definition of a subgroup.

Definition 6.2. If $(G, \circ)$ is a group and H is a subset of G such that $(H, \circ)$ is also a group, then $(H, \circ)$ is a *subgroup* of $(G, \circ)$.

In all of the above examples, each element of the group is also a one-to-one function defined on a set. Further, the group operation is in fact function composition. To isolate the essential features of this situation, we define the following.

Definition 6.3. Let $(G, \circ)$ be a group and let $\mathcal{X}$ be a set. The group $(G, \circ)$ *acts on the left* of $\mathcal{X}$ if to each pair $(g, x) \in G \times \mathcal{X}$, there corresponds a unique element of $\mathcal{X}$, denoted by gx, such that

(i) $g_1(g_2x) = (g_1 \circ g_2)x$.

(ii) $ex = x$.

The content of Definition 6.3 is that there is a function on $G \times \mathcal{X}$ to $\mathcal{X}$ whose value at (g, x) is denoted by gx and under this mapping, (g_1, g_2x) and $(g_1 \circ g_2, x)$ are sent into the same element. Furthermore, (e, x) is mapped to x. Thus each $g \in G$ can be thought of as a one-to-one onto function from $\mathcal{X}$ to $\mathcal{X}$ and the group operation in G is function composition. To make this claim precise, for each $g \in G$, define t_g on $\mathcal{X}$ to $\mathcal{X}$ by $t_g(x) = gx$.

Proposition 6.1. Suppose G acts on the left of $\mathcal{X}$. Then each t_g is a one-to-one onto function from $\mathcal{X}$ to $\mathcal{X}$ and:

(i) $t_{g_1}t_{g_2} = t_{g_1 \circ g_2}$.
(ii) $t_g^{-1} = t_{g^{-1}}$.

Proof. To show t_g is onto, consider $x \in \mathcal{X}$. Then $t_g(g^{-1}x) = g(g^{-1}x) = (g \circ g^{-1})x = ex = x$ where (i) and (ii) of Definition 6.3 have been used. Thus t_g is onto. If $t_g(x_1) = t_g(x_2)$, then $gx_1 = gx_2$ so

$$x_1 = ex_1 = (g^{-1} \circ g)x_1 = g^{-1}(gx_1) = g^{-1}(gx_2)$$
$$= (g^{-1} \circ g)x_2 = ex_2 = x_2.$$

Thus t_g is one-to-one. Assertion (i) follows immediately from (i) of Definition 6.3. Since t_e is the identity function on $\mathcal{X}$ and (i) implies that

$$t_g t_{g^{-1}} = t_{g^{-1}} t_g = t_e,$$

we have $t_{g^{-1}} = t_g^{-1}$. □

Henceforth, we dispense with t_g and simply regard each g as a function on $\mathcal{X}$ to $\mathcal{X}$ where function composition is group composition and e is the identity function on $\mathcal{X}$. All of the examples considered thus far are groups of functions on a vector space to itself and the group operation is defined to be function composition. In particular, $Gl(V)$ is the set of all one-to-one onto linear transformations of V to V and the group operation is function composition. In the next example, the motivation for the definition of the group operation is provided by thinking of each group element as a function.

◆ **Example 6.4.** Let V be an n-dimensional vector space and consider the set $Al(V)$ that is the collection of all pairs (A, x) with $A \in Gl(V)$ and $x \in V$. Each pair (A, x) defines a one-to-one onto function from V to V by

$$(A, x)v = Av + x, \qquad v \in V.$$

The composition of (A_1, x_1) and (A_2, x_2) is

$$(A_1, x_1)(A_2, x_2)v = (A_1, x_1)(A_2v + x_2) = A_1A_2v + A_1x_2 + x_1$$
$$= (A_1A_2, A_1x_2 + x_1)v.$$

Also, $(I, 0) \in Al(V)$ is the identity function on V and the inverse of (A, x) is $(A^{-1}, -A^{-1}x)$. It is now an easy matter to verify that $Al(V)$ is a group where the group operation in $Al(V)$ is

$$(A_1, x_1)(A_2, x_2) \equiv (A_1A_2, A_1x_2 + x_1).$$

This group $Al(V)$ is called the *affine group* of V. When V is the coordinate space R^n, $Al(V)$ is denoted by Al_n. ◆

An interesting and useful subgroup of $Al(V)$ is given in the next example.

◆ **Example 6.5.** Suppose V is a finite dimensional vector space and let M be a subspace of V. Let H be the collection of all pairs (A, x) where $x \in M$, $A(M) \subseteq M$, and $(A, x) \in Al(V)$. The group operation in H is that inherited from $Al(V)$. It is a routine calculation to show that H is a subgroup of $Al(V)$. As a particular case, suppose that V is R^n and M is the m-dimensional subspace of R^n consisting of those vectors $x \in R^n$ whose last $n - m$ coordinates are zero. An $n \times n$ matrix $A \in Gl_n$ satisfies $AM \subseteq M$ iff

$$A = \begin{pmatrix} A_{11} & A_{12} \\ 0 & A_{22} \end{pmatrix}$$

where A_{11} is $m \times m$ and nonsingular, A_{12} is $m \times (n - m)$, and A_{22} is $(n - m) \times (n - m)$ and nonsingular. Thus H consists of all pairs (A, x) where $A \in Gl_n$ has the above form and x has its last $n - m$ coordinates zero. ◆

◆ **Example 6.6.** In this example, we consider two finite groups that arise naturally in statistical problems. Consider the space R^n and let P be an $n \times n$ matrix that permutes the coordinates of a vector $x \in R^n$. Thus in each row and in each column of P, there is a single element that is one and the remaining elements are zero. Conversely, any such matrix permutes the coordinates of vectors in R^n. The set $\mathcal{P}_n$ of all such matrices is called the group of *permutation matrices*. It is clear that $\mathcal{P}_n$ is a group under matrix multiplication and $\mathcal{P}_n$ has $n!$ elements. Also, let $\mathfrak{D}_n$ be the set of all $n \times n$ diagonal matrices whose diagonal elements are plus or minus one. Obviously, $\mathfrak{D}_n$ is a group under matrix multiplication and $\mathfrak{D}_n$ has 2^n elements. The group $\mathfrak{D}_n$ is called the group of *sign changes* on R^n. A bit of reflection shows that both $\mathcal{P}_n$ and $\mathfrak{D}_n$ are subgroups of $\mathcal{O}_n$. Now, let

H be the set

$$H = \{PD | P \in \mathcal{P}_n, D \in \mathcal{D}_n\}.$$

The claim is that H is a group under matrix multiplication. To see this, first note that for $P \in \mathcal{P}_n$ and $D \in \mathcal{D}_n$, PDP' is an element of $\mathcal{D}_n$. Thus if P_1D_1 and P_2D_2 are in H, then

$$P_1D_1P_2D_2 = P_1P_2P_2'D_1P_2D_2 = P_3D_3 \in H$$

where $P_3 = P_1P_2$ and $D_3 = P_2'D_1P_2D_2$. Also,

$$(PD)^{-1} = DP' = P'PDP' \in H.$$

Therefore H is a group and clearly has $2^n n!$ elements. ◆

Suppose that G is a group and H is a subgroup of G. The quotient space G/H, to be defined next, is often a useful representation of spaces that arise in later considerations. The subgroup H of G defines an equivalence relation in G by $g_1 \simeq g_2$ iff $g_2^{-1}g_1 \in H$. That $\simeq$ is an equivalence relation is easily verified using the assumption that H is a subgroup of G. Also, it is not difficult to show that $g_1 \simeq g_2$ iff the set $g_1H = \{g_1h | h \in H\}$ is equal to the set g_2H. Thus the set of points in G equivalent to g_1 is the set g_1H.

Definition 6.4. If H is a subgroup of G, the quotient space G/H is defined to be the set whose elements are gH for $g \in G$.

The quotient space G/H is obviously the set of equivalence classes (defined by H) of elements of G. Under certain conditions on H, the quotient space G/H is in fact a group under a natural definition of a group operation.

Definition 6.5. A subgroup H of G is called a *normal subgroup* if $g^{-1}Hg = H$ for all $g \in G$.

When H is a normal subgroup of G, and $g_iH \in G/H$ for $i = 1, 2$, then

$$g_1Hg_2H \equiv \{g | g = g_1h_1g_2h_2;\ h_1, h_2 \in H\}$$

$$= g_1g_2g_2^{-1}Hg_2H = g_1g_2HH = g_1g_2H$$

since $HH = H$.

Proposition 6.2. When H is a normal subgroup of G, the quotient space G/H is a group under the operation

$$(g_1H)(g_2H) \equiv g_1g_2H.$$

Proof. This is a routine calculation and is left to the reader. □

◆ **Example 6.7.** Let $Al(V)$ be the affine group of the vector space V. Then

$$H \equiv \{(I, x)|x \in V\}$$

is easily shown to be a subgroup of G, since $(I, x_1)(I, x_2) = (I, x_1 + x_2)$. To show H is normal in $Al(V)$, consider $(A, x) \in Al(V)$ and $(I, x_0) \in H$. Then

$$\begin{aligned}(A, x)^{-1}(I, x_0)(A, x) &= (A^{-1}, -A^{-1}x)(A, x + x_0)\\ &= \left(I, A^{-1}x + A^{-1}x_0 - A^{-1}x\right)\\ &= \left(I, A^{-1}x_0\right),\end{aligned}$$

which is an element of H. Thus $g^{-1}Hg \subseteq H$ for all $g \in Al(V)$. But if $(I, x_0) \in H$ and $(A, x) \in Al(V)$, then

$$(A, x)^{-1}(I, Ax_0)(A, x) = (I, x_0)$$

so $g^{-1}Hg = H$, for $g \in Al(V)$. Therefore, H is normal in $Al(V)$. To describe the group $Al(V)/H$, we characterize the equivalence relation defined by H. For $(A_i, x_i) \in Al(V)$, $i = 1, 2$,

$$\begin{aligned}(A_1, x_1)^{-1}(A_2, x_2) &= \left(A_1^{-1}, -A_1^{-1}x_1\right)(A_2, x_2)\\ &= \left(A_1^{-1}A_2, A_1^{-1}x_2 - A_1^{-1}x_1\right)\end{aligned}$$

is an element of H iff $A_1^{-1}A_2 = I$ or $A_1 = A_2$. Thus (A_1, x_1) is equivalent to (A_2, x_2) iff $A_1 = A_2$. From each equivalence class, select the element $(A, 0)$. Then it is clear that the quotient group $Al(V)/H$ can be identified with the group

$$K = \{(A, 0)|A \in Gl(V)\}$$

where the group operation is

$$(A_1, 0)(A_2, 0) = (A_1 A_2, 0). \qquad \blacklozenge$$

Now, suppose the group G acts on the left of the set $\mathfrak{X}$. We say G acts *transitively* on $\mathfrak{X}$ if, for each x_1 and x_2 in $\mathfrak{X}$, there exists a $g \in G$ such that $gx_1 = x_2$. When G acts transitively on $\mathfrak{X}$, we want to show that there is a natural one-to-one correspondence between $\mathfrak{X}$ and a certain quotient space. Fix an element $x_0 \in \mathfrak{X}$ and let

$$H = \{h | hx_0 = x_0, h \in G\}.$$

The subgroup H of G is called the *isotropy subgroup* of x_0. Now, define the function τ on G/H to $\mathfrak{X}$ by $\tau(gH) = gx_0$.

Proposition 6.3. The function τ is one-to-one and onto. Further,

$$\tau(g_1 gH) = g_1 \tau(gH).$$

Proof. The definition of τ clearly makes sense as $ghx_0 = gx_0$ for all $h \in H$. Also, τ is an onto function since G acts transitively on $\mathfrak{X}$. If $\tau(g_1 H) = \tau(g_2 H)$, then $g_1 x_0 = g_2 x_0$ so $g_2^{-1} g_1 \in H$. Therefore, $g_1 H = g_2 H$ so τ is one-to-one. The rest is obvious. □

If H is any subgroup of G, then the group G acts transitively on $\mathfrak{X} \equiv G/H$ where the group action is

$$g_1(gH) \equiv g_1 gH.$$

Thus we have a complete description of the spaces $\mathfrak{X}$ that are acted on transitively by G. Namely, these spaces are simply relabelings of the quotient spaces G/H where H is a subgroup of G. Further, the action of g on $\mathfrak{X}$ corresponds to the action of G on the quotient space described in Proposition 6.3. A few examples illustrate these ideas.

◆ **Example 6.8.** Take the set $\mathfrak{X}$ to be $\mathfrak{F}_{p,n}$—the set of $n \times p$ real matrices Ψ that satisfy $\Psi'\Psi = I_p$, $1 \leqslant p \leqslant n$. The group $G = \mathcal{O}_n$ of all $n \times n$ orthogonal matrices acts on $\mathfrak{F}_{p,n}$ by matrix multiplication. That is, if $\Gamma \in \mathcal{O}_n$ and $\Psi \in \mathfrak{F}_{p,n}$, then $\Gamma\Psi$ is the matrix product of Γ and Ψ. To show that this group action is transitive, consider Ψ_1 and Ψ_2 in $\mathfrak{F}_{p,n}$. Then, the columns of Ψ_1 form a set of p orthonormal vectors in R^n as do the columns of Ψ_2. By Proposition 1.30, there exists an $n \times n$ orthogonal matrix Γ that maps the columns of Ψ_1 into the columns of Ψ_2. Thus $\Gamma\Psi_1 = \Psi_2$ so $\mathcal{O}_n$ is transitive on $\mathcal{F}_{p,n}$. A convenient choice of $x_0 \in \mathcal{F}_{p,n}$ to define the map τ is

$$x_0 = \begin{pmatrix} I_p \\ 0 \end{pmatrix}$$

where 0 is a block of $(n-p) \times p$ zeroes. It is not difficult to show that the subgroup $H = \{\Gamma | \Gamma x_0 = x_0, \Gamma \in \mathcal{O}_n\}$ is

$$H = \left\{ \Gamma | \Gamma = \begin{pmatrix} I_p & 0 \\ 0 & \Gamma_{22} \end{pmatrix}, \Gamma_{22} \in \mathcal{O}_{(n-p)} \right\}.$$

The function τ is

$$\tau(\Gamma H) = \Gamma x_0 = \Gamma \begin{pmatrix} I_p \\ 0 \end{pmatrix},$$

which is the $n \times p$ matrix consisting of the first p columns of Γ. This gives an obvious representation of $\mathcal{F}_{p,n}$. ◆

◆ **Example 6.9.** Let $\mathcal{X}$ be the set of all $p \times p$ positive definite matrices and let $G = Gl_p$. The transitive group action is given by $A(x) = AxA'$ where A is a $p \times p$ nonsingular matrix, $x \in \mathcal{X}$, and A' is the transpose of A. Choose $x_0 \in \mathcal{X}$ to be I_p. Obviously, $H = \mathcal{O}_p$ and the map τ is given by

$$\tau(AH) = A(x_0) = AA'.$$

The reader should compare this example with the assertion of Proposition 1.31. ◆

◆ **Example 6.10.** In this example, take $\mathcal{X}$ to be the set of all $n \times p$ real matrices of rank p, $p \leqslant n$. Consider the group G defined by

$$G = \{g | g = \Gamma \otimes T, \Gamma \in \mathcal{O}_n, T \in G_T^+\}$$

where G_T^+ is the group of all $p \times p$ lower triangular matrices with positive diagonal elements. Of course, $\otimes$ denotes the Kronecker product and group composition is

$$(\Gamma_1 \otimes T_1)(\Gamma_2 \otimes T_2) = (\Gamma_1\Gamma_2) \otimes (T_1T_2).$$

The action of G on $\mathfrak{X}$ is

$$(\Gamma \otimes T)X = \Gamma X T', \qquad X \in \mathfrak{X}.$$

To show G acts transitively on $\mathfrak{X}$, consider $X_1, X_2 \in \mathfrak{X}$ and write $X_i = \Psi_i U_i$, where $\Psi_i \in \mathfrak{F}_{p,n}$ and $U_i \in G_U^+$, $i = 1, 2$ (see Proposition 5.2). From Example 6.8, there is a $\Gamma \in \mathcal{O}_n$ such that $\Gamma\Psi_1 = \Psi_2$. Let $T' = U_1^{-1}U_2$ so

$$\Gamma X_1 T' = \Gamma \Psi_1 U_1 U_1^{-1} U_2 = \Psi_2 U_2 = X_2.$$

Choose $X_0 \in \mathfrak{X}$ to be

$$X_0 = \begin{pmatrix} I_p \\ 0 \end{pmatrix}$$

as in Example 6.8. Then the equation $(\Gamma \otimes T)X_0 = X_0$ implies that

$$I_p = X_0'X_0 = ((\Gamma \otimes T)X_0)'(\Gamma \otimes T)X_0 = TX_0'\Gamma'\Gamma X_0 T' = TT'$$

so $T = I_p$ by Proposition 5.4. Then the equation $(\Gamma \otimes I_p)X_0 = X_0$ is exactly the equation occurring in Example 6.8 for elements of the subgroup H. Thus for this example,

$$H = \left\{ \Gamma \otimes I_p | \Gamma = \begin{pmatrix} I_p & 0 \\ 0 & \Gamma_{22} \end{pmatrix}, \Gamma_{22} \in \mathcal{O}_{n-p} \right\}.$$

Therefore,

$$\tau((\Gamma \otimes T)H) = (\Gamma \otimes T)X_0 = \Gamma \begin{pmatrix} I_p \\ 0 \end{pmatrix} T'$$

is the representation for elements of $\mathfrak{X}$. Obviously,

$$\Gamma \begin{pmatrix} I_p \\ 0 \end{pmatrix} \equiv \Psi \in \mathfrak{F}_{p,n}$$

and the representation of elements of $\mathfrak{X}$ via the map τ is precisely the representation established in Proposition 5.2. This representation of $\mathfrak{X}$ is used on a number of occasions. ◆

6.2. INVARIANT MEASURES AND INTEGRALS

Before beginning a discussion of invariant integrals on locally compact topological groups, we first outline the basic results of integration theory on locally compact topological spaces. Consider a set $\mathcal{X}$ and let $\mathcal{J}$ be a Hausdorff topology for $\mathcal{X}$.

Definition 6.6. The topological space $(\mathcal{X}, \mathcal{J})$ is a *locally compact* space if for each $x \in \mathcal{X}$, there exists a compact neighborhood of x.

Most of the groups introduced in the examples of the previous section are subsets of the space R^m, for some m, and when these groups are given the topology of R^m, they are locally compact spaces. The verification of this is not difficult and is left to the reader. If $(\mathcal{X}, \mathcal{J})$ is a locally compact space, $\mathcal{K}(\mathcal{X})$ denotes the set of all continuous real-valued functions that have compact support. Thus $f \in \mathcal{K}(\mathcal{X})$ if f is a continuous and there is a compact set K such that $f(x) = 0$ if $x \notin K$. It is clear that $\mathcal{K}(\mathcal{X})$ is a real vector space with addition and scalar multiplication being defined in the obvious way.

Definition 6.7. A real-valued function J defined on $\mathcal{K}(\mathcal{X})$ is called an *integral* if:

(i) $J(\alpha_1 f_1 + \alpha_2 f_2) = \alpha_1 J(f_1) + \alpha_2 J(f_2)$ for $\alpha_1, \alpha_2 \in R$ and $f_1, f_2 \in \mathcal{K}(\mathcal{X})$.

(ii) $J(f) \geqslant 0$ if $f \geqslant 0$, $f \in \mathcal{K}(\mathcal{X})$.

An integral J is simply a linear function on $\mathcal{K}(\mathcal{X})$ that has the additional property that $J(f)$ is nonnegative when $f \geqslant 0$. Let $\mathcal{B}(\mathcal{X})$ be the σ-algebra generated by the compact subsets of $\mathcal{X}$. If μ is a measure on $\mathcal{B}(\mathcal{X})$ such that $\mu(K) < +\infty$ for each compact set K, it is clear that

$$J(f) \equiv \int_{\mathcal{X}} f(x)\mu(dx)$$

defines an integral on $\mathcal{K}(\mathcal{X})$. Such measures μ are called *Radon measures.* Conversely, given an integral J, there is a measure μ on $\mathcal{B}(\mathcal{X})$ such that $\mu(K) < +\infty$ for all compact sets K and

$$J(f) = \int_{\mathcal{X}} f(x)\mu(dx)$$

for $f \in \mathcal{K}(\mathcal{X})$. For a proof of this result, see Segal and Kunze (1978,

Chapter 5). In the special case when $(\mathcal{X}, \mathcal{J})$ is a σ-compact space—that is, $\mathcal{X} = \cup_1^\infty K_i$ where K_i is compact—then the correspondence between integrals J and measures μ that satisfy $\mu(K) < +\infty$ for K compact is one-to-one (see Segal and Kunze, 1978). All of the examples considered here are σ-compact spaces and we freely identify integrals with Radon measures and vice versa.

Now, assume $(\mathcal{X}, \mathcal{J})$ is a σ-compact space. If an integral J on $\mathcal{K}(\mathcal{X})$ corresponds to a Radon measure μ on $\mathcal{B}(\mathcal{X})$, then J has a natural extension to the class of all $\mathcal{B}(\mathcal{X})$-measurable and μ-integrable functions. Namely, J is extended by the equation

$$J(f) = \int_{\mathcal{X}} f(x)\mu(dx)$$

for all f for which the right-hand side is defined. Obviously, the extension of J is unique and is determined by the values of J on $\mathcal{K}(\mathcal{X})$. In many of the examples in this chapter, we use J to denote both an integral on $\mathcal{K}(\mathcal{X})$ and its extension. With this convention, J is defined for any $\mathcal{B}(\mathcal{X})$ measurable function that is μ-integrable where μ corresponds to J.

Suppose G is a group and $\mathcal{J}$ is a topology on G.

Definition 6.8. Given the topology $\mathcal{J}$ on G, $(G, \mathcal{J})$ is a *topological group* if the mapping $(x, y) \to xy^{-1}$ is continuous from $G \times G$ to G. If $(G, \mathcal{J})$ is a topological group and $(G, \mathcal{J})$ is a locally compact topological space, $(G, \mathcal{J})$ is called a *locally compact topological group*

In what follows, all groups under consideration are locally compact topological groups. Examples of such groups include the vector space R^n, the general linear group Gl_n, the affine group Al_n, and G_T^+. The verification that these groups are locally compact topological groups with the Euclidean space topology is left to the reader.

If $(G, \mathcal{J})$ is a locally compact topological group, $\mathcal{K}(G)$ denotes the real vector space of all continuous functions on G that have compact support. For $s \in G$ and $f \in \mathcal{K}(G)$, the *left translate* of f by s, denoted by sf, is defined by $(sf)(x) \equiv f(s^{-1}x)$, $x \in G$. Clearly, $sf \in \mathcal{K}(G)$ for all $s \in G$. Similarly, the *right translate* of $f \in \mathcal{K}(G)$, denoted by fs, is $(fs)(x) \equiv f(xs^{-1})$ and $fs \in \mathcal{K}(G)$.

Definition 6.9. An integral $J \neq 0$ on $\mathcal{K}(G)$ is *left invariant* if $J(sf) = J(f)$ for all $f \in \mathcal{K}(G)$ and $s \in G$. An integral $J \neq 0$ on $\mathcal{K}(G)$ is *right invariant* if $J(fs) = J(f)$ for all $f \in \mathcal{K}(G)$ and $s \in G$.

The basic properties of left and right invariant integrals are summarized in the following two results.

Theorem 6.1. If G is a locally compact topological group, then there exist left and right invariant integrals on $\mathcal{K}(G)$. If J_1 and J_2 are left (right) invariant integrals on $\mathcal{K}(G)$, then $J_2 = cJ_1$ for some positive constant c.

Proof. See Nachbin (1965, Section 4, Chapter 2).

Theorem 6.2. Suppose that

$$J(f) \equiv \int f(x)\mu(dx)$$

is a left invariant integral on $\mathcal{K}(G)$. Then there exists a unique continuous function Δ_r mapping G into $(0, \infty)$ such that

$$\int f(xs^{-1})\mu(dx) = \Delta_r(s)\int f(x)\mu(dx)$$

for all $s \in G$ and $f \in \mathcal{K}(G)$. The function Δ_r, called the *right-hand modulus* of G, also satisfies:

(i) $\Delta_r(st) = \Delta_r(s)\Delta_r(t)$, $s, t \in G$.
(ii) $\int f(x^{-1})\mu(dx) = \int f(x)\Delta_r(x^{-1})\mu(dx)$.

Further, the integral

$$J_1(f) = \int f(x)\Delta_r(x^{-1})\mu(dx)$$

is right invariant.

Proof. See Nachbin (1965, Section 5, Chapter 2).

The two results above establish the existence and uniqueness of right and left invariant integrals and show how to construct right invariant integrals from left invariant integrals via the right-hand modulus Δ_r. The right-hand modulus is a continuous homomorphism from G into $(0, \infty)$—that is, Δ_r is continuous and satisfies $\Delta_r(st) = \Delta_r(s)\Delta_r(t)$, for $s, t \in G$. (The definition of a homomorphism from one group to another group is given shortly.)

Before presenting examples of invariant integrals, it is convenient to introduce relatively left (and right) invariant integrals. Proposition 6.4, given below, provides a useful method for constructing invariant integrals from relatively invariant integrals.

Definition 6.10. A nonzero integral J on $\mathcal{K}(G)$ given by

$$J(f) = \int f(x)m(dx), \qquad f \in \mathcal{K}(G),$$

is called *relatively left invariant* if there exists a function χ on G to $(0, \infty)$ such that

$$\int f(s^{-1}x)m(dx) = \chi(s)\int f(x)m(dx)$$

for all $s \in G$ and $f \in \mathcal{K}(G)$. The function χ is the *multiplier* for J.

It can be shown that any multiplier χ is continuous (see Nachbin, 1965). Further, if J is relatively left invariant with multiplier χ, then for $s, t \in G$ and $f \in \mathcal{K}(G)$,

$$\begin{aligned}\chi(st)\int f(x)m(dx) &= \int f((st)^{-1}x)m(dx) = \int (tf)(s^{-1}x)m(dx)\\ &= \chi(s)\int (tf)(x)m(dx) = \chi(s)\int f(t^{-1}x)m(dx)\\ &= \chi(s)\chi(t)\int f(x)m(dx).\end{aligned}$$

Thus $\chi(st) = \chi(s)\chi(t)$. Hence all multipliers are continuous and are homomorphisms from G into $(0, \infty)$. For any such homomorphism χ, it is clear that $\chi(e) = 1$ and $\chi(s^{-1}) = 1/\chi(s)$. Also, $\chi(G) = \{\chi(s) | s \in G\}$ is a subgroup of the group $(0, \infty)$ with multiplication as the group operation.

Proposition 6.4. Let χ be a continuous homomorphism on G to $(0, \infty)$.

(i) If $J(f) = \int f(x)\mu(dx)$ is left invariant on $\mathcal{K}(G)$, then

$$J_1(f) \equiv \int f(x)\chi(x)\mu(dx)$$

is a relatively left invariant integral on $\mathcal{K}(G)$ with multiplier χ.

(ii) If $J_1(f) = \int f(x)m(dx)$ is relatively left invariant with multiplier χ, then

$$J(f) \equiv \int f(x)\chi(x^{-1})m(dx)$$

is a left invariant integral.

Proof. The proof is a calculation. For (i),

$$J_1(sf) = \int (sf)(x)\chi(x)\mu(dx) = \int f(s^{-1}x)\chi(ss^{-1}x)\mu(dx)$$

$$= \chi(s)\int f(s^{-1}x)\chi(s^{-1}x)\mu(dx) = \chi(s)\int f(x)\chi(x)\mu(dx)$$

$$= \chi(s)J_1(f).$$

Thus J_1 is relatively left invariant with multiplier χ. For (ii),

$$J(sf) = \int f(s^{-1}x)\chi(x^{-1})m(dx) = \int f(s^{-1}x)\chi(s^{-1}sx^{-1})m(dx)$$

$$= \chi(s^{-1})\int f(s^{-1}x)\chi\left((s^{-1}x)^{-1}\right)m(dx)$$

$$= \chi(s^{-1})\chi(s)\int f(x)\chi(x^{-1})m(dx)$$

$$= \int f(x)\chi(x^{-1})m(dx) = J(f).$$

Thus J is a left invariant integral and the proof is complete. □

If J is a relatively left invariant integral with multiplier χ, say

$$J(x) = \int f(x)m(dx),$$

the measure m is also called relatively left invariant with multiplier χ. A nonzero integral J_1 on $\mathcal{K}(G)$ is *relatively right invariant* with multiplier χ if $J_1(fs) = \chi(s)J_1(f)$. Using the results given above, if J_1 is relatively right invariant with multiplier χ, then J_1 is relatively left invariant with multiplier

χ/Δ_r where Δ_r is the right-hand modulus of G. Thus all relatively right and left invariant integrals can be constructed from a given relatively left (or right) invariant integral once all the continuous homomorphisms are known. Also, if a relatively left invariant measure m can be found and its multiplier χ calculated, then a left invariant measure is given by m/χ according to Proposition 6.4. This observation is used in the examples below.

◆ **Example 6.11.** Consider the group Gl_n of all nonsingular $n \times n$ matrices. Let ds denote Lebesgue measure on Gl_n. Since $Gl_n = \{s|\det(s) \neq 0\}$, Gl_n is a nonempty open subset of n^2-dimensional Euclidean space and hence has positive Lebesgue measure. For $f \in \mathcal{K}(Gl_n)$, let

$$J(f) = \int f(t)\, dt.$$

To find a left invariant measure on Gl_n, it is now shown that $J(sf) = |\det(s)|^n J(f)$ so J is relatively left invariant with multiplier $\chi(s) = |\det(s)|^n$. From Proposition 5.10, the Jacobian of the transformation $g(t) = st$, $s \in Gl_n$, is $|\det(s)|^n$. Thus

$$J(sf) = \int f(s^{-1}t)\, dt = |\det(s)|^n \int f(t)\, dt = |\det(s)|^n J(f).$$

From Proposition 6.4, it follows that the measure

$$\mu(dt) = \frac{dt}{|\det(t)|^n}$$

is a left invariant measure on Gl_n. A similar Jacobian argument shows that μ is also right invariant, so the right-hand modulus of Gl_n is $\Delta_r \equiv 1$. To construct all of the relatively invariant measures on Gl_n, it is necessary that the continuous homomorphisms χ be characterized. For each $\alpha \in R$, let

$$\chi_\alpha(s) = |\det(s)|^\alpha, \qquad s \in Gl_n.$$

Obviously, each χ_α is a continuous homomorphism. However, it can be shown (see the problems at the end of this chapter) that if χ is a continuous homomorphism of Gl_n into $(0, \infty)$, then $\chi = \chi_\alpha$ for some $\alpha \in R$. Hence every relatively invariant measure on Gl_n is given by

$$m(dt) = c\chi_\alpha(t)\frac{dt}{\chi_n(t)}$$

where c is a positive constant and $\alpha \in R$. ◆

A group G for which $\Delta_r = 1$ is called *unimodular*. Clearly, all commutative groups are unimodular as a left invariant integral is also right invariant. In the following example, we consider the group G_T^+, which is not unimodular, but G_T^+ is a subgroup of the unimodular group Gl_n.

◆ **Example 6.12.** Let G_T^+ be the group of all $n \times n$ lower triangular matrices with positive diagonal elements. Thus G_T^+ is a nonempty open subset of $[n(n+1)/2]$-dimensional Euclidean space so G_T^+ has positive Lebesgue measure. Let dt denote $[n(n+1)/2]$-dimensional Lebesgue measure restricted to G_T^+. Consider the integral

$$J(f) \equiv \int f(t)\, dt$$

defined on $\mathcal{K}(G_T^+)$. The Jacobian of the transformation $g(t) = st$, $s \in G_T^+$, is equal to

$$\chi_0(s) \equiv \prod_{i=1}^{n} s_{ii}^{i}$$

where s has diagonal elements $s_{11}, \ldots, s_{nn}$ (see Proposition 5.13). Thus

$$J(sf) = \int f(s^{-1}t)\, dt = \chi_0(s) \int f(t)\, dt = \chi_0(s) J(f).$$

Hence J is relatively left invariant with multiplier χ_0 so the measure

$$\mu(dt) \equiv \frac{dt}{\prod_{i=1}^{n} t_{ii}^{i}} = \frac{dt}{\chi_0(t)}$$

is left invariant. To compute the right-hand modulus Δ_r for G_T^+, let

$$J_1(f) = \int f(t)\mu(dt)$$

so J_1 is left invariant. Then

$$J_1(fs) = \int f(ts^{-1})\mu(dt) = \int f(ts^{-1}) \frac{dt}{\prod_1^n t_{ii}^{i}} = \int f(ts^{-1}) \frac{dt}{\chi_0(t)}$$

$$= \int f(ts^{-1}) \frac{\chi_0(s^{-1})}{\chi_0(ts^{-1})} dt = \chi_0(s^{-1}) \int \frac{f(ts^{-1})}{\chi_0(ts^{-1})} dt.$$

By Proposition 5.14, the Jacobian of the transform $g(t) = ts$ is

$$\chi_1(s) = \prod_{i=1}^{n} s_{ii}^{n-i+1}.$$

Therefore,

$$J_1(fs) = \chi_0(s^{-1}) \int \frac{f(ts^{-1})}{\chi_0(ts^{-1})} dt = \chi_0(s^{-1})\chi_1(s) \int f(t) \frac{dt}{\chi_0(t)}$$

$$= \frac{\chi_1(s)}{\chi_0(s)} J_1(f).$$

By Theorem 6.2,

$$\Delta_r(s) = \frac{\chi_1(s)}{\chi_0(s)} = \prod_{i=1}^{n} s_{ii}^{n-2i+1}$$

is the right-hand modulus for G_T^+. Therefore, the measure

$$\nu(dt) \equiv \frac{\mu(dt)}{\Delta_r(t)} = \frac{dt}{\chi_0(t)\Delta_r(t)} = \frac{dt}{\prod_{i=1}^{n} t_{ii}^{n-i+1}}$$

is right invariant. As in the previous example, a description of the relatively left invariant measures is simply a matter of describing all the continuous homomorphisms on G_T^+. For each vector $c \in R^n$ with coordinates $c_1, \ldots, c_n$, let

$$\chi_c(t) \equiv \prod_{i=1}^{n} (t_{ii})^{c_i}$$

where $t \in G_T^+$ has diagonal elements $t_{11}, \ldots, t_{nn}$. It is easy to verify that χ_c is a continuous homomorphism on G_T^+. It is known that if χ is a continuous homomorphism on G_T^+, then χ is given by χ_c for some $c \in R^n$ (see Problems 6.4 and 6.9). Thus every relatively left invariant measure on G_T^+ has the form

$$m(dt) = k\chi_c(t) \frac{dt}{\chi_0(t)}$$

for some positive constant k and some vector $c \in R^n$. ◆

The following two examples deal with the affine group and a subgroup of Gl_n related to the group introduced in Example 6.5.

◆ **Example 6.13.** Consider the group Al_n of all affine transformations on R^n. An element of Al_n is a pair (s, x) where $s \in Gl_n$ and $x \in R^n$. Recall that the group operation in Al_n is

$$(s_1, x_1)(s_2, x_2) = (s_1 s_2, s_1 x_2 + x_1)$$

so

$$(s, x)^{-1} = (s^{-1}, -s^{-1}x).$$

Let $ds\,dx$ denote Lebesgue measure restricted to Al_n. In order to construct a left invariant measure on Al_n, it is shown that the integral

$$J(f) \equiv \int f(t, y)\,dt\,dy$$

is relatively left invariant with multiplier

$$\chi_0(s, x) = |\det(s)|^{n+1}.$$

For $(s, x) \in Al_n$,

$$J((s, x)f) = \int f((s, x)^{-1}(t, y))\,dt\,dy$$

$$= \int f((s^{-1}, -s^{-1}x)(t, y))\,dt\,dy$$

$$= \int f(s^{-1}t, s^{-1}y - s^{-1}x)\,dt\,dy$$

$$= |\det(s)| \int f(s^{-1}t, u)\,dt\,du.$$

The last equality follows from the change of variable $u = s^{-1}y - sx$, which has a Jacobian $|\det(s)|$. As in Example 6.11,

$$\int_{Gl_n} f(s^{-1}t, u)\,dt = |\det(s)|^n \int_{Gl_n} f(t, u)\,dt$$

for each fixed $u \in R^n$. Thus

$$J((s, x)f) = |\det(s)|^{n+1} \int f(t, u)\,dt\,du = |\det(s)|^{n+1} J(f)$$

$$= \chi_0(s, x) J(f)$$

so J is relatively left invariant with multiplier χ_0. Hence the measure

$$\mu(ds, du) \equiv \frac{ds\,du}{\chi_0(s, u)} = \frac{ds\,du}{|\det(s)|^{n+1}}$$

is left invariant. To find the right-hand modulus of Al_n, let

$$J_1(f) = \int f(t, u)\frac{dt\,du}{\chi_0(t, u)}$$

be a left invariant integral. Then using an argument similar to that above, we have

$$\begin{aligned}
J_1(f(s, x)) &= \int f((t, u)(s, x)^{-1})\frac{dt\,du}{\chi_0(t, u)} \\
&= \int f((t, u)(s^{-1}, -sx))\frac{dt\,du}{\chi_0(t, u)} \\
&= \int f(ts^{-1}, u - ts^{-1}x)\frac{dt\,du}{|\det(t)|^{n+1}} \\
&= \int f(ts^{-1}, u)\frac{dt\,du}{|\det(t)|^{n+1}} \\
&= |\det(s^{-1})|^{n+1}\int f(ts^{-1}, u)\frac{dt\,du}{|\det(ts^{-1})|^{n+1}} \\
&= |\det(s^{-1})|^{n+1}|\det(s)|^{n}\int f(t, u)\frac{dt\,du}{|\det(t)|^{n+1}} \\
&= |\det(s)|^{-1}J_1(f).
\end{aligned}$$

Thus $\Delta_r(s, x) = |\det(s)|^{-1}$ so a right invariant measure on Al_n is

$$\nu(ds, du) = \frac{1}{\Delta_r(s, u)}\mu(ds, du) = \frac{ds\,du}{|\det(s)|^{n}}.$$

Now, suppose that χ is a continuous homomorphism on Al_n. Since

$$(s, x) = (s, 0)(e, s^{-1}x) = (e, x)(s, 0)$$

where e is the $n \times n$ identity matrix, χ must satisfy the equation

$$\chi(s, x) = \chi(s, 0)\chi(e, s^{-1}x) = \chi(s, 0)\chi(e, x)$$

Thus for all $s \in Gl_n$,

$$\chi(e, x) = \chi(e, s^{-1}x).$$

Letting s^{-1} converge to the zero matrix, the continuity of χ implies that

$$\chi(e, x) = \chi(e, 0) = 1$$

since $(e, 0)$ is the identity in Al_n. Therefore,

$$\chi(s, x) = \chi(s, 0), \qquad s \in Gl_n.$$

However,

$$\chi((s_1, 0)(s_2, 0)) = \chi((s_1 s_2), 0) = \chi(s_1, 0)\chi(s_2, 0)$$

so χ is a continuous homomorphism on Gl_n. But every continuous homomorphism on Gl_n is given by $s \to |\det(s)|^\alpha$ for some real α. In summary, χ is a continuous homomorphism on Al_n iff

$$\chi(s, x) = |\det(s)|^\alpha$$

for some real number α. Thus we have a complete description of all the relatively invariant integrals on Al_n. ◆

◆ **Example 6.14.** In this example, the group G consists of all the $n \times n$ nonsingular matrices s that have the form

$$s = \begin{pmatrix} s_{11} & s_{12} \\ 0 & s_{22} \end{pmatrix}; \qquad s_{11} \in Gl_p, \qquad s_{22} \in Gl_q$$

where $p + q = n$. Let M be the subspace of R^n consisting of those vectors whose last q coordinates are zero. Then G is the subgroup of Gl_n consisting of those elements s that satisfy $s(M) \subseteq M$. Let $ds_{11}\, ds_{12}\, ds_{22}$ denote Lebesgue measure restricted to G when G is regarded as a subset of $(p^2 + q^2 + pq)$-dimensional Euclidean space. Since G is a nonempty open subset of this space, G has positive Lebesgue measure. As in previous examples, it is shown that the integral

$$J(f) \equiv \int f(t)\, dt_{11}\, dt_{12}\, dt_{22}$$

is relatively left invariant. For $s \in G$,

$$J(sf) = \int f(s^{-1}t)\, dt_{11}\, dt_{12}\, dt_{22}.$$

A bit of calculation shows that

$$\begin{pmatrix} s_{11} & s_{12} \\ 0 & s_{22} \end{pmatrix}^{-1} = \begin{pmatrix} s_{11}^{-1} & -s_{11}^{-1}s_{12}s_{22}^{-1} \\ 0 & s_{22}^{-1} \end{pmatrix}$$

and

$$s^{-1}t = \begin{pmatrix} s_{11}^{-1}t_{11} & s_{11}^{-1}t_{12} - s_{11}^{-1}s_{12}s_{22}^{-1}t_{22} \\ 0 & s_{22}^{-1}t_{22} \end{pmatrix}.$$

Let

$$u_{11} = s_{11}^{-1}t_{11}, \qquad u_{22} = s_{22}^{-1}t_{22}$$

$$u_{12} = s_{11}^{-1}t_{12} - s_{11}^{-1}s_{12}s_{22}^{-1}t_{22}.$$

The Jacobian of this transformation is

$$\chi_0(s) \equiv |\det(s_{11})|^p |\det(s_{22})|^q |\det(s_{11})|^q = |\det(s_{11})|^n |\det(s_{22})|^q.$$

Therefore,

$$J(sf) = \chi_0(s) J(f)$$

so the measure

$$\mu(dt_{11}, dt_{12}, dt_{22}) \equiv \frac{dt_{11}\, dt_{12}\, dt_{22}}{|\det(t_{11})|^n |\det(t_{22})|^q}$$

is left invariant. Setting

$$J_1(f) \equiv \int f(t) \mu(dt_{11}, dt_{12}, dt_{22}),$$

a calculation similar to that above yields

$$J_1(fs) = \Delta_r(s)J_1(f)$$

where

$$\Delta_r(s) = |\det s_{11}|^{-q}|\det s_{22}|^{p}.$$

Thus Δ_r is the right-hand modulus of G and the measure

$$\nu(dt_{11}, dt_{12}, dt_{22}) \equiv \frac{\mu(dt_{11}, dt_{12}, dt_{22})}{\Delta_r(t)} = \frac{dt_{11}\, dt_{12}\, dt_{22}}{|\det(t_{11})|^{p}|\det(t_{22})|^{n}}$$

is right invariant. For $\alpha, \beta \in R$, let

$$\chi_{\alpha\beta}(s) \equiv |\det(s_{11})|^{\alpha}|\det(s_{22})|^{\beta}.$$

Clearly, $\chi_{\alpha\beta}$ is a continuous homomorphism of G into $(0, \infty)$. Conversely, it is not too difficult to show that every continuous homomorphism of G into $(0, \infty)$ is equal to $\chi_{\alpha\beta}$ for some $\alpha, \beta \in R$. Again, this gives a complete description of all the relatively invariant integrals on G. ◆

In the four examples above, the same argument was used to derive the left and right invariant measures, the modular function, and all of the relatively invariant measures. Namely, the group G had positive Lebesgue measure when regarded as a subset of an obvious Euclidean space. The integral on $\mathcal{K}(G)$ defined by Lebesgue measure was relatively left invariant with a multiplier that we calculated. Thus a left invariant measure on G was simply Lebesgue measure divided by the multiplier. From this, the right-hand modulus and a right invariant measure were easily derived. The characterization of the relatively invariant integrals amounted to finding all the solutions to the functional equation $\chi(st) = \chi(s)\chi(t)$ where χ is a continuous function on G to $(0, \infty)$. Of course, the above technique can be applied to many other matrix groups—for example, the matrix group considered in Example 6.5. However, there are important matrix groups for which this argument is not available because the group has Lebesgue measure zero in the "natural" Euclidean space of which the group is a subset. For example, consider the group of $n \times n$ orthogonal matrices $\mathcal{O}_n$. When regarded as a subset of n^2-dimensional Euclidean space, $\mathcal{O}_n$ has Lebesgue measure zero. But, without a fairly complicated parameterization of $\mathcal{O}_n$, it is not possible to regard $\mathcal{O}_n$ as a set of positive Lebesgue measure of some Euclidean space.

For this reason, we do not demonstrate directly the existence of an invariant measure on $\mathcal{O}_n$ in this chapter. In the following chapter, a probabilistic proof of the existence of an invariant measure on $\mathcal{O}_n$ is given.

The group $\mathcal{O}_n$, as well as other groups to be considered later, are in fact compact topological groups. A basic property of such groups is given next.

Proposition 6.5. Suppose G is a locally compact topological group. Then G is compact iff there exists a left invariant probability measure on G.

Proof. See Nachbin (1965, Section 5, Chapter 2). □

The following result shows that when G is compact, left invariant measures are right invariant measures and all relatively invariant measures are in fact invariant.

Proposition 6.6. If G is compact and χ is a continuous homomorphism on G to $(0, \infty)$, then $\chi(s) = 1$ for all $s \in G$.

Proof. Since χ is continuous and G is compact, $\chi(G) = \{\chi(s)|s \in G\}$ is a compact subset of $(0, \infty)$. Since χ is a homomorphism, $\chi(G)$ is a subgroup of $(0, \infty)$. However, the only compact subgroup of $(0, \infty)$ is $\{1\}$. Thus $\chi(s) = 1$ for all $s \in G$. □

The nonexistence of nontrivial continuous homomorphisms on compact groups shows that all compact groups are unimodular. Further, all relatively invariant measures are invariant. Whenever G is compact, the invariant measure on G is always taken to be a probability measure.

6.3. INVARIANT MEASURES ON QUOTIENT SPACES

In this section, we consider the existence and uniqueness of invariant integrals on spaces that are acted on transitively by a group. Throughout this section, $\mathcal{X}$ is a locally compact Hausdorff space and $\mathcal{K}(\mathcal{X})$ denotes the set of continuous functions on $\mathcal{X}$ that have compact support. Also, G is a locally compact topological group that acts on the left of $\mathcal{X}$.

Definition 6.11. The group G acts *topologically* on $\mathcal{X}$ if the function from $G \times \mathcal{X}$ to $\mathcal{X}$ given by $(g, x) \to gx$ is continuous. When G acts topologically on $\mathcal{X}$, $\mathcal{X}$ is a *left homogeneous space* if for each $x \in \mathcal{X}$, the function π_x on G to $\mathcal{X}$ defined by $\pi_x(g) = gx$ is continuous, open, and onto $\mathcal{X}$.

The assumption that each π_x is an onto function is just another way to say that G acts transitively on $\mathcal{X}$. Also, it is not difficult to show that if, for one $x \in \mathcal{X}$, π_x is continuous, open, and onto $\mathcal{X}$, then for all x, π_x is continuous, open, and onto $\mathcal{X}$. To describe the structure of left homogeneous spaces $\mathcal{X}$, fix an element $x_0 \in \mathcal{X}$ and let

$$H_0 = \{g | gx_0 = x_0, g \in G\}.$$

That H_0 is a closed subgroup of G is easily verified. Further, the function τ considered in Proposition 6.3 is now one-to-one, onto, and τ and τ^{-1} are both continuous. Thus we have a one-to-one, onto, bicontinuous mapping between $\mathcal{X}$ and the quotient space G/H_0 endowed with the quotient topology. Conversely, let H be a closed subgroup of G and take $\mathcal{X} = G/H$ with the quotient topology. The group G acts on G/H in the obvious way $(g(g_1H) = gg_1H)$ and it is easily verified that G/H is a left homogeneous space (see Nachbin 1965, Section 3, Chapter 3). Thus we have a complete description of the left homogeneous spaces (up to relabelings by τ) as quotient spaces G/H where H is a closed subgroup of G.

In the notation above, let $\mathcal{X}$ be a left homogeneous space.

Definition 6.12. A nonzero integral J on $\mathcal{K}(\mathcal{X})$

$$J(f) = \int f(x)m(dx), \qquad f \in \mathcal{K}(\mathcal{X})$$

is *relatively invariant* with multiplier χ if, for each $s \in G$,

$$\int f(s^{-1}x)m(dx) = \chi(s)\int f(x)m(dx)$$

for all $f \in \mathcal{K}(\mathcal{X})$.

For $f \in \mathcal{K}(\mathcal{X})$, the function sf given by $(sf)(x) = f(s^{-1}x)$ is the *left translate* of f by $s \in G$. Thus an integral J on $\mathcal{K}(\mathcal{X})$ is relatively invariant with multiplier χ if $J(sf) = \chi(s)J(f)$. For such an integral,

$$\chi(st)J(f) = J((st)f) = J(s(tf)) = \chi(s)J(tf) = \chi(s)\chi(t)J(f)$$

so $\chi(st) = \chi(s)\chi(t)$. Also, any multiplier χ is continuous, which implies that a multiplier is a continuous homomorphism of G into the multiplicative group $(0, \infty)$.

◆ **Example 6.15.** Let $\mathcal{X}$ be the set of all $p \times p$ positive definite matrices. The group $G = Gl_p$ acts transitively on $\mathcal{X}$ as shown in Example 6.9. That $\mathcal{X}$ is a left homogeneous space is easily verified. For $\alpha \in R$, define the measure m_α by

$$m_\alpha(dx) = (\det(x))^{\alpha/2} \frac{dx}{(\det(x))^{(p+1)/2}}$$

where dx is Lebesgue measure on $\mathcal{X}$. Let $J_\alpha(f) \equiv \int f(x) m_\alpha(dx)$. For $s \in Gl_p$, $s(x) = sxs'$ is the group action on $\mathcal{X}$. Therefore,

$$\begin{aligned} J_\alpha(sf) &= \int f(s^{-1}(x)) m_\alpha(dx) \\ &= \int f(s^{-1}xs'^{-1})(\det(x))^{\alpha/2} \frac{dx}{(\det(x))^{(p+1)/2}} \\ &= |\det(s)|^\alpha \int f(s^{-1}xs'^{-1}) \det(s^{-1}xs'^{-1})^{\alpha/2} \frac{dx}{(\det(x))^{(p+1)/2}} \\ &= |\det(s)|^\alpha \int f(x)(\det(x))^{\alpha/2} \frac{dx}{(\det(x))^{(p+1)/2}}. \end{aligned}$$

The last equality follows from the change of variable $x = sys'$, which has a Jacobian equal to $|\det(s)|^{p+1}$ (see Proposition 5.11). Hence

$$J_\alpha(sf) = |\det(s)|^\alpha J(f)$$

for all $s \in Gl_p$, $f \in \mathcal{K}(\mathcal{X})$, and J_α is relatively invariant with multiplier $\chi_\alpha(s) = |\det(s)|^\alpha$. For this example, it has been shown that for every continuous homomorphism χ on G, there is a relatively invariant integral with multiplier χ. That this is not the case in general is demonstrated in future examples. ◆

The problem of the existence and uniqueness of relatively invariant integrals on left homogeneous spaces $\mathcal{X}$ is completely solved in the following result due to Weil (see Nachbin, 1965, Section 4, Chapter 3). Recall that x_0 is a fixed element of $\mathcal{X}$ and

$$H_0 = \{g | gx_0 = x_0, g \in G\}$$

is a closed subgroup of G. Let Δ_r denote the right-hand modulus of G and let Δ_r^0 denote the right-hand modulus of H_0.

Theorem 6.3. In the notation above:

(i) If $J(f) = \int f(x)m(dx)$ is relatively invariant with multiplier χ, then

$$\Delta_r^0(h) = \chi(h)\Delta_r(h) \quad \text{for all } h \in H_0.$$

(ii) If χ is a continuous homomorphism of G to $(0, \infty)$ that satisfies $\Delta_r^0(h) = \chi(h)\Delta_r(h)$, $h \in H_0$, then a relatively invariant integral with multiplier χ exists.

(iii) If J_1 and J_2 are relatively invariant with the same multiplier, then there exists a constant $c > 0$ such that $J_2 = cJ_1$.

Before turning to applications of Theorem 6.3, a few general comments are in order. If the subgroup H_0 is compact, then $\Delta_r^0(h) = 1$ for all $h \in H_0$. Since the restrictions of χ and of Δ_r to H_0 are both continuous homomorphisms on H_0, $\Delta_r(h) = \chi(h) = 1$ for all $h \in H_0$ as H_0 is compact. Thus when H_0 is compact, any continuous homomorphism χ is a multiplier for a relatively invariant integral and the description of all the relatively invariant integrals reduces to finding all the continuous homomorphisms of G. Further, when G is compact, then only an invariant integral on $\mathcal{K}(X)$ can exist as $\chi \equiv 1$ is the only continuous homomorphism. When G and H are not compact, the situation is a bit more complicated. Both Δ_r and Δ_r^0 must be calculated and then, the continuous homomorphisms χ on G to $(0, \infty)$ that satisfy (ii) of Theorem 6.3 must be found. Only then do we have a description of the relatively invariant integrals on $\mathcal{K}(X)$. Of course, the condition for the existence of an invariant integral ($\chi \equiv 1$) is that $\Delta_r^0(h) = \Delta_r(h)$ for all $h \in H_0$.

If J is a relatively invariant integral (with multiplier χ) given by

$$J(f) = \int f(x)m(dx), \qquad f \in \mathcal{K}(\mathcal{X}),$$

then the measure m is called relatively invariant with multiplier χ. In Example 6.15, it was shown that for each $\alpha \in R$, the measure m_α was relatively invariant under Gl_p with multiplier χ_α. Theorem 6.3 implies that any relatively invariant measure on the space of $p \times p$ positive definite matrices is equal to a positive constant times an m_α for some $\alpha \in R$. We now proceed with further examples.

◆ **Example 6.16.** Let $\mathcal{X} = \mathcal{F}_{p,n}$ and let $G = \mathcal{O}_n$. It was shown in Example 6.8 that $\mathcal{O}_n$ acts transitively on $\mathcal{F}_{p,n}$. The verification that

$\mathfrak{F}_{p,n}$ is a left homogeneous space is left to the reader. Since $\mathcal{O}_n$ is compact, Theorem 6.3 implies that there is a unique probability measure μ on $\mathfrak{F}_{p,n}$ that is invariant under the action of $\mathcal{O}_n$ on $\mathfrak{F}_{p,n}$. Also, any relatively invariant measure on $\mathfrak{F}_{p,n}$ will be equal to a positive constant times μ. The distribution μ is sometimes called the *uniform distribution* on $\mathfrak{F}_{p,n}$. When $p = 1$, then

$$\mathfrak{F}_{1,n} = \{x | x \in R^n, \|x\| = 1\},$$

which is the rim of the unit sphere in R^n. The uniform distribution on $\mathfrak{F}_{1,n}$ is just surface Lebesgue measure normalized so that it is a probability measure. When $p = n$, then $\mathfrak{F}_{n,n} = \mathcal{O}_n$ and μ is the uniform distribution on the orthogonal group. A different argument, probabilistic in nature, is given in the next chapter, which also establishes the existence of the uniform distribution on $\mathfrak{F}_{p,n}$. ◆

◆ **Example 6.17.** Take $\mathfrak{X} = R^p - \{0\}$ and let $G = Gl_p$. The action of Gl_p on $\mathfrak{X}$ is that of a matrix acting on a vector and this action is obviously transitive. The verification that $\mathfrak{X}$ is a left homogeneous space is routine. Consider the integral

$$J(f) = \int f(x)\,dx, \qquad f \in \mathcal{K}(\mathfrak{X})$$

where dx is Lebesgue measure on $\mathfrak{X}$. For $s \in Gl_p$, it is clear that $J(sf) = |\det(s)| J(f)$ so J is relatively invariant with multiplier $\chi_1(s) = |\det(s)|$. We now show that J is the only relatively invariant integral on $\mathcal{K}(\mathfrak{X})$. This is done by proving that χ_1 is the only possible multiplier for relatively invariant integrals on $\mathcal{K}(X)$. A convenient choice of $x_0 \in \mathfrak{X}$ is $x_0 = \varepsilon_1$ where $\varepsilon_1' = (1, 0, \ldots, 0)$. Then

$$H_0 = \{h | h\varepsilon_1 = \varepsilon_1, h \in Gl_p\}.$$

A bit of reflection shows that $h \in H_0$ iff

$$h = \begin{pmatrix} 1 & h_{12} \\ 0 & h_{22} \end{pmatrix}$$

where $h_{22} \in Gl_{(p-1)}$ and h_{12} is $1 \times (p - 1)$. A calculation similar to that in Example 6.14 yields

$$\mu(dh_{12}, dh_{22}) = \frac{dh_{12}\,dh_{22}}{|\det(h_{22})|^{p-1}}$$

as a left invariant measure on H_0. Then the integral

$$J_1(f) \equiv \int f(h)\mu(dh_{12}, dh_{22})$$

is left invariant on $\mathcal{K}(H_0)$ and a standard Jacobian argument yields

$$J_1(fh) = \Delta_r^0(h) J_1(f), \qquad f \in \mathcal{K}(H_0)$$

where

$$\Delta_r^0(h) = |\det(h_{22})|, \qquad h \in H_0.$$

Every continuous homomorphism on Gl_p has the form $\chi_\alpha(s) = |\det(s)|^\alpha$ for some $\alpha \in R$. Since $\Delta_r = 1$ for Gl_p, χ_α can be a multiplier for an invariant integral iff

$$\Delta_r^0(h) = \chi_\alpha(h), \qquad h \in H_0.$$

But $\Delta_r^0(h) = |\det(h_{22})|$ and for $h \in H_0$, $\chi_\alpha(h) = |\det(h_{22})|^\alpha$ so the only value for α for which χ_α can be a multiplier is $\alpha = 1$. Further, the integral J is relatively invariant with multiplier χ_1. Thus Lebesgue measure on $\mathcal{X}$ is the only (up to a positive constant) relatively invariant measure on $\mathcal{X}$ under the action of Gl_p. ◆

Before turning to the next example, it is convenient to introduce the direct product of two groups. If G_1 and G_2 are groups, the *direct product* of G_1 and G_2, denoted by $G \equiv G_1 \times G_2$, is the group consisting of all pairs (g_1, g_2) with $g_i \in G_i$, $i = 1, 2$, and group operation

$$(g_1, g_2)(h_1, h_2) \equiv (g_1 h_1, g_2 h_2).$$

If e_i is the identity in G_i, $i = 1, 2$, then (e_1, e_2) is the identity in G and $(g_1, g_2)^{-1} = (g_1^{-1}, g_2^{-1})$. When G_1 and G_2 are locally compact topological groups, then $G_1 \times G_2$ is a locally compact topological group when endowed with the product topology. The next two results describe all the continuous homomorphisms and relatively left invariant measures on $G_1 \times G_2$ in terms of continuous homomorphisms and relatively left invariant measures on G_1 and G_2.

Proposition 6.7. Suppose G_1 and G_2 are locally compact topological groups. Then χ is a continuous homomorphism on $G_1 \times G_2$ iff $\chi((g_1, g_2)) = \chi_1(g_1)\chi_2(g_2)$, $(g_1, g_2) \in G_1 \times G_2$, where χ_i is a continuous homomorphism on G_i, $i = 1, 2$.

Proof. If $\chi((g_1, g_2)) = \chi_1(g_1)\chi_2(g_2)$, clearly χ is a continuous homomorphism on $G_1 \times G_2$. Conversely, since $(g_1, g_2) = (g_1, e_2)(e_1, g_2)$, if χ is a continuous homomorphism on $G_1 \times G_2$, then

$$\chi((g_1, g_2)) = \chi(g_1, e_2)\chi(e_1, g_2).$$

Setting $\chi_1(g_1) = \chi(g_1, e_2)$ and $\chi_2(g_2) = \chi(e_1, g_2)$, the desired result follows. □

Proposition 6.8. Suppose χ is a continuous homomorphism on $G_1 \times G_2$ with $\chi(g_1, g_2) = \chi_1(g_1)\chi_2(g_2)$ where χ_i is a continuous homomorphism on G_i, $i = 1, 2$. If m is a relatively left invariant measure with multiplier χ, then there exist relatively left invariant measures m_i on G_i with multipliers χ_i, $i = 1, 2$, and m is product measure $m_1 \times m_2$. Conversely, if m_i is a relatively left invariant measure on G_i with multiplier χ_i, $i = 1, 2$, then $m_1 \times m_2$ is a relatively left invariant measure on $G_1 \times G_2$ with multiplier χ, which satisfies $\chi(g_1, g_2) = \chi_1(g_1)\chi_2(g_2)$.

Proof. This result is a direct consequence of Fubini's Theorem and the existence and uniqueness of relatively left invariant integrals. □

The following example illustrates many of the results presented in this chapter and has a number of applications in multivariate analysis. For example, one of the derivations of the Wishart distribution is quite easy given the results of this example.

◆ **Example 6.18.** As in Example 6.10, $\mathfrak{X}$ is the set of all $n \times p$ matrices with rank p and G is the direct product group $\mathcal{O}_n \times G_T^+$. The action of $(\Gamma, T) \in \mathcal{O}_n \times G_T^+$ on $\mathfrak{X}$ is

$$(\Gamma, T)X \equiv (\Gamma \otimes T)X = \Gamma X T', \qquad X \in \mathfrak{X}.$$

Since $\mathfrak{X} = \{X | X \in \mathcal{L}_{p, n}, \det(X'X) > 0\}$, $\mathfrak{X}$ is a nonempty open subset of $\mathcal{L}_{p,n}$. Let dX be Lebesgue measure on $\mathcal{X}$ and define a measure on $\mathcal{X}$ by

$$m(dX) = \frac{dX}{(\det(X'X))^{n/2}}.$$

Using Proposition 5.10, it is an easy calculation to show that the integral

$$J(f) \equiv \int f(X) m(dX)$$

is invariant—that is, $J((\Gamma, T)f) = J(f)$ for $(\Gamma, T) \in \mathcal{O}_n \times G_T^+$ and $f \in \mathcal{K}(\mathcal{X})$. However, it takes a bit more work to characterize all the relatively invariant measures on $\mathcal{X}$. First, it was shown in Example 6.10 that, if X_0 is

$$X_0 = \begin{pmatrix} I_p \\ 0 \end{pmatrix} \in \mathcal{X},$$

then $H_0 = \{(\Gamma, T) | (\Gamma, T)X_0 = X_0\}$ is a closed subgroup of $\mathcal{O}_n$ and hence is compact. By Theorem 6.3, every continuous homomorphism on $\mathcal{O}_n \times G_T^+$ is the multiplier for a relatively invariant integral. But every continuous homomorphism χ on $\mathcal{O}_n \times G_T^+$ has the form $\chi(\Gamma, T) = \chi_1(\Gamma)\chi_2(T)$ where χ_1 and χ_2 are continuous homomorphisms on $\mathcal{O}_n$ and G_T^+. Since $\mathcal{O}_n$ is compact, $\chi_1 = 1$. From Example 6.12,

$$\chi_2(T) = \prod_{i=1}^{p} (t_{ii})^{c_i} \equiv \chi_c(T)$$

where $c \in R^p$ has coordinates $c_1, \ldots, c_p$. Now that all the possible multipliers have been described, we want to exhibit the relatively invariant integrals on $\mathcal{K}(\mathcal{X})$. To this end, consider the space $\mathcal{Y} = \mathcal{F}_{p,n} \times G_U^+$ so points in $\mathcal{Y}$ are (Ψ, U) where Ψ is an $n \times p$ linear isometry and U is a $p \times p$ upper triangular matrix in G_U^+. The group $\mathcal{O}_n \times G_T^+$ acts transitively on $\mathcal{Y}$ under the group action

$$(\Gamma, T)(\Psi, U) \equiv (\Gamma\Psi, UT').$$

Let μ_0 be the unique probability measure on $\mathcal{F}_{p,n}$ that is $\mathcal{O}_n$-invariant and let ν_r be the particular right invariant measure on the group G_U^+

given by

$$\nu_r(dU) = \frac{dU}{\prod_{i=1}^{p} u_{ii}^{i}}.$$

Obviously, the integral

$$J_1(f) \equiv \iint f(\Psi, U)\mu_0(d\Psi)\nu_r(dU)$$

is invariant under the action of $\mathcal{O}_n \times G_T^+$ on $\mathfrak{F}_{p,n} \times G_U^+$, $f \in \mathcal{K}(\mathfrak{F}_{p,n} \times G_U^+)$. Consider the integral

$$J_2(f) \equiv \iint f(\Psi, U)\chi_c(U')\mu_0(d\Psi)\nu_r(dU)$$

defined on $\mathcal{K}(\mathfrak{F}_{p,n} \times G_U^+)$ where χ_c is a continuous homomorphism on G_T^+. The claim is that $J_2((\Gamma, T)f) = \chi_c(T)J_2(f)$ so J_2 is relatively invariant with multiplier χ_c. To see this, compute as follows:

$$\begin{aligned} J_2((\Gamma, T)f) &= \iint f\big((\Gamma, T)^{-1}(\Psi, U)\big)\chi_c(U')\mu_0(d\Psi)\nu_r(dU) \\ &= \iint f(\Gamma'\Psi, UT'^{-1})\chi_c(TT^{-1}U')\mu_0(d\Psi)\nu_r(dU) \\ &= \chi_c(T)\iint f(\Gamma'\Psi, UT'^{-1})\chi_c((UT'^{-1})')\mu_0(d\Psi)\nu_r(dU) \\ &= \chi_c(T)J_2(f). \end{aligned}$$

The last equality follows from the invariance of μ_0 and ν_r. Thus all the relatively invariant integrals on $\mathcal{K}(\mathfrak{F}_{p,n} \times G_U^+)$ have been explicitly described. To do the same for $\mathcal{K}(\mathfrak{X})$, the basic idea is to move the integral J_2 over to $\mathcal{K}(\mathfrak{X})$. It was mentioned earlier that the map ϕ_0 on $\mathfrak{F}_{p,n} \times G_U^+$ to $\mathfrak{X}$ given by

$$\phi_0(\Psi, U) = \Psi U \in \mathfrak{X}$$

is one-to-one, onto, and satisfies

$$\phi_0((\Gamma, T)(\Psi, U)) = (\Gamma, T)\phi_0(\Psi, U),$$

for group elements (Γ, T). For $f \in \mathcal{K}(\mathfrak{X})$, consider the integral

$$J_3(f) \equiv \int\int f(\phi_0(\Psi, U))\mu_0(d\Psi)\nu_r(dU).$$

Then for $(\Gamma, \mathrm{T}) \in \mathcal{O}_n \times G_T^+$,

$$J_3((\Gamma, T)f) = \int\int f\big((\Gamma, T)^{-1}\phi_0(\Psi, U)\big)\mu_0(d\Psi)\nu_r(dU)$$

$$= \int\int f\big((\Gamma', T^{-1})\phi_0(\mu, U)\big)\mu_0(d\Psi)\nu_r(dU)$$

$$= \int\int f\big(\phi_0(\Gamma'\Psi, UT'^{-1})\big)\mu_0(d\Psi)\nu_r(dU) = J_3(f)$$

since μ_0 and ν_r are invariant. Therefore, J_3 is an invariant integral on $\mathcal{K}(\mathfrak{X})$. Since J is also an invariant integral on $\mathcal{K}(\mathfrak{X})$, Theorem 6.3 shows that there is a positive constant k such that

$$J(f) = kJ_3(f), \qquad f \in \mathcal{K}(\mathfrak{X}).$$

More explicitly, we have the equation

$$\int f(X)\frac{dX}{|X'X|^{n/2}} = k\int\int f(\Psi U)\mu_0(d\Psi)\nu_r(dU)$$

for all $f \in \mathcal{K}(\mathfrak{X})$. This equation is a formal way to state the very nontrivial fact that the measure m on $\mathfrak{X}$ gets transformed into the measure $k(\mu_0 \times \nu_r)$ on $\mathcal{F}_{p,n} \times G_U^+$ under the mapping ϕ_0^{-1}. To evaluate the constant k, it is sufficient to find one particular function so that both sides of the above equality can be evaluated. Consider

$$f_0(X) = |X'X|^{n/2}(2\pi)^{-np/2}\exp\left[-\tfrac{1}{2}\operatorname{tr}(X'X)\right].$$

Clearly,

$$\int f_0(X)\frac{dX}{|X'X|^{n/2}} = 1$$

so

$$\frac{1}{k} = \iint f_0(\Psi U)\mu_0(d\Psi)\nu_r(dU)$$

$$= (2\pi)^{-np/2}\int |U'U|^{n/2}\exp\left[-\frac{1}{2}\operatorname{tr} U'U\right]\nu_r(dU)$$

$$= (2\pi)^{-np/2}\int \prod_1^p u_{ii}^{n-i}\exp\left[-\frac{1}{2}\sum_{i\leqslant j} u_{ij}^2\right] dU$$

$$= (2\pi)^{-np/2}2^{-p}c(n, p).$$

The last equality follows from the result in Example 5.1, where $c(n, p)$ is defined. Therefore,

$$(6.1)\quad \int f(X)\frac{dX}{|X'X|^{n/2}} = \frac{(2\pi)^{np/2}2^p}{c(n, p)}\iint f(\Psi U)\mu_0(d\Psi)\nu_r(dU).$$

It is now an easy matter to derive all the relatively invariant integrals on $\mathcal{K}(\mathcal{X})$. Let χ_c be a given continuous homomorphism on G_T^+. For each $X \in \mathcal{X}$, let $U(X)$ be the unique element in G_U^+ such that $X = \Psi U(X)$ for some $\Psi \in \mathcal{F}_{p,n}$ (see Proposition 5.2). It is clear that $U(\Gamma X T') = U(X)T'$ for $\Gamma \in \mathcal{O}_n$ and $T \in G_T^+$. We have shown that

$$J_2(f) = \iint f(\Psi, U)\chi_c(U')\mu_0(d\Psi)\nu_r(dU)$$

is relatively invariant with multiplier χ_c on $\mathcal{K}(\mathcal{F}_{p,n} \times G_U^+)$. For $h \in \mathcal{K}(X)$, define an integral J_4 by

$$J_4(h) = \iint h(\Psi U)\chi_c(U')\mu_0(d\Psi)\nu_r(dU).$$

Clearly, J_4 is relatively invariant with multiplier χ_c since $J_4(h) = J_2(\tilde{h})$ where $\tilde{h}(\Psi, U) \equiv h(\Psi U)$. Now, we move J_4 over to $\mathcal{X}$ by (6.1). In (6.1), take $f(X) = h(X)\chi_c(U'(X))$ so $f(\Psi U) = h(\Psi U)\chi_c(U')$. Thus the integral

$$J_5(h) = \int h(X)\chi_c(U'(X))\frac{dX}{|X'X|^{n/2}}$$

is relatively invariant with multiplier χ_c. Of course, any relatively invariant integral with multiplier χ_c on $\mathcal{K}(\mathcal{X})$ is equal to a positive constant times J_5. ◆

6.4. TRANSFORMATIONS AND FACTORIZATIONS OF MEASURES

The results of Example 6.18 describe how an invariant measure on the set of $n \times p$ matrices is transformed into an invariant measure on $\mathcal{F}_{p,n} \times G_U^+$ under a particular mapping. The first problem to be discussed in this section is an abstraction of this situation. The notion of a group homomorphism plays a role in what follows.

Definition 6.13. Let G and H be groups. A function η from G onto H is a *homomorphism* if:

(i) $\eta(g_1 g_2) = \eta(g_1)\eta(g_2)$, $g_1, g_2 \in G$.
(ii) $\eta(g) = (\eta(g))^{-1}$, $g \in G$.

When there is a homomorphism from G to H, H is called a *homomorphic image* of G.

For notational convenience, a homomorphic image of G is often denoted by $\bar{G}$ and the value of the homomorphism at g is $\bar{g}$. In this case, $\overline{g_1 g_2} = \bar{g}_1 \bar{g}_2$ and $\overline{g^{-1}} = \bar{g}^{-1}$. Also, if e is the identity in G, then $\bar{e}$ is the identity in $\bar{G}$. Suppose $\mathcal{X}$ and $\mathcal{Y}$ are locally compact spaces, and G and $\bar{G}$ are locally compact topological groups that act topologically on $\mathcal{X}$ and $\mathcal{Y}$, respectively. It is assumed that $\bar{G}$ is a homomorphic image of G.

Definition 6.14. A measurable function ϕ from $\mathcal{X}$ onto $\mathcal{Y}$ is called *equivariant* if $\phi(gx) = \bar{g}\phi(x)$ for all $g \in G$ and $x \in \mathcal{X}$.

Now, consider an integral

$$J(f) = \int f(x)\mu(dx), \qquad f \in \mathcal{K}(\mathcal{X}),$$

which is invariant under the action of G on $\mathcal{X}$, that is

$$J(gf) \equiv \int f(g^{-1}x)\mu(dx) = \int f(x)\mu(dx) = J(f)$$

for $g \in G$ and $f \in \mathcal{K}(\mathcal{X})$. Given an equivariant function ϕ from $\mathcal{X}$ to $\mathcal{Y}$, there is a natural measure ν induced on $\mathcal{Y}$. Namely, if B is a measurable subset of $\mathcal{Y}$, $\nu(B) \equiv \mu(\phi^{-1}(B))$. The result below shows that under a regularity condition on ϕ, the measure ν defines an invariant (under $\bar{G}$) integral on $\mathcal{K}(\mathcal{Y})$.

Proposition 6.9. If ϕ is an equivariant function from $\mathcal{X}$ onto $\mathcal{Y}$ that satisfies $\mu(\phi^{-1}(K)) < +\infty$ for all compact sets $K \subseteq \mathcal{Y}$, then the integral

$$J_1(f) \equiv \int f(y)\nu(dy), \qquad f \in \mathcal{K}(\mathcal{Y})$$

is invariant under $\bar{G}$.

Proof. First note that J_1 is well defined and finite since $\mu(\phi^{-1}(K)) < +\infty$ for all compact sets $K \subseteq \mathcal{Y}$. From the definition of the measure ν, it follows immediately that

$$J_1(f) = \int f(y)\nu(dy) = \int f(\phi(x))\mu(dx), \qquad f \in \mathcal{K}(\mathcal{Y}).$$

Using the equivariance of ϕ and the invariance of μ, we have

$$J_1(\bar{g}f) = \int f(\bar{g}^{-1}y)\nu(dy) = \int f(\bar{g}^{-1}\phi(x))\mu(dx)$$

$$= \int f(\phi(g^{-1}x))\mu(dx) = \int f(\phi(x))\mu(dx) = J_1(f)$$

so J_1 is invariant under $\bar{G}$. □

Before presenting some applications of Proposition 6.9, a few remarks are in order. The groups G and $\bar{G}$ are not assumed to act transitively on $\mathcal{X}$ and $\mathcal{Y}$, respectively. However, if $\bar{G}$ does act transitively on $\mathcal{Y}$ and if $\mathcal{Y}$ is a left homogeneous space, then the measure ν is uniquely determined up to a positive constant. Thus if we happen to know an invariant measure on $\mathcal{Y}$, the identity

$$\int f(y)\nu(dy) = \int f(\phi(x))\mu(dx), \qquad f \in \mathcal{K}(\mathcal{Y})$$

relates the G-invariant measure μ to the $\bar{G}$-invariant measure ν. It was this line of reasoning that led to (6.1) in Example 6.18. We now consider some further examples.

◆ **Example 6.19.** As in Example 6.18, let $\mathcal{X}$ be the set of all $n \times p$ matrices of rank p, and let $\mathcal{Y}$ be the space $\mathcal{S}_p^+$ of $p \times p$ positive definite matrices. Consider the map ϕ on $\mathcal{X}$ to $\mathcal{S}_p^+$ defined by

$$\phi(X) = X'X, \qquad X \in \mathcal{X}.$$

The group $\mathcal{O}_n \times Gl_p$ acts on $\mathcal{X}$ by

$$(\Gamma, A)X = (\Gamma \otimes A)X = \Gamma X A'$$

and the measure

$$\mu(dX) = \frac{dX}{|X'X|^{n/2}}$$

is invariant under $\mathcal{O}_n \times Gl_p$. Further,

$$\phi((\Gamma, A)X) = AX'XA' = A\phi(X)A',$$

and this defines an action of Gl_p on $\mathcal{S}_p^+$. It is routine to check that the mapping

$$(\Gamma, A) \to A \equiv \overline{(\Gamma, A)}$$

is a homomorphism. Obviously,

$$\phi((\Gamma, A)X) = \overline{(\Gamma, A)}\,\phi(X)$$

since the action of Gl_p on $\mathcal{S}_p^+$ is

$$A(S) = ASA'; \qquad S \in \mathcal{S}_p^+, \qquad A \in Gl_p.$$

Since Gl_p acts transitively on $\mathcal{S}_p^+$, the invariant measure

$$\nu_1(dS) = \frac{dS}{|S|^{(p+1)/2}}$$

is unique up to a positive constant. The remaining assumption to verify in order to apply Proposition 6.9 is that $\phi^{-1}(K)$ has finite μ measure for compact sets $K \subseteq \mathcal{S}_p^+$. To do this, we show that

$\phi^{-1}(K)$ is compact in $\mathfrak{X}$. Recall that the mapping h on $\mathcal{F}_{p,n} \times S_p^+$ onto $\mathfrak{X}$ given by

$$h(\Psi, S) = \Psi S \in \mathfrak{X}$$

is one-to-one and is obviously continuous. Given the compact set $K \subseteq \mathbb{S}_p^+$, let

$$K_1 = \{S | S \in \mathbb{S}_p^+, S^2 \in K\}.$$

Then K_1 is compact so $\mathcal{F}_{p,n} \times K_1$ is a compact subset of $\mathcal{F}_{p,n} \times \mathbb{S}_p^+$. It is now routine to show that

$$\phi^{-1}(K) = \{X | X'X \in K\} = h(\mathcal{F}_{p,n} \times K_1),$$

which is compact since h is continuous and the continuous image of a compact set is compact. By Proposition 6.9, we conclude that the measure $\nu = \mu \circ \phi^{-1}$ is invariant under Gl_p and satisfies

$$\int_{\mathfrak{X}} f(X'X) \frac{dX}{|X'X|^{n/2}} = \int_{\mathbb{S}_p^+} f(S) \nu(dS),$$

for all $f \in \mathcal{K}(\mathbb{S}_p^+)$. Since ν is invariant under Gl_p, $\nu = c\nu_1$ where c is a positive constant. Thus we have the identity

$$\int f(X'X) \frac{dX}{|X'X|^{n/2}} = c \int f(S) \frac{dS}{|S|^{(p+1)/2}}. \tag{6.2}$$

To find the constant c, it is sufficient to evaluate both sides of (6.2) for a particular function f_0. For f_0, take the function

$$f_0(S) = (\sqrt{2\pi})^{-np} |S|^{n/2} \exp[-\tfrac{1}{2} \operatorname{tr} S],$$

so

$$f_0(X'X) = (\sqrt{2\pi})^{-np} |X'X|^{n/2} \exp[-\tfrac{1}{2} \operatorname{tr} X'X].$$

Clearly, the left-hand side of (6.2) integrates to one and this yields the equation

$$c \int (\sqrt{2\pi})^{-np} |S|^{(n-p-1)/2} \exp[-\tfrac{1}{2} \operatorname{tr} S] \, dS = 1.$$

The result of Example 5.1 gives

$$c\left(\sqrt{2\pi}\right)^{-np} c(n, p) = 1$$

so

$$c = \frac{\left(\sqrt{2\pi}\right)^{np}}{c(n, p)} = \left(\sqrt{2\pi}\right)^{np} \omega(n, p).$$

In conclusion, the identity

$$\text{(6.3)} \qquad \int_{\mathcal{X}} f(X'X) \frac{dX}{|X'X|^{n/2}} = \left(\sqrt{2\pi}\right)^{np} \omega(n, p) \int_{\mathcal{S}_p^+} f(S) \frac{dS}{|S|^{(p+1)/2}}$$

has been established for all $f \in \mathcal{K}(\mathcal{S}_p^+)$, and thus for all measurable f for which either side exists. ◆

◆ **Example 6.20.** Again let $\mathcal{X}$ be the set of $n \times p$ matrices of rank p so the group $\mathcal{O}_n \times G_T^+$ acts on $\mathcal{X}$ by

$$(\Gamma, T)X \equiv (\Gamma \otimes T)X = \Gamma X T'.$$

Each element $X \in \mathcal{X}$ has a unique representation $X = \Psi U$ where $\Psi \in \mathcal{F}_{p,n}$ and $U \in G_U^+$. Define ϕ on $\mathcal{X}$ onto G_U^+ by defining $\phi(X)$ to be the unique element $U \in G_U^+$ such that $X = \Psi U$ for some $\Psi \in \mathcal{F}_{p,n}$. If $\phi(X) - U$, then $\phi((\Gamma, T)X) = UT'$, since when $X = \Psi U$, $(\Gamma, T)X = \Gamma \Psi U T'$. This implies that UT' is the unique element in G_U^+ such that $X = (\Gamma\Psi)UT'$ as $\Gamma\Psi \in \mathcal{F}_{p,n}$. The mapping $(\Gamma, T) \to T \equiv \overline{(\Gamma, T)}$ is clearly a homomorphism of (Γ, T) onto G_T^+ and the action of G_T^+ on G_U^+ is

$$T(U) \equiv UT'; \qquad U \in G_U^+, T \in G_T^+.$$

Therefore, $\phi((\Gamma, T)X) = \overline{(\Gamma, T)}\phi(X)$ so ϕ is equivariant. The measure

$$\mu(dX) = \frac{dX}{|X'X|^{n/2}}$$

is $\mathcal{O}_n \times G_T^+$ invariant. To show that $\phi^{-1}(K)$ has finite μ measure when $K \subseteq G_U^+$ is compact, note that $h(\Psi, U) \equiv \Psi U$ is a continuous function on $\mathcal{F}_{p,n} \times G_U^+$ onto $\mathcal{X}$. It is easily verified that

$$\phi^{-1}(K) = h\left(\mathcal{F}_{p,n} \times K\right).$$

But $\mathcal{F}_{p,n} \times K$ is compact, which shows that $\phi^{-1}(K)$ is compact since h is continuous. Thus $\mu(\phi^{-1}(K)) < +\infty$. Proposition 6.9 shows that $\nu \equiv \mu \circ \phi^{-1}$ is a G_T^+-invariant measure on G_U^+ and we have the identity

$$\int_{\mathfrak{X}} f(\phi(X)) \frac{dX}{|X'X|^{n/2}} = \int_{G_U^+} f(U)\nu(dU)$$

for all $f \in \mathcal{K}(G_U^+)$. However, the measure

$$\nu_1(dU) \equiv \frac{dU}{\prod_{i=1}^{p} u_{ii}^{i}}$$

is a right invariant measure on G_U^+, and therefore, ν_1 is invariant under the transitive action of G_T^+ on G_U^+. The uniqueness of invariant measures implies that $\nu = c\nu_1$ for some positive constant c and

$$\int_{\mathfrak{X}} f(\phi(X)) \frac{dX}{|X'X|^{n/2}} = c\int_{G_U^+} f(U) \frac{dU}{\prod_1^p u_{ii}^i}.$$

The constant c is evaluated by choosing f to be

$$f(U) = (\sqrt{2\pi})^{-np} |U'U|^{n/2} \exp[-\tfrac{1}{2} \operatorname{tr} U'U].$$

Since $(\phi(X))'\phi(X) = X'X$,

$$f(\phi(X)) = (\sqrt{2\pi})^{-np} |X'X|^{n/2} \exp[-\tfrac{1}{2} \operatorname{tr} X'X]$$

and

$$\int f(\phi(X)) \frac{dX}{|X'X|^{n/2}} = 1.$$

Therefore,

$$1 = c(\sqrt{2\pi})^{-np} \int_{G_U^+} |U'U|^{n/2} \exp\left[-\frac{1}{2} \operatorname{tr} U'U\right] \frac{dU}{\prod_1^p u_{ii}^i}$$

$$= c(\sqrt{2\pi})^{-np} \int_{G_U^+} \prod_1^p u_{ii}^{n-i} \exp\left[-\frac{1}{2} \operatorname{tr} U'U\right] dU$$

$$= c(\sqrt{2\pi})^{-np} 2^{-p} c(n, p)$$

where $c(n, p)$ is defined in Example 5.1. This yields the identity

$$\int f(\phi(X)) \frac{dX}{|X'X|^{n/2}} = 2^p (\sqrt{2\pi})^{np} \omega(n, p) \int f(U) \frac{dU}{\prod_1^p u_{ii}^i}$$

for all $f \in \mathcal{K}(G_U^+)$. In particular, when $f(U) = f_1(U'U)$, we have

$$\int f_1(X'X) \frac{dX}{|X'X|^{n/2}} = 2^p (\sqrt{2\pi})^{np} \omega(n, p) \int f_1(U'U) \frac{dU}{\prod_1^p u_{ii}^i} \tag{6.4}$$

whenever either integral exists. Combining this with (6.3) yields the identity

$$\int_{\mathcal{S}_p^+} f(S) \frac{dS}{|S|^{(p+1)/2}} = 2^p \int_{G_p^+} f(U'U) \frac{dU}{\prod_1^p u_{ii}^i} \tag{6.5}$$

for all measurable f for which either integral exists. Setting $T = U'$ in (6.5) yields the assertion of Proposition 5.18. ◆

The final topic in this chapter has to do with the factorization of a Radon measure on a product space. Suppose $\mathcal{X}$ and $\mathcal{Y}$ are locally compact and σ-compact Hausdorff spaces and assume that G is a locally compact topological group that acts on $\mathcal{X}$ in such a way that $\mathcal{X}$ is a homogeneous space. It is also assumed that μ_1 is a G-invariant Radon measure on $\mathcal{X}$ so the integral

$$J_1(f_1) \equiv \int f_1(x) \mu_1(dx), \qquad f_1 \in \mathcal{K}(\mathcal{X})$$

is G-invariant, and is unique up to a positive constant.

Proposition 6.10. Assume the conditions above on $\mathcal{X}$, $\mathcal{Y}$, G, and J_1. Define G acting on the locally compact and σ-compact space $\mathcal{X} \times \mathcal{Y}$ by $g(x, y) = (gx, y)$. If m is a G-invariant Radon measure on $\mathcal{X} \times \mathcal{Y}$, then $m = \mu_1 \times \nu$ for some Radon measure ν on $\mathcal{Y}$.

Proof. By assumption, the integral

$$J(f) \equiv \iint_{\mathcal{Y}\mathcal{X}} f(x, y) m(dx, dy), \qquad f \in \mathcal{K}(\mathcal{X} \times \mathcal{Y})$$

satisfies

$$J(gf) = \iint_{\mathcal{Y}\mathcal{X}} f(g^{-1}x, y)m(dx, dy) = J(f).$$

For $f_2 \in \mathcal{K}(\mathcal{Y})$ and $f_1 \in \mathcal{K}(\mathcal{X})$, the product $f_1 f_2$, defined by $(f_1 f_2)(x, y) = f_1(x) f_2(y)$, is in $\mathcal{K}(\mathcal{X} \times \mathcal{Y})$ and

$$J(f_1 f_2) = \iint_{\mathcal{Y}\mathcal{X}} f_1(x) f_2(y) m(dx, dy).$$

Fix $f_2 \in \mathcal{K}(\mathcal{Y})$ such that $f_2 \geqslant 0$ and let

$$H(f_1) \equiv \iint_{\mathcal{Y}\mathcal{X}} f_1(x) f_2(y) m(dx, dy), \qquad f_1 \in \mathcal{K}(\mathcal{X}).$$

Since $J(gf) = J(f)$, it follows that

$$H(gf_1) = H(f_1) \quad \text{for } g \in G \text{ and } f_1 \in \mathcal{K}(\mathcal{X}).$$

Therefore H is a G-invariant integral on $\mathcal{K}(\mathcal{X})$. Hence there exists a non-negative constant $c(f_2)$ depending on f_2 such that

$$H(f_1) = c(f_2) J_1(f_1)$$

and $c(f_2) = 0$ iff $H(f_1) = 0$ for all $f_1 \in \mathcal{K}(\mathcal{X})$. For an arbitrary $f_2 \in \mathcal{K}(\mathcal{Y})$, write $f_2 = f_2^+ - f_2^-$ where $f_2^+ = \max(f_2, 0)$ and $f_2^- = \max(-f_2, 0)$ are in $\mathcal{K}(\mathcal{Y})$. For such an f_2, it is easy to show

$$J(f_1 f_2) = c(f_2^+) J_1(f_1) - c(f_2^-) J_1(f_1) = (c(f_2^+) - c(f_2^-)) J_1(f_1).$$

Thus defining c on $\mathcal{K}(\mathcal{Y})$ by $c(f_2) = c(f_2^+) - c(f_2^-)$, it is easy to show that c is an integral on $\mathcal{K}(\mathcal{Y})$. Hence

$$c(f_2) = \int_{\mathcal{Y}} f_2(y) \nu(dy)$$

for some Radon measure ν. Therefore,

$$\iint_{\mathcal{Y}\mathcal{X}} f_1(x) f_2(y) m(dx, dy) = \iint_{\mathcal{Y}\mathcal{X}} f_1(x) f_2(y) \mu_1(dx) \nu(dy).$$

A standard approximation argument now implies that m is the product measure $\mu_1 \times \nu$. □

Proposition 6.10 provides one technique for establishing the stochastic independence of two random vectors. This technique is used in the next chapter. The one application of Proposition 6.10 given here concerns the space of positive definite matrices.

◆ **Example 6.21.** Let $\mathcal{Z}$ be the set of all $p \times p$ positive definite matrices that have distinct eigenvalues. That $\mathcal{Z}$ is an open subset of $\mathcal{S}_p^+$ follows from the fact that the eigenvalues of $S \in \mathcal{S}_p^+$ are continuous functions of the elements of the matrix $\mathcal{S}$. Thus $\mathcal{Z}$ has nonzero Lebesgue measure in $\mathcal{S}_p^+$. Also, let $\mathcal{Y}$ be the set of $p \times p$ diagonal matrices Y with diagonal elements $y_1, \dots, y_p$ that satisfy $y_1 > y_2 > \cdots > y_p$. Further, let $\mathcal{X}$ be the quotient space $\mathcal{O}_p/\mathcal{D}_p$ where $\mathcal{D}_p$ is the group of sign changes introduced in Example 6.6. We now construct a natural one-to-one onto map from $\mathcal{X} \times \mathcal{Y}$ to $\mathcal{Z}$. For $X \in \mathcal{X}$, $X = \Gamma\mathcal{D}_p$ for some $\Gamma \in \mathcal{O}_p$. Define ϕ by

$$\phi(X, Y) = \Gamma Y \Gamma', \qquad X = \Gamma\mathcal{D}_p,\ Y \in \mathcal{Y}.$$

To verify that ϕ is well defined, suppose that $X = \Gamma_1\mathcal{D}_p = \Gamma_2\mathcal{D}_p$. Then

$$\phi(X, Y) = \Gamma_1 Y \Gamma_1' = \Gamma_2\Gamma_2'\Gamma_1 Y \Gamma_1'\Gamma_2\Gamma_2' = \Gamma_2 Y \Gamma_2'$$

since $\Gamma_2'\Gamma_1 \in \mathcal{D}_p$ and every element $D \in \mathcal{D}_p$ satisfies $DYD = Y$ for all $Y \in \mathcal{Y}$. It is clear that $\phi(X, Y)$ has ordered eigenvalues $y_1 > y_2 > \cdots > y_p > 0$, the diagonal elements of Y. Clearly, the function ϕ is onto and continuous. To show ϕ is one-to-one, first note that, if Y is any element of $\mathcal{Y}$, then the equation

$$\Gamma Y \Gamma' = Y, \qquad \Gamma \in \mathcal{O}_p$$

implies that $\Gamma \in \mathcal{D}_p$ ($\Gamma Y \Gamma' = Y$ implies that $\Gamma Y = Y\Gamma$ and equating the elements of these two matrices shows that Γ must be diagonal so $\Gamma \in \mathcal{D}_p$). If

$$\phi(X_1, Y_1) = \phi(X_2, Y_2),$$

then $Y_1 = Y_2$ by the uniqueness of eigenvalues and the ordering of the diagonal elements of $Y \in \mathcal{Y}$. Thus

$$\Gamma_1 Y_1 \Gamma_1' = \Gamma_2 Y_1 \Gamma_2'$$

when

$$\phi(X_1, Y_1) = \phi(X_2, Y_1).$$

Therefore,

$$\Gamma_2'\Gamma_1 Y_1 \Gamma_1' \Gamma_2 = Y_1,$$

which implies that $\Gamma_2'\Gamma_1 \in \mathfrak{D}_p$. Since $X_i = \Gamma_i \mathfrak{D}_p$ for $i = 1, 2$, this shows that $X_1 = X_2$ and that ϕ is one-to-one. Therefore, ϕ has an inverse and the spectral theorem for matrices specifies just what ϕ^{-1} is. Namely, for $Z \in \mathfrak{Z}$, let $y_1 > \cdots > y_p > 0$ be the ordered eigenvalues of Z and write Z as

$$Z = \Gamma Y \Gamma', \qquad \Gamma \in \mathcal{O}_p$$

where $Y \in \mathcal{Y}$ has diagonal elements $y_1 > \cdots > y_p > 0$. The problem is that $\Gamma \in \mathcal{O}_p$ is not unique since

$$\Gamma Y \Gamma' = \Gamma D Y D \Gamma' \quad \text{for } D \in \mathfrak{D}_p.$$

To obtain uniqueness, we simply have "quotiented out" the subgroup $\mathfrak{D}_p$ in order that ϕ^{-1} be well defined. Now, let

$$\mu(dZ) = dZ$$

be Lebesgue measure on $\mathfrak{Z}$ and consider $\nu = \mu \circ \phi$—the induced measure on $\mathfrak{X} \times \mathcal{Y}$. The problem is to obtain some information about the measure ν. Since ϕ is continuous, ν is a Radon measure on $\mathfrak{X} \times \mathcal{Y}$, and ν satisfies

$$\iint f(X, Y)\nu(dX, dY) = \int f(\phi^{-1}(Z))\, dZ$$

for $f \in \mathcal{K}(\mathfrak{X} \times \mathcal{Y})$. The claim is that the measure ν is invariant under the action of $\mathcal{O}_p$ on $\mathfrak{X} \times \mathcal{Y}$ defined by

$$\Gamma(X, Y) = (\Gamma X, Y).$$

To see this, we have

$$\iint f(\Gamma'(X, Y))\nu(dX, dY) = \int f(\Gamma'\phi^{-1}(Z))\, dZ.$$

But a bit of reflection shows that $\Gamma'\phi^{-1}(Z) = \phi^{-1}(\Gamma' Z \Gamma)$. Since the Jacobian of the transformation $\Gamma' Z \Gamma$ is equal to one, it follows that ν is $\mathcal{O}_p$-invariant. By Proposition 6.10, the measure ν is a product measure $\nu_1 \times \nu_2$ where ν_1 is an $\mathcal{O}_p$-invariant measure on $\mathcal{X}$. Since $\mathcal{O}_p$ is compact and $\mathcal{X}$ is compact, the measure ν_1 is finite and we take $\nu_1(\mathcal{X}) = 1$ as a normalization. Therefore,

$$\int f(\phi^{-1}(Z))\, dZ = \int\int f(X, Y)\nu_1(dX)\nu_2(dY)$$

for all $f \in \mathcal{K}(\mathcal{X} \times \mathcal{Y})$. Setting $h = f\phi^{-1}$ yields

$$\int h(Z)\, dZ = \int\int h(\phi(X, Y))\nu_1(dX)\nu_2(dY)$$

for $h \in \mathcal{K}(\mathcal{Z})$. In particular, if $h \in \mathcal{K}(\mathcal{Z})$ satisfies $h(Z) = h(\Gamma Z \Gamma')$ for all $\Gamma \in \mathcal{O}_p$ and $Z \in \mathcal{Z}$, then $h(\phi(X, Y)) = h(Y)$ and we have the identity

$$\int h(Z)\, dZ = \int h(Y)\nu_2(dY).$$

It is quite difficult to give a rigorous derivation of the measure ν_2 without the theory of differential forms. In fact, it is not obvious that ν_2 is absolutely continuous with respect to Lebesgue measure on $\mathcal{Y}$. The subject of this example is considered again in later chapters. ◆

PROBLEMS

1. Let M be a proper subspace for V and set

$$G(M) = \{g | g \in Gl(V),\ g(M) = M\}$$

where $g(M) = \{x | x = gv \text{ for some } v \in M\}$.

(i) Show that $g(M) = M$ iff $g(M) \subseteq M$ for $g \in Gl(V)$ and show that $G(M)$ is a group.

Now, assume $V = R^p$ and, for $x \in R^p$, write $x = \binom{y}{z}$ with $y \in R^q$ and $z \in R^r$, $q + r = p$. Let $M = \{x | x = \binom{y}{0},\ y \in R^q\}$.

(ii) For $g \in Gl_p$, partition g as

$$g = \begin{pmatrix} g_{11} & g_{12} \\ g_{21} & g_{22} \end{pmatrix}, \qquad g_{11} \text{ is } q \times q.$$

Show that $g \in G(M)$ iff $g_{11} \in Gl_q$, $g_{22} \in Gl_r$, and $g_{21} = 0$. For such g show that

$$g^{-1} = \begin{pmatrix} g_{11}^{-1} & -g_{11}^{-1}g_{12}g_{22}^{-1} \\ 0 & g_{22}^{-1} \end{pmatrix}.$$

(iii) Verify that $G_1 = \{g \in G(M) | g_{11} = I_q, g_{12} = 0\}$ and $G_2 = \{g \in G(M) | g_{22} = I_r\}$ are subgroups of $G(M)$ and G_2 is a normal subgroup of $G(M)$.

(iv) Show that $G_1 \cap G_2 = \{I\}$ and show that each g can be written uniquely as $g = hk$ with $h \in G_1$ and $k \in G_2$. Conclude that, if $g_i = h_i k_i$, $i = 1, 2$, then $g_1 g_2 = h_3 k_3$, where $h_3 = h_1 h_2$ and $k_3 = h_2^{-1} k_1 h_2 k_2$, is the unique representation of $g_1 g_2$ with $h_3 \in G_1$ and $k_3 \in G_2$.

2. Let $G(M)$ be as in Problem 1. Does $G(M)$ act transitively on $V - \{0\}$? Does $G(M)$ act transitively on $V \cap M^c$ where M^c is the complement of the set M in V?

3. Show that $\mathcal{O}_n$ is a compact subset of R^m with $m = n^2$. Show that $\mathcal{O}_n$ is a topological group when $\mathcal{O}_n$ has the topology inherited from R^m. If χ is a continuous homomorphism from $\mathcal{O}_n$ to the multiplicative group $(0, \infty)$, show that $\chi(\Gamma) = 1$ for all $\Gamma \in \mathcal{O}_n$.

4. Suppose χ is a continuous homomorphism on $(0, \infty)$ to $(0, \infty)$. Show that $\chi(x) = x^\alpha$ for some real number α.

5. Show that $\mathcal{O}_n$ is a compact subgroup of Gl_n and show that G_U^+ (of dimension $n \times n$) is a closed subgroup of Gl_n. Show that the uniqueness of the representation $A = \Gamma U$ ($A \in Gl_n$, $\Gamma \in \mathcal{O}_n$, $U \in G_U^+$) is equivalent to $\mathcal{O}_n \cap G_U^+ = \{I_n\}$. Show that neither $\mathcal{O}_n$ nor G_U^+ is a normal subgroup of Gl_n.

6. Let $(V, (\cdot, \cdot))$ be an inner product space.

(i) For fixed $v \in V$, show that χ defined by $\chi(x) = \exp[(v, x)]$ is a continuous homomorphism on V to $(0, \infty)$. Here V is a group under addition.

(ii) If χ is a continuous homomorphism on V, show that $\chi(x) = \log\chi(x)$ is a linear function on V. Conclude that $\chi(x) = \exp[(v, x)]$ for some $v \in V$.

7. Suppose χ is a continuous homomorphism defined on Gl_n to $(0, \infty)$. Using the steps outlined below, show that $\chi(A) = |\det A|^\alpha$ for some real α.

(i) First show that $\chi(\Gamma) = 1$ for $\Gamma \in \mathcal{O}_n$.

(ii) Write $A = \Gamma D \Delta$ with $\Gamma, \Delta \in \mathcal{O}_n$ and D diagonal with positive diagonals $\lambda_1, \dots, \lambda_n$. Show that $\chi(A) = \chi(D)$.

(iii) Next, write $D = \prod D_i(\lambda_i)$ where $D_i(c)$ is diagonal with all diagonal elements equal to one except the ith diagonal element, which is c. Conclude that $\chi(D) = \prod \chi(D_i(\lambda_i))$.

(iv) Show that $D_i(c) = PD_1(c)P'$ for some permutation matrix $P \in \mathcal{O}_n$. Using this, show that $\chi(D) = \chi(D_1(\lambda))$ where $\lambda = \prod \lambda_i$.

(v) For $\lambda \in (0, \infty)$, set $\xi(\lambda) = \chi(D_1(\lambda))$ and show that ξ is a continuous homomorphism on $(0, \infty)$ to $(0, \infty)$ so $\xi(\lambda) = \lambda^\beta$ for some real β. Now, complete the proof of $\chi(A) = |\det A|^\alpha$.

8. Let $\mathfrak{X}$ be the set of all rank r orthogonal projections on R^n to R^n $(1 \leqslant r \leqslant n - 1)$.

(i) Show that $\mathcal{O}_n$ acts transitively on $\mathfrak{X}$ via the action $x \to \Gamma x \Gamma'$, $\Gamma \in \mathcal{O}_n$. For

$$x_0 = \begin{pmatrix} I_r & 0 \\ 0 & 0 \end{pmatrix} \in \mathfrak{X}.$$

what is the isotropy subgroup? Show that the representation of x in this case is $x = \psi\psi'$ where $\psi : n \times r$ consists of the first r columns of $\Gamma \in \mathcal{O}_n$.

(ii) The group $\mathcal{O}_r$ acts on $\mathfrak{F}_{r,n}$ by $\psi \to \psi\Delta'$, $\Delta \in \mathcal{O}_r$. This induces an equivalence relation on $\mathfrak{F}_{r,n}$ ($\psi_1 \cong \psi_2$ iff $\psi_1 = \psi_2\Delta'$ for some $\Delta \in \mathcal{O}_r$), and hence defines a quotient space. Show that the map $[\psi] \to \psi\psi'$ defines a one-to-one onto map from this quotient space to $\mathfrak{X}$. Here $[\psi]$ is the equivalence class of ψ.

9. Following the steps outlined below, show that every continuous homomorphism on G_T^+ to $(0, \infty)$ has the form $\chi(T) = \prod_1^p (t_{ii})^{c_i}$ where $T: p \times p$ has diagonal elements $t_{11}, \dots, t_{pp}$ and $c_1, \dots, c_p$ are real numbers.

(i) Let

$$G_1 = \left\{ T | T = \begin{pmatrix} T_{11} & 0 \\ 0 & 1 \end{pmatrix}, T_{11} : (p-1) \times (p-1) \right\}$$

and

$$G_2 = \left\{ T | T = \begin{pmatrix} I_{p-1} & 0 \\ T_{21} & t_{pp} \end{pmatrix} \right\}.$$

Show that G_1 and G_2 are subgroups of G_T^+ and G_2 is normal. Show that every T has a unique representation as $T = hk$ with $h \in G_1$, $k \in G_2$.

(ii) An induction assumption yields $\chi(h) = \prod_1^{p-1}(t_{ii})^{c_i}$. Also for $T = hk$, $\chi(T) = \chi(h)\chi(k)$.

(iii) Show that $\chi(k) = (t_{pp})^{c_p}$ for some real c_p.

10. Evaluate the integral $I_\gamma = \int |X'X|^\gamma \exp[-\frac{1}{2}\operatorname{tr} X'X]\, dX$ where X ranges over all $n \times p$ matrices of rank p. In particular, for what values of γ is this integral finite?

11. In the notation of Problems 1 and 2, find all of the relatively invariant integrals on $R^p \cap M^c$ under the action of $G(M)$.

12. In R^n, let $\mathcal{X} = \{x | x \in R^n, x \notin \operatorname{span}\{e\}\}$. Also, let $S_{n-1}(e) = \{x | \|x\| = 1, x \in R^n, x'e = 0\}$ and let $\mathcal{Y} = R^1 \times (0, \infty) \times S_{n-1}(e)$. For $x \in \mathcal{X}$, set $\bar{x} = n^{-1}e'x$ and set $s^2(x) = \Sigma(x_i - \bar{x})^2$. Define a mapping τ on $\mathcal{X}$ to $\mathcal{Y}$ by $\tau(x) = \{\bar{x}, s, (x - \bar{x}e)/s\}$.

(i) Show that τ is one-to-one, onto and find τ^{-1}. Let $\mathcal{O}_n(e) = \{\Gamma | \Gamma \in \mathcal{O}_n, \Gamma e = e\}$ and consider a group G defined by $G = \{(a, b, \Gamma) | a \in (0, \infty),\ b \in R^1,\ \Gamma \in \mathcal{O}_n(e)\}$ with group composition given by $(a_1, b_1, \Gamma_1)(a_2, b_2, \Gamma_2) = (a_1a_2, a_1b_2 + b_1, \Gamma_1\Gamma_2)$. Define G acting on $\mathcal{X}$ and $\mathcal{Y}$ by $(a, b, \Gamma)x = a\Gamma x + be$, $x \in \mathcal{X}$, $(a, b, \Gamma)(u, v, w) = (au + b, av, \Gamma w)$ for $(u, v, w) \in \mathcal{Y}$.

(ii) Show that $\tau(gx) = g\tau(x)$, $g \in G$.

(iii) Show that the measure $\mu(dx) = dx/s^n$ is an invariant measure on $\mathcal{X}$.

(iv) Let $\gamma(dw)$ be the unique $\mathcal{O}_n(e)$ invariant probability measure on $S_{n-1}(e)$. Show that the measure

$$\nu(d(u, v, w)) = du \frac{dv}{v^2} \gamma(dw)$$

is an invariant measure on $\mathcal{Y}$.

(v) Prove that $\int_{\mathcal{X}} f(x)\mu(dx) = k\int_{\mathcal{Y}} f(\tau^{-1}(y))\nu(dy)$ for all integrable f where k is a fixed constant. Find k.

(vi) Suppose a random vector $X \in \mathcal{X}$ has a density (with respect to dx) given by

$$f(x) = \frac{1}{\sigma^n} h\left(\frac{\|x - \delta e\|^2}{\sigma^2}\right), \qquad x \in \mathcal{X}$$

where $\delta \in R^1$ and $\sigma > 0$ are parameters. Find the joint density of $\bar{X}$ and s.

13. Let $\mathcal{X} = R^n - \{0\}$ and consider $X \in \mathcal{X}$ with an $\mathcal{O}_n$-invariant distribution. Define ϕ on $\mathcal{X}$ to $(0, \infty) \times \mathcal{F}_{1,n}$ by $\phi(x) = (\|x\|, x/\|x\|)$. The group $\mathcal{O}_n$ acts on $(0, \infty) \times \mathcal{F}_{1,n}$ by $\Gamma(u, v) = (u, \Gamma v)$. Show that $\phi(\Gamma x) = \Gamma\phi(x)$ and use this to prove that:

(i) $\|X\|$ and $X/\|X\|$ are independent.

(ii) $X/\|X\|$ has a uniform distribution on $\mathcal{F}_{1,n}$.

14. Let $\mathcal{X} = \{x \in R^n | x_i \neq x_j \text{ for all } i \neq j\}$ and let $\mathcal{Y} = \{y \in R^n | y_1 < y_2 < \cdots < y_n\}$. Also, let $\mathcal{P}_n$ be the group of $n \times n$ permutation matrices so $\mathcal{P}_n \subseteq \mathcal{O}_n$ and $\mathcal{P}_n$ acts on $\mathcal{X}$ by $x \to gx$.

(i) Show that the map $\phi(g, y) = gy$ is one-to-one and onto from $\mathcal{P}_n \times \mathcal{Y}$ to $\mathcal{X}$. Describe ϕ^{-1}.

(ii) Let $X \in \mathcal{X}$ be a random vector such that $\mathcal{L}(X) = \mathcal{L}(gX)$ for $g \in \mathcal{P}_n$. Write $\phi^{-1}(X) = (P(X), Y(X))$ where $P(X) \in \mathcal{P}_n$ and $Y(X) \in \mathcal{Y}$. Show that $P(X)$ and $Y(X)$ are independent and that $P(X)$ has a uniform distribution on $\mathcal{P}_n$.

NOTES AND REFERENCES

1. For an alternative to Nachbin's treatment of invariant integrals, see Segal and Kunze (1978).

2. Proposition 6.10 is the Radon measure version of a result due to Farrell (see Farrell, 1976). The extension of Proposition 6.10 to relatively invariant integrals that are unique up to constant is immediate—the proof of Proposition 6.10 is valid.

3. For the form of the measure ν_2 in Example 6.21, see Deemer and Olkin (1951), Farrell (1976), or Muirhead (1982).

CHAPTER 7

First Applications of Invariance

We now begin to reap some of the benefits of the labor of Chapter 6. The one unifying notion throughout this chapter is that of a group of transformations acting on a space. Within this framework independence and distributional properties of random vectors are discussed and a variety of structural problems are considered. In particular, invariant probability models are introduced and the invariance of likelihood ratio tests and maximum likelihood estimators is established. Further, maximal invariant statistics are discussed in detail.

7.1. LEFT $\mathcal{O}_n$ INVARIANT DISTRIBUTIONS ON $n \times p$ MATRICES

The main concern of this section is conditions under which the two matrices Ψ and U in the decomposition $X = \Psi U$ (see Example 6.20) are stochastically independent when X is a random $n \times p$ matrix. Before discussing this problem, a useful construction of the uniform distribution on $\mathcal{F}_{p,n}$ is presented. Throughout this section, $\mathcal{X}$ denotes the space of $n \times p$ matrices of rank p so $n \geqslant p$. First, a technical result.

Proposition 7.1. Let $X \in \mathcal{L}_{p,n}$ have a normal distribution with mean zero and $\text{Cov}(X) = I_n \otimes I_p$. Then $P\{X \in \mathcal{X}\} = 1$ and the complement of $\mathcal{X}$ in $\mathcal{L}_{p,n}$ has Lebesgue measure zero.

Proof. Let $X_1, \ldots, X_p$ denote the p columns of X. Thus $X_1, \ldots, X_p$ are independent random vectors in R^n and $\mathcal{L}(X_i) = N(0, I_n)$, $i = 1, \ldots, p$. It is shown that $P\{X \in \mathcal{X}^c\} = 0$. To say that $X \in \mathcal{X}^c$ is to say that, for some index i,

$$X_i \in \operatorname{span}\{X_j | j \neq i\}.$$

Therefore,

$$P\{X \in \mathfrak{X}^c\} = P\left\{\bigcup_{i=1}^{p} \left[X_i \in \operatorname{span}\{X_j | j \neq i\}\right]\right\}$$

$$\leqslant \sum_1^p P\{X_i \in \operatorname{span}\{X_j | j \neq i\}\}.$$

However, X_i is independent of the set of random vectors $\{X_j | j \neq i\}$ and the probability of any subspace M of dimension less than n is zero. Since $p \leqslant n$, the subspace span $\{X_j | j \neq i\}$ has dimension less than n. Thus conditioning on X_j for $j \neq i$, we have

$$P\{X_i \in \operatorname{span}\{X_j | j \neq i\}\} = \mathcal{E}P\{X_i \in \operatorname{span}\{X_j | j \neq i\} | X_j,\ j \neq i\} = 0.$$

Hence $P\{X \in \mathfrak{X}^c\} = 0$. Since $\mathfrak{X}^c$ has probability zero under the normal distribution on $\mathcal{L}_{p,n}$ and since the normal density function with respect to Lebesgue measure is strictly positive on $\mathcal{L}_{p,n}$, it follows the $\mathfrak{X}^c$ has Lebesgue measure zero. □

If $X \in \mathcal{L}_{p,n}$ is a random vector that has a density with respect to Lebesgue measure, the previous result shows that $P\{X \in \mathfrak{X}\} = 1$ since $\mathfrak{X}^c$ has Lebesgue measure zero. In particular, if $X \in \mathcal{L}_{p,n}$ has a normal distribution with a nonsingular covariance, then $P\{X \in \mathfrak{X}\} = 1$, and we often restrict such normal distributions to $\mathfrak{X}$ in order to insure that X has rank p. For many of the results below, it is assumed that X is a random vector in $\mathfrak{X}$, and in applications X is a random vector in $\mathcal{L}_{p,n}$, which has been restricted to $\mathfrak{X}$ after it has been verified that $\mathfrak{X}^c$ has probability zero under the distribution of X.

Proposition 7.2. Suppose $X \in \mathfrak{X}$ has a normal distribution with $\mathcal{L}(X) = N(0, I_n \otimes I_p)$. Let $X_1, \dots, X_p$ be the columns of X and let $\Psi \in \mathfrak{F}_{p,n}$ be the random matrix whose p columns are obtained by applying the Gram–Schmidt orthogonalization procedure to $X_1, \dots, X_p$. Then Ψ has the uniform distribution on $\mathfrak{F}_{p,n}$, that is, the distribution of Ψ is the unique probability measure on $\mathfrak{F}_{p,n}$ that is invariant under the action of $\mathcal{O}_n$ on $\mathfrak{F}_{p,n}$ (see Example 6.16).

Proof. Let Q be the probability distribution of Ψ of $\mathcal{F}_{p,n}$. It must be verified that

$$Q(\Gamma B) = Q(B), \qquad \Gamma \in \mathcal{O}_n$$

for all Borel sets B of $\mathcal{F}_{p,n}$. If $\Gamma \in \mathcal{O}_n$, it is clear that $\mathcal{L}(\Gamma X) = \mathcal{L}(X)$. Also, it is not difficult to verify that Ψ, which we now write as a function of X, say $\Psi(X)$, satisfies

$$\Psi(\Gamma X) = \Gamma \Psi(X), \qquad \Gamma \in \mathcal{O}_n.$$

This follows by looking at the Gram–Schmidt Procedure, which defined the columns of Ψ. Thus

$$Q(B) = P\{\Psi(X) \in B\} = P\{\Psi(\Gamma X) \in B\} = P\{\Gamma\Psi(X) \in B\}$$

$$= P\{\Psi(X) \in \Gamma' B\} = Q(\Gamma' B)$$

for all $\Gamma \in \mathcal{O}_n$. The second equality above follows from the observation that $\mathcal{L}(X) = \mathcal{L}(\Gamma X)$. Hence Q is an $\mathcal{O}_n$-invariant probability measure on $\mathcal{F}_{p,n}$ and the uniqueness of such a measure shows that Q is what was called the uniform distribution on $\mathcal{F}_{p,n}$. □

Now, consider the two spaces $\mathcal{X}$ and $\mathcal{F}_{p,n} \times G_U^+$. Let ϕ be the function on $\mathcal{X}$ to $\mathcal{F}_{p,n} \times G_U^+$ that maps X into the unique pair (Ψ, U) such that $X = \Psi U$. Obviously, $\phi^{-1}(\Psi, U) = \Psi U \in \mathcal{X}$.

Definition 7.1. If $X \in \mathcal{X}$ is a random vector with a distribution P, then P is *left invariant under* $\mathcal{O}_n$ if $\mathcal{L}(X) = \mathcal{L}(\Gamma X)$ for all $\Gamma \in \mathcal{O}_n$.

The remainder of this section is devoted to a characterization of the $\mathcal{O}_n$-left invariant distributions on $\mathcal{X}$. It is shown that, if $X \in \mathcal{X}$ has an $\mathcal{O}_n$-left invariant distribution, then for $\phi(X) = (\Psi, U) \in \mathcal{F}_{p,n} \times G_U^+$, Ψ and U are stochastically independent and Ψ has a uniform distribution on $\mathcal{F}_{p,n}$. This assertion and its converse are given in the following proposition.

Proposition 7.3. Suppose $X \in \mathcal{X}$ is a random vector with an $\mathcal{O}_n$-left invariant distribution P and write $(\Psi, U) = \phi(X)$. Then Ψ and U are stochastically independent and Ψ has a uniform distribution on $\mathcal{F}_{p,n}$. Conversely, if $\Psi \in \mathcal{F}_{p,n}$ and $U \in G_U^+$ are independent and if Ψ has a uniform distribution on $\mathcal{F}_{p,n}$, then $X = \Psi U$ has an $\mathcal{O}_n$-left invariant distribution on $\mathcal{X}$.

Proof. The joint distribution Q of (Ψ, U) is determined by

$$Q(B_1 \times B_2) = P\big(\phi^{-1}(B_1 \times B_2)\big)$$

where B_1 is a Borel subset of $\mathfrak{F}_{p,n}$ and B_2 is a Borel subset of G_U^+. Also,

$$\iint f(\Psi, U) Q(d\Psi, dU) = \int f(\phi(X)) P(dX)$$

for any Borel measurable function that is integrable. The group $\mathcal{O}_n$ acts on the left of $\mathfrak{F}_{p,n} \times G_U^+$ by

$$\Gamma(\Psi, U) = (\Gamma\Psi, U)$$

and it is clear that

$$\phi(\Gamma X) = \Gamma\phi(X) \quad \text{for } X \in \mathfrak{X}, \Gamma \in \mathcal{O}_n.$$

We now show that Q is invariant under this group action and apply Proposition 6.10. For $\Gamma \in \mathcal{O}_n$,

$$\begin{aligned}\iint f(\Gamma(\Psi, U)) Q(d\Psi, dU) &= \int f(\Gamma\phi(X)) P(dX) = \int f(\phi(\Gamma X)) P(dX)\\ &= \int f(\phi(X)) P(dX)\\ &= \iint f(\Psi, U) Q(d\Psi, dU).\end{aligned}$$

Therefore, Q is $\mathcal{O}_n$-invariant and, by Proposition 6.10, Q is a product measure $Q_1 \times Q_2$ where Q_1 is taken to be the uniform distribution on $\mathfrak{F}_{p,n}$. That Q_2 is a probability measure is clear since Q is a probability measure. The first assertion has been established. For the converse, let Q_1 and Q_2 be the distributions of Ψ and U so Q_1 is the uniform distribution on $\mathfrak{F}_{p,n}$ and $Q_1 \times Q_2$ is the joint distribution of (Ψ, U) in $\mathfrak{F}_{p,n} \times G_U^+$. The distribution P of $X = \Psi U = \phi^{-1}(\Psi, U)$ is determined by the equation

$$\int f(X) P(dX) = \iint f(\phi^{-1}(\Psi, U)) Q_1(d\Psi) Q_2(dU),$$

for all integrable f. To show P is $\mathcal{O}_n$-left invariant, it must be verified that

$$\int f(\Gamma X) P(dX) = \int f(X) P(dX)$$

for all integrable f and $\Gamma \in \mathcal{O}_n$. But

$$\int f(\Gamma X)P(dX) = \iint f(\Gamma\phi^{-1}(\Psi, U))Q_1(d\Psi)Q_2(dU)$$

$$= \iint f(\phi^{-1}(\Gamma\Psi, U))Q_1(d\Psi)Q_2(dU)$$

$$= \iint f(\phi^{-1}(\Psi, U))Q_1(d\Psi)Q_2(dU) = \int f(X)P(dX)$$

where the next to the last equality follows from the $\mathcal{O}_n$-invariance of Q_1. Thus P is $\mathcal{O}_n$-left invariant. □

When $p = 1$, Proposition 7.3 is interesting. In this case $\mathcal{X} = R^n - \{0\}$ and the $\mathcal{O}_n$-left invariant distributions on $\mathcal{X}$ are exactly the orthogonally invariant distributions on R^n that have no probability at $0 \in R^n$. If $X \in R^n - \{0\}$ has an orthogonally invariant distribution, then $\Psi = X/\|X\| \in \mathcal{F}_{1,n}$ is independent of $U = \|X\|$ and Ψ has a uniform distribution on $\mathcal{F}_{1,n}$.

There is an analogue of Proposition 7.3 for the decomposition of $X \in \mathcal{X}$ into (Ψ, A) where $\Psi \in \mathcal{F}_{p,n}$ and $A \in \mathcal{S}_p^+$ (see Proposition 5.5).

Proposition 7.4. Suppose $X \in \mathcal{X}$ is a random vector with an $\mathcal{O}_n$-left invariant distribution and write $\phi(X) = (\Psi, A)$ where $\Psi \in \mathcal{F}_{p,n}$ and $A \in \mathcal{S}_p^+$ are the unique matrices such that $X = \Psi A$. Then Ψ and A are independent and Ψ has a uniform distribution on $\mathcal{F}_{p,n}$. Conversely, if $\Psi \in \mathcal{F}_{p,n}$ and $A \in \mathcal{S}_p^+$ are independent and if Ψ has a uniform distribution on $\mathcal{F}_{p,n}$, then $X = \Psi A$ has an $\mathcal{O}_n$-left invariant distribution on $\mathcal{X}$.

Proof. The proof is essentially the same as that of Proposition 7.3 and is left to the reader. □

Thus far, it has been shown that if $X \in \mathcal{X}$ has an $\mathcal{O}_n$-left invariant distribution for $X = \Psi U$, Ψ and U are independent and Ψ has a uniform distribution. However, nothing has been said about the distribution of $U \in G_U^+$. The next result gives the density function of U with respect to the right invariant measure

$$\nu_r(dU) = \frac{dU}{\prod_1^p u_{ii}^i}$$

in the case that X has a density of a special form.

Proposition 7.5. Suppose $X \in \mathfrak{X}$ has a distribution P given by a density function

$$f_0(X'X), \qquad X \in \mathfrak{X}$$

with respect to the measure

$$\mu(dX) = \frac{dX}{|X'X|^{n/2}}$$

on $\mathfrak{X}$. Then the density function of U (with respect to ν_r) in the representation $X = \Psi U$ is

$$g_0(U) = 2^p(\sqrt{2\pi})^{np}\omega(n, p) f_0(U'U).$$

Proof. If $X \in \mathfrak{X}$, $U(X)$ denotes the unique element of G_U^+ such that $X = \Psi U(X)$ for some $\Psi \in \mathfrak{F}_{p,n}$. To show g_0 is the density function of U, it is sufficient to verify that

$$\int h(U(X)) f_0(X'X)\mu(dX) = \int h(U) g_0(U)\nu_r(dU)$$

for all integrable functions h. Since $X'X = U'(X)U(X)$, the results of Example 6.20 show that

$$\int h(U(X)) f_0(U'(X)U(X))\mu(dX) = c\int h(U) f_0(U'U)\nu_r(dU)$$

where $c = 2^p(\sqrt{2\pi})^{np}\omega(n, p)$. Since $g_0(U) = cf_0(U'U)$, g_0 is the density of U. □

A similar argument gives the density of $S = X'X$.

Proposition 7.6. Suppose $X \in \mathfrak{X}$ has distribution P given by a density function

$$f_0(X'X), \qquad X \in \mathfrak{X}$$

with respect to the measure μ. Then the density of $S = X'X$ is

$$g_0(S) = (\sqrt{2\pi})^{np}\omega(n, p) f_0(S)$$

with respect to the measure

$$\nu(dS) = \frac{dS}{|S|^{(p+1)/2}}.$$

Proof. With the notation $S(X) = X'X$, it is sufficient to verify that

$$\int h(S(X)) f_0(X'X)\mu(dX) = \int h(S) g_0(S)\nu(dS)$$

for all integrable functions h. Combining the identities (6.4) and (6.5), we have

$$\int h(S(X)) f_0(X'X)\mu(dX) = c\int h(S) f_0(S)\nu(dS)$$

where $c = (\sqrt{2\pi})^{np}\omega(n, p)$. Since $g_0 = cf_0$, the proof is complete. □

When $X \in \mathcal{X}$ has the density assumed in Propositions 7.5 and 7.6, it is clear that the distribution of X is $\mathcal{O}_n$-left invariant. In this case, for $X = \Psi U$, Ψ and U are independent, Ψ has a uniform distribution on $\mathcal{F}_{p,n}$, and U has the density given in Proposition 7.5. Thus the joint distribution of Ψ and U has been completely described. Similar remarks apply to the situation treated in Proposition 7.6. The reader has probably noticed that the distribution of $S = X'X$ was derived rather than the distribution of A in the representation $X = \Psi A$ for $\Psi \in \mathcal{F}_{p,n}$ and $A \in \mathcal{S}_p^+$. Of course, $S = A^2$ so A is the unique positive definite square root of S. The reason for giving the distribution of S rather than that of A is quite simple—the distribution of A is substantially more complicated than that of S and harder to derive.

In the following example, we derive the distributions of U and S when $X \in \mathcal{X}$ has a nonsingular $\mathcal{O}_n$-left invariant normal distribution.

◆ **Example 7.1.** Suppose $X \in \mathcal{X}$ has a normal distribution with a nonsingular covariance and also assume that $\mathcal{L}(X) = \mathcal{L}(\Gamma X)$ for all $\Gamma \in \mathcal{O}_n$. Thus $\mathcal{E}X = \Gamma\mathcal{E}X$ for all $\Gamma \in \mathcal{O}_n$, which implies that $\mathcal{E}X = 0$. Also, Cov(X) must satisfy Cov$((\Gamma \otimes I_p)X) = $ Cov(X) since $\mathcal{L}(X) = \mathcal{L}((\Gamma \otimes I_p)X)$. From Proposition 2.19, this implies that

$$\text{Cov}(X) = I_n \otimes \Sigma$$

for some positive definite Σ as Cov(X) is assumed to be nonsingular. In summary, if X has a normal distribution in $\mathcal{X}$ that is $\mathcal{O}_n$-left invariant, then

$$\mathcal{L}(X) = N(0, I_n \otimes \Sigma).$$

Conversely, if X is normal with mean zero and $\text{Cov}(X) = I_n \otimes \Sigma$, then $\mathcal{L}(X) = \mathcal{L}(\Gamma X)$ for all $\Gamma \in \mathcal{O}_n$. Now that the $\mathcal{O}_n$-left invariant normal distributions on $\mathfrak{X}$ have been described, we turn to the distribution of $S = X'X$ and U described in Propositions 7.5 and 7.6. When $\mathcal{L}(X) = N(0, I_n \otimes \Sigma)$, the density function of X with respect to the measure $\mu(dX) = dX/|X'X|^{n/2}$ is

$$f_0(X) = (\sqrt{2\pi})^{-np}|\Sigma^{-1}X'X|^{n/2}\exp[-\tfrac{1}{2}\text{tr}\,\Sigma^{-1}X'X].$$

Therefore, the density of S with respect to the measure

$$\nu(dS) = \frac{dS}{|S|^{(p+1)/2}}, \qquad S \in \mathcal{S}_p^+$$

is given by

$$g_0(S) = \omega(n, p)|\Sigma^{-1}S|^{n/2}\exp[-\tfrac{1}{2}\,\text{tr}\,\Sigma^{-1}S]$$

according to Proposition 7.6. This density is called the *Wishart density* with parameters Σ, p, and n. Here, p is the dimension of S and n is called the degrees of freedom. When S has such a density function, we write $\mathcal{L}(S) = W(\Sigma, p, n)$, which is read "the distribution of S is Wishart with parameters Σ, p, and n." A slightly more general definition of the Wishart distribution is given in the next chapter, where a thorough discussion of the Wishart distribution is presented. A direct application of Proposition 7.5 yields the density

$$g_1(U) = 2^p\omega(n, p)|\Sigma^{-1}U'U|^{n/2}\exp[-\tfrac{1}{2}\,\text{tr}\,\Sigma^{-1}U'U]$$

with respect to measure

$$\nu_r(dU) = \frac{dU}{\prod_1^p u_{ii}^i}$$

when $X = \Psi U$, $\Psi \in \mathcal{F}_{p,n}$, and $U \in G_U^+$. Here, the nonzero elements of U are u_{ij}, $1 \leqslant i \leqslant j \leqslant p$. When $\Sigma = I_p$, g_1 becomes

$$\begin{aligned} g_1(U)\nu_r(dU) &= 2^p\omega(n, p)\prod_{i=1}^p u_{ii}^i\exp[-\tfrac{1}{2}\,\text{tr}U'U]\nu_r(dU) \\ &= 2^p\omega(n, p)\prod_{i=1}^p u_{ii}^{n-i}\exp\left[-\tfrac{1}{2}\sum_{1\leqslant j} u_{ij}^2\right] dU. \end{aligned}$$

In G_U^+, the diagonal elements of U range between 0 and ∞ and the elements above the diagonal range between $-\infty$ and $+\infty$. Writing the density above as

$$g_1(U)\nu_r(dU) = 2^p\omega(n,p)\prod_1^p\left(u_{ii}^{n-i}\exp\left[-\tfrac{1}{2}u_{ii}^2\right]du_{ii}\right)$$

$$\times\prod_{i<j}\left(\exp\left[-\tfrac{1}{2}u_{ij}^2\right]du_{ij}\right),$$

we see that this density factors into a product of functions that are, when normalized by a constant, density functions. It is clear by inspection that

$$\mathcal{L}(u_{ij}) = N(0,1) \quad \text{for } i<j.$$

Further, a simple change of variable shows that

$$\mathcal{L}\left(u_{ii}^2\right) = \chi_{n-i+1}^2, \qquad i = 1,\dots,p.$$

Thus when $\Sigma = I_p$, the nonzero elements of U are independent, the elements above the diagonal are all $N(0,1)$, and the square of the ith diagonal element has a chi-square distribution with $n-i+1$ degrees of freedom. This result is sometimes useful for deriving the distribution of functions of $S = U'U$. ◆

7.2. GROUPS ACTING ON SETS

Suppose $\mathcal{X}$ is a set and G is a group that acts on the left of $\mathcal{X}$ according to Definition 6.3. The group G defines a natural equivalence relation between elements of $\mathcal{X}$—namely, write $x_1 \simeq x_2$ if there exists a $g \in G$ such that $x_1 = gx_2$. It is easy to check that $\simeq$ is in fact an equivalence relation. Thus the group G partitions the set $\mathcal{X}$ into disjoint equivalence classes, say

$$\mathcal{X} = \bigcup_{\alpha\in A}\mathcal{X}_\alpha,$$

where A is an index set and the equivalence classes $\mathcal{X}_\alpha$ are disjoint. For each $x \in \mathcal{X}$, the set $\{gx|g \in G\}$ is the *orbit* of x under the action of G. From the definition of the equivalence relation, it is clear that, if $x \in \mathcal{X}_\alpha$, then $\mathcal{X}_\alpha$ is just the orbit of x. Thus the decomposition of $\mathcal{X}$ into equivalence classes is simply a decomposition of $\mathcal{X}$ into disjoint orbits and two points are equivalent iff they are in the same orbit.

Definition 7.2. Suppose G acts on the left of $\mathcal{X}$. A function f on $\mathcal{X}$ to $\mathcal{Y}$ is *invariant* if $f(x) = f(gx)$ for all $x \in \mathcal{X}$ and $g \in G$. The function f is *maximal invariant* if f is invariant and $f(x_1) = f(x_2)$ implies that $x_1 = gx_2$ for some $g \in G$.

Obviously, f is invariant iff f is constant on each orbit in X. Also, f is maximal invariant iff it is constant on each orbit and takes different values on different orbits.

Proposition 7.7. Suppose f maps $\mathcal{X}$ onto $\mathcal{Y}$ and f is maximal invariant. Then h, mapping $\mathcal{X}$ into $\mathcal{Z}$, is invariant iff $h(x) = k(f(x))$ for some function k mapping $\mathcal{Y}$ into $\mathcal{Z}$.

Proof. If $h(x) = k(f(x))$, then h is invariant as f is invariant. Conversely, suppose h is invariant. Given $y \in \mathcal{Y}$, the set $\{x | f(x) = y\}$ is exactly one orbit in $\mathcal{X}$ since f is maximal invariant. Let $z \in \mathcal{Z}$ be the value of h on this orbit (h is invariant), and define $k(y) = z$. Obviously, k is well defined and $k(f(x)) = h(x)$. □

Proposition 7.7 is ordinarily paraphrased by saying that a function is invariant iff it is a function of a maximal invariant. Once a maximal invariant function has been constructed, then all the invariant functions are known—namely, they are functions of the maximal invariant function. If the group G acts transitively on $\mathcal{X}$, then there is just one orbit and the only invariant functions are the constants. We now turn to some examples.

◆ **Example 7.2.** Let $\mathcal{X} = R^n - \{0\}$ and let $G = \mathcal{O}_n$ act on $\mathcal{X}$ as a group of matrices acts on a vector space. Given $x \in \mathcal{X}$, it is clear that the orbit of x is $\{y | \|y\| = \|x\|\}$. Let $S_r = \{x | \|x\| = r\}$, so

$$\mathcal{X} = \bigcup_{r>0} S_r$$

is the decomposition of $\mathcal{X}$ into equivalence classes. The real number $r > 0$ indexes the orbits. That $f(x) = \|x\|$ is a maximal invariant function follows from the invariance of f and the fact that f takes a different value on each orbit. Thus a function is invariant under the action of G on $\mathcal{X}$ iff it is a function of $\|x\|$. Now, consider the space $S_1 \times (0, \infty)$ and define the function ϕ on $\mathcal{X}$ to $S_1 \times (0, \infty)$ by

$\phi(x) = (x/\|x\|, \|x\|)$. Obviously, ϕ is one-to-one, onto, and $\phi^{-1}(u, r) = ru$ for $(u, r) \in S_1 \times (0, \infty)$. Further, the group action on $\mathfrak{X}$ corresponds to the group action on $S_1 \times (0, \infty)$ given by

$$\Gamma(u, r) \equiv (\Gamma u, r), \qquad \Gamma \in \mathcal{O}_n.$$

In other words, $\phi(\Gamma x) = \Gamma\phi(x)$ so ϕ is an equivariant function (see Definition 6.14). Since $\mathcal{O}_n$ acts transitively on S_1, a function h on $S_1 \times (0, \infty)$ is invariant iff $h(u, r)$ does not depend on u. For this example, the space $\mathfrak{X}$ has been mapped onto $S_1 \times (0, \infty)$ by ϕ so that the group action on $\mathfrak{X}$ corresponds to a special group action on $S_1 \times (0, \infty)$—namely, $\mathcal{O}_n$ acts transitively on S_1 and is the identity on $(0, \infty)$. The whole point of introducing $S_1 \times (0, \infty)$ is that the function $h_0(u, r) = r$ is obviously a maximal invariant function due to the special way in which $\mathcal{O}_n$ acts on $S_1 \times (0, \infty)$. To say it another way, the orbits in $S_1 \times (0, \infty)$ are $S_1 \times \{r\}$, $r > 0$, so the product space structure provides a convenient way to index the orbits and hence to give a maximal invariant function. This type of product space structure occurs in many other examples. ◆

The following example provides a useful generalization of the example above.

◆ **Example 7.3.** Suppose $\mathfrak{X}$ is the space of all $n \times p$ matrices of rank $p, p \leqslant n$. Then $\mathcal{O}_n$ acts on the left of $\mathfrak{X}$ by matrix multiplication. The first claim is that $f_0(X) = X'X$ is a maximal invariant function. That f_0 is invariant is clear, so assume that $f_0(X_1) = f_0(X_2)$. Thus $X_1'X_1 = X_2'X_2$ and, by Proposition 1.31, there exists a $\Gamma \in \mathcal{O}_n$ such that $\Gamma X_1 = X_2$. This proves that f_0 is a maximal invariant. Now, the question is: where did f_0 come from? To answer this question, recall that each $X \in \mathfrak{X}$ has a unique representation as $X = \Psi A$ where $\Psi \in \mathfrak{F}_{p,n}$ and $A \in \mathcal{S}_p^+$. Let ϕ denote the map that sends X into the pair $(\Psi, A) \in \mathfrak{F}_{p,n} \times \mathcal{S}_p^+$ such that $X = \Psi A$. The group $\mathcal{O}_n$ acts on $\mathfrak{F}_{p,n} \times \mathcal{S}_p^+$ by

$$\Gamma(\Psi, A) = (\Gamma\Psi, A)$$

and ϕ satisfies

$$\phi(\Gamma X) = \Gamma\phi(X).$$

It is clear that $h_0(\Psi A) \equiv A$ is a maximal invariant function on $\mathfrak{F}_{p,n} \times \mathcal{S}_p^+$ under the action of $\mathcal{O}_n$ since $\mathcal{O}_n$ acts transitively on $\mathfrak{F}_{p,n}$.

Also, the orbits in $\mathcal{F}_{p,n} \times \mathcal{S}_p^+$ are $\mathcal{F}_{p,n} \times \{A\}$ for $A \in \mathcal{S}_p^+$. It follows immediately from the equivariance of ϕ that

$$\phi^{-1}\left(\mathcal{F}_{p,n} \times \{A\}\right) = \{X | X = \Psi A \text{ for some } \Psi \in \mathcal{F}_{p,n}\}$$

are the orbits in $\mathcal{X}$ under the action of $\mathcal{O}_n$. Thus we have a convenient indexing of the orbits in $\mathcal{X}$ given by A. A maximal invariant function on $\mathcal{X}$ must be a one-to-one function of an orbit index—namely, $A \in \mathcal{S}_p^+$. However, $f_0(X) = X'X = A^2$ when

$$X \in \{X | X = \Psi A, \quad \text{for some } \Psi \in \mathcal{F}_{p,n}\}.$$

Since A is the unique positive definite square root of $A^2 = X'X$, we have explicitly shown why f_0 is a one-to-one function of the orbit index A. A similar orbit indexing in $\mathcal{X}$ can be given by elements $U \in G_U^+$ by representing each $X \in \mathcal{X}$ as $X = \Psi U$, $\Psi \in \mathcal{F}_{p,n}$, and $U \in G_U^+$. The details of this are left to the reader. ◆

◆ **Example 7.4.** In this example, the set $\mathcal{X}$ is $(R^p - \{0\}) \times \mathcal{S}_p^+$. The group Gl_p acts on the left of $\mathcal{X}$ in the following manner:

$$A(y, S) \equiv (Ay, ASA')$$

for $(y, S) \in \mathcal{X}$ and $A \in Gl_p$. A useful method for finding a maximal invariant function is to consider a point $(y, S) \in \mathcal{X}$ and then "reduce" (y, S) to a convenient representative in the orbit of (y, S). The orbit of (y, S) is $\{A(y, S) | A \in Gl_p\}$. To reduce a given point (y, S) by $A \in Gl_p$, first choose $A = \Gamma S^{-1/2}$ where $\Gamma \in \mathcal{O}_p$ and $S^{-1/2}$ is the inverse of the positive definite square root of S. Then

$$ASA' = \Gamma S^{-1/2} S S^{-1/2} \Gamma' = \Gamma\Gamma' = I_p$$

and

$$A(y, S) = \left(\Gamma S^{-1/2} y, I\right),$$

which is in the orbit of (y, S). Since $S^{-1/2}y$ and $\|S^{-1/2}y\|\varepsilon_1$ have the same length ($\varepsilon_1' = (1, 0, \ldots, 0)$), we can choose $\Gamma \in \mathcal{O}_p$ such that

$$\Gamma S^{-1/2} y = \|S^{-1/2} y\| \varepsilon_1.$$

Therefore, for each $(y, S) \in \mathcal{X}$, the point

$$\left(\|S^{-1/2} y\| \varepsilon_1, I_p\right)$$

is in the orbit of (y, S). Let

$$f_0(y, S) = y'S^{-1}y = \|S^{-1/2}y\|^2.$$

The above reduction argument suggests, but does not prove, that f_0 is maximal invariant. However, the reduction argument does provide a method for checking that f_0 is maximal invariant. First, f_0 is invariant. To show f_0 is maximal invariant, if $f_0(y_1, S_1) = f_0(y_2, S_2)$, we must show there exists an $A \in Gl_p$ such that $A(y_1, S_1) = (y_2, S_2)$. From the reduction argument, there exists $A_i \in Gl_p$ such that

$$A_i(y_i, S_i) = \left(\|S_i^{-1/2}y_i\|\varepsilon_1, I_p\right), \qquad i = 1, 2.$$

Since $f_0(y_1, S_1) = f_0(y_2, S_2)$,

$$\|S_1^{-1/2}y_1\| = \|S_2^{-1/2}y_2\|$$

and this shows that

$$A_1(y_1, S_1) = A_2(y_2, S_2).$$

Setting $A = A_2^{-1}A_1$, we see that $A(y_1, S_1) = (y_2, S_2)$ so f_0 is maximal invariant. As in the previous two examples, it is possible to represent $\mathcal{X}$ as a product space where a maximal invariant is obvious. Let

$$\mathcal{Y} = \left\{(u, S)|u \in R^p, S \in \mathcal{S}_p^+, u'S^{-1}u = 1\right\}.$$

Then Gl_p acts on the left of $\mathcal{Y}$ by

$$A(u, S) \equiv (Au, ASA').$$

The reduction argument used above shows that the action of Gl_p is transitive on $\mathcal{Y}$. Consider the map ϕ from $\mathcal{X}$ to $\mathcal{Y} \times (0, \infty)$ given by

$$\phi(x, S) = \left(\left(\frac{x}{(x'S^{-1}x)^{1/2}}, S\right), x'S^{-1}x\right).$$

The group action of Gl_p on $\mathcal{Y} \times (0, \infty)$ is

$$A((u, S), r) \equiv (A(u, S), r)$$

and a maximal invariant function is

$$f_1((u, S), r) = r$$

since Gl_p is transitive on $\mathcal{Y}$. Clearly, ϕ is a one-to-one onto function and satisfies

$$\phi(A(x, S)) = A\phi(x, S).$$

Thus $f_1(\phi(x, S)) = x'S^{-1}x$ is maximal invariant. ◆

In the three examples above, the space $\mathcal{X}$ has been represented as a product space $\mathcal{Y} \times \mathcal{Z}$ in such a way that the group action on $\mathcal{X}$ corresponds to a group action on $\mathcal{Y} \times \mathcal{Z}$—namely,

$$g(y, z) = (gy, z)$$

and G acts transitively on $\mathcal{Y}$. Thus it is obvious that

$$f_1(y, z) = z$$

is maximal invariant for G acting on $\mathcal{Y} \times \mathcal{Z}$. However, the correspondence ϕ, a one-to-one onto mapping, satisfies

$$\phi(gx) = g\phi(x) \quad \text{for } g \in G, x \in \mathcal{X}.$$

The conclusion is that $f_1(\phi(x))$ is a maximal invariant function on $\mathcal{X}$. A direct proof in the present generality is easy. Since

$$f_1(\phi(gx)) = f_1(g\phi(x)) = f_1(\phi(x)),$$

$f_1(\phi(x))$ is invariant. If $f_1(\phi(x_1)) = f_1(\phi(x_2))$, then there is a $g \in G$ such that $g\phi(x_1) = \phi(x_2)$ since f_1 is maximal invariant on $\mathcal{Y} \times \mathcal{Z}$. But $g\phi(x_1) = \phi(gx_1) = \phi(x_2)$, so $gx_1 = x_2$ as ϕ is one-to-one. Thus $f_1(\phi(x))$ is maximal invariant. In the next example, a maximal invariant function is easily found but the product space representation in the form just discussed is not available.

◆ **Example 7.5.** The group $\mathcal{O}_p$ acts on $\mathcal{S}_p^+$ by

$$\Gamma(S) = \Gamma S \Gamma', \qquad \Gamma \in \mathcal{O}_p.$$

A maximal invariant function is easily found using a reduction argument similar to that given in Example 7.4. From the spectral theorem for matrices, every $S \in \mathbb{S}_p^+$ can be written in the form $S = \Gamma_1 D \Gamma_1'$ where $\Gamma_1 \in \mathcal{O}_p$ and D is a diagonal matrix whose diagonal elements are the ordered eigenvalues of S, say

$$\lambda_1(S) \geqslant \lambda_2(S) \geqslant \cdots \geqslant \lambda_p(S).$$

Thus $\Gamma_1' S \Gamma_1 = D$, which shows that D is in the orbit of S. Let f_0 on $\mathbb{S}_p^+$ to R^p be defined by: $f_0(S)$ is the vector of ordered eigenvalues of S. Obviously, f_0 is $\mathcal{O}_p$-invariant and, to show f_0 is maximal invariant, suppose $f_0(S_1) = f_0(S_2)$. Then S_1 and S_2 have the same eigenvalues and we have

$$S_i = \Gamma_i D \Gamma_i', \qquad i = 1, 2$$

where D is the diagonal matrix of eigenvalues of S_i, $i = 1, 2$. Thus $\Gamma_2 \Gamma_1' S_1 (\Gamma_2 \Gamma_1')' = S_2$, so f_0 is maximal invariant. To describe the technical difficulty when we try to write $\mathbb{S}_p^+$ as a product space, first consider the case $p = 2$. Then $\mathbb{S}_2^+ = \mathcal{X}_1 \cup \mathcal{X}_2$ where

$$\mathcal{X}_1 = \{S | S \in \mathbb{S}_2^+, \lambda_1(S) = \lambda_2(S)\}$$

and

$$\mathcal{X}_2 = \{S | S \in \mathbb{S}_2^+, \lambda_1(S) > \lambda_2(S)\}.$$

That $\mathcal{O}_2$ acts on both $\mathcal{X}_1$ and $\mathcal{X}_2$ is clear. The function ϕ_1 defined on $\mathcal{X}_1$ by $\phi_1(S) = \lambda_1(S) \in (0, \infty)$ is maximal invariant and ϕ_1 establishes a one-to-one correspondence between $\mathcal{X}_1$ and $(0, \infty)$. For $\mathcal{X}_2$, define ϕ_2 by

$$\phi_2(S) = \begin{pmatrix} \lambda_1(S) & 0 \\ 0 & \lambda_2(S) \end{pmatrix}$$

so ϕ_2 is a maximal invariant function and takes values in the set $\mathcal{Y}$ of all 2×2 diagonal matrices with diagonal elements y_1 and y_2, $y_1 > y_2 > 0$. Let $\mathcal{D}_2$ be the subgroup of $\mathcal{O}_2$ consisting of those diagonal matrices with ± 1 for each diagonal element. The argument given in Example 6.21 shows that the mapping constructed there establishes a one-to-one onto correspondence between $\mathcal{X}_2$ and $(\mathcal{O}_2/\mathcal{D}_2) \times \mathcal{Y}$, and $\mathcal{O}_2$ acts on $(\mathcal{O}_2/\mathcal{D}_2) \times \mathcal{Y}$ by

$$\Gamma(z, y) = (\Gamma z, y); \qquad (z, y) \in (\mathcal{O}_2/\mathcal{D}_2) \times \mathcal{Y}.$$

Further, ϕ satisfies

$$\phi(\Gamma(z, y)) = \Gamma(\phi(z, y)).$$

Thus for $p = 2$, $\mathcal{S}_2^+$ has been decomposed into $\mathfrak{X}_1$ and $\mathfrak{X}_2$, which are both invariant under $\mathcal{O}_2$. The action of $\mathcal{O}_2$ on $\mathfrak{X}_1$ is trivial in that $\Gamma x = x$ for all $x \in \mathfrak{X}_1$ and a maximal invariant function on $\mathfrak{X}_1$ is the identity function. Also, $\mathfrak{X}_2$ was decomposed into a product space where $\mathcal{O}_2$ acted transitively on the first component of the product space and trivially on the second component. From this decomposition, a maximal invariant function was obvious. Similar decompositions for $p > 2$ can be given for $\mathcal{S}_p^+$, but the number of component spaces increases. For example, when $p = 3$, let $\lambda_1(S) \geqslant \lambda_2(S) \geqslant \lambda_3(S)$ denote the ordered eigenvalues of $S \in \mathcal{S}_3^+$. The relevant decomposition for $\mathcal{S}_3^+$ is

$$\mathcal{S}_3^+ = \mathfrak{X}_1 \cup \mathfrak{X}_2 \cup \mathfrak{X}_3 \cup \mathfrak{X}_4$$

where

$$\mathfrak{X}_1 = \{S|\lambda_1(S) = \lambda_2(S) = \lambda_3(S)\}$$

$$\mathfrak{X}_2 = \{S|\lambda_1(S) = \lambda_2(S) > \lambda_3(S)\}$$

$$\mathfrak{X}_3 = \{S|\lambda_1(S) > \lambda_2(S) = \lambda_3(S)\}$$

$$\mathfrak{X}_4 = \{S|\lambda_1(S) > \lambda_2(S) > \lambda_3(S)\}.$$

Each of the four components is acted on by $\mathcal{O}_3$ and can be written as a product space with the structure described previously. The details of this are left to the reader. In some situations, it is sufficient to consider the subset $\mathfrak{Z}$ of $\mathcal{S}_p^+$ where

$$\mathfrak{Z} = \{S|\lambda_1(S) > \lambda_2(S) > \cdots > \lambda_p(S)\}.$$

The argument given in Example 6.21 shows how to write $\mathfrak{Z}$ as a product space so that a maximal invariant function is obvious under the action of $\mathcal{O}_p$ on $\mathfrak{Z}$. ◆

Further examples of maximal invariants are given as the need arises. We end this section with a brief discussion of equivariant functions. Recall (see Definition 6.14) that a function ϕ on $\mathfrak{X}$ onto $\mathfrak{Y}$ is called equivariant if $\phi(gx) = \bar{g}\phi(x)$ where G acts on $\mathfrak{X}$, $\bar{G}$ acts on $\mathfrak{Y}$, and $\bar{G}$ is a homomorphic image of G. If $\bar{G} = \{\bar{e}\}$ consists only of the identity, then equivariant functions are invariant under G. In this case, we have a complete description of all the equivariant functions—namely, a function is equivariant iff it is a function of a maximal invariant function on $\mathcal{X}$. In the general case when $\bar{G}$ is not the trivial group, a useful description of all the equivariant functions appears to be rather difficult. However, there is one special case when the equivariant functions can be characterized.

Assume that G acts transitively on $\mathcal{X}$ and $\bar{G}$ acts transitively on $\mathcal{Y}$, where $\bar{G}$ is a homomorphic image of G. Fix $x_0 \in \mathcal{X}$ and let

$$H_0 = \{g | gx_0 = x_0\}.$$

The subgroup H_0 of G is called the *isotropy subgroup* of x_0. Also, fix $y_0 \in \mathcal{Y}$ and let

$$K_0 = \{\bar{g} | \bar{g}y_0 = y_0\}$$

be the isotropy subgroup of y_0.

Proposition 7.8. In order that there exist an equivariant function ϕ on $\mathcal{X}$ to $\mathcal{Y}$ such that $\phi(x_0) = y_0$, it is necessary and sufficient that $\bar{H}_0 \subseteq K_0$. Here $\bar{H}_0 \subseteq \bar{G}$ is the image of H_0 under the given homomorphism.

Proof. First, suppose that ϕ is equivariant and satisfies $\phi(x_0) = y_0$. Then, for $g \in H_0$,

$$\phi(x_0) = \phi(gx_0) = \bar{g}\phi(x_0) = \bar{g}y_0 = y_0$$

so $\bar{g} \in K_0$. Thus $\bar{H}_0 \subseteq K_0$. Conversely, suppose that $\bar{H}_0 \subseteq K_0$. For $x \in \mathcal{X}$, the transitivity of G on $\mathcal{X}$ implies that $x = gx_0$ for some g. Define ϕ on $\mathcal{X}$ to $\mathcal{Y}$ by

$$\phi(x) = \bar{g}y_0 \quad \text{where } x = gx_0.$$

It must be shown that ϕ is well defined and is equivariant. If $x = g_1x_0 = g_2x_0$, then $g_2^{-1}g_1 \in H_0$ so $\overline{g_2^{-1}g_1} \in K_0$. Thus

$$\phi(x) = \bar{g}_1 y_0 = \bar{g}_2 y_0$$

since

$$\bar{g}_2^{-1}\bar{g}_1 y_0 = \overline{g_2^{-1}g_1}\, y_0 = y_0.$$

Therefore ϕ is well defined and is onto $\mathcal{Y}$ since $\bar{G}$ acts transitively on $\mathcal{Y}$. That ϕ is equivariant is easily checked. □

The proof of Proposition 7.8 shows that an equivariant function is determined by its value at one point when G acts transitively on $\mathcal{X}$. More precisely, if ϕ_1 and ϕ_2 are equivariant functions on $\mathcal{X}$ such that $\phi_1(x_0) = \phi_2(x_0)$ for some $x_0 \in \mathcal{X}$, then $\phi_1(x) = \phi_2(x)$ for all x. To see this, write $x = gx_0$ so

$$\phi_1(x) = \phi_1(gx_0) = \bar{g}\phi_1(x_0) = \bar{g}\phi_2(x_0) = \phi_2(gx_0) = \phi_2(x).$$

Thus to characterize all the equivariant functions, it is sufficient to determine the possible values of $\phi(x_0)$ for some fixed $x_0 \in \mathcal{X}$. The following example illustrates these ideas.

◆ **Example 7.6.** Suppose $\mathcal{X} = \mathcal{Y} = \mathcal{S}_p^+$ and $G = \bar{G} = Gl_p$ where the homomorphism is the identity. The action of Gl_p on $\mathcal{S}_p^+$ is

$$A(S) = ASA'; \qquad A \in Gl_p,\, S \in \mathcal{S}_p^+.$$

To characterize the equivariant functions, pick $x_0 = I_p \in \mathcal{S}_p^+$. An equivariant function ϕ must satisfy

$$\phi(I_p) = \phi(\Gamma\Gamma') = \Gamma\phi(I_p)\Gamma'$$

for all $\Gamma \in \mathcal{O}_p$. By Proposition 2.13, a matrix $\phi(I_p)$ satisfies this equation iff $\phi(I_p) = kI_p$ for some real constant k. Since $\phi(I_p) \in \mathcal{S}_p^+$, $k > 0$. Thus

$$\phi(I_p) = kI_p, \qquad k > 0$$

and for $S \in \mathcal{S}_p^+$,

$$\phi(S) = \phi(S^{1/2}S^{1/2}) = S^{1/2}\phi(I_p)S^{1/2} = kS.$$

Therefore, every equivariant function has the form $\phi(S) = kS$ for some $k > 0$. ◆

Further applications of the above ideas occur in the following sections after it is shown that, under certain conditions, maximum likelihood estimators are equivariant functions.

7.3. INVARIANT PROBABILITY MODELS

Invariant probability models provide the mathematical framework in which the connection between statistical problems and invariance can be studied. Suppose $(\mathfrak{X}, \mathfrak{B})$ is a measurable space and G is a group of transformations acting on $\mathfrak{X}$ such that each $g \in G$ is a one-to-one onto measurable function from $\mathfrak{X}$ to $\mathfrak{X}$. If P is a probability measure on $(\mathfrak{X}, \mathfrak{B})$ and $g \in G$, the probability measure gP on $(\mathfrak{X}, \mathfrak{B})$ is defined by

$$(gP)(B) \equiv P(g^{-1}B); \qquad g \in G, \quad B \in \mathfrak{B}.$$

It is easily verified that $(g_1 g_2)P = g_1(g_2 P)$ so the group G acts on the space of all probability measures defined on $(\mathfrak{X}, \mathfrak{B})$.

Definition 7.3. Let $\mathfrak{P}$ be a set of probability measures defined on $(\mathfrak{X}, \mathfrak{B})$. The set $\mathfrak{P}$ is *invariant* under G if for each $P \in \mathfrak{P}$, $gP \in \mathfrak{P}$ for all $g \in G$. Sets of probability measures $\mathfrak{P}$ are called *probability models*, and when $\mathfrak{P}$ is invariant under G, we speak of a *G-invariant probability model.*

If $X \in \mathfrak{X}$ is a random vector with $\mathcal{L}(X) = P$, then $\mathcal{L}(gX) = gP$ for $g \in G$ since

$$\Pr\{gX \in B\} = \Pr\{X \in g^{-1}B\} = P\{g^{-1}B\} = (gP)(B).$$

Thus $\mathfrak{P}$ is invariant under G iff whenever $\mathcal{L}(X) \in \mathfrak{P}$, $\mathcal{L}(gX) \in \mathfrak{P}$ for all $g \in G$.

There are a variety of ways to construct invariant probability models from other invariant probability models. For example, if $\mathfrak{P}_\alpha$, $\alpha \in A$, are G-invariant probability models, it is clear that

$$\bigcup_{\alpha \in A} \mathfrak{P}_\alpha \quad \text{and} \quad \bigcap_{\alpha \in A} \mathfrak{P}_\alpha$$

are both G-invariant. Now, given $(\mathfrak{X}, \mathfrak{B})$ and a G-invariant probability model $\mathfrak{P}$, form the product space

$$\mathfrak{X}^{(n)} = \mathfrak{X} \times \mathfrak{X} \times \cdots \times \mathfrak{X}$$

and the product σ-algebra $\mathfrak{B}^{(n)}$ on $\mathfrak{X}^{(n)}$. For $P \in \mathfrak{P}$, define $P^{(n)}$ on $\mathfrak{B}^{(n)}$ by first defining

$$P^{(n)}(B_1 \times B_2 \times \cdots \times B_n) = \prod_{i=1}^{n} P(B_i)$$

where $B_i \in \mathfrak{B}$. Once $P^{(n)}$ is defined on sets of the form $B_1 \times \cdots \times B_n$, its extension to $\mathfrak{B}^{(n)}$ is unique. Also, define G acting on $\mathfrak{X}^{(n)}$ by

$$g(x_1,\ldots,x_n) \equiv (gx_1,\ldots,gx_n)$$

for $x = (x_1,\ldots,x_n) \in \mathfrak{X}^{(n)}$.

Proposition 7.9. Let $\mathfrak{P}^{(n)} = \{P^{(n)} | P \in \mathfrak{P}\}$. Then $\mathfrak{P}^{(n)}$ is a G-invariant probability model on $(\mathfrak{X}^{(n)}, \mathfrak{B}^{(n)})$ when $\mathfrak{P}$ is G-invariant.

Proof. It must be shown that $gP^{(n)} \in \mathfrak{P}^{(n)}$ for $g \in G$ and $P^{(n)} \in \mathfrak{P}^{(n)}$. However, $P^{(n)}$ is the product measure

$$P^{(n)} = P \times P \times \cdots \times P; \qquad P \in \mathfrak{P}$$

and $P^{(n)}$ is determined by its values on sets of the form $B_1 \times \cdots \times B_n$. But

$$(gP)^{(n)}(B_1 \times \cdots \times B_n) = P^{(n)}\left(g^{-1}B_1 \times g^{-1}B_2 \times \cdots \times g^{-1}B_n\right)$$
$$= \prod_1^n P\left(g^{-1}B_i\right) = \prod_1^n (gP)(B_i)$$

where the first equality follows from the definition of the action of G on $\mathfrak{X}^{(n)}$. Then $gP^{(n)}$ is the product measure

$$gP^{(n)} = (gP) \times (gP) \times \cdots \times (gP),$$

which is in $\mathfrak{P}^{(n)}$ as $gP \in \mathfrak{P}$. □

For an application of Proposition 7.9, suppose X is a random vector with $\mathcal{L}(X) \in \mathfrak{P}$ where $\mathfrak{P}$ is a G-invariant probability model on $\mathfrak{X}$. If $X_1,\ldots,X_n$ are independent and identically distributed with $\mathcal{L}(X_i) \in \mathfrak{P}$, then the random vector

$$Y = (X_1,\ldots,X_n) \in \mathfrak{X}^{(n)}$$

has distribution $P^{(n)} \in \mathfrak{P}^{(n)}$ when $\mathcal{L}(X_i) = P$, $i = 1,\ldots,n$. Thus $\mathfrak{P}^{(n)}$ is a G-invariant probability model for Y.

In most applications, probability models $\mathfrak{P}$ are described in the form $\mathfrak{P} = \{P_\theta | \theta \in \theta\}$ where θ is a *parameter* and Θ is the *parameter space*. When discussing indexed families of probability measures, the term "parameter space" is used only in the case that the indexing is one-to-one—that is,

$P_{\theta_1} = P_{\theta_2}$ implies that $\theta_1 = \theta_2$. Now, suppose $\mathcal{P} = \{P_\theta | \theta \in \Theta\}$ is G-invariant. Then for each $g \in G$ and $\theta \in \Theta$, $gP_\theta \in \mathcal{P}$, so $gP_\theta = P_{\theta'}$ for some unique $\theta' \in \Theta$. Define a function $\bar{g}$ on Θ to Θ by

$$gP_\theta = P_{\bar{g}\theta}, \quad \theta \in \Theta.$$

In other words, $\bar{g}\theta$ is the unique point in Θ that satisfies the above equation.

Proposition 7.10. Each $\bar{g}$ is a one-to-one onto function from Θ to Θ. Let $\bar{G} = \{\bar{g} | g \in G\}$. Then $\bar{G}$ is a group under the group operation of function composition and the mapping $g \to \bar{g}$ is a group homomorphism from G to $\bar{G}$, that is:

(i) $\overline{g_1 g_2} = \bar{g}_1 \bar{g}_2$.
(ii) $\overline{g^{-1}} = \bar{g}^{-1}$.

Proof. To show that $\bar{g}$ is one-to-one, suppose $\bar{g}\theta_1 = \bar{g}\theta_2$. Then

$$gP_{\theta_1} = P_{\bar{g}\theta_1} = P_{\bar{g}\theta_2} = gP_{\theta_2},$$

which implies that $P_{\theta_1} = P_{\theta_2}$ so $\theta_1 = \theta_2$. The verification that $\bar{g}$ is onto goes as follows. If $\theta \in \Theta$, let $\theta' = \overline{g^{-1}}\theta$. Then

$$P_{\bar{g}\theta'} = gP_{\theta'} = g(g^{-1}P_\theta) = (gg^{-1})P_\theta = P_\theta,$$

so $\bar{g}\theta' = \theta$. Equations (i) and (ii) follow by calculations similar to those above. This shows that $\bar{G}$ is the homomorphic image of G and $\bar{G}$ is a group. □

An important special case of a G-invariant parametric model is the following. Suppose G acts on $(\mathcal{X}, \mathcal{B})$ and assume that ν is a σ-finite measure on $(\mathcal{X}, \mathcal{B})$ that is relatively invariant with multiplier χ, that is,

$$\int f(g^{-1}x)\nu(dx) = \chi(g)\int f(x)\nu(dx), \qquad g \in G$$

for all integrable functions f. Assume that $\mathcal{P} = \{P_\theta | \theta \in \Theta\}$ is a parametric model and

$$P_\theta(B) = \int I_B(x)p(x|\theta)\nu(dx)$$

for all measurable sets B. Thus $p(\cdot|\theta)$ is a density for P_θ with respect to ν. If

$\mathcal{P}$ is G-invariant, then

$$gP_\theta = P_{\bar{g}\theta} \qquad \text{for } g \in G, \theta \in \Theta.$$

Therefore,

$$\begin{aligned} gP_\theta(B) = P_\theta(g^{-1}B) &= \int I_B(gx)p(x|\theta)\nu(dx) \\ &= \int I_B(gx)p(g^{-1}gx|\theta)\nu(dx) \\ &= \chi(g^{-1})\int I_B(x)p(g^{-1}x|\theta)\nu(dx) \\ &= P_{\bar{g}\theta}(B) = \int I_B(x)p(x|\bar{g}\theta)\nu(dx) \end{aligned}$$

for all measurable sets B. Thus the density p must satisfy

$$\chi(g^{-1})p(g^{-1}x|\theta) = p(x|\bar{g}\theta) \qquad \text{a.e. } (\nu)$$

or, equivalently,

$$p(x|\theta) = p(gx|\bar{g}\theta)\chi(g) \qquad \text{a.e. } (\nu).$$

It should be noted that the null set where the above equality does not hold may depend on both θ and g. However, in most applications, a version of the density is available so the above equality is valid everywhere. This leads to the following definition.

Definition 7.4. The family of densities $\{p(\cdot|\theta)|\theta \in \Theta\}$ with respect to the relatively invariant measure ν with multiplier χ is $(G - \bar{G})$-invariant if

$$p(x|\theta) = p(gx|\bar{g}\theta)\chi(g)$$

for all x, θ, and g.

It is clear that if a family of densities is $(G - \bar{G})$-invariant where $\bar{G}$ is a homomorphic image of G that acts on Θ, then the family of probability measures defined by these densities is a G-invariant probability model. A few examples illustrate these notions.

◆ **Example 7.7.** Let $\mathfrak{X} = R^n$ and suppose $f(\|x\|^2)$ is a density with respect to Lebesgue measure on R^n. For $\mu \in R^n$ and $\Sigma \in \mathbb{S}_p^+$, set

$$p(x|\mu, \Sigma) = |\Sigma|^{-1/2} f((x-\mu)'\Sigma^{-1}(x-\mu)).$$

For each μ and Σ, $p(\cdot|\mu, \Sigma)$ is a density on R^n. The affine group Al_n acts on R^n by $(A, b)x = Ax + b$ and Lebesgue measure is relatively invariant with multiplier

$$\chi(A, b) = |\det(A)|$$

where $(A, b) \in Al_n$. Consider the parameter space $R^n \times \mathbb{S}_p^+$ and the family of densities

$$\{p(\cdot|\mu, \Sigma) | (\mu, \Sigma) \in R^n \times \mathbb{S}_p^+\}.$$

The group Al_n acts on the parameter space $R^n \times \mathbb{S}_p^+$ by

$$(A, b)(\mu, \Sigma) = (A\mu + b, A\Sigma A').$$

It is now verified that the family of densities above is $(G - \overline{G})$-invariant where $G = \overline{G} = Al_n$. For $(A, b) \in Al_n$,

$$\begin{aligned}
p((A, b)x | (A, b)(\mu, \Sigma)) &= p(Ax + b | (A\mu + b, A\Sigma A')) \\
&= |A\Sigma A'|^{-1/2} f\big((Ax + b - A\mu - b)' \\
&\qquad \times (A\Sigma A')^{-1}(Ax + b - A\mu - b)\big) \\
&= |\det A|^{-1} |\Sigma|^{-1/2} f((x-\mu)'\Sigma^{-1}(x-\mu)) \\
&= \frac{1}{\chi(A, b)} p(x | (\mu, \Sigma)).
\end{aligned}$$

Therefore, the parametric model determined by the family of densities is Al_n-invariant. ◆

A useful method for generating a G-invariant probability model on a measurable space $(\mathfrak{X}, \mathfrak{B})$ is to consider a fixed probability measure P_0 on $(\mathfrak{X}, \mathfrak{B})$ and set

$$\mathfrak{P} = \{gP_0 | g \in G\}.$$

Obviously, $\mathfrak{P}$ is G-invariant. However, in many situations, the group G does not serve as a parameter space for $\mathcal{P}$ since $g_1P_0 = g_2P_0$ does not necessarily imply that $g_1 = g_2$. For example, consider $\mathcal{X} = R^n$ and let P_0 be given by

$$P_0(B) = \int_{R^n} I_B(x) f(\|x\|^2)\, dx$$

where $f(\|x\|^2)$ is the density on R^n of Example 7.7. Also, let $G = Al_n$. To obtain the density of gP_0, suppose X is a random vector with $\mathcal{L}(X) = P_0$. For $g = (A, b) \in Al_n$, $(A, b)X = AX + b$ has a density given by

$$p(x|b, AA') = |\det(AA')|^{-1/2} f\big((x-b)'(AA')^{-1}(x-b)\big)$$

and this is the density of $(A, b)P_0$. Thus the parameter space for

$$\mathcal{P} = \{(A, b)P_0 | (A, b) \in Al_n\}$$

is $R^n \times \mathcal{S}_n^+$. Of course, the reason that Al_n is not a parameter space for $\mathcal{P}$ is that

$$(\Gamma, 0)P_0 = P_0$$

for all $n \times n$ orthogonal matrices Γ. In other words, P_0 is an orthogonally invariant probability on R^n.

Some of the linear models introduced in Chapter 4 provide interesting examples of parametric models that are generated by groups of transformations.

◆ **Example 7.8.** Consider an inner product space $(V, [\cdot, \cdot])$ and let P_0 be a probability measure on V so that if $\mathcal{L}(X) = P_0$, then $\mathcal{E}X = 0$ and $\text{Cov}(X) = I$. Given a subspace M of V, form the group G whose elements consist of pairs (a, x) with $a > 0$ and $x \in M$. The group operation is

$$(a_1, x_1)(a_2, x_2) \equiv (a_1a_2, a_1x_2 + x_1).$$

The probability model $\mathcal{P} = \{gP_0 | g \in G\}$ consists of all the distributions of $(a, x)X = aX + x$ where $\mathcal{L}(X) = P_0$. Clearly,

$$\mathcal{E}(aX + x) = x \quad \text{and} \quad \text{Cov}(aX + x) = a^2 I.$$

Therefore, if $\mathcal{L}(Y) \in \mathcal{P}$, then $\mathcal{E}Y \in M$ and $\text{Cov}(Y) = \sigma^2 I$ for some $\sigma^2 > 0$, so $\mathcal{P}$ is a linear model for Y. For this particular example the group G is a parameter space for $\mathcal{P}$. This linear model is generated by G in the sense that $\mathcal{P}$ is obtained by transforming a fixed probability measure P_0 by elements of G. ◆

An argument similar to that in Example 7.8 shows that the multivariate linear model introduced in Example 4.4 is also generated by a group of transformations.

◆ **Example 7.9.** Let $\mathcal{L}_{p,n}$ be the linear space of real $n \times p$ matrices with the usual inner product $\langle \cdot , \cdot \rangle$ on $\mathcal{L}_{p,n}$. Assume that P_0 is a probability measure on $\mathcal{L}_{p,n}$ so that, if $\mathcal{L}(X) = P_0$, then $\mathcal{E}X = 0$ and $\text{Cov}(X) = I_n \otimes I_p$. To define a regression subspace M, let Z be a fixed $n \times k$ real matrix and set

$$M = \{y | y = ZB, B \in \mathcal{L}_{p,k}\}.$$

Obviously, M is a subspace of $\mathcal{L}_{p,n}$. Consider the group G whose elements are pairs (A, y) with $A \in Gl_p$ and $y \in M$. Then G acts on $\mathcal{L}_{p,n}$ by

$$(A, y)x = xA' + y = (I_n \otimes A)x + y,$$

and the group operation is

$$(A_1, y_1)(A_2, y_2) = (A_1 A_2, y_2 A_1' + y_1).$$

The probability model $\mathcal{P} = \{gP_0 | g \in G\}$ consists of the distributions of $(A, y)X = (I_n \otimes A)X + y$ where $\mathcal{L}(X) = P_0$. Since

$$\mathcal{E}((I_n \otimes A)X + y) = y \in M$$

and

$$\text{Cov}((I_n \otimes A)X + y) = I_n \otimes AA',$$

if $\mathcal{L}(Y) \in \mathcal{P}$, then $\mathcal{E}Y \in M$ and $\text{Cov}(Y) = I_n \otimes \Sigma$ for some $p \times p$ positive definite matrix Σ. Thus $\mathcal{P}$ is a multivariate linear model as described in Example 4.4. If $p > 1$, the group G is not a parameter space for $\mathcal{P}$, but G does generate $\mathcal{P}$. ◆

Most of the probability models discussed in later chapters are examples of probability models generated by groups of transformations. Thus these models are G-invariant and this invariance can be used in a variety of ways.

First, invariance can be used to give easy derivations of maximum likelihood estimators and to suggest test statistics in some situations. In addition, distributional and independence properties of certain statistics are often best explained in terms of invariance.

7.4. THE INVARIANCE OF LIKELIHOOD METHODS

In this section, it is shown that under certain conditions maximum likelihood estimators are equivariant functions and likelihood ratio tests are invariant functions. Throughout this section, G is a group of transformations that act measurably on $(\mathfrak{X}, \mathfrak{B})$ and ν is a σ-finite relatively invariant measure on $(\mathfrak{X}, \mathfrak{B})$ with multiplier χ. Suppose that $\mathcal{P} = \{P_\theta | \theta \in \Theta\}$ is a G-invariant parametric model such that each P_θ has a density $p(\cdot|\theta)$, which satisfies

$$p(x|\theta) = p(gx|\bar{g}\theta)\chi(g)$$

for all $x \in \mathfrak{X}$, $\theta \in \Theta$, and $g \in G$. The group $\bar{G} = \{\bar{g} | g \in G\}$ is the homomorphic image of G described in Proposition 7.10. In the present context, a point estimator of θ, say t, mapping $\mathfrak{X}$ into Θ, is equivariant (see Definition 6.14) if

$$t(gx) = \bar{g}t(x), \qquad g \in G, \quad x \in \mathfrak{X}.$$

Proposition 7.11. Given the $(G - \bar{G})$-invariant family of densities $\{p(\cdot|\theta) | \theta \in \Theta\}$, assume there exists a unique function $\hat{\theta}$ mapping $\mathfrak{X}$ into Θ that satisfies

$$\sup_{\theta \in \Theta} p(x|\theta) = p(x|\hat{\theta}(x)).$$

Then $\hat{\theta}$ is an equivariant function—that is,

$$\hat{\theta}(gx) = \bar{g}\hat{\theta}(x), \qquad x \in \mathfrak{X}, \quad g \in G.$$

Proof. By assumption, $\hat{\theta}(gx)$ is the unique point in Θ that satisfies

$$\sup_{\theta \in \Theta} p(gx|\theta) = p(gx|\hat{\theta}(gx)).$$

But

$$p(gx|\theta) = \chi(g^{-1})p(x|\bar{g}^{-1}\theta)$$

so

$$\sup_{\theta} p(gx|\theta) = \chi(g^{-1}) \sup_{\theta} p(x|\bar{g}^{-1}\theta) = \chi(g^{-1}) \sup_{\theta} p(x|\theta)$$

$$= \chi(g^{-1}) p(x|\hat{\theta}(x)) = p(gx|\bar{g}\hat{\theta}(x)).$$

Thus

$$p(gx|\hat{\theta}(gx)) = p(gx|\bar{g}\hat{\theta}(x))$$

and, by the uniqueness assumption,

$$\hat{\theta}(gx) = \bar{g}\hat{\theta}(x). \qquad \square$$

Of course, the estimator $\hat{\theta}(x)$ whose existence and uniqueness is assumed in Proposition 7.11 is the maximum likelihood estimator of θ. That $\hat{\theta}$ is an equivariant function is useful information about the maximum likelihood estimator, but the above result does not indicate how to use invariance to find the maximum likelihood estimator. The next result rectifies this situation.

Proposition 7.12. Let $\{p(\cdot|\theta)|\theta \in \Theta\}$ be a $(G - \bar{G})$-invariant family of densities on $(\mathcal{X}, \mathcal{B})$. Fix a point $x_0 \in \mathcal{X}$ and let $\mathcal{O}_{x_0}$ be the orbit of x_0. Assume that

$$\sup_{\theta \in \Theta} p(x_0|\theta) = p(x_0|\theta_0)$$

and that θ_0 is unique. For $x \in \mathcal{O}_{x_0}$, define $\hat{\theta}(x)$ by

$$\hat{\theta}(x) = \bar{g}_x\theta_0 \qquad \text{where } x = g_x x_0.$$

Then $\hat{\theta}$ is well defined on $\mathcal{O}_{x_0}$ and satisfies

(i) $\hat{\theta}(gx) = \bar{g}\hat{\theta}(x)$, $x \in \mathcal{O}_{x_0}$.
(ii) $\sup_{\theta \in \Theta} p(x|\theta) = p(x|\hat{\theta}(x))$, $x \in \mathcal{O}_{x_0}$.

Furthermore, $\hat{\theta}$ is unique.

Proof. The density $p(\cdot|\theta)$ satisfies

$$p(y|\theta) = p(gy|\bar{g}\theta)\chi(g)$$

where χ is a multiplier on G. To show $\hat{\theta}$ is well defined on $\mathcal{O}_{x_0}$, it must be verified that if $x = g_x x_0 = h_x x_0$, then $\bar{g}_x\theta_0 = \bar{h}_x\theta_0$. Set $k = h_x^{-1}g_x$ so $kx_0 = x_0$ and we need to show that $\bar{k}\theta_0 = \theta_0$. But

$$\begin{aligned} p(x_0|\theta_0) &= \sup_{\theta\in\Theta} p(kx_0|\theta) = \chi(k^{-1})\sup_{\theta} p(x_0|\bar{k}^{-1}\theta) \\ &= \chi(k^{-1})\sup_{\theta} p(x_0|\theta) = \chi(k^{-1})p(x_0|\theta_0) \\ &= p(kx_0|\bar{k}\theta_0) = p(x_0|\bar{k}\theta_0). \end{aligned}$$

By the uniqueness assumption, $\bar{k}\theta_0 = \theta_0$ so $\hat{\theta}$ is well defined on $\mathcal{O}_{x_0}$. To establish (i), if $x = g_x x_0$, then $gx = (gg_x)x_0$ so

$$\hat{\theta}(gx) = \overline{gg_x}\theta_0 = \bar{g}(\bar{g}_x\theta_0) = \bar{g}\hat{\theta}(x).$$

For (ii), $x = g_x x_0$ so

$$\begin{aligned} \sup_{\theta} p(x|\theta) &= \sup_{\theta} p(g_x x_0|\theta) = \chi(g_x^{-1})\sup_{\theta} p(x_0|\bar{g}_x^{-1}\theta_0) \\ &= \chi(g_x^{-1})\sup_{\theta} p(x_0|\theta) = \chi(g_x^{-1})p(x_0|\theta_0) \\ &= p(g_x x_0|\bar{g}_x\theta_0) = p(x|\hat{\theta}(x)). \end{aligned}$$

To establish the uniqueness of $\hat{\theta}$, fix $x \in \mathcal{O}_{x_0}$ and consider $\theta_1 \neq \bar{g}_x\theta_0$. Then

$$\begin{aligned} p(x|\theta_1) &= p(g_x x_0|\bar{g}_x\bar{g}_x^{-1}\theta_1) = \chi(g_x^{-1})p(x_0|\bar{g}_x^{-1}\theta_1) \\ &< \chi(g_x^{-1})p(x_0|\theta_0) = p(x|\hat{\theta}(x)). \end{aligned}$$

The strict inequality follows from the uniqueness assumption concerning θ_0. □

In applications, Proposition 7.12 is used as follows. From each orbit in the sample space $\mathfrak{X}$, we pick a convenient point x_0 and show that $p(x_0|\theta)$ is uniquely maximized at θ_0. Then for other points x in this orbit, write $x = g_x x_0$ and set $\hat{\theta}(x) = \bar{g}_x\theta_0$. The function $\hat{\theta}$ is then the maximum likelihood estimator of θ and is equivariant. In some situations, there is only one orbit in $\mathfrak{X}$ so this method is relatively easy to apply.

◆ **Example 7.10.** Consider $\mathfrak{X} = \Theta = \mathbb{S}_p^+$ and let

$$p(S|\Sigma) = \omega(n, p)|\Sigma^{-1}S|^{n/2}\exp\left[-\tfrac{1}{2}\operatorname{tr}\Sigma^{-1}S\right]$$

for $S \in \mathbb{S}_p^+$ and $\Sigma \in \mathbb{S}_p^+$. The constant $\omega(n, p)$, $n \geqslant p$, was defined in Example 5.1. That $p(\cdot|\Sigma)$ is a density with respect to the measure

$$\nu(dS) = \frac{dS}{|S|^{(p+1)/2}}$$

follows from Example 5.1. The group Gl_p acts on $\mathbb{S}_p^+$ by

$$A(S) \equiv ASA'$$

for $A \in Gl_p$ and $S \in \mathbb{S}_p^+$ and the measure ν is invariant. Also, it is clear that the density $p(\cdot|\Sigma)$ satisfies

$$p(ASA'|A\Sigma A') = p(S|\Sigma).$$

To find the maximum likelihood estimator of $\Sigma \in \mathbb{S}_p^+$, we apply the technique described above. Consider the point $I_p \in \mathbb{S}_p^+$ and note that the orbit of I_p under the action of Gl_p is $\mathbb{S}_p^+$ so in this case there is only one orbit. Thus to apply Proposition 7.12, it must be verified that

$$\sup_{\Sigma \in \mathbb{S}_p^+} p(I_p|\Sigma) = p(I_p|\Sigma_0)$$

where Σ_0 is unique. Taking the logarithm of $p(I_p|\Sigma)$ and ignoring the constant term, we have

$$\sup_{\Sigma \in \mathbb{S}_p^+}\left[\frac{n}{2}\log|\Sigma^{-1}| - \frac{1}{2}\operatorname{tr}\Sigma^{-1}\right] = \sup_{B \in \mathbb{S}_p^+}\left[\frac{n}{2}\log|B| - \frac{1}{2}\operatorname{tr}B\right]$$

$$= \sup_{\lambda_i > 0}\left[\frac{n}{2}\sum_1^p \log\lambda_i - \frac{1}{2}\sum_1^p \lambda_i\right]$$

where $\lambda_1, \ldots, \lambda_p$ are the eigenvalues of $B = \Sigma^{-1} \in \mathbb{S}_p^+$. However, for $\lambda > 0$, $n\log\lambda - \lambda$ is a strictly concave function of λ and is uniquely maximized at $\lambda = n$. Thus the function

$$\frac{n}{2}\sum_1^n \log \lambda_i - \frac{1}{2}\sum_1^n \lambda_i$$

is uniquely maximized at $\lambda_1 = \cdots = \lambda_n = n$, which means that

$$\frac{n}{2}\log|B| - \frac{1}{2}\operatorname{tr} B$$

is uniquely maximized at $B = nI$. Therefore,

$$\sup_{\Sigma \in \mathbb{S}_p} p(I_p|\Sigma) = p\left(I_p|\frac{1}{n}I_p\right)$$

and $(1/n)I_p$ is the unique point in $\mathbb{S}_p^+$ that achieves this supremum. To find the maximum likelihood estimator of Σ, say $\hat{\Sigma}(S)$, write $S = AA'$ for $A \in Gl_p$. Then

$$\hat{\Sigma}(S) = A\left(\frac{1}{n}I_p\right) = \frac{1}{n}AI_pA' = \frac{1}{n}S.$$

In summary,

$$\hat{\Sigma} = \frac{1}{n}S$$

is the unique maximum likelihood estimator of Σ and

$$\sup_{\Sigma \in \mathbb{S}_p^+} p(S|\Sigma) = p\left(S|\frac{1}{n}S\right)$$

$$= \omega(n, p)\left|\left(\frac{1}{n}S\right)^{-1}S\right|^{n/2}\exp\left[-\frac{1}{2}\operatorname{tr}\left(\frac{1}{n}S\right)^{-1}S\right]$$

$$= \omega(n, p)n^{np/2}\exp\left[-\frac{np}{2}\right].$$

The results of this example are used later to derive the maximum likelihood estimator of a covariance matrix in a variety of multivariate normal models. ◆

We now turn to the invariance of likelihood ratio tests. First, invariant testing problems need to be defined. Let $\mathcal{P} = \{P_\theta | \theta \in \Theta\}$ be a parametric probability model on $(\mathfrak{X}, \mathfrak{B})$ and suppose that G acts measurably on $\mathfrak{X}$. Let Θ_0 and Θ_1 be two disjoint subsets of Θ. On the basis of an observation vector $X \in \mathfrak{X}$ with $\mathcal{L}(X) \in \mathcal{P}_0 \cup \mathcal{P}_1$ where

$$\mathcal{P}_i = \{P_\theta | \theta \in \Theta_i\}, \qquad i = 0, 1,$$

suppose it is desired to test the hypothesis

$$H_0 : \mathcal{L}(X) \in \mathcal{P}_0$$

against the alternative

$$H_1 : \mathcal{L}(X) \in \mathcal{P}_1.$$

Definition 7.5. The above hypothesis testing problem is invariant under G if $\mathcal{P}_0$ and $\mathcal{P}_1$ are both G-invariant probability models.

Now suppose that $\mathcal{P}_0 = \{P_\theta | \theta \in \Theta_0\}$ and $\mathcal{P}_1 = \{P_\theta | \theta \in \Theta_1\}$ are disjoint families of probability measures on $(\mathfrak{X}, \mathfrak{B})$ such that each P has a density $p(\cdot|\theta)$ with respect to a σ-finite measure ν. Consider

$$\Lambda(x) = \frac{\sup_{\theta \in \Theta_0} p(x|\theta)}{\sup_{\theta \in \Theta_0 \cup \Theta_1} p(x|\theta)}.$$

For testing the null hypothesis that $\mathcal{L}(X) \in \mathcal{P}_0$ versus the alternative that $\mathcal{L}(X) \in \mathcal{P}_1$, the test that rejects the null hypothesis iff $\Lambda(x) < k$, where k is chosen to control the level of the test, is commonly called the *likelihood ratio test*.

Proposition 7.13. Given the family of densities $\{p(\cdot|\theta) | \theta \in \Theta_0 \cup \Theta_1\}$, assume the testing problem for $\mathcal{L}(X) \in \mathcal{P}_0$ versus $\mathcal{L}(X) \in \mathcal{P}_1$ is invariant under a group G and suppose that

$$p(x|\theta) = p(gx|\bar{g}\theta)\chi(g)$$

for some multiplier χ. Then the likelihood ratio

$$\Lambda(x) \equiv \frac{\sup_{\theta \in \Theta_0} p(x|\theta)}{\sup_{\theta \in \Theta_0 \cup \Theta_1} p(x|\theta)}$$

is an invariant function.

Proof. It must be shown that $\Lambda(x) = \Lambda(gx)$ for $x \in \mathfrak{X}$ and $g \in G$. For $g \in G$,

$$\Lambda(gx) = \frac{\sup_{\theta \in \Theta_0} p(gx|\theta)}{\sup_{\theta \in \Theta_0 \cup \Theta_1} p(gx|\theta)} = \frac{\sup_{\theta \in \Theta_0} \chi(g^{-1}) p(x|\bar{g}^{-1}\theta)}{\sup_{\theta \in \Theta_0 \cup \Theta_1} \chi(g^{-1}) p(x|\bar{g}^{-1}\theta)}$$

$$= \frac{\sup_{\theta \in \Theta_0} p(x|\theta)}{\sup_{\theta \in \Theta_0 \cup \Theta_1} p(x|\theta)} = \Lambda(x).$$

The next to the last equality follows from the positivity of χ and the invariance of Θ_0 and $\Theta_0 \cup \Theta_1$. □

For invariant testing problems, Proposition 7.13 shows that that test function determined by Λ, namely

$$\phi_0(x) = \begin{cases} 1 & \text{if } \Lambda(x) < k \\ 0 & \text{if } \Lambda(x) \geqslant k, \end{cases}$$

is an invariant function. More generally, any test function ϕ is invariant if $\phi(x) = \phi(gx)$ for all $x \in \mathfrak{X}$ and $g \in G$. The whole point of the above discussion is to show that, when attention is restricted to invariant tests for invariant testing problems, the likelihood ratio test is never excluded from consideration. Furthermore, if a particular invariant test has been shown to have an optimal property among invariant tests, then this test has been compared to the likelihood ratio test. Illustrations of these comments are given later in this section when we consider testing problems for the multivariate normal distribution.

Comments similar to those above apply to equivariant estimators. Suppose $\{p(\cdot|\theta)|\theta \in \Theta\}$ is a $(G - \bar{G})$-invariant family of densities and satisfies

$$p(x|\theta) = p(gx|\bar{g}\theta)\chi(g)$$

for some multiplier χ. If the conditions of Proposition 7.12 hold, then an equivariant maximum likelihood estimator exists. Thus if an equivariant estimator t with some optimal property (relative to the class of all equivariant estimators) has been found, then this property holds when t is compared to the maximum likelihood estimator. The Pitman estimator, derived in the next example, is an illustration of this situation.

◆ **Example 7.11.** Let f be a density on R^p with respect to Lebesgue measure and consider the translation family of densities $\{p(\cdot|\theta)|\theta \in R^p\}$ defined by

$$p(x|\theta) = f(x - \theta), \qquad x, \theta \in R^p.$$

For this example, $\mathfrak{X} = \Theta = G = R^p$ and the group action is

$$g(x) = x + g, \qquad x, g \in R^p.$$

It is clear that

$$p(gx|g\theta) = p(x|\theta),$$

so the family of densities is invariant and the multiplier is unity. It is assumed that

$$\int_{R^p} xf(x)\,dx = 0 \quad \text{and} \quad \int \|x\|^2 f(x)\,dx < +\infty.$$

Initially, assume we have one observation X with $\mathcal{L}(X) \in \{p(\cdot|\theta)|\theta \in R^p\}$. The problem is to estimate the parameter θ. As a measure of how well an estimator t performs, consider

$$R(t,\theta) \equiv \mathcal{E}_\theta \|t(X) - \theta\|^2.$$

If $t(X)$ is close to θ on the average, then $R(t,\theta)$ should be small. We now want to show that, if t is an equivariant estimator of θ, then

$$R(t,\theta) = R(t,0)$$

and the equivariant estimator $t_0(X) = X$ minimizes $R(t,0)$ over all equivariant estimators. If t is an equivariant estimator, then

$$t(x+g) = t(x) + g$$

so, with $g = -x$,

$$t(x) = x + t(0).$$

Therefore, every equivariant estimator has the form $t(x) = x + c$ where $c \in R^p$ is a constant. Conversely, any such estimator $t(x) =$

$x + c$ is equivariant. For $t(x) = x + c$,

$$R(t,\theta) = \mathcal{E}_\theta\|t(X) - \theta\|^2 = \int \|x + c - \theta\|^2 f(x-\theta)\,dx$$

$$= \int \|x + c\|^2 f(x)\,dx = R(t,0).$$

To minimize $R(t,0)$ over all equivariant t, the integral

$$\int \|x + c\|^2 f(x)\,dx$$

must be minimized by an appropriate choice of c. But

$$\mathcal{E}\|X + c\|^2 = \mathcal{E}\|X - \mathcal{E}(X)\|^2 + \|\mathcal{E}(X) + c\|^2$$

so

$$c = -\mathcal{E}(X) = \int x f(x)\,dx = 0$$

minimizes the above integral. Hence $t_0(X) = X$ minimizes $R(t,0)$ over all equivariant estimators. Now, we want to generalize this result to the case when $X_1,\ldots, X_n$ are independent and identically distributed with $\mathcal{L}(X_i) \in \{p(\cdot|\theta)|\theta \in R^p\}$, $i = 1,\ldots, p$. The argument is essentially the same as when $n = 1$. An estimator t is equivariant if

$$t(x_1 + g,\ldots, x_n + g) = t(x_1,\ldots, x_n) + g$$

so, setting $g = -x_1$,

$$t(x_1,\ldots, x_n) = x_1 + t(0, x_2 - x_1,\ldots, x_n - x_1).$$

Conversely, if

$$t(x_1,\ldots, x_n) = x_1 + \Psi(x_2 - x_1,\ldots, x_n - x_1)$$

then t is equivariant. Here, Ψ is some measurable function taking values in R^p. Thus a complete description of the equivariant estimators has been given. For such an estimator,

$$R(t,\theta) = \mathcal{E}_\theta \|t(X_1,\ldots,X_n) - \theta\|^2 = \mathcal{E}_\theta \|t(X_1 - \theta,\ldots,X_n - \theta)\|^2$$

$$= \mathcal{E}_0 \|t(X_1,\ldots,X_n)\|^2 = R(t,0).$$

To minimize $R(t,0)$, we need to choose the function Ψ to minimize

$$R(t,0) = \mathcal{E}_0 \|X_1 + \Psi(X_2 - X_1,\ldots,X_n - X_1)\|^2.$$

Let $U_i = X_i - X_1$, $i = 2,\ldots,n$. Then

$$R(t,0) = \mathcal{E}_0 \|X_1 + \Psi(U_2,\ldots,U_n)\|^2$$

$$= \mathcal{E}\{\mathcal{E}\{\|X_1 + \Psi(U_2,\ldots,U_n)\|^2 | U_2,\ldots,U_n\}\}.$$

However, conditional on $(U_2,\ldots,U_n) \equiv U$,

$$\mathcal{E}(\|X_1 + \Psi(U)\|^2 | U) = \mathcal{E}(\|X_1 - \mathcal{E}(X_1|U) + \mathcal{E}(X_1|U) + \Psi(U)\|^2 | U)$$

$$= \mathcal{E}(\|X_1 - \mathcal{E}(X_1|U)\|^2 | U) + \|\mathcal{E}(X_1|U) + \Psi(U)\|^2.$$

Thus it is clear that

$$\Psi_0(U) \equiv -\mathcal{E}(X_1|U)$$

minimizes $R(t,0)$. Hence the equivariant estimator

$$t_0(X_1,\ldots,X_n) = X_1 - \mathcal{E}_0(X_1|X_2 - X_1,\ldots,X_n - X_1)$$

satisfies

$$R(t_0,\theta) = R(t_0,0) \leqslant R(t,0) = R(t,\theta)$$

for all $\theta \in R^p$ and all equivariant estimators t. The estimator t_0 is commonly called the Pitman estimator. ◆

7.5. DISTRIBUTION THEORY AND INVARIANCE

When a family of distributions is invariant under a group of transformations, useful information can often be obtained about the distribution of invariant functions by using the invariance. For example, some of the results in Section 7.1 are generalized here.

The first result shows that the distribution of an invariant function depends invariantly on a parameter. Suppose $(\mathcal{X}, \mathcal{B})$ is a measurable space acted on measurably by a group G. Consider an invariant probability model $\mathcal{P} = \{P_\theta | \theta \in \Theta\}$ and let $\bar{G}$ be the induced group of transformations on Θ. Thus

$$gP_\theta = P_{\bar{g}\theta}, \qquad \theta \in \Theta, \quad g \in G.$$

A measurable mapping τ on $(\mathcal{X}, \mathcal{B})$ to $(\mathcal{Y}, \mathcal{C})$ induces a family of distributions on $(\mathcal{Y}, \mathcal{C})$, $\{Q_\theta | \theta \in \Theta\}$ given by

$$Q_\theta(C) \equiv P_\theta(\tau^{-1}(C)), \qquad C \in \mathcal{C}, \quad \theta \in \Theta.$$

Proposition 7.14. If τ is G-invariant, then $Q_\theta = Q_{\bar{g}\theta}$ for $\theta \in \Theta$ and $\bar{g} \in \bar{G}$.

Proof. For each $C \in \mathcal{C}$, it must be shown that

$$Q_\theta(C) = Q_{\bar{g}\theta}(C)$$

or, equivalently, that

$$P_\theta(\tau^{-1}(C)) = P_{\bar{g}\theta}(\tau^{-1}(C)).$$

But

$$P_{\bar{g}\theta}(\tau^{-1}(C)) = (gP_\theta)(\tau^{-1}(C)) = P_\theta(g^{-1}\tau^{-1}C)) = P_\theta((\tau g)^{-1}(C)).$$

Since $\tau g = \tau$ as τ is invariant,

$$Q_{\bar{g}\theta}(C) = P_\theta((\tau g)^{-1}(C)) = P_\theta(\tau^{-1}(C)) = Q_\theta(C). \qquad \square$$

An alternative formulation of Proposition 7.14 is useful. If $\mathcal{L}(X) \in \{P_\theta | \theta \in \Theta\}$ and if τ is G-invariant, then the induced distribution of $Y = \tau(X)$, which is Q_θ, satisfies $Q_\theta = Q_{\bar{g}\theta}$. In other words, the distribution of an invariant function depends only on a maximal invariant parameter. By definition, a maximal invariant parameter is any function defined on Θ that is maximal invariant under the action of $\bar{G}$ on Θ. Of course, Θ is usually not a parameter space for the family $\{Q_\theta | \theta \in \Theta\}$ as $Q_\theta = Q_{\bar{g}\theta}$, but any maximal $\bar{G}$-invariant function on Θ often serves as a parameter index for the distribution of $Y = \tau(X)$.

◆ **Example 7.12.** In this example, we establish a property of the distribution of the bivariate sample correlation coefficient. Consider a family of densities $p(\cdot|\mu, \Sigma)$ on R^2 given by

$$p(x|\mu, \Sigma) = |\Sigma|^{-1/2} f_0((x - \mu)'\Sigma^{-1}(x - \mu))$$

where $\mu \in R^2$ and $\Sigma \in \mathcal{S}_2^+$. Hence

$$\int_{R^2} f_0(\|x\|^2)\, dx = 1$$

and it is assumed that

$$\int_{R^2} \|x\|^2 f_0(\|x\|^2)\, dx < +\infty.$$

Since the distribution on R^2 determined by f_0 is orthogonally invariant, if $Z \in R^2$ has density $f_0(\|x\|^2)$, then

$$\mathcal{E}Z = 0 \quad \text{and} \quad \text{Cov}(Z) = cI_2$$

for some $c > 0$ (see Proposition 2.13). Also, $Z_1 = \Sigma^{1/2}Z + \mu$ has density $p(\cdot|\mu, \Sigma)$ when Z has density $f_0(\|x\|^2)$. Thus

$$\mathcal{E}Z_1 = \mu \quad \text{and} \quad \text{Cov}(Z_1) = c\Sigma.$$

The group Al_2 acts on R^2 by

$$(A, b)x = Ax + b$$

and it is clear that the family of distributions, say $\mathcal{P} = \{P_{\mu, \Sigma} | (\mu, \Sigma) \in \mathcal{R}^2 \times \mathcal{S}_2^+\}$, having the densities $p(\cdot|\mu, \Sigma)$, $\mu \in R^2$, $\Sigma \in \mathcal{S}_2^+$, is invariant under this group action. Lebesgue measure on R^2 is relatively invariant with multiplier

$$\chi(A, b) = |\det(A)|$$

and

$$p(x|\mu, \Sigma) = p((A, b)x | A\mu + b, A\Sigma A')\chi(A, b).$$

Obviously, the group action on the parameter space is

$$(A, b)(\mu, \Sigma) = (A\mu + b, A\Sigma A')$$

and

$$(A, b)P_{\mu, \Sigma} = P_{(A, B)(\mu, \Sigma)}; \qquad P_{\mu, \Sigma} \in \mathcal{P}.$$

Now, let $X_1, \ldots, X_n$, $n \geqslant 3$, be a random sample with $\mathcal{L}(X_i) \in \mathcal{P}$ so the probability model for the random sample is Al_2-invariant by Proposition 7.9. Consider $\bar{X} = (1/n)\Sigma_1^n X_i$ and $S = \Sigma_1^n (X_i - \bar{X})(X_i - \bar{X})'$ so $\bar{X}$ is the sample mean and S is the sample covariance matrix (not normalized). Obviously, $S = S(X_1, \ldots, X_n)$ is a function of $X_1, \ldots, X_n$ and

$$S(AX_1 + b, \ldots, AX_n + b) = AS(X_1, \ldots, X_n)A'.$$

That is, S is an equivariant function on $(R^2)^n$ to $\mathbb{S}_2^+$ where the group action on $\mathbb{S}_2^+$ is

$$(A, b)(S) = ASA'.$$

Writing $S \in \mathbb{S}_2^+$ as

$$S = \begin{pmatrix} s_{11} & s_{12} \\ s_{21} & s_{22} \end{pmatrix}, \qquad s_{12} = s_{21},$$

the sample correlation coefficient is

$$r = \frac{s_{12}}{\sqrt{s_{11}s_{22}}}.$$

Also, the population correlation coefficient is

$$\rho = \frac{\sigma_{12}}{\sqrt{\sigma_{11}\sigma_{22}}}$$

when the distribution under consideration is $P_{\mu, \Sigma}$, and

$$\Sigma = \begin{pmatrix} \sigma_{11} & \sigma_{12} \\ \sigma_{21} & \sigma_{22} \end{pmatrix}.$$

Now, given that the random sample is from $P_{\mu, \Sigma}$, the question is: how does the distribution of r depend on (μ, Σ)? To show that the distribution of r depends only on ρ, we use an invariance argument. Let G be the subgroup of Al_2 defined by

$$G = \left\{(A, b) | (A, b) \in Al_2, A = \begin{pmatrix} a_{11} & 0 \\ 0 & a_{22} \end{pmatrix}, a_{ii} > 0, i = 1, 2\right\}.$$

For $(A, b) \in G$, a bit of calculation shows that $r = r(X_1, \ldots, X_n) =$

$r(AX_1 + b, \ldots, AX_n + b)$ so r is a G-invariant function of $X_1, \ldots, X_n$. By Proposition 7.14, the distribution of r, say $Q_{\mu, \Sigma}$, satisfies

$$Q_{\mu, \Sigma} = Q_{(A, b)(\mu, \Sigma)}, \qquad (A, b) \in G.$$

Thus $Q_{\mu, \Sigma}$ depends on μ, Σ only through a maximal invariant function on the parameter space $R^2 \times \mathbb{S}_2^+$ under the action of G. Of course, the action of G is

$$(A, b)(\mu, \Sigma) = (A\mu + b, A\Sigma A'), \qquad (A, b) \in G.$$

We now claim that

$$\rho = \rho(\mu, \Sigma) = \frac{\sigma_{12}}{\sqrt{\sigma_{11}\sigma_{22}}}$$

is a maximal G-invariant function. To see this, consider $(\mu, \Sigma) \in R^2 \times \mathbb{S}_2^+$. By choosing

$$A = \begin{pmatrix} \sigma_{11}^{-1/2} & 0 \\ 0 & \sigma_{22}^{-1/2} \end{pmatrix}$$

and $b = -A\mu$, $(A, b) \in G$ and

$$(A, b)(\mu, \Sigma) = \left(\begin{pmatrix} 0 \\ 0 \end{pmatrix}, \begin{pmatrix} 1 & \rho \\ \rho & 1 \end{pmatrix} \right),$$

so this point is in the orbit of (μ, Σ) and an orbit index is ρ. Thus ρ is maximal invariant and the distribution of r depends only on (μ, Σ) through the maximal invariant function ρ. Obviously, the distribution of r also depends on the function f_0, but f_0 was considered fixed in this discussion. ◆

Proposition 7.14 asserts that the distribution of an invariant function depends only on a maximal invariant parameter, but this result is not especially useful if the exact distribution of an invariant function is desired. The remainder of the section is concerned with using invariance arguments, when G is compact, to derive distributions of maximal invariants and to characterize the G-invariant distributions.

First, we consider the distribution of a maximal invariant function when a compact topological group G acts measurably on a space $(\mathcal{X}, \mathcal{B})$. Suppose that μ_0 is a σ-finite G-invariant measure on $(\mathcal{X}, \mathcal{B})$ and f is a density with respect to μ_0. Let τ be a measurable mapping from $(\mathcal{X}, \mathcal{B})$ onto $(\mathcal{Y}, \mathcal{C})$. Then τ induces a measure on $(\mathcal{Y}, \mathcal{C})$, say ν_0, given by

$$\nu_0(C) = \mu_0(\tau^{-1}(C))$$

and the equation

$$\int_{\mathcal{X}} h(\tau(x))\mu_0(dx) = \int_{\mathcal{Y}} h(y)\nu_0(dy)$$

holds for all integrable functions h on $(\mathcal{Y}, \mathcal{C})$. Since the group G is compact, there exists a unique probability measure, say δ, that is left and right invariant.

Proposition 7.15. Suppose the mapping τ from $(\mathcal{X}, \mathcal{B})$ onto $(\mathcal{Y}, \mathcal{C})$ is maximal invariant under the action of G on $\mathcal{X}$. If $X \in \mathcal{X}$ has density f with respect to μ_0, then the density of $Y = \tau(X)$ with respect to ν_0 is given by

$$q(\tau(x)) = \int_G f(gx)\delta(dg).$$

Proof. First, the integral

$$\int f(gx)\delta(dg)$$

is a G-invariant function of x and thus can be written as a function of the maximal invariant τ. This defines the function q on $\mathcal{Y}$. To show that q is the density of Y, it suffices to show that

$$\mathcal{E}k(Y) = \int_{\mathcal{Y}} k(y)q(y)\nu_0(dy)$$

for all bounded measurable functions k. But

$$\mathcal{E}k(Y) = \mathcal{E}k(\tau(X)) = \int_{\mathcal{X}} k(\tau(x))f(x)\mu_0(dx)$$

$$= \int_{\mathcal{X}} k(\tau(x))f(gx)\mu_0(dx).$$

The last equality holds since μ_0 is G-invariant and τ is G-invariant. Since δ is a probability measure

$$\mathcal{E}k(Y) = \int_G \int_{\mathfrak{X}} k(\tau(x)) f(gx) \mu_0(dx) \delta(dg).$$

Using Fubini's Theorem, the definition of q and the relationship between μ_0 and ν_0, we have

$$\mathcal{E}k(Y) = \int_{\mathfrak{X}} k(\tau(x)) q(\tau(x)) \mu_0(dx) = \int_{\mathfrak{Y}} k(y) q(y) \nu_0(dy). \qquad \square$$

In most situations, the compact group G will be the orthogonal group or some subgroup of the orthogonal group. Concrete applications of Proposition 7.15 involve two separate steps. First, the function q must be calculated by evaluating

$$\int_G f(gx) \delta(dg).$$

Also, given μ_0 and the maximal invariant τ, the measure ν_0 must be found.

◆ **Example 7.13.** Take $\mathfrak{X} = R^n$ and let μ_0 be Lebesgue measure. The orthogonal group $\mathcal{O}_n$ acts on R^n and a maximal invariant function is $\tau(x) = \|x\|^2$ so $\mathfrak{Y} = [0, \infty)$. If a random vector $X \in R^n$ has a density f with respect to Lebesgue measure, Proposition 7.15 tells us how to find the density of $Y = \|X\|^2$ with respect to the measure ν_0. To find ν_0, consider the particular density

$$f_0(x) = (\sqrt{2\pi})^{-n} \exp\left[-\tfrac{1}{2}\|x\|^2\right].$$

Thus $\mathcal{L}(X) = N(0, I_n)$, so $\mathcal{L}(Y) = \chi^2_n$ and the density of Y with respect to Lebesgue measure dy on $[0, \infty)$ is

$$p(y) = \frac{y^{n/2-1}\exp\left[-\tfrac{1}{2}y\right]}{2^{n/2}\Gamma(n/2)}.$$

Therefore,

$$p(y)\, dy = q_0(y) \nu_0(dy)$$

where

$$q_0(\tau(x)) = \int_{\mathcal{O}_n} f_0(\Gamma x) \delta(d\Gamma).$$

Since $f_0(\Gamma x) = f_0(x)$, the integration of f_0 over $\mathcal{O}_n$ is trivial and

$$q_0(y) = (\sqrt{2\pi})^{-n}\exp[-\tfrac{1}{2}y].$$

Solving for $\nu_0(dy)$, we have

$$\nu_0(dy) = \frac{(2\pi)^{n/2}}{2^{n/2}\Gamma(n/2)}y^{n/2-1}\,dy = \frac{[\Gamma(\frac{1}{2})]^n}{\Gamma(n/2)}y^{n/2-1}\,dy$$

since $\Gamma(\frac{1}{2}) = \sqrt{\pi}$. Now that ν_0 has been found, consider a general density f on R^n. Then

$$q(\tau(x)) = \int_{\mathcal{O}_n} f(\Gamma x)\delta(d\Gamma)$$

and $q(y)$ is the density of $Y = \|X\|^2$ with respect to ν_0. When the density f is given by

$$f(x) = h(\|x\|^2),$$

then it is clear that

$$q(y) = h(y), \qquad y \in [0, \infty)$$

so the distribution of Y has been found in this case. The noncentral chi-square distribution of $Y = \|X\|^2$ provides an interesting example where the integration over $\mathcal{O}_n$ is not trivial. Suppose $\mathcal{L}(X) = N(\mu, I_n)$ so

$$\begin{aligned} f(x) &= (\sqrt{2\pi})^{-n}\exp[-\tfrac{1}{2}\|x-\mu\|^2] \\ &= (\sqrt{2\pi})^{-n}\exp[-\tfrac{1}{2}(\|x\|^2 - 2x'\mu + \|\mu\|^2)]. \end{aligned}$$

Thus

$$\begin{aligned} q(\tau(x)) = (\sqrt{2\pi})^{-n}&\exp[-\tfrac{1}{2}\|\mu\|^2]\exp[-\tfrac{1}{2}\|x\|^2] \\ &\times \int_{\mathcal{O}_n}\exp[(\Gamma x)'\mu]\delta(d\Gamma). \end{aligned}$$

Since x and $\|x\|\varepsilon_1$ have the same length, $x = \|x\|\Gamma_1\varepsilon_1$ for some $\Gamma_1 \in \mathcal{O}_n$ where ε_1 is the first standard unit vector in R^n. Similarly,

$\mu = \|\mu\|\Gamma_2\varepsilon_1$ for some $\Gamma_2 \in \mathcal{O}_n$. Setting $\lambda = \|\mu\|^2$,

$$q(y) = (\sqrt{2\pi})^{-n}\exp[-\tfrac{1}{2}\lambda]\exp[-\tfrac{1}{2}y]$$
$$\times \int_{\mathcal{O}_n} \exp[\sqrt{y\lambda}\,(\Gamma\Gamma_1\varepsilon_1)'\Gamma_2\varepsilon_1]\delta(d\Gamma).$$

Thus to evaluate q, we need to calculate

$$H(u) = \int_{\mathcal{O}_n} \exp[u\varepsilon_1'\Gamma_1'\Gamma'\Gamma_2\varepsilon_1]\delta(d\Gamma).$$

Since δ is left and right invariant,

$$H(u) = \int_{\mathcal{O}_n} \exp[u\varepsilon_1'\Gamma'\varepsilon_1]\delta(d\Gamma) = \int_{\mathcal{O}_n} \exp[u\gamma_{11}]\delta(d\Gamma)$$

where γ_{11} is the $(1,1)$ element of Γ. The representation of the uniform distribution on $\mathcal{O}_n$ given in Proposition 7.2 shows that when Γ is uniform on $\mathcal{O}_n$, then

$$\mathcal{L}(\gamma_{11}) = \mathcal{L}\left(\frac{Z_1}{\|Z\|}\right)$$

where $\mathcal{L}(Z) = N(0, I_n)$ and Z_1 is the first coordinate of Z. Expanding the exponential in a power series, we have

$$H(u) = \sum_{j=0}^{\infty} \frac{1}{j!} \int_{\mathcal{O}_n} u^j \gamma_{11}^j \delta(d\Gamma) = \sum_{j=0}^{\infty} \frac{u^j}{j!} \mathcal{E}\left(\frac{Z_1}{\|Z\|}\right)^j.$$

Thus the moments of $U_1 \equiv Z_1/\|Z\|$ need to be found. Obviously, $\mathcal{L}(U_1) = \mathcal{L}(-U_1)$, so all odd moments of U_1 are zero. Also, $U_1^2 = Z_1^2/(Z_1^2 + \Sigma_2^n Z_i^2)$, which has a beta distribution with parameters $\frac{1}{2}$ and $(n-1)/2$. Therefore,

$$a_j \equiv \mathcal{E}(U_1^2)^j = \frac{\Gamma(n/2)\Gamma(j+\frac{1}{2})}{\Gamma(n/2+j)\Gamma(\frac{1}{2})}$$

so

$$H(u) = \sum_{j=0}^{\infty} \frac{a_j u^{2j}}{(2j)!}.$$

Hence

$$q(y) = (\sqrt{2\pi})^{-n}\exp\left[-\frac{1}{2}\lambda\right]\exp\left[-\frac{1}{2}y\right]\sum_{j=0}^{\infty}\lambda^j y^j \frac{a_j}{(2j)!}$$

is the density of Y with respect to the measure ν_0. A bit of algebra and some manipulation with the gamma function shows that

$$q(y)\nu_0(dy) = \left\{\sum_{j=0}^{\infty}\frac{\exp[-\frac{1}{2}\lambda]}{j!}(\lambda/2)^j h_{n+2j}(y)\right\} dy$$

where

$$h_m(y) = \frac{y^{m/2-1}\exp[-\frac{1}{2}y]}{2^{m/2}\Gamma(m/2)}$$

is the density of a χ^2_m distribution. This is the expression for the density of the noncentral chi-square distribution discussed in Chapter 3. ◆

◆ **Example 7.14.** In this example, we derive the density function of the order statistic of a random vector $X \in R^n$. Suppose X has a density f with respect to Lebesgue measure and let $X_1, \ldots, X_n$ be the coordinates of X. Consider the space $\mathcal{Y} \subseteq R^n$ defined by

$$\mathcal{Y} = \{y | y \in R^n,\ y_1 \leqslant y_2 \leqslant \cdots \leqslant y_n\}.$$

The order statistic of X is the random vector $Y \in \mathcal{Y}$ consisting of the ordered values of the coordinates of X. More precisely, Y_1 is the smallest coordinate of X, Y_2 is the next smallest coordinate of X, and so on. Thus $Y = \tau(X)$ where τ maps each $x \in R^n$ into the ordered coordinates of x—say $\tau(x) \in \mathcal{Y}$. To derive the density function of Y, we show that Y is a maximal invariant under a compact group operating on R^n and then apply Proposition 7.15. Let G be the group of all one-to-one onto functions from $\{1, 2, \ldots, n\}$ to $\{1, 2, \ldots, n\}$—that is, G is the permutation group of $\{1, 2, \ldots, n\}$. Of course, the group operation is function composition, the group inverse is function inverse, and G has $n!$ elements. The group G acts on the left of R^n in the following way. For $x \in R^n$ and $\pi \in G$, define $\pi x \in R^n$ to have ith coordinate $x(\pi^{-1}(i))$. Thus the ith coordinate of πx is the $\pi^{-1}(i)$ coordinate of x, so

$$(\pi x)(i) \equiv x(\pi^{-1}(i)).$$

The reason for the inverse on π in this definition is so that G acts on the left of R^n—that is,

$$(\pi_1\pi_2)x = \pi_1(\pi_2 x).$$

It is routine to verify that the function τ on $\mathcal{X}$ to $\mathcal{Y}$ is a maximal invariant under the action of G on R^n. Also, Lebesgue measure, say l, is invariant so Proposition 7.15 is applicable as G is a finite group and hence compact. Obviously, the density q of $Y = \tau(X)$ is

$$q(\tau(x)) = \frac{1}{n!}\sum_{\pi} f(\pi x) = \frac{1}{n!}\sum_{\pi} f(\pi\tau(x))$$

so

$$q(y) = \frac{1}{n!}\sum_{\pi} f(\pi y)$$

for $y \in \mathcal{Y}$. To derive the measure ν_0 on $\mathcal{Y}$, consider a measurable subset $C \subseteq \mathcal{Y}$. Then

$$\tau^{-1}(C) = \bigcup_{\pi \in G} (\pi C)$$

and

$$\nu_0(C) = l(\tau^{-1}(C)) = l\Big(\bigcup_{\pi \in G}(\pi C)\Big) = \sum_{\pi \in G} l(\pi(C)) = n!\,l(C).$$

The third equality follows since $(\pi_1 C) \cap (\pi_2 C)$ has Lebesgue measure zero for $\pi_1 \neq \pi_2$ as the boundary of $\mathcal{Y}$ in R^n has Lebesgue measure zero. Thus ν_0 is just $n!$ times l restricted to $\mathcal{Y}$. Therefore, the density of the order statistic Y, with respect to ν_0 restricted to $\mathcal{Y}$, is

$$q(y) = \frac{1}{n!}\sum_{\pi} f(\pi y).$$

When f is invariant under permutations, as is the case when $X_1, \ldots, X_n$ are independent and identically distributed, we have

$$q(y) = f(y), \qquad y \in \mathcal{Y}. \qquad \blacklozenge$$

The next example is an extension of Example 7.13 and is related to the results in Proposition 7.6.

◆ **Example 7.15.** Suppose X is a random vector in $\mathcal{L}_{p,n}$, $n \geqslant p$, which has a density f with respect to Lebesgue measure dx on $\mathcal{L}_{p,n}$. Let τ map $\mathcal{L}_{p,n}$ onto the space of $p \times p$ positive semidefinite matrices, say $\bar{\mathcal{S}}_p^+$, by $\tau(x) = x'x$. The problem in this example is to derive the density of $S = \tau(X) = X'X$. The compact group $\mathcal{O}_n$ acts on $\mathcal{L}_{p,n}$ and a group element $\Gamma \in \mathcal{O}_n$ sends x into Γx. It follows immediately from Proposition 1.31 that τ is a maximal invariant function under the action of $\mathcal{O}_n$ on $\mathcal{L}_{p,n}$. Since dx is invariant under $\mathcal{O}_n$, Proposition 7.15 shows that the density of S is

$$q(\tau(x)) = \int_{\mathcal{O}_n} f(\Gamma x)\mu(d\Gamma)$$

with respect to the measure ν_0 on $\bar{\mathcal{S}}_p^+$ induced by dx and τ. To find the measure ν_0, we argue as in Example 7.13. Consider the particular density

$$f_0(x) = (\sqrt{2\pi})^{-np}\exp[-\tfrac{1}{2}\operatorname{tr}(x'x)]$$

on $\mathcal{L}_{p,n}$ so $\mathcal{L}(X) = N(0, I_n \otimes I_p)$. For this f_0, the density of S is

$$q_0(S) = q_0(\tau(x)) = \int_{\mathcal{O}_n} f_0(\Gamma x)\mu(d\Gamma) = (\sqrt{2\pi})^{-np}\exp[-\tfrac{1}{2}\operatorname{tr}(S)]$$

with respect to ν_0. However, by Propostion 7.6, the density of S with respect to $dS/|S|^{(p+1)/2}$ is

$$q_1(S) = \omega(n, p)|S|^{n/2}\exp[-\tfrac{1}{2}\operatorname{tr}(S)].$$

Therefore,

$$q_1(S)\frac{dS}{|S|^{(p+1)/2}} = q_0(S)\nu_0(dS)$$

so

$$\omega(n, p)|S|^{(n-p-1)/2}\exp[-\tfrac{1}{2}\operatorname{tr}(S)]\,dS$$
$$= (\sqrt{2\pi})^{-np}\exp[-\tfrac{1}{2}\operatorname{tr}(S)]\nu_0(dS),$$

which shows that

$$\nu_0(dS) = (\sqrt{2\pi})^{np}\omega(n, p)|S|^{(n-p-1)/2}\,dS.$$

In the above argument, we have ignored the set of Lebesgue measure zero where $x \in \mathcal{L}_{p,n}$ has rank less than p. The justification for this is left to the reader. Now that ν_0 has been found, the density of S for a general density f is obtained by calculating

$$q(\tau(x)) = \int_{\mathcal{O}_n} f(\Gamma x)\mu(d\Gamma).$$

When $f(x) = h(x'x)$, then $f(\Gamma x) = h(x'x) = h(\tau(x))$ and $q(S) = h(S)$. In this case, the integration over $\mathcal{O}_n$ is trivial. Another example where the integration over $\mathcal{O}_n$ is not trivial is given in the next chapter when we discuss the noncentral Wishart distribution. ◆

As motivation for the next result of this section, consider the situation discussed in Proposition 7.3. This result gives a characterization of the $\mathcal{O}_n$-left invariant distributions by representing each of these distributions as a product measure where one measure is a fixed $\mathcal{O}_n$-invariant distribution and the other measure is arbitrary. The decomposition of the space $\mathcal{X}$ into the product space $\mathcal{F}_{p,n} \times G_U^+$ provided the framework in which to state this representation of $\mathcal{O}_n$-left invariant distributions. In some situations, this product space structure is not available (see Example 7.5) but a product measure representation for $\mathcal{O}_n$-invariant distributions can be obtained. It is established below that, under some mild regularity conditions, such a representation can be given for probability measures that are invariant under any compact topological group that acts on the sample space. We now turn to the technical details.

In what follows, G is a compact topological group that acts measurably on a measure space $(\mathcal{X}, \mathcal{B})$ and P is a G-invariant probability measure on $(\mathcal{X}, \mathcal{B})$. The unique invariant probability measure on G is denoted by μ and the symbol $U \in G$ denotes a random variable with values in G and distribution μ. The σ-algebra for G is the Borel σ-algebra of open sets so U is a measurable function defined on some probability space with induced distribution μ. Since G acts on $\mathcal{X}$, $\mathcal{X}$ can be written as a disjoint union of orbits, say

$$\mathcal{X} = \bigcup_{\alpha \in \mathcal{A}} \mathcal{X}_\alpha$$

where $\mathcal{A}$ is an index set for the orbits and $\mathcal{X}_\alpha \cap \mathcal{X}_{\alpha'} = \phi$ if $\alpha \neq \alpha'$. Let x_α be a fixed element of $\mathcal{X}_\alpha = \{gx_\alpha | g \in G\}$. Also, set

$$\mathcal{Y} = \{x_\alpha | \alpha \in A\} \subseteq \mathcal{X}$$

and assume that $\mathcal{Y}$ is a measurable subset of $\mathcal{X}$. The function τ defined on $\mathcal{X}$ to $\mathcal{Y}$ by

$$\tau(x) = x_\alpha \qquad \text{if } x \in \mathcal{X}_\alpha$$

is obviously a maximal invariant function under the action of G on $\mathcal{X}$. It is assumed that τ is a measurable function from $\mathcal{X}$ to $\mathcal{Y}$ where $\mathcal{Y}$ has the σ-algebra inherited from $\mathcal{X}$. A subset $B_1 \subseteq \mathcal{Y}$ is measurable iff $B_1 = \mathcal{Y} \cap B$ for some $B \in \mathcal{B}$. If $X \in \mathcal{X}$ has distribution P, then the maximal invariant $Y = \tau(X)$ has the induced distribution Q defined by

$$Q(B_1) = P(\tau^{-1}(B_1))$$

for measurable subsets $B_1 \subseteq \mathcal{Y}$. What we would like to show is that P is represented by the product measure $\mu \times Q$ on $G \times \mathcal{Y}$ in the following sense. If $Y \in \mathcal{Y}$ has the distribution Q and is independent of $U \in G$, then the random variable $Z = UY \in \mathcal{X}$ has the distribution P. In other words, $\mathcal{L}(X) = \mathcal{L}(UY)$ where U and Y are independent. Here, UY means the group element U operating on the point $Y \in \mathcal{X}$. The intuitive argument that suggests this representation is the following. The distribution of X, conditional on $\tau(X) = x_\alpha$, should be G-invariant on $\mathcal{X}_\alpha$ as the distribution of X is G-invariant. But G acts transitively on $\mathcal{X}_\alpha$ and, since G is compact, there should be a unique invariant probability distribution on $\mathcal{X}_\alpha$ that is induced by μ on G. In other words, conditional on $\tau(X) = x_\alpha$, X should have the same distribution as Ux_α where U is "uniform" on G. The next result makes all of this precise.

Proposition 7.16. Consider $\mathcal{X}$, $\mathcal{Y}$, and G to be as above with their respective σ-algebras. Assume that the mapping h on $G \times \mathcal{Y}$ to $\mathcal{X}$ given by $h(g, y) = gy$ is measurable.

(i) If $U \in G$ and $Y \in \mathcal{Y}$ are independent with $\mathcal{L}(U) = \mu$ and $\mathcal{L}(Y) = Q$, then the distribution of $X = UY$ is a G-invariant distribution on $\mathcal{X}$.

(ii) If $X \in \mathcal{X}$ has a G-invariant distribution, say P, let the maximal invariant $Y = \tau(X)$ have an induced distribution Q on $\mathcal{Y}$. Let $U \in G$ have the distribution μ and be independent of X. Then $\mathcal{L}(X) = \mathcal{L}(UY)$.

Proof. For the proof of (i), it suffices to show that

$$\mathcal{E}f(X) = \mathcal{E}f(gX)$$

for all integrable functions f and all $g \in G$. But

$$\mathcal{E}f(gX) = \mathcal{E}f(g(UY)) = \mathcal{E}f((gU)Y) = \mathcal{E}\mathcal{E}[f((gU)Y)|Y]$$
$$= \mathcal{E}\mathcal{E}[f(UY)|Y] = \mathcal{E}f(UY) = \mathcal{E}f(X).$$

In the above calculation, we have used the assumption that U and Y are independent, so conditional on Y, $\mathcal{L}(U) = \mathcal{L}(gU)$ for $g \in G$.

To prove (ii) it suffices to show that

$$\mathcal{E}f(X) = \mathcal{E}f(UY)$$

for all integrable f. Since the distribution of X is G-invariant

$$\mathcal{E}f(X) = \mathcal{E}f(gX), \quad g \in G.$$

Therefore,

$$\mathcal{E}f(X) = \mathcal{E}_U\mathcal{E}_X f(UX),$$

as U and X are independent. Thus

$$\int_{\mathfrak{X}} f(x)P(dx) = \int_G\int_{\mathfrak{X}} f(gx)P(dx)\mu(dg) = \int_{\mathfrak{X}}\int_G f(gx)\mu(dg)P(dx).$$

However, for $x \in \mathfrak{X}_\alpha$, there exists an element $k \in G$ such that $x = kx_\alpha$. Using the definition of τ and the right invariance of μ, we have

$$\int_G f(gx)\mu(dg) = \int_G f(gkx_\alpha)\mu(dg) = \int_G f(gx_\alpha)\mu(dg)$$
$$= \int_G f(g\tau(x))\mu(dg).$$

Hence

$$\int_{\mathfrak{X}} f(x)P(dx) = \int_{\mathfrak{X}}\int_G f(g\tau(x))\mu(dg)P(dx) = \int_{\mathfrak{Y}}\int_G f(gy)\mu(dg)Q(dy)$$

where the second equality follows from the definition of the induced measure Q. In terms of the random variables,

$$\mathcal{E}f(X) = \mathcal{E}_U\mathcal{E}_Y f(UY)$$

where U and Y are independent as U and X are independent. □

The technical advantage of Proposition 7.16 over the method discussed in Section 7.1 is that the space $\mathfrak{X}$ is not assumed to be in one-to-one correspondence with the product space $G \times \mathfrak{Y}$. Obviously, the mapping h on $G \times \mathfrak{Y}$ to $\mathfrak{X}$ is onto, but h will ordinarily not be one-to-one.

◆ **Example 7.16.** In this example, take $\mathfrak{X} = \mathbb{S}_p$, the set of all $p \times p$ symmetric matrices. The group $G = \mathcal{O}_p$ acts on $\mathbb{S}_p$ by

$$\Gamma(S) = \Gamma S \Gamma', \quad S \in \mathbb{S}_p, \quad \Gamma \in \mathcal{O}_p.$$

For $S \in \mathbb{S}_p$, let

$$\tau(S) = Y = \begin{pmatrix} y_1 & & & \\ & y_2 & & \\ & & \ddots & \\ & & & y_p \end{pmatrix} \in \mathbb{S}_p$$

where $y_1 \geqslant \cdots \geqslant y_p$ are ordered eigenvalues of S and the off-diagonal elements of Y are zero. Also, let $\mathfrak{Y} = \{Y | Y = \tau(S),\ S \in \mathbb{S}_p\}$. The spectral theorem shows that τ is a maximal invariant function under the action of $\mathcal{O}_p$ and the elements of $\mathfrak{Y}$ index the orbits in $\mathbb{S}_p$. The measurability assumptions of Proposition 7.16 are easily verified, so every $\mathcal{O}_n$-invariant distribution on $\mathbb{S}_p$, say P, has the representation given by

$$\int_{\mathbb{S}_p} f(S) P(dS) = \int_{\mathcal{O}_p} \int_{\mathfrak{Y}} f(\Gamma Y \Gamma') Q(dY) \mu(d\Gamma)$$

where μ is the uniform distribution on $\mathcal{O}_p$ and Q is the induced distribution of Y. In terms of random variables, if $\mathcal{L}(S) = P$ and $\mathcal{L}(\Gamma S \Gamma') = \mathcal{L}(S)$ for all $\Gamma \in \mathcal{O}_p$, then

$$\mathcal{L}(S) = \mathcal{L}(\Psi \tau(S) \Psi')$$

where Ψ is uniform on $\mathcal{O}_p$ and is independent of the matrix of eigenvalues of S. As a particular case, consider the probability measure P_0 on $\mathbb{S}_p^+ \subseteq \mathbb{S}_p$ with the Wishart density

$$p_0(S) = \omega(p, n) |S|^{(n-p-1)/2} \exp\left[-\tfrac{1}{2} \operatorname{tr} S\right] I(S)$$

where $n \geqslant p$, $I(S) = 1$ if $S \in \mathbb{S}_p^+$ and is zero otherwise. That p_0 is a density on $\mathcal{S}_p$ with respect to Lebesgue measure dS on $\mathcal{S}_p$ follows from Example 5.1. Also, p_0 is $\mathcal{O}_p$-invariant since dS is $\mathcal{O}_p$-invariant and $p_0(\Gamma S \Gamma') = p_0(S)$ for all $S \in \mathcal{S}_p$ and $\Gamma \in \mathcal{O}_p$. Thus the above results are applicable to this particular Wishart distribution. ◆

The final example of this section deals with the singular value decomposition of a random $n \times p$ matrix.

◆ **Example 7.17.** The compact group $\mathcal{O}_n \times \mathcal{O}_p$ acts on the space $\mathcal{L}_{p,n}$ by

$$(\Gamma, \Delta)X \equiv \Gamma X \Delta'; \qquad (\Gamma, \Delta) \in \mathcal{O}_n \times \mathcal{O}_p, \quad X \in \mathcal{L}_{p,n}.$$

For definiteness, we take $p \leqslant n$. Define τ on $\mathcal{L}_{p,n}$ by

$$\tau(X) = \begin{pmatrix} \lambda_1 & & & 0 \\ & \lambda_2 & & \\ & & \ddots & \\ 0 & & & \lambda_p \\ \hline & & 0 & \end{pmatrix}$$

where $\lambda_1 \geqslant \cdots \geqslant \lambda_p \geqslant 0$ and $\lambda_1^2, \ldots, \lambda_p^2$ are the ordered eigenvalues of $X'X$. Let $\mathcal{Y} \subseteq \mathcal{L}_{p,n}$ be the range of τ so $\mathcal{Y}$ is a closed subset of $\mathcal{L}_{p,n}$. It is clear that $\tau(\Gamma X \Delta') = \tau(X)$ for $\Gamma \in \mathcal{O}_n$ and $\Delta \in \mathcal{O}_p$ so τ is invariant. That τ is a maximal invariant follows easily from the singular value decomposition theorem. Thus the elements of $\mathcal{Y}$ index the orbits in $\mathcal{L}_{p,n}$ and every $X \in \mathcal{L}_{p,n}$ can be written as

$$X = \Gamma y \Delta' = (\Gamma, \Delta)y$$

for some $y \in \mathcal{Y}$ and $(\Gamma, \Delta) \in \mathcal{O}_n \times \mathcal{O}_p$. The measurability assumptions of Proposition 7.16 are easily checked. Thus if P is an $(\mathcal{O}_n \times \mathcal{O}_p)$-invariant probability measure on $\mathcal{L}_{p,n}$ and $\mathcal{L}(X) = P$, then

$$\mathcal{L}(X) = \mathcal{L}(\Gamma Y \Delta')$$

where (Γ, Δ) has a uniform distribution on $\mathcal{O}_n \times \mathcal{O}_p$, Y has a distribution Q induced by τ and P, and Y and (Γ, Δ) are independent. However, we can say a bit more. Since $\mathcal{O}_n \times \mathcal{O}_p$ is a product group, the unique invariant probability measure on $\mathcal{O}_n \times \mathcal{O}_p$ is the product measure $\mu_1 \times \mu_2$ where $\mu_1(\mu_2)$ is the unique invariant probability measure on $\mathcal{O}_n(\mathcal{O}_p)$. Thus Γ and Δ are independent and each is uniform in its respective group. In summary,

$$\mathcal{L}(X) = \mathcal{L}(\Gamma Y \Delta').$$

where Γ, Y, and Δ are mutually independent with the distributions given above. As a particular case, consider the density

$$f_0(X) = (\sqrt{2\pi})^{-np} \exp[-\tfrac{1}{2}\operatorname{tr}(X'X)]$$

with respect to Lebesgue measure on $\mathcal{L}_{p,n}$. Since $f_0(\Gamma X \Delta') = f_0(X)$ and Lebesgue measure is $(\mathcal{O}_n \times \mathcal{O}_p)$-invariant, the probability measure defined by f_0 is $(\mathcal{O}_n \times \mathcal{O}_p)$-invariant. Therefore, when $\mathcal{L}(X) = N(0, I_n \times I_p)$, X has the same distribution as $\Gamma Y \Delta'$ where Γ and Δ are uniform and Y has the induced distribution Q on $\mathcal{Y}$. ◆

7.6. INDEPENDENCE AND INVARIANCE

Considerations that imply the stochastic independence of an invariant function and an equivariant function are the subject of this section. To motivate the abstract discussion to follow, we begin with the familiar random sample from a univariate normal distribution. Consider $X \in \mathcal{X}$ with $\mathcal{L}(X) = N(\mu e, \sigma^2 I_n)$ where $\mu \in R$, $\sigma^2 > 0$, and e is the vector of ones in R^n. The set $\mathcal{X}$ is $R^n - \operatorname{span}\{e\}$ and the reason for choosing this as the sample space is to guarantee that $\Sigma_1^n (x_i - \bar{x})^2 > 0$ for $x \in \mathcal{X}$. The coordinates of X, say $X_1, \ldots, X_n$, are independent and $\mathcal{L}(X_i) = N(\mu, \sigma^2)$ for $i = 1, \ldots, n$. When μ and σ^2 are unknown parameters, the statistic $t(X) = (s, \bar{X})$ where

$$\bar{X} = \frac{1}{n}\sum_1^n X_i, \qquad s^2 = \sum_1^n (X_i - \bar{X})^2$$

is minimal sufficient and complete. The reason for using s rather than s^2 in the definition of $t(X)$ is based on invariance considerations. The affine group Al_1 acts on $\mathcal{X}$ by

$$(a, b)x \equiv ax + be$$

for $(a, b) \in Al_1$. Let G be the subgroup of Al_1 given by $G = \{(a, b) | (a, b) \in Al_1, a > 0\}$ so G also acts on $\mathcal{X}$.

The probability model for $X \in \mathcal{X}$ is generated by G in the sense that if $Z \in \mathcal{X}$ and $\mathcal{L}(Z) = N(0, I_n)$,

$$\mathcal{L}((a, b)Z) = \mathcal{L}(aZ + be) = N\big(be, a^2 I_n\big).$$

Thus the set of distributions $\mathcal{P} = \{N(\mu e, \sigma^2 I_n) | \mu \in R, \sigma^2 > 0\}$ is obtained from an $N(0, I_n)$ distribution by a group operation. For this example, the group G serves as a parameter space for $\mathcal{P}$. Further, the statistic t takes its values in G and satisfies

$$t((a, b)X) = (a, b)(s, \bar{X}),$$

that is, t evaluated at $(a, b)X = aX + be$ is the same as the group element (a, b) composed with the group element $(s, \bar{X})$. Thus t is an equivariant function defined on $\mathcal{X}$ to G and G acts on both $\mathcal{X}$ and G. Now, which functions of X, say $h(X)$, might be independent of $t(X)$? Intuitively, since $t(X)$ is sufficient, $t(X)$ "contains all the information in X about the parameters." Thus if $h(X)$ has a distribution that does not depend on the parameter value (such an $h(X)$ will be called *ancillary*), there is some reason to believe that $h(X)$ and $t(X)$ might be independent. However, the group structure given above provides a method for constructing ancillary statistics. If h is an invariant function of X, then the distribution of h is an invariant function of the parameter (μ, σ^2). But the group G acts transitively on the parameter space (i.e., G), so any invariant function will be ancillary. Also, h is invariant iff h is a function of a maximal invariant statistic. This suggests that a maximal invariant statistic will be independent of $t(X)$. Consider the statistic

$$Z(X) = (t(X))^{-1}X = \frac{X - \bar{X}e}{s},$$

where the inverse on $t(X)$ denotes the group inverse in G. The verification that $Z(X)$ is maximal invariant partially justifies choosing t to have values in G. For $(a, b) \in G$,

$$\begin{aligned} Z((a, b)X) &= (t((a, b)X))^{-1}(a, b)X = ((a, b)t(X))^{-1}(a, b)X \\ &= (t(X))^{-1}(a, b)^{-1}(a, b)X = (t(X))^{-1}X = Z(X), \end{aligned}$$

so Z is invariant. Also, if

$$(t(X))^{-1}X = Z(X) = Z(Y) = (t(Y))^{-1}Y,$$

then

$$Y = \left[t(Y)(t(X))^{-1}\right]X,$$

so X and Y are in the same orbit. Thus Z is maximal invariant and is an ancillary statistic. That $Z(X)$ and $t(X)$ are stochastically independent for each value of μ and σ^2 follows from Basu's Theorem given in the Appendix. The whole purpose of this discussion was to show that sufficiency coupled with the invariance suggested the independence of $Z(X)$ and $t(X)$. The role of the equivariance of t is not completely clear, but it is essential in the more abstract treatment that follows.

Let P_0 be a fixed probability on $(\mathfrak{X}, \mathfrak{B})$ and suppose that G is a group that acts measurably on $(\mathfrak{X}, \mathfrak{B})$. Consider a measurable function t on $(\mathfrak{X}, \mathfrak{B})$ to $(\mathfrak{Y}, \mathcal{C}_1)$ and assume that $\bar{G}$ is a homomorphic image of G that acts transitively on $(\mathfrak{Y}, \mathcal{C}_1)$ and that

$$t(gx) = \bar{g}t(x); \qquad x \in \mathfrak{X}, \quad g \in G.$$

Thus t is an equivariant function. For technical reasons that become apparent later, it is assumed that $\bar{G}$ is a locally compact and σ-compact topological group endowed with the Borel σ-algebra. Also, the mapping $(\bar{g}, y) \to \bar{g}y$ from $\bar{G} \times \mathfrak{Y}$ to $\mathfrak{Y}$ is assumed to be jointly measurable.

Now, let h be a measurable function on $(\mathfrak{X}, \mathfrak{B})$ to $(\mathfrak{Z}, \mathcal{C}_2)$, which is G-invariant. If $X \in \mathfrak{X}$ and $\mathcal{L}(X) = P_0$, we want to find conditions under which $Y \equiv t(X)$ and $Z \equiv h(X)$ are stochastically independent. The following informal argument, which is made precise later, suggests the conditions needed. To show that Y and Z are independent, it is sufficient to verify that, for all bounded measurable functions f on $(\mathfrak{Z}, \mathcal{C}_2)$,

$$H(y) = \mathcal{E}_{P_0}(f(h(X))|t(X) = y)$$

is constant for $y \in \mathfrak{Y}$. That this condition is sufficient follows by integrating H with respect to the induced distribution of Y, say Q_0. More precisely, if k is a bounded function on $(\mathfrak{Y}, \mathcal{C}_1)$ and $H(y) = H(y_0)$ for $y \in \mathfrak{Y}$, then

$$\begin{aligned}
\mathcal{E}_{P_0}[k(t(X))f(h(X))] &= \int \mathcal{E}_{P_0}[k(t(X))f(h(X))|t(X) = y]Q_0(dy) \\
&= \int k(y)\mathcal{E}_{P_0}[f(h(X))|t(X) = y]Q_0(dy) \\
&= \int k(y)H(y)Q_0(dy) = H(y_0)\int k(y)Q_0(dy) \\
&= \mathcal{E}_{P_0}f(h(X))\mathcal{E}_{P_0}k(t(X)),
\end{aligned}$$

and this implies independence. The assumption that H is constant justifies the next to the last equality while the last equality follows from

$$H(y_0) = \int H(y)Q_0(dy) = \mathcal{E}_{P_0} f(h(X))$$

when H is constant. Thus under what conditions will H be constant? Since $\bar{G}$ acts transitively on $\mathcal{Y}$, if H is $\bar{G}$-invariant, then H must be constant and conversely. However,

$$\begin{aligned} H(\bar{g}^{-1}y) &= \mathcal{E}_{P_0}[f(h(X))|t(X) = \bar{g}^{-1}y] = \mathcal{E}_{P_0}[f(h(X))|\bar{g}t(X) = y] \\ &= \mathcal{E}_{P_0}[f(h(X))|t(gX) = y] = \mathcal{E}_{P_0}[f(h(gX))|t(gX) = y] \\ &= \mathcal{E}_{gP_0}[f(h(X))|t(X) = y]. \end{aligned}$$

The equivariance of t and the invariance of h justify the third and fourth equalities while the last equality is a consequence of $\mathcal{L}(gX) = gP_0$ when $\mathcal{L}(X) = P_0$. Now, if $t(X)$ is a sufficient statistic for the family $\mathcal{P} = \{gP_0 | g \in G\}$, then the last member of the above string of equalities is just $H(y)$. Under this sufficiency assumption, $H(y) = H(\bar{g}^{-1}y)$ so H is invariant and hence is a constant. The technical problem with this argument is caused by the nonuniqueness of conditional expectations. The conclusion that $H(y) = H(\bar{g}^{-1}y)$ should really be $H(y) = H(\bar{g}^{-1}y)$ except for $y \in N_g$ where N_g is a set of Q_0 measure zero. Since this null set can depend on g, even the conclusion that H is a constant a.e. (Q_0) is not justified without some further work. Once these technical problems are overcome, we prove that, if $t(X)$ is sufficient for $\{gP_0 | g \in G\}$, then for each $g \in G$, $h(X)$ and $t(X)$ are stochastically independent when $\mathcal{L}(X) = gP_0$.

The first gap to fill concerns almost invariant functions.

Definition 7.6. Let $(\mathcal{X}_1, \mathcal{B}_1)$ be a measurable space that is acted on measurably by a group G_1. If μ is a σ-finite measure on $(\mathcal{X}_1, \mathcal{B}_1)$ and f is a real-valued Borel measurable function, f is *almost G_1-invariant* if for each $g \in G_1$, the set $N_g = \{x | f(x) \neq f(gx)\}$ has μ measure zero.

The following result shows that under certain conditions, an almost G_1-invariant function is equal a.e. (μ) to a G_1-invariant function.

Proposition 7.17. Suppose that G_1 acts measurably on $(\mathcal{X}_1, \mathcal{B}_1)$ and that G_1 is a locally compact and σ-compact topological group with the Borel σ-algebra. Assume that the mapping $(g, x) \to gx$ from $G_1 \times \mathcal{X}_1$ to $\mathcal{X}_1$ is measurable. If μ is a σ-finite measure on $(\mathcal{X}_1, \mathcal{B}_1)$ and f is a bounded almost G_1-invariant function, then there exists a measurable invariant function f_1 such that $f = f_1$ a.e. (μ).

Proof. This follows from Theorem 4, p. 227 of Lehmann (1959) and the proof is not repeated here. □

The next technical problem has to do with conditional expectations.

Proposition 7.18. In the notation introduced earlier, suppose $(\mathcal{X}, \mathcal{B})$ and $(\mathcal{Y}, \mathcal{C}_1)$ are measurable spaces acted on by groups G and $\overline{G}$ where $\overline{G}$ is a homomorphic image of G. Assume that τ is an equivariant function from $\mathcal{X}$ to $\mathcal{Y}$. Let P_0 be a probability measure on $(\mathcal{X}, \mathcal{B})$ and let Q_0 be the induced distribution of $\tau(X)$ when $\mathcal{L}(X) = P_0$. If f is a bounded G-invariant function on $\mathcal{X}$, let

$$H(y) \equiv \mathcal{E}_{P_0}(f(X)|\tau(X) = y),$$

and

$$H_1(y) = \mathcal{E}_{gP_0}(f(X)|\tau(X) = y).$$

Then $H_1(\bar{g}y) = H(y)$ a.e. (Q_0) for each fixed $\bar{g} \in G$.

Proof. The conditional expectations are well defined since f is bounded. $H(y)$ is the unique a.e. (Q_0) function that satisfies the equation

$$\int_{\mathcal{Y}} k(y)H(y)Q_0(dy) = \int_{\mathcal{X}} k(\tau(x))f(x)P_0(dx)$$

for all bounded measurable k. The probability measure gP_0 satisfies the equation

$$\int_{\mathcal{X}} f_1(x)(gP_0)(dx) = \int_{\mathcal{X}} f_1(gx)P_0(dx)$$

for all bounded f_1. Since τ is equivariant, this implies that if $\mathcal{L}(X) = gP_0$, then $\mathcal{L}(\tau(X)) = \bar{g}Q_0$. Using this, the invariance of f, and the characterizing property of conditional expectation, we have for all bounded k,

$$\begin{aligned}\int_{\mathcal{Y}} H(y)k(y)Q_0(dy) &= \int_{\mathcal{X}} f(x)k(\tau(x))P_0(dx)\\ &= \int f(gx)k(\bar{g}^{-1}\tau(gx))P_0(dx)\\ &= \int f(x)k(\bar{g}^{-1}\tau(x))(gP_0)(dx)\\ &= \int H_1(y)k(\bar{g}^{-1}y)(\bar{g}Q_0)(dy)\\ &= \int H_1(\bar{g}y)k(\bar{g}^{-1}\bar{g}y)Q_0(dy)\\ &= \int H_1(\bar{g}y)k(y)Q_0(dy).\end{aligned}$$

Since the first and the last terms in this equality are equal for all bounded k, we have that $H(y) = H_1(\bar{g}y)$ a.e. (Q_0). □

With the technical problems out of the way, the main result of this section can be proved.

Proposition 7.19. Consider measurable spaces $(\mathcal{X}, \mathcal{B})$ and $(\mathcal{Y}, \mathcal{C}_1)$, which are acted on measurably by groups G and $\bar{G}$ where $\bar{G}$ is a homomorphic image of G. It is assumed that the conditions of Proposition 7.17 hold for the group $\bar{G}$ and the space $(\mathcal{Y}, \mathcal{C}_1)$, and that $\bar{G}$ acts transitively on $\mathcal{Y}$. Let τ on $\mathcal{X}$ to $\mathcal{Y}$ be measurable and equivariant. Also let $(\mathcal{Z}, \mathcal{C}_2)$ be a measurable space and let h be a G-invariant measurable function from $\mathcal{X}$ to $\mathcal{Z}$. For a random variable $X \in \mathcal{X}$ with $\mathcal{L}(X) = P_0$, set $Y = \tau(X)$ and $Z = h(X)$ and assume that $\tau(X)$ is a sufficient statistic for the family $\{gP_0 | g \in G\}$ of distributions on $(\mathcal{X}, \mathcal{B})$. Under these assumptions, Y and Z are independent when $\mathcal{L}(X) = gP_0$, $g \in G$.

Proof. First we prove that Y and Z are independent when $\mathcal{L}(X) = P_0$. Fix a bounded measurable function f on Z and let

$$H_g(y) = \mathcal{E}_{gP_0}(f(h(X)) | \tau(X) = y).$$

Since $\tau(X)$ is a sufficient statistic, there is a measurable function H on $\mathcal{Y}$ such that

$$H_g(y) = H(y) \qquad \text{for } y \notin N_g$$

where N_g is a set of $\bar{g}Q_0$-measure zero. Thus $(\bar{g}Q_0)(N_g) = Q_0(\bar{g}^{-1}N_g) = 0$.

Let e denote the identity in G. We now claim that H is a Q_0 almost G-invariant function. By Proposition 7.18, $H_e(y) = H_g(\bar{g}y)$ a.e. (Q_0). However, $H(y) = H_e(y)$ a.e. Q_0 and $H_g(\bar{g}y) = H(\bar{g}y)$ for $\bar{g}y \notin N_g$, where $Q_0(\bar{g}^{-1}N_g) = 0$. Thus $H_g(\bar{g}y) = H(\bar{g}y)$ a.e. Q_0, and this implies that $H(y) = H(\bar{g}y)$ a.e. Q_0. Therefore, there exists a $\bar{G}$-invariant measurable function, say $\tilde{H}$, such that $H = \tilde{H}$ a.e. Q_0. Since $\bar{G}$ is transitive on $\mathcal{Y}$, $\tilde{H}$ must be a constant, so H is a constant a.e. Q_0. Therefore,

$$H_e(y) = \mathcal{E}_{P_0}(f(h(X))|\tau(X) = y)$$

is a constant a.e. Q_0 and, as noted earlier, this implies that $Z = h(X)$ and $Y = \tau(X)$ are independent when $\mathcal{L}(X) = P_0$. When $\mathcal{L}(X) = g_1 P_0$, let $\tilde{P}_0 = g_1 P_0$ and note that $\{gP_0|g \in G\} = \{g\tilde{P}_0|g \in G\}$ so $\tau(X)$ is sufficient for $\{g\tilde{P}_0|g \in G\}$. The argument given for P_0 now applies for $\tilde{P}_0$. Thus Z and Y are independent when $\mathcal{L}(X) = g_1 P_0$. □

A few comments concerning this result are in order. Since G acts transitively on $\{gP_0|g \in G\}$ and $Z = h(X)$ is G-invariant, the distribution of Z is the same under each gP_0, $g \in G$. In other words, Z is an ancillary statistic. Basu's Theorem, given in the Appendix, asserts that a sufficient statistic, whose induced family of distributions is complete, is independent of an ancillary statistic. Although no assumptions concerning invariance are made in the statement of Basu's Theorem, most applications are to problems where invariance is used to show a statistic is ancillary. In Proposition 7.19, the completeness assumption of Basu's Theorem has been replaced by the invariance assumptions and, most particularly, by the assumption that the group $\bar{G}$ acts transitively on the space $\mathcal{Y}$.

◆ **Example 7.18.** The normal distribution example at the beginning of this section provided a situation where the sample mean and sample variance are independent of a scale and translation invariant statistic. We now consider a generalization of that situation. Let $\mathcal{X} = R^n - (\text{span}\{e\})$ where e is the vector of ones in R^n and suppose that a random vector $X \in \mathcal{X}$ has a density $f(\|x\|^2)$ with respect to Lebesgue measure dx on $\mathcal{X}$. The group G in the example at the beginning of this section acts on $\mathcal{X}$ by

$$(a, b)x = ax + be, \qquad (a, b) \in G.$$

Consider the statistic $t(X) = (s, \bar{X})$ where

$$\bar{X} = \frac{1}{n}\sum_1^n X_i \qquad \text{and} \qquad s^2 = \sum_1^n (X_i - \bar{X})^2.$$

Then t takes values in G and satisfies

$$t((a,b)X) = (a,b)t(X)$$

for $(a, b) \in G$. It is shown that $t(X)$ and the G-invariant statistic

$$Z(X) \equiv (t(X))^{-1}X = \frac{X - \bar{X}e}{s}$$

are independent. The verification that $Z(X)$ is invariant goes as follows:

$$\begin{aligned} Z((a,b)X) &= (t((a,b)X))^{-1}(a,b)X \\ &= ((a,b)t(X))^{-1}(a,b)X = (t(X))^{-1}X = Z(X). \end{aligned}$$

To apply Proposition 7.19, let P_0 be the probability measure with density $f(\|x\|^2)$ on $\mathcal{X}$ and let $G = \bar{G} = \mathcal{Y}$. Thus $t(X)$ is equivariant and $Z(X)$ is invariant. The sufficiency of $t(X)$ for the parametric family $\{gP_0 | g \in G\}$ is established by using the factorization theorem. For $(a, b) \in G$, it is not difficult to show that $(a, b)P_0$ has a density $k(x|a, b)$ with respect to dx given by

$$k(x|a,b) = \frac{1}{a^n} f\left(\left\|\frac{x - be}{a}\right\|^2\right), \qquad x \in \mathcal{X}.$$

Since

$$\left\|\frac{x - be}{a}\right\|^2 = \frac{1}{a^2}\left(\sum_1^n x_i^2 - 2b\sum_1^n x_i + nb^2\right),$$

the density $k(x|a, b)$ is a function of Σx_i^2 and Σx_i so the pair $(\Sigma X_i^2, \Sigma X_i)$ is a sufficient statistic for the family $\{gP_0 | g \in G\}$. However, $t(X) = (s, \bar{X})$ is a one-to-one function of $(\Sigma X_i^2, \Sigma X_i)$ so $t(X)$ is a sufficient statistic. The remaining assumptions of Proposition 7.19 are easily verified. Therefore, $t(X)$ and $Z(X)$ are independent under each of the measures $(a, b)P_0$ for (a, b) in G. ◆

Before proceeding with the next example, an extension of Proposition 7.1 is needed.

Proposition 7.20. Consider the space $\mathcal{L}_{p,n}$, $n \geqslant p$, and let Q be an $n \times n$ rank k orthogonal projection. If $k \geqslant p$, then the set

$$B = \{X | X \in \mathcal{L}_{p,n}, \operatorname{rank}(QX) < p\}$$

has Lebesgue measure zero.

Proof. Let $X \in \mathcal{L}_{p,n}$ be a random vector with $\mathcal{L}(X) = N(0, I_n \otimes I_p) = P_0$. It suffices to show that $P_0(B) = 0$ since P_0 and Lebesgue measure are absolutely continuous with respect to each other. Also, write Q as

$$Q = \Gamma' D \Gamma, \qquad \Gamma \in \mathcal{O}_n$$

where

$$D = \begin{pmatrix} I_k & 0 \\ 0 & 0 \end{pmatrix}.$$

Since

$$\operatorname{rank}(\Gamma' D \Gamma X) = \operatorname{rank}(D \Gamma X)$$

and $\mathcal{L}(\Gamma X) = \mathcal{L}(X)$, it suffices to show that

$$P_0(\operatorname{rank}(DX) < p) = 0.$$

Now, partition X as

$$X = \begin{pmatrix} X_1 \\ X_2 \end{pmatrix}, \qquad X_1 : k \times p$$

so

$$DX = \begin{pmatrix} X_1 \\ 0 \end{pmatrix} \in \mathcal{L}_{p,n}.$$

Thus $\operatorname{rank}(DX) = \operatorname{rank}(X_1)$. Since $k \geqslant p$ and $\mathcal{L}(X_1) = N(0, I_k \otimes I_p)$, Proposition 7.1 implies that X_1 has rank p with probability one. Thus $P_0(B) = 0$ so B has Lebesgue measure zero. □

◆ **Example 7.19.** This is generalization of Example 7.18 and deals with the general multivariate linear model discussed in Chapter 4.

As in Example 4.4, let M be a linear subspace of $\mathcal{L}_{p,n}$ defined by

$$M = \{x | x \in \mathcal{L}_{p,n}, x = ZB, B \in \mathcal{L}_{p,k}\}$$

where Z is a fixed $n \times k$ matrix of rank k. For reasons that are apparent in a moment, it is assumed that $n - k \geqslant p$. The orthogonal projection onto M relative to the natural inner product $\langle \cdot , \cdot \rangle$ on $\mathcal{L}_{p,n}$ is $P_M = P_z \otimes I_p$ where

$$P_z = Z(Z'Z)^{-1}Z'$$

is a rank k orthogonal projection on R^n. Also, $Q_M \equiv Q_z \otimes I_p$, where $Q_z = I_n - P_z$ is the orthogonal projection onto $M^{\perp}$ and Q_z is a rank $n - k$ orthogonal projection on R^n. For this example, the sample space $\mathcal{X}$ is

$$\mathcal{X} = \{x | x \in \mathcal{L}_{p,n}, \operatorname{rank}(Q_z x) = p\}.$$

Since $n - k \geqslant p$, Proposition 7.20 implies that the complement of $\mathcal{X}$ has Lebesgue measure zero in $\mathcal{L}_{p,n}$. In this example, the group G has elements that are pairs (T, u) with $T \in G_T^+$ where T is $p \times p$ and $u \in M$. The group operation is

$$(T_1, u_1)(T_2, u_2) = (T_1T_2, u_1 + u_2T_1')$$

and the action of G on $\mathcal{X}$ is

$$(T, u)x = xT' + u.$$

For this example, $\mathcal{Y} = \overline{G} = G$ and t on $\mathcal{X}$ to G is defined by

$$t(x) = (T(x), P_M x) \in G$$

where $T(x)$ is the unique element in G_T^+ such that $x'Q_z x = T(x)T'(x)$. The assumption that $n - k \geqslant p$ insures that $x'Q_z x$ has rank p. It is now routine to verify that

$$t((T, u)x) = (T, u)t(x)$$

for $x \in \mathcal{X}$ and $(T, u) \in G$. Using this relationship, the function

$$h(x) \equiv (t(x))^{-1}x$$

is easily shown to be a maximal invariant under the action of G on $\mathfrak{X}$. Now consider a random vector $X \in \mathfrak{X}$ with $\mathcal{L}(X) = P_0$ where P_0 has a density $f(\langle x, x\rangle)$ with respect to Lebesgue measure on $\mathfrak{X}$. We apply Proposition 7.19 to show that $t(X)$ and $h(X)$ are independent under gP_0 for each $g \in G$. Since $\mathcal{Y} = \overline{G} = G$, t is an equivariant function and $\overline{G}$ acts transitively on $\mathcal{Y}$. The measurability assumptions of Proposition 7.19 are easily checked. It remains to show that $t(X)$ is a sufficient statistic for the family $\{gP_0 | g \in G\}$. For $g = (T, \mu) \in G$, gP_0 has a density given by

$$p(x|(T, \mu)) = |TT'|^{-n/2} f\left(\langle (x - \mu)(TT')^{-1}, x - \mu\rangle\right).$$

Letting $\Sigma = TT'$ and arguing as in Example 4.4, it follows that

$$\begin{aligned}\langle (x - \mu)\Sigma^{-1}, x - \mu\rangle &= \langle (P_M x - \mu)\Sigma^{-1}, P_M x - \mu\rangle \\ &\quad + \operatorname{tr}\left(\Sigma^{-1} x' Q_z x\right)\end{aligned}$$

since $\mu \in M$. Therefore, the density $p(x|(T, \mu))$ is a function of the pair $(x'Q_z x, P_M x)$ so this pair is a sufficient statistic for the family $\{gP_0 | g \in G\}$. However, $T(x)$ is a one-to-one function of $x'Q_z x$ so

$$t(x) = (T(x), P_M x)$$

is also a sufficient statistic. Thus Proposition 7.19 implies that $t(X)$ and $h(X)$ are stochastically independent under each probability measure gP_0 for $g \in G$. Of course, the choice of f that motivated this example is

$$f(w) = (\sqrt{2\pi})^{-np} \exp\left[-\tfrac{1}{2} w\right]$$

so that P_0 is the probability measure of a $N(0, I_n \otimes I_p)$ distribution on $\mathfrak{X}$.

One consequence of Proposition 7.19 is that the statistic $h(X)$ is ancillary. But for the case at hand, we now describe the distribution of $h(X)$ and show that its distribution does not even depend on the particular density f used to define P_0. Recall that

$$\begin{aligned}h(x) &= (t(x))^{-1} x = (x - P_M x)(T'(x))^{-1} = (Q_M x)(T'(x))^{-1} \\ &= (Q_z x)(T'(x))^{-1}\end{aligned}$$

where $T(x)T'(x) = x'Q_z x$ and $T(x) \in G_T^+$. Fix $x \in \mathfrak{X}$ and set

$\Psi = (Q_z x)(T'(x))^{-1}$. First note that

$$\Psi'\Psi = (T(x))^{-1}x'Q_z x(T'(x))^{-1} = I_p$$

so Ψ is a linear isometry. Let N be the orthogonal complement in R^n of the linear subspace spanned by the columns of the matrix Z. Clearly, $\dim(N) = n - k$ and the range of Ψ is contained in N since Q_z is the orthogonal projection onto N. Therefore, Ψ is an element of the space

$$\mathfrak{F}_p(N) = \{\Psi | \Psi'\Psi = I_p, \text{range}(\Psi) \subseteq N\}.$$

Further, the group

$$H = \{\Gamma | \Gamma \in \mathcal{O}_n, \Gamma(N) = N\}$$

is compact and acts transitively on $\mathfrak{F}_p(N)$ under the group action

$$\Psi \to \Gamma\Psi, \qquad \Psi \in \mathfrak{F}_p(N), \qquad \Gamma \in H.$$

Now, we return to the original problem of describing the distribution of $W = h(X)$ when $\mathcal{L}(X) = P_0$. The above argument shows that $W \in \mathfrak{F}_p(N)$. Since the compact group H acts transitively on $\mathfrak{F}_p(N)$, there is a unique invariant probability measure ν on $\mathfrak{F}_p(N)$. It will be shown that $\mathcal{L}(W) = \nu$ by proving $\mathcal{L}(\Gamma W) = \mathcal{L}(W)$ for all $\Gamma \in H$. It is not difficult to verify that $\Gamma Q_z = Q_z \Gamma$ for $\Gamma \in H$. Since $\mathcal{L}(\Gamma X) = \mathcal{L}(X)$ and $T(\Gamma X) = T(X)$, we have

$$\begin{aligned}
\mathcal{L}(\Gamma W) &= \mathcal{L}(\Gamma h(X)) = \mathcal{L}\left(\Gamma Q_z X(T'(X))^{-1}\right) \\
&= \mathcal{L}\left(Q_z \Gamma X(T'(\Gamma X))^{-1}\right) = \mathcal{L}\left(Q_z X(T'(X))^{-1}\right) \\
&= \mathcal{L}(h(X)) = \mathcal{L}(W).
\end{aligned}$$

Therefore, the distribution of W is H-invariant so $\mathcal{L}(W) = \nu$. ◆

Further applications of Proposition 7.19 occur in the next three chapters. In particular, this result is used to derive the distribution of the determinant of a certain matrix that arises in testing problems for the general linear model.

PROBLEMS

1. Suppose the random $n \times p$ matrix $X \in \mathcal{X}$ ($\mathcal{X}$ as in Section 7.1) has a density given by $f(x) = k|x'x|^{\gamma}\exp[-\frac{1}{2}\operatorname{tr} x'x]$ with respect to dx. The constant k depends on n, p, and γ (see Problem 6.10). Derive the density of $S = X'X$ and the density of U in the representation $X = \Psi U$ with $U \in G_T^+$ and $\Psi \in \mathfrak{F}_{p,n}$.

2. Suppose $X \in \mathcal{X}$ has an $\mathcal{O}_n$-left invariant distribution. Let $P(X) = X(X'X)^{-1}X'$ and $S(X) = X'X$. Prove that $P(X)$ and $S(X)$ are independent.

3. Let Q be an $n \times n$ non-negative definite matrix of rank r and set $A = \{x | x \in \mathcal{L}_{p,n}, x'Qx \text{ has rank } p\}$. Show that, if $r \geqslant p$, then A^c has Lebesgue measure zero.

4. With $\mathcal{X}$ as in Section 7.1, $\mathcal{O}_n \times Gl_p$ acts on $\mathcal{X}$ by $x \to \Gamma x A'$ for $\Gamma \in \mathcal{O}_n$ and $A \in Gl_p$. Also, $\mathcal{O}_n \times Gl_p$ acts on $\mathcal{S}_p^+$ by $S \to ASA'$. Show that $\phi(x) = kx'x$ is equivariant for each constant $k > 0$. Are these the only equivariant functions?

5. The permutation group $\mathcal{P}_n$ acts on R^n via matrix multiplication $x \to gx$, $g \in \mathcal{P}_n$. Let $\mathcal{Y} = \{y | y \in R^n, y_1 \leqslant y_2 \leqslant \cdots \leqslant y_n\}$. Define $f: R^n \to \mathcal{Y}$ by $f(x)$ is the vector of ordered values of the set $\{x_1, \ldots, x_n\}$ with multiple values listed.
   (i) Show f is a maximal invariant.
   (ii) Set $I_0(u) = 1$ if $u \geqslant 0$ and 0 if $u < 0$. Define $F_n(t) = n^{-1}\Sigma_1^n I_0(t - x_i)$ for $t \in R^1$. Show F_n is also a maximal invariant.

6. Let M be a proper subspace of the inner product space $(V, (\cdot, \cdot))$. Let A_0 be defined by $A_0 x = -x$ for $x \in M$ and $A_0 x = x$ for $x \in M^{\perp}$.
   (i) Verify that the set of pairs (B, y), with $y \in M$ and B either A_0 or A_0^2, forms a subgroup of the affine group $Al(V)$. Let G be this group.
   (ii) Show that G acts on M and on V.
   (iii) Suppose $t: V \to M$ is equivariant ($t(Bx + y) = Bt(x) + y$ for $(B, y) \in G$ and $x \in V$). Prove that $t(x) = P_M x$.

7. Let M be a subspace of R^n ($M \neq R^n$) so the complement of $\mathcal{X} = R^n \cap M^c$ has Lebesgue measure zero. Suppose $X \in \mathcal{X}$ has a density given by

$$p(x|\mu, \sigma) = \frac{1}{\sigma^n} f_0\left(\frac{\|x - \mu\|^2}{\sigma^2}\right)$$

where $\mu \in M$ and $\sigma > 0$. Assume that $\int \|x\|^2 f_0(\|x\|^2)\, dx < +\infty$. For $a > 0$, $\Gamma \in \mathcal{O}_n(M)$, and $b \in M$, the affine transformation $(a, \Gamma, b)x = a\Gamma x + b$ acts on $\mathcal{X}$.

(i) Show that the probability model for X ($\mu \in M$, $\sigma > 0$) is invariant under the above affine transformations. What is the induced group action on (μ, σ^2)?

(ii) Show that the only equivariant estimator of μ is $P_M x$. Show that any equivariant estimator of σ^2 has the form $k\|Qx\|^2$ for some $k > 0$.

8. With $\mathcal{X}$ as in Section 7.1, suppose f is a function defined on Gl_p to $[0, \infty)$, which satisfies $f(AB) = f(BA)$ and

$$\int_{\mathcal{X}} f(x'x) \frac{dx}{|x'x|^{n/2}} = 1.$$

(i) Show that $f(x'x\Sigma^{-1})$, $\Sigma \in \mathcal{S}_p^+$, is a density on $\mathcal{X}$ with respect to $dx/|x'x|^{n/2}$ and under this density, the covariance (assuming it exists) is $cI_n \otimes \Sigma$ where $c > 0$.

(ii) Show that the family of distributions of (i) indexed by $\Sigma \in \mathcal{S}_p^+$ is invariant under the group $\mathcal{O}_n \times Gl_p$ acting on $\mathcal{X}$ by $(\Gamma, A)x = \Gamma x A'$. Also, show that $\overline{(\Gamma, A)}\Sigma = A\Sigma A'$.

(iii) Show that the equivariant estimators of Σ all have the form $kX'X$, $k > 0$.

Now, assume that

$$\sup_{C \in \mathcal{S}_p^+} f(C) = f(C_0)$$

where $C_0 \in \mathcal{S}_p^+$ is unique.

(iv) Show $C_0 = \alpha I_p$ for some $\alpha > 0$.

(v) Find the maximum likelihood estimator of Σ (expressed in terms of X and α in (iv)).

9. In an inner product space $(V, (\cdot, \cdot))$, suppose X has a distribution P_0.

(i) Show that $\mathcal{L}(\|X\|) = \mathcal{L}(\|Y\|)$ whenever $\mathcal{L}(Y) = gP_0$, $g \in \mathcal{O}(V)$.

(ii) In the special case that $\mathcal{L}(X) = \mathcal{L}(\mu + Z)$ where μ is a fixed vector and Z has an $\mathcal{O}(V)$-invariant distribution, how does the distribution of $\|X\|$ depend on μ?

10. Under the assumptions of Problem 4.5, use an invariance argument to show that the distribution of F depends on (μ, σ^2) only through the parameter $(\|\mu\|^2 - \|P_\omega \mu\|^2)/\sigma^2$. What happens when $\mu \in \omega$?

11. Suppose $X_1, \ldots, X_n$ is a random sample from a distribution on R^p $(n > p)$ with density $p(x|\mu, \Sigma) = |\Sigma|^{-1/2} f((x-\mu)'\Sigma^{-1}(x-\mu))$ where $\mu \in R^p$ and $\Sigma \in \mathcal{S}_p^+$. The parameter $\theta = \det(\Sigma)$ is sometimes called the *population generalized variance*. The *sample generalized variance* is $V = \det((1/n)S)$ where $S = \Sigma_1^n (x_i - \bar{x})(x_i - \bar{x})'$. Show that the distribution of V depends on (μ, Σ) only through θ.

12. Assume the conditions under which Proposition 7.16 was proved. Given a probability Q on $\mathcal{Y}$, let $\bar{Q}$ denote the extension of Q to $\mathcal{X}$—that is, $\bar{Q}(B) = Q(B \cap \mathcal{Y})$ for $B \in \mathcal{B}$. For $g \in G$, $g\bar{Q}$ is defined in the usual way—$(g\bar{Q})(B) = \bar{Q}(g^{-1}B)$.

(i) Assume that P is a probability measure on $\mathcal{X}$ and

$$P = \int_G g\bar{Q}\mu(dg), \tag{7.1}$$

that is,

$$P(B) = \int_G (g\bar{Q})(B)\mu(dg); \qquad B \in \mathcal{B}.$$

Show that P is G-invariant.

(ii) If P is G-invariant, show that (7.1) holds for some Q.

13. Under the assumptions used to prove Proposition 7.16, let $\mathcal{P}$ be all the G-invariant distributions. Prove that $\tau(X)$ is a sufficient statistic for the family $\mathcal{P}$.

14. Suppose $X \in R^n$ has coordinates $X_1, \ldots, X_n$ that are i.i.d. $N(\mu, 1)$, $\mu \in R^1$. Thus the parameter space for the distributions of X is the additive group $G = R^1$. The function $t: R^n \to G$ given by $t(x) = \bar{x}$ gives a complete sufficient statistic for the model for X. Also, G acts on R^n by $gx = x + ge$ where $e \in R^n$ is the vector of ones.

(i) Show that $t(gx) = gt(x)$ and that $Z(X) = (t(X))^{-1}X$ is an ancillary statistic. Here, $(t(X))^{-1}$ means the group inverse of $t(X) \in G$ so $(t(X))^{-1}X = X - \bar{X}e$. What is the distribution of $Z(X)$?

(ii) Suppose we want to find a minimum variance unbiased estimator (MVUE) of $h(\mu) = \mathcal{E}_\mu f(X_1)$ where f is a given function. The Rao–Blackwell Theorem asserts that the MVUE is $\mathcal{E}(f(X_1)|t(X) = t)$. Show that this conditional expectation is

$$\int_{-\infty}^{\infty} f(z + t)\frac{1}{\sqrt{2\pi\delta}}\exp\left[-\frac{z^2}{2\delta^2}\right]dz$$

where $\delta^2 = \text{var}(X_1 - \bar{X}) = (n-1)/n$. Evaluate this for $f(x) = 1$ if $x \leqslant u_0$ and $f(x) = 0$ if $x > u_0$.

(iii) What is the MVUE of the parametric function $(\sqrt{2\pi})^{-1}\exp[-\frac{1}{2}(x_0 - \mu)^2]$ where x_0 is a fixed number?

15. Using the notation, results, and assumptions of Example 7.18, find an unbiased estimator based on $t(X)$ of the parametric function $h(a, b) = ((a, b)P_0)(X_1 \leqslant u_0)$ where u_0 is a fixed number and X_1 is the first coordinate of X. Express the answer in terms of the distribution of Z_1 —the first coordinate of Z. What is this distribution? In the case that P_0 is the $N(0, I_n)$ distribution, show this gives a MVUE for $h(a, b)$.

16. This problem contains an abstraction of the technique developed in Problems 14 and 15. Under the conditions used to prove Proposition 7.19, assume the space $(\mathcal{Y}, \mathcal{C}_1)$ is $(G, \mathcal{B}_G)$ and $\bar{G} = G$. The equivariance assumption on τ then becomes $\tau(gx) = g \circ \tau(x)$ since $\tau(x) \in G$. Of course, $\tau(X)$ is assumed to be a sufficient statistic for $\{gP_0 | g \in G\}$.

(i) Let $Z(X) = (\tau(X))^{-1}X$ where $(\tau(X))^{-1}$ is the group inverse of $\tau(X)$. Show that $Z(X)$ is a maximal invariant and $Z(X)$ is ancillary. Hence Proposition 7.19 applies.

(ii) Let Q_0 denote the distribution of Z when $\mathcal{L}(X)$ is one of the distributions gP_0, $g \in G$. Show that a version of the conditional expectation $\mathcal{E}(f(X)|\tau(X) = g)$ is $\mathcal{E}_{Q_0}f(gZ)$ for any bounded measurable f.

(iii) Apply the above to the case when P_0 is $N(0, I_n \otimes I_p)$ on $\mathcal{X}$ (as in Section 7.1) and take $G = G_T^+$. The group action is $x \to xT'$ for $x \in \mathcal{X}$ and $T \in G_T^+$. The map τ is $\tau(X) = T$ in the representation $X = \Psi T'$ with $\Psi \in \mathcal{F}_{p,n}$ and $T \in G_T^+$. What is Q_0?

(iv) When $X \in \mathcal{X}$ is $N(0, I_n \otimes \Sigma)$ with $\Sigma \in \mathcal{S}_p^+$, use (iii) to find a MVUE of the parametric function

$$(\sqrt{2\pi})^{-p}|\Sigma|^{-1/2}\exp\left[-\tfrac{1}{2}u_0'\Sigma^{-1}u_0\right]$$

where u_0 is a fixed vector in R^p.

NOTES AND REFERENCES

1. For some material related to Proposition 7.3, see Dawid (1978). The extension of Proposition 7.3 to arbitrary compact groups (Proposition 7.16) is due to Farrell (1962). A related paper is Das Gupta (1979).

2. If G acts on $\mathcal{X}$ and t is a function from $\mathcal{X}$ onto $\mathcal{Y}$, it is natural to ask if we can define a group action on $\mathcal{Y}$ (using t and G) so that t becomes equivariant. The obvious thing to do is to pick $y \in \mathcal{Y}$, write $y = t(x)$, and then define gy to be $t(gx)$. In order that this definition make sense it is necessary (and sufficient) that whenever $t(x) = t(\tilde{x})$, then $t(gx) = t(g\tilde{x})$ for all $g \in G$. When this condition holds, it is easy to show that G then acts on $\mathcal{Y}$ via the above definition and t is equivariant. For some further discussion, see Hall, Wijsman, and Ghosh (1965).

3. Some of the early work on invariance by Stein, and Hunt and Stein, first appeared in print in the work of other authors. For example, the famous Hunt–Stein Theorem given in Lehmann (1959) was established in 1946 but was never published. This early work laid the foundation for much of the material in this chapter. Other early invariance works include Hotelling (1931), Pitman (1939), and Peisakoff (1950). The paper by Kiefer (1957) contains a generalization of the Hunt–Stein Theorem. For some additional discussion on the development of invariance arguments, see Hall, Wijsman, and Ghosh (1965).

4. Proposition 7.15 is probably due to Stein, but I do not know a reference.

5. Make the assumptions on $\mathcal{X}$, $\mathcal{Y}$, and G that lead to Proposition 7.16, and note that $\mathcal{Y}$ is just a particular representation of the quotient space $\mathcal{X}/G$. If ν is any σ-finite G-invariant measure on $\mathcal{X}$, let δ be the measure on $\mathcal{Y}$ defined by

$$\delta(C) = \nu(\tau^{-1}(C)), \qquad C \subseteq \mathcal{Y}.$$

Then (see Lehmann, 1959, p. 39),

$$\int_{\mathcal{X}} h(\tau(x))\nu(dx) = \int_{\mathcal{Y}} h(y)\delta(dy)$$

for all measurable functions h. The proof of Proposition 7.16 shows that for any ν-integrable function f, the equation

$$(7.2) \qquad \int_{\mathcal{X}} f(x)\nu(dx) = \int_{\mathcal{Y}}\int_G f(gy)\mu(dg)\delta(dy)$$

holds. In an attempt to make sense of (7.2) when G is not compact, let

μ_r denote a right invariant measure on G. For $f \in \mathcal{K}(\mathcal{X})$, set

$$\hat{f}(x) = \int_G f(gx)\mu_r(dg).$$

Assuming this integral is well defined (it may not be in certain examples —e.g., $\mathcal{X} = R^n - \{0\}$ and $G = Gl_n$), it follows that $\hat{f}(hx) = \hat{f}(x)$ for $h \in G$. Thus $\hat{f}$ is invariant and can be regarded as a function on $\mathcal{Y} = \mathcal{X}/G$. For any measure δ on $\mathcal{Y}$, write $\int \hat{f}\, d\delta$ to mean the integral of $\hat{f}$, expressed as a function of y, with respect to the measure δ. In this case, the right-hand side of (7.2) becomes

$$J(f) = \int\left(\int_G f(gx)\mu_r(dg)\right) d\delta = \int \hat{f}\, d\delta.$$

However, for $h \in G$, it is easy to show

$$J(hf) = \Delta_r^{-1}(h)J(f)$$

so J is a relatively invariant integral. As usual, Δ_r is the right-hand modulus of G. Thus the left-hand side of (7.2) must also be relatively invariant with multiplier Δ_r^{-1}. The argument thus far shows that when μ in (7.2) is replaced by μ_r (this choice looks correct so that the inside integral defines an invariant function), the resulting integral J is relatively invariant with multiplier Δ_r^{-1}. Hence the only possible measures ν for which (7.2) can hold must be relatively invariant with multiplier Δ_r^{-1}. However, given such a ν, further assumptions are needed in order that (7.2) hold for some δ (when G is not compact and μ is replaced by μ_r). Some examples where (7.2) is valid for noncompact groups are given in Stein (1956), but the first systematic account of such a result is Wijsman (1966), who uses some Lie group theory. A different approach due to Schwarz is reported in Farrell (1976). The description here follows Andersson (1982) most closely.

6. Proposition 7.19 is a special case of a result in Hall, Wijsman, and Ghosh (1965). Some version of this result was known to Stein but never published by him. The development here is a modification of that which I learned from Bondesson (1977).

CHAPTER 8

The Wishart Distribution

The Wishart distribution arises in a natural way as a matrix generalization of the chi-square distribution. If $X_1, \ldots, X_n$ are independent with $\mathcal{L}(X_i) = N(0, 1)$, then $\Sigma_1^n X_i^2$ has a chi-square distribution with n degrees of freedom. When the X_i are random vectors rather than real-valued random variables say $X_i \in R^p$ with $\mathcal{L}(X_i) = N(0, I_p)$, one possible way to generalize the above sum of squares is to form the $p \times p$ positive semidefinite matrix $S = \Sigma_1^n X_i X_i'$. Essentially, this representation of S is used to define a Wishart distribution. As with the definition of the multivariate normal distribution, our definition of the Wishart distribution is not in terms of a density function and allows for Wishart distributions that are singular. In fact, most of the properties of the Wishart distribution are derived without reference to densities by exploiting the representation of the Wishart in terms of normal random vectors. For example, the distribution of a partitioned Wishart matrix is obtained by using properties of conditioned normal random vectors.

After formally defining the Wishart distribution, the characteristic function and convolution properties of the Wishart are derived. Certain generalized quadratic forms in normal random vectors are shown to have Wishart distributions and the basic decomposition of the Wishart into submatrices is given. The remainder of the chapter is concerned with the noncentral Wishart distribution in the rank one case and certain distributions that arise in connection with likelihood ratio tests.

8.1. BASIC PROPERTIES

The Wishart distribution, or more precisely, the family of Wishart distributions, is indexed by a $p \times p$ positive semidefinite symmetric matrix Σ, by a dimension parameter p, and by a degrees of freedom parameter n. Formally, we have the following definition.

Definition 8.1. A random $p \times p$ symmetric matrix S has a Wishart distribution with parameters Σ, p, and n if there exist independent random vectors $X_1, \dots, X_n$ in R^p such that $\mathcal{L}(X_i) = N(0, \Sigma)$, $i = 1, \dots, n$ and

$$\mathcal{L}(S) = \mathcal{L}\left(\sum_1^n X_i X_i'\right).$$

In this case, we write $\mathcal{L}(S) = W(\Sigma, p, n)$.

In the above definition, p and n are positive integers and Σ is a $p \times p$ positive semidefinite matrix. When $p = 1$, it is clear that the Wishart distribution is just a chi-square distribution with n degrees of freedom and scale parameter $\Sigma \geqslant 0$. When $\Sigma = 0$, then $X_i = 0$ with probability one, so $S = 0$ with probability one. Since $\Sigma_1^n X_i X_i'$ is positive semidefinite, the Wishart distribution has all of its mass on the set of positive semidefinite matrices. In an abuse of notation, we often write

$$S = \sum_1^n X_i X_i'$$

when $\mathcal{L}(S) = W(\Sigma, p, n)$. As distributional questions are the primary concern in this chapter, this abuse causes no technical problems. If $X \in \mathcal{L}_{p,n}$ has rows $X_1', \dots, X_n'$, it is clear that $\mathcal{L}(X) = N(0, I_n \otimes \Sigma)$ and $X'X = \Sigma_1^n X_i X_i'$. Thus if $\mathcal{L}(S) = W(\Sigma, p, n)$, then $\mathcal{L}(S) = \mathcal{L}(X'X)$ where $\mathcal{L}(X) = N(0, I_n \otimes \Sigma)$ in $\mathcal{L}_{p,n}$. Also, the converse statement is clear. Some further properties of the Wishart distribution follow.

Proposition 8.1. If $\mathcal{L}(S) = W(\Sigma, p, n)$ and A is an $r \times p$ matrix, then $\mathcal{L}(ASA') = W(A\Sigma A', r, n)$.

Proof. Since $\mathcal{L}(S) = W(\Sigma, p, n)$,

$$\mathcal{L}(S) = \mathcal{L}(X'X)$$

where $\mathcal{L}(X) = N(0, I_n \otimes \Sigma)$ in $\mathcal{L}_{p,n}$. Thus $\mathcal{L}(ASA') = \mathcal{L}(AX'XA') = \mathcal{L}[((I_n \otimes A)X)'(I_n \otimes A)X]$. But $Y = (I_n \otimes A)X$ satisfies $\mathcal{L}(Y) = N(0, I_n \otimes (A\Sigma A'))$ in $\mathcal{L}_{r,n}$ and $\mathcal{L}(Y'Y) = \mathcal{L}(ASA')$. The conclusion follows from the definition of the Wishart distribution. □

One consequence of Proposition 8.1 is that, for fixed p and n, the family of distributions $\{W(\Sigma, p, n)|\Sigma \geqslant 0\}$ can be generated from the $W(I_p, p, n)$ distribution and $p \times p$ matrices. Here, the notation $\Sigma \geqslant 0$ ($\Sigma > 0$) means that Σ is positive semidefinite (positive definite). To see this, if $\mathcal{L}(S) = W(I_p, p, n)$ and $\Sigma = AA'$, then

$$\mathcal{L}(ASA') = W(AA', p, n) = W(\Sigma, p, n).$$

In particular, the family $\{W(\Sigma, p, n)|\Sigma > 0\}$ is generated by the $W(I_p, p, n)$ distribution and the group Gl_p acting on $\mathcal{S}_p$ by $A(S) \equiv ASA'$. Many proofs are simplified by using the above representation of the Wishart distribution. The question of the nonsingularity of the Wishart distribution is a good example. If $\mathcal{L}(S) = W(\Sigma, p, n)$, then S has a *nonsingular Wishart distribution* if S is positive definite with probability one.

Proposition 8.2. Suppose $\mathcal{L}(S) = W(\Sigma, p, n)$. Then S has a nonsingular Wishart distribution iff $n \geqslant p$ and $\Sigma > 0$. If S has a nonsingular Wishart distribution, then S has a density with respect to the measure $\nu(dS) = dS/|S|^{(p+1)/2}$ given by

$$p(S|\Sigma) = \omega(n, p)|\Sigma^{-1}S|^{n/2}\exp\left[-\tfrac{1}{2}\operatorname{tr}\Sigma^{-1}S\right].$$

Here, $\omega(p, n)$ is the Wishart constant defined in Example 5.1.

Proof. Represent the $W(\Sigma, p, n)$ distribution as $\mathcal{L}(AS_1A')$ where $\mathcal{L}(S_1) = W(I_p, p, n)$ and $AA' = \Sigma$. Obviously, the rank of A is the rank of Σ and $\Sigma > 0$ iff rank of Σ is p. If $n < p$, then by Proposition 7.1, if $\mathcal{L}(X_i) = N(0, I_p)$, $i = 1, \ldots, n$, the rank of $\Sigma_1^n X_i X_i'$ is n with probability one. Thus $S_1 = \Sigma_1^n X_i X_i'$ has rank n, which is less than p, and $S = AS_1A'$ has rank less than p with probability one. Also, if the rank of Σ is $r < p$, then A has rank r so AS_1A' has rank at most r no matter what n happens to be. Therefore, if $n < p$ or if Σ is singular, then S is singular with probability one. Now, consider the case when $n \geqslant p$ and Σ is positive definite. Then $S_1 = \Sigma_1^n X_i X_i'$ has rank p with probability one by Proposition 7.1, and A has rank p. Therefore, $S = AS_1A'$ has rank p with probability one.

When $\Sigma > 0$, the density of $X \in \mathcal{L}_{p,n}$ is

$$f(X) = (\sqrt{2\pi})^{-np}|\Sigma|^{-n/2}\exp\left[-\tfrac{1}{2}\operatorname{tr}\Sigma^{-1}X'X\right]$$

when $\mathcal{L}(X) = N(0, I_n \otimes \Sigma)$. When $n \geqslant p$, it follows from Proposition 7.6 that the density of S with respect to $\nu(dS)$ is $p(S|\Sigma)$. □

Recall that the natural inner product on $\mathcal{S}_p$, when $\mathcal{S}_p$ is regarded as a subspace of $\mathcal{L}_{p,p}$, is

$$\langle S_1, S_2 \rangle = \operatorname{tr} S_1 S_2, \qquad S_i \in \mathcal{S}_p, \quad i = 1, 2.$$

The mean vector, covariance, and characteristic function of a Wishart distribution on the inner product space $(\mathcal{S}_p, \langle \cdot , \cdot \rangle)$ are given next.

Proposition 8.3. Suppose $\mathcal{L}(S) = W(\Sigma, p, n)$ on $(\mathcal{S}_p, \langle \cdot , \cdot \rangle)$. Then

(i) $\mathcal{E}S = n\Sigma$.
(ii) $\operatorname{Cov}(S) = 2n\Sigma \otimes \Sigma$.
(iii) $\phi(A) \equiv \mathcal{E}\exp[i\langle A, S\rangle] = |I_p - 2i\Sigma A|^{-n/2}$.

Proof. To prove (i) write $S = \Sigma_1^n X_i X_i'$ where $\mathcal{L}(X_i) = N(0, \Sigma)$, and $X_1, \ldots, X_n$ are independent. Since $\mathcal{E}X_i X_i' = \Sigma$, it is clear that $\mathcal{E}S = n\Sigma$. For (ii), the independence of $X_1, \ldots, X_n$ implies that

$$\operatorname{Cov}(S) = \operatorname{Cov}\left(\sum_1^n X_i X_i'\right) = \sum_1^n \operatorname{Cov}(X_i X_i') = n \operatorname{Cov}(X_1 X_1')$$

$$= n \operatorname{Cov}(X_1 \Box X_1)$$

where $X_1 \Box X_1$ is the outer product of X_1 relative to the standard inner product on R^p. Since $\mathcal{L}(X_1) = \mathcal{L}(CZ)$ where $\mathcal{L}(Z) = N(0, I_p)$ and $CC' = \Sigma$, it follows from Proposition 2.24 that $\operatorname{Cov}(X_1 \Box X_1) = 2\Sigma \otimes \Sigma$. Thus (ii) holds. To establish (iii), first write $C'AC = \Gamma D \Gamma'$ where $A \in \mathcal{S}_p$, $CC' = \Sigma$, $\Gamma \in \mathcal{O}_n$, and D is a diagonal matrix with diagonal entries $\lambda_1, \ldots, \lambda_p$. Then

$$\phi(A) = \mathcal{E}\exp[i \operatorname{tr}(AS)] = \mathcal{E}\exp\left[i \operatorname{tr} A\left(\sum_1^n X_j X_j'\right)\right]$$

$$= \mathcal{E}\prod_{j=1}^n \exp\left[i \operatorname{tr} A X_j X_j'\right] = \prod_{j=1}^n \mathcal{E}\exp\left[i \operatorname{tr} A X_j X_j'\right]$$

$$[\mathcal{E}\exp[i \operatorname{tr} A X_1 X_1']]^n = [\mathcal{E}\exp[i X_1' A X_1]]^n \equiv (\xi(A))^n.$$

Again, $\mathcal{L}(X_1) = \mathcal{L}(CZ)$ where $\mathcal{L}(Z) = N(0, I_p)$. Also, $\mathcal{L}(\Gamma Z) = \mathcal{L}(Z)$ for

$\Gamma \in \mathcal{O}_p$. Therefore,

$$\xi(A) = \mathcal{E} \exp[iX_1'AX_1] = \mathcal{E} \exp[iZ'C'ACZ]$$

$$= \mathcal{E} \exp[iZ'DZ] = \mathcal{E} \exp\left[i \sum_{j=1}^{p} \lambda_j Z_j^2\right]$$

where $Z_1, \ldots, Z_p$ are the coordinates of Z. Since $Z_1, \ldots, Z_p$ are independent with $\mathcal{L}(Z_j) = N(0,1)$, Z_j^2 has a χ_1^2 distribution and we have

$$\xi(A) = \prod_{j=1}^{p} \mathcal{E} \exp\left[i\lambda_j Z_j^2\right] = \prod_{j=1}^{p} (1 - 2i\lambda_j)^{-1/2}$$

$$= |I_p - 2iD|^{-1/2} = |I_p - 2i\Gamma D\Gamma'|^{-1/2} = |I_p - 2iC'AC|^{-1/2}$$

$$= |I_p - 2iCC'A|^{-1/2} = |I_p - 2i\Sigma A|^{-1/2}.$$

The next to the last equality is a consequence of Proposition 1.35. Thus (iii) holds. □

Proposition 8.4. If $\mathcal{L}(S_i) = W(\Sigma, p, n_i)$ for $i = 1, 2$ and if S_1 and S_2 are independent, then $\mathcal{L}(S_1 + S_2) = W(\Sigma, p, n_1 + n_2)$.

Proof. An application of (iii) of Proposition 8.3 yields this convolution result. Specifically,

$$\phi(A) = \mathcal{E} \exp[i\langle A, S_1 + S_2\rangle] = \prod_{j=1}^{2} \mathcal{E} \exp i\langle A, S_j\rangle$$

$$= \prod_{j=1}^{2} |I_p - 2i\Sigma A|^{-n_j/2} = |I_p - 2i\Sigma A|^{-(n_1+n_2)/2}.$$

The uniqueness of characteristic functions shows that $\mathcal{L}(S_1 + S_2) = W(\Sigma, p, n_1 + n_2)$. □

It should be emphasized that $\langle \cdot, \cdot \rangle$ is not what we might call the standard inner product on $\mathcal{S}_p$ when $\mathcal{S}_p$ is regarded as a $[p(p+1)/2]$-dimensional coordinate space. For example, if $p = 2$, and $S, T \in \mathcal{S}_p$, then

$$\langle S, T\rangle = \operatorname{tr} ST = s_{11}t_{11} + s_{22}t_{22} + 2s_{12}t_{12}$$

while the three-dimensional coordinate space inner product between S and

T would be $s_{11}t_{11} + s_{22}t_{22} + s_{12}t_{12}$. In this connection, equation (ii) of Proposition 8.3 means that

$$\begin{aligned}\operatorname{cov}(\langle A, S\rangle, \langle B, S\rangle) &= 2n\langle A, (\Sigma \otimes \Sigma)B\rangle \\ &= 2n\langle A, \Sigma B \Sigma\rangle = 2n \operatorname{tr}(A\Sigma B\Sigma),\end{aligned}$$

that is, (ii) depends on the inner product $\langle \cdot\, , \cdot \rangle$ on $\mathcal{S}_p$ and is not valid for other inner products.

In Chapter 3, quadratic forms in normal random vectors were shown to have chi-square distributions under certain conditions. Similar results are available for generalized quadratic forms and the Wishart distribution. The following proposition is not the most general possible, but suffices in most situations.

Proposition 8.5. Consider $X \in \mathcal{L}_{p,n}$ where $\mathcal{L}(X) = N(\mu, Q \otimes \Sigma)$. Let $S = X'PX$ where P is $n \times n$ and positive semidefinite, and write $P = A^2$ with A positive semidefinite. If AQA is a rank k orthogonal projection and if $P\mu = 0$, then

$$\mathcal{L}(S) = W(\Sigma, p, k).$$

Proof. With $Y = AX$, it is clear that $S = Y'Y$ and

$$\mathcal{L}(Y) = N(A\mu, (AQA) \otimes \Sigma).$$

Since $\mathfrak{N}(A) = \mathfrak{N}(P)$ and $P\mu = 0$, $A\mu = 0$ so

$$\mathcal{L}(Y) = N(0, (AQA) \otimes \Sigma).$$

By assumption, $B = AQA$ is a rank k orthogonal projection. Also, $S = Y'Y = Y'BY + Y'(I - B)Y$, and $\mathcal{L}((I - B)Y) = N(0, 0 \otimes \Sigma)$ so $Y'(I - B)Y$ is zero with probability one. Thus it remains to show that if $\mathcal{L}(Y) = N(0, B \otimes \Sigma)$ where B is a rank k orthogonal projection, then $S = Y'BY$ has a $W(\Sigma, p, k)$ distribution. Without loss of generality (make an orthogonal transformation),

$$B = \begin{pmatrix} I_k & 0 \\ 0 & 0 \end{pmatrix} : n \times n.$$

Partitioning Y into $Y_1 : k \times p$ and $Y_2 : (n - k) \times p$, it follows that $S = Y_1'Y_1$

and

$$\mathcal{L}(Y_1) = N(0, I_k \otimes \Sigma).$$

Thus $\mathcal{L}(S) = W(\Sigma, p, k)$. □

◆ **Example 8.1.** We again return to the multivariate normal linear model introduced in Example 4.4. Consider $X \in \mathcal{L}_{p,n}$ with

$$\mathcal{L}(X) = N(\mu, I_n \otimes \Sigma)$$

where μ is an element of the subspace $M \subseteq \mathcal{L}_{p,n}$ defined by

$$M = \{x | x \in \mathcal{L}_{p,n}, x = ZB, B \in \mathcal{L}_{p,k}\}.$$

Here, Z is an $n \times k$ matrix of rank k and it is assumed that $n - k \geqslant p$. With $P_z = Z(Z'Z)^{-1}Z'$, $P_M = P_z \otimes I_p$ is the orthogonal projection onto M and $Q_M = Q_z \otimes I_p$, $Q_z = I - P_z$, is the orthogonal projection onto $M^\perp$. We know that

$$\hat{\mu} = P_M X = (P_z \otimes I_p) X = P_z X$$

is the maximum likelihood estimator of μ. As demonstrated in Example 4.4, the maximum likelihood estimator of Σ is found by maximizing

$$p(x|\hat{\mu}, \Sigma) = |\Sigma|^{-n/2} \exp\left[-\tfrac{1}{2} \operatorname{tr} \Sigma^{-1} x' Q_z x\right].$$

Since $n - k \geqslant p$, $x'Q_z x$ has rank p with probability one. When $X'Q_z X$ has rank p, Example 7.10 shows that

$$\hat{\Sigma} = \frac{1}{n} X' Q_z X$$

is the maximum likelihood estimator of Σ. The conditions of Proposition 8.5 are easily checked to verify that $S \equiv X'Q_z X$ has a $W(\Sigma, p, n-k)$ distribution. In summary, for the multivariate linear model, $\hat{\mu} = P_M X$ and $\hat{\Sigma} = n^{-1} X' Q_z X$ are the maximum likelihood estimators of μ and Σ. Further, $\hat{\mu}$ and $\hat{\Sigma}$ are independent and

$$\mathcal{L}(\hat{\mu}) = N(\mu, P_z \otimes \Sigma)$$

$$\mathcal{L}(n\hat{\Sigma}) = W(\Sigma, p, n-k).$$ ◆

8.2. PARTITIONING A WISHART MATRIX

The partitioning of the Wishart distribution considered here is motivated partly by the transformation described in Proposition 5.8. If $\mathcal{L}(S) = W(\Sigma, p, n)$ where $n \geqslant p$, partition S as

$$S = \begin{pmatrix} S_{11} & S_{12} \\ S_{21} & S_{22} \end{pmatrix}$$

where $S_{21} = S_{12}'$ and let

$$S_{11\cdot 2} = S_{11} - S_{12}S_{22}^{-1}S_{21}.$$

Here, S_{ij} is $p_i \times p_j$ for $i, j = 1, 2$ so $p_1 + p_2 = p$. The primary result of this section describes the joint distribution of $(S_{11\cdot 2}, S_{21}, S_{22})$ when Σ is nonsingular. This joint distribution is derived by representing the Wishart distribution in terms of the normal distribution. Since $\mathcal{L}(S) = W(\Sigma, p, n)$, $S = X'X$ where $\mathcal{L}(X) = N(0, I_n \otimes \Sigma)$. Discarding a set of Lebesgue measure zero, X is assumed to take values in $\mathcal{X}$, the set of all $n \times p$ matrices of rank p. With

$$X = (X_1, X_2), \qquad X_i : n \times p_i, \quad i = 1, 2,$$

it is clear that

$$S_{ij} = X_i'X_j \qquad \text{for } i, j = 1, 2.$$

Thus

$$S_{11\cdot 2} = X_1'X_1 - X_1'X_2(X_2'X_2)^{-1}X_2'X_1 = X_1'QX_1$$

where

$$Q = I_n - X_2(X_2'X_2)^{-1}X_2' \equiv I_n - P$$

is an orthogonal projection of rank $n - p_2$ for each value of X_2 when $X \in \mathcal{X}$. To obtain the desired result for the Wishart distribution, it is useful to first give the joint distribution of (QX_1, PX_1, X_2).

Proposition 8.6. The joint distribution of (QX_1, PX_1, X_2) can be described as follows. Conditional on X_2, QX_1 and PX_1 are independent with

$$\mathcal{L}(QX_1 | X_2) = N(0, Q \otimes \Sigma_{11\cdot 2})$$

and

$$\mathcal{L}(PX_1|X_2) = N(X_2\Sigma_{22}^{-1}\Sigma_{21}, P \otimes \Sigma_{11\cdot 2}).$$

Also,

$$\mathcal{L}(X_2) = N(0, I_n \otimes \Sigma_{22}).$$

Proof. From Example 3.1, the conditional distribution of X_1 given X_2, say $\mathcal{L}(X_1|X_2)$, is

$$\mathcal{L}(X_1|X_2) = N(X_2\Sigma_{22}^{-1}\Sigma_{21}, I_n \otimes \Sigma_{11\cdot 2}).$$

Thus conditional on X_2, the random vector

$$W \equiv \begin{pmatrix} Q \otimes I_{p_1} \\ P \otimes I_{p_1} \end{pmatrix} X_1 = \begin{pmatrix} QX_1 \\ PX_1 \end{pmatrix} : (2n) \times p_1$$

is a linear transformation of X_1. Thus W has a normal distribution with mean vector

$$\begin{pmatrix} Q \otimes I_{p_1} \\ P \otimes I_{p_1} \end{pmatrix} X_2\Sigma_{22}^{-1}\Sigma_{21} = \begin{pmatrix} 0 \\ X_2\Sigma_{22}^{-1}\Sigma_{21} \end{pmatrix}$$

since $QX_2 = 0$ and $PX_2 = X_2$. Also, using the calculational rules for partitioned linear transformations, the covariance of W is

$$\begin{pmatrix} Q \otimes I_{p_1} \\ P \otimes I_{p_1} \end{pmatrix} (I_n \otimes \Sigma_{11\cdot 2})(Q \otimes I_{p_1}, P \otimes I_{p_1}) = \begin{pmatrix} Q & 0 \\ 0 & P \end{pmatrix} \otimes \Sigma_{11\cdot 2}$$

since $QP = 0$. The conditional independence and conditional distribution of QX_1 and PX_1 follow immediately. That X_2 has the claimed marginal distribution is obvious. □

Proposition 8.7. Suppose $\mathcal{L}(S) = W(\Sigma, p, n)$ with $n \geqslant p$ and $\Sigma > 0$. Partition S into S_{ij}, $i, j = 1, 2$, where S_{ij} is $p_i \times p_j$, $p_1 + p_2 = p$, and partition Σ similarly. With $S_{11\cdot 2} = S_{11} - S_{12}S_{22}^{-1}S_{21}$, $S_{11\cdot 2}$ and (S_{21}, S_{22}) are stochastically independent. Further,

$$\mathcal{L}(S_{11\cdot 2}) = W(\Sigma_{11\cdot 2}, p_1, n - p_2)$$

and conditional on S_{22},

$$\mathcal{L}(S_{21}|S_{22}) = N(S_{22}\Sigma_{22}^{-1}\Sigma_{21}, S_{22} \otimes \Sigma_{11\cdot 2}).$$

The marginal distribution of S_{22} is $W(\Sigma_{22}, p_2, n)$.

Proof. In the notation of Proposition 8.6, consider $X \in \mathcal{X}$ with $\mathcal{L}(X) = N(0, I_n \otimes \Sigma)$ and $S = X'X$. Then $S_{ij} = X_i'X_j$ for $i, j = 1, 2$ and $S_{11\cdot 2} = X_1'QX_1$. Since $PX_2 = X_2$ and $S_{21} = X_2'X_1$, we see that $S_{21} = (PX_2)'X_1 = X_2'PX_1$, and conditional on X_2,

$$\mathcal{L}(S_{21}|X_2) = N(X_2'X_2\Sigma_{22}^{-1}\Sigma_{21}, (X_2'X_2) \otimes \Sigma_{11\cdot 2}).$$

To show that $S_{11\cdot 2}$ and (S_{21}, S_{22}) are independent, it suffices to show that

$$\mathcal{E}f(S_{11\cdot 2})h(S_{21}, S_{22}) = \mathcal{E}f(S_{11\cdot 2})\mathcal{E}h(S_{21}, S_{22})$$

for bounded measurable functions f and h with the appropriate domains of definition. Using Proposition 8.6, we argue as follows. For fixed X_2, QX_1 and PX_1 are independent so $S_{11\cdot 2} = X_1'QQX_1$ and $S_{21} = X_2'PX_1$ are conditionally independent. Also,

$$\mathcal{L}(QX_1|X_2) = N(0, Q \otimes \Sigma_{11\cdot 2})$$

and Q is a rank $n - p_2$ orthogonal projection. By Proposition 8.5, $\mathcal{L}(X_1'QX_1|X_2) = W(\Sigma_{11\cdot 2}, p_1, n - p_2)$ for each X_2 so $X_1'QX_1$ and X_2 are independent. Conditioning on X_2, we have

$$\begin{aligned}
\mathcal{E}f(S_{11\cdot 2})h(S_{21}, S_{22}) &= \mathcal{E}\mathcal{E}[f(X_1'QX_1)h(X_2'PX_1, X_2'X_2)|X_2] \\
&= \mathcal{E}[\mathcal{E}(f(X_1'QX_1)|X_2)\mathcal{E}(h(X_2'PX_1, X_2'X_2)|X_2)] \\
&= \mathcal{E}[\mathcal{E}f(X_1'QX_1)\mathcal{E}(h(X_2'PX_1, X_2'X_2)|X_2)] \\
&= \mathcal{E}f(X_1'QX_1)\mathcal{E}[\mathcal{E}(h(X_2'PX_1, X_2'X_2)|X_2)] \\
&= \mathcal{E}f(X_1'QX_1)\mathcal{E}h(X_2'PX_1, X_2'X_2) \\
&= \mathcal{E}f(S_{11\cdot 2})\mathcal{E}h(S_{21}, S_{22}).
\end{aligned}$$

Therefore, $S_{11\cdot 2}$ and (S_{21}, S_{22}) are stochastically independent. To describe the joint distribution of S_{21} and S_{22}, again condition on X_2. Then

$$\mathcal{L}(S_{21}|X_2) = N(X_2'X_2\Sigma_{22}^{-1}\Sigma_{21}, (X_2'X_2)\otimes\Sigma_{11\cdot 2})$$

and this conditional distribution depends on X_2 only through $S_{22} = X_2'X_2$. Thus

$$\mathcal{L}(S_{21}|S_{22}) = N(S_{22}\Sigma_{22}^{-1}\Sigma_{21}, S_{22}\otimes\Sigma_{11\cdot 2}).$$

That S_{22} has the claimed marginal distribution is obvious. □

By simply permuting the indices in Proposition 8.7, we obtain the following proposition.

Proposition 8.8. With the notation and assumptions of Proposition 8.7, let $S_{22\cdot 1} = S_{22} - S_{21}S_{11}^{-1}S_{12}$. Then $S_{22\cdot 1}$ and (S_{11}, S_{12}) are stochastically independent and

$$\mathcal{L}(S_{22\cdot 1}) = W(\Sigma_{22\cdot 1}, p_2, n - p_1).$$

Conditional on S_{11},

$$\mathcal{L}(S_{12}|S_{11}) = N(S_{11}\Sigma_{11}^{-1}\Sigma_{12}, S_{11}\otimes\Sigma_{22\cdot 1})$$

and the marginal distribution of S_{11} is $W(\Sigma_{11}, p_1, n)$.

Proposition 8.7 is one of the most useful results for deriving distributions of functions of Wishart matrices. Applications occur in this and the remaining chapters. For example, the following assertion provides a simple proof of the distribution of Hotelling's-T^2, discussed in the next chapter.

Proposition 8.9. Suppose S_0 has a nonsingular Wishart distribution, say $W(\Sigma, p, n)$, and let A be an $r \times p$ matrix of rank r. Then

$$\mathcal{L}\left((AS_0^{-1}A')^{-1}\right) = W\left((A\Sigma^{-1}A')^{-1}, r, n - p + r\right).$$

Proof. First, an invariance argument shows that it is sufficient to consider the case when $\Sigma = I$. More precisely, write $\Sigma = B^2$ with $B > 0$ and let $C = AB^{-1}$. With $S = B^{-1}S_0B^{-1}$, $\mathcal{L}(S) = W(I, p, n)$ and the assertion is that

$$\mathcal{L}\left((CS^{-1}C')^{-1}\right) = W\left((CC')^{-1}, r, n - p + r\right).$$

Now, let $\Psi = C'(CC')^{-1/2}$, so the assertion becomes

$$\mathcal{L}\left((\Psi'S^{-1}\Psi)^{-1}\right) = W(I_r, r, n - p + r).$$

However, Ψ is $p \times r$ and satisfies $\Psi'\Psi = I_r$—that is, Ψ is a linear isometry. Since $\mathcal{L}(\Gamma'S\Gamma) = \mathcal{L}(S)$ for all $\Gamma \in \mathcal{O}_p$,

$$\mathcal{L}\left((\Psi'\Gamma'S^{-1}\Gamma\Psi)^{-1}\right) = \mathcal{L}\left((\Psi'S^{-1}\Psi)^{-1}\right).$$

Choose Γ so that

$$\Gamma\Psi = \begin{pmatrix} I_r \\ 0 \end{pmatrix} : p \times r.$$

For this choice of Γ, the matrix $(\Psi'\Gamma'S^{-1}\Gamma\Psi)^{-1}$ is just the inverse of the $r \times r$ upper left corner of S^{-1}, and this matrix is

$$S_{11} - S_{12}S_{22}^{-1}S_{21} \equiv V$$

where V is $r \times r$. By Proposition 8.7,

$$\mathcal{L}(V) = W(I_r, r, n - p + r)$$

since $\mathcal{L}(S) = W(I, p, n)$. This establishes the assertion of the proposition. □

When $r = 1$ in Proposition 8.9, the matrix A' is nonzero vector, say $A' = a \in R^p$. In this case,

$$\mathcal{L}\left(\frac{a'\Sigma^{-1}a}{a'S^{-1}a}\right) = \chi^2_{n-p+1}$$

when $\mathcal{L}(S) = W(\Sigma, p, n)$. Another decomposition result for the Wishart distribution, which is sometimes useful, follows.

Lemma 8.10. Suppose S has a nonsingular Wishart distribution, say $\mathcal{L}(S) = W(\Sigma, p, n)$, and let $S = TT'$ where $T \in G_T^+$. Then the density of T with respect to the left invariant measure $\nu(dT) = dT/\prod t_{ii}^i$ is

$$p(T|\Sigma) = 2^p\omega(n, p)|\Sigma^{-1}TT'|^{n/2}\exp\left[-\tfrac{1}{2}\operatorname{tr}\Sigma^{-1}TT'\right].$$

If S and T are partitioned as

$$S = \begin{pmatrix} S_{11} & S_{12} \\ S_{21} & S_{22} \end{pmatrix}, \qquad T = \begin{pmatrix} T_{11} & 0 \\ T_{21} & T_{22} \end{pmatrix}$$

where S_{ij} is $p_i \times p_j$, $p_1 + p_2 = p$, then $S_{11} = T_{11}T'_{11}$, $S_{12} = T_{11}T'_{21}$, and $S_{22\cdot 1} = T_{22}T'_{22}$. Further, the pair (T_{11}, T_{21}) is independent of T_{22} and

$$\mathcal{L}(T'_{21}|T_{11}) = N\left(T'_{11}\Sigma_{11}^{-1}\Sigma_{12}, I_{p_1} \otimes \Sigma_{22\cdot 1}\right).$$

Proof. The expression for the density of T is a consequence of Proposition 7.5, and a bit of algebra shows that $S_{11} = T_{11}T'_{11}$, $S_{12} = T_{11}T'_{21}$, and $S_{22\cdot 1} = T_{22}T'_{22}$. The independence of (T_{11}, T_{21}) and T_{22} follows from Proposition 8.8 and the fact that the mapping between S and T is one-to-one and onto. Also,

$$\mathcal{L}(S_{12}|S_{11}) = N\left(S_{11}\Sigma_{11}^{-1}\Sigma_{12}, S_{11} \otimes \Sigma_{22\cdot 1}\right).$$

Since S_{11} and T_{11} are one-to-one functions of each other and $S_{12} = T_{11}T'_{21}$,

$$\mathcal{L}(T_{11}T'_{21}|T_{11}) = N\left(T_{11}T'_{11}\Sigma_{11}^{-1}\Sigma_{12}, T_{11}T'_{11} \otimes \Sigma_{22\cdot 1}\right).$$

Thus

$$\mathcal{L}(T'_{21}|T_{11}) = N\left(T'_{11}\Sigma_{11}^{-1}\Sigma_{12}, I_{p_1} \otimes \Sigma_{22\cdot 1}\right),$$

as

$$T'_{21} = \left(T_{11}^{-1} \otimes I_{p_2}\right)(T_{11}T'_{21})$$

and T_{11} is fixed. □

Proposition 8.11. Suppose S has a nonsingular Wishart distribution with $\mathcal{L}(S) = W(\Sigma, p, n)$ and assume that Σ is diagonal with diagonal elements $\sigma_{11}, \dots, \sigma_{pp}$. If $S = TT'$ with $T \in G_T^+$, then the random variables $\{t_{ij}|i \geqslant j\}$ are mutually independent and

$$\mathcal{L}(t_{ij}) = N(0, \sigma_{ii}) \qquad \text{for } i > j$$

and

$$\mathcal{L}(t_{ii}^2) = \sigma_{ii}\chi^2_{n-i+1}, \qquad i = 1, \dots, p.$$

Proof. First, partition S, Σ, and T as

$$S = \begin{pmatrix} S_{11} & S_{12} \\ S_{21} & S_{22} \end{pmatrix}, \qquad \Sigma = \begin{pmatrix} \sigma_{11} & 0 \\ 0 & \Sigma_{22} \end{pmatrix}, \qquad T = \begin{pmatrix} t_{11} & 0 \\ T_{21} & T_{22} \end{pmatrix}$$

where S_{11} is 1×1. Since $\Sigma_{12} = 0$, the conditional distribution of T_{21} given T_{11} does not depend on T_{11} and Σ_{22} has diagonal elements $\sigma_{22}, \ldots, \sigma_{pp}$. It follows from Proposition 8.10 that t_{11}, T_{21}', and T_{22} are mutually independent and

$$\mathcal{L}(T_{21}') = N(0, \Sigma_{22}).$$

The elements of T_{21} are $t_{21}, t_{31}, \ldots, t_{p1}$, and since Σ_{22} is diagonal, these are independent with

$$\mathcal{L}(t_{i1}) = N(0, \sigma_{ii}), \qquad i = 2, \ldots, p.$$

Also,

$$\mathcal{L}(t_{11}^2) = \sigma_{11}\chi_n^2$$

and

$$\mathcal{L}(S_{22\cdot 1}) = \mathcal{L}(T_{22}T_{22}') = W(\Sigma_{22}, p - 1, n - 1).$$

The conclusion of the proposition follows by an induction argument on the dimension parameter p. □

When $\mathcal{L}(S) = W(\Sigma, p, n)$ is a nonsingular Wishart distribution, the random variable $|S|$ is called the *generalized variance*. The distribution of $|S|$ is easily derived using Proposition 8.11. First, write $\Sigma = B^2$ with $B > 0$ and let $S_1 = B^{-1}SB^{-1}$. Then $\mathcal{L}(S_1) = W(I, p, n)$ and $|S| = |\Sigma||S_1|$. Also, if $TT' = S_1$, $T \in G_T^+$, then $\mathcal{L}(t_{ii}^2) = \chi_{n-i+1}^2$ for $i = 1, \ldots, p$, and $t_{11}, \ldots, t_{pp}$ are mutually independent. Thus

$$\mathcal{L}(|S|) = \mathcal{L}(|\Sigma||S_1|) = \mathcal{L}(|\Sigma||TT'|) = \mathcal{L}\left(|\Sigma|\prod_{i=1}^{p} t_{ii}^2\right).$$

Therefore, the distribution of $|S|$ is the same as the constant $|\Sigma|$ times a product of p independent chi-square random variables with $n - i + 1$ degrees of freedom for $i = 1, \ldots, p$.

8.3. THE NONCENTRAL WISHART DISTRIBUTION

Just as the Wishart distribution is a matrix generalization of the chi-square distribution, the noncentral Wishart distribution is a matrix analog of the noncentral chi-square distribution. Also, the noncentral Wishart distribution arises in a natural way in the study of distributional properties of test statistics in multivariate analysis.

Definition 8.2. Let $X \in \mathcal{L}_{p,n}$ have a normal distribution $N(\mu, I_n \otimes \Sigma)$. A random matrix $S \in \mathcal{S}_p$ has a noncentral Wishart distribution with parameters Σ, p, n, and $\Delta \equiv \mu'\mu$ if $\mathcal{L}(S) = \mathcal{L}(X'X)$. In this case, we write $\mathcal{L}(S) = W(\Sigma, p, n; \Delta)$.

In this definition, it is not obvious that the distribution of $X'X$ depends on μ only through $\Delta = \mu'\mu$. However, an invariance argument establishes this. The group $\mathcal{O}_n$ acts on $\mathcal{L}_{p,n}$ by sending x into Γx for $x \in \mathcal{L}_{p,n}$ and $\Gamma \in \mathcal{O}_n$. A maximal invariant under this action is $x'x$. When $\mathcal{L}(X) = N(\mu, I_n \otimes \Sigma)$, $\mathcal{L}(\Gamma X) = N(\Gamma\mu, I_n \otimes \Sigma)$ and we know the distribution of $X'X$ depends only on a maximal invariant parameter. But the group action on the parameter space is $(\mu, \Sigma) \to (\Gamma\mu, \Sigma)$ and a maximal invariant is obviously $(\mu'\mu, \Sigma)$. Thus the distribution of $X'X$ depends only on $(\mu'\mu, \Sigma)$.

When $\Delta = 0$, the noncentral Wishart distribution is just the $W(\Sigma, p, n)$ distribution. Let $X_1', \dots, X_n'$ be the rows of X in the above definition so $X_1, \dots, X_n$ are independent and $\mathcal{L}(X_i) = N(\mu_i, \Sigma)$ where $\mu_1', \dots, \mu_n'$ are the rows of μ. Obviously,

$$\mathcal{L}(X_i X_i') = W(\Sigma, p, 1; \Delta_i)$$

where $\Delta_i = \mu_i \mu_i'$. Thus $S_i = X_i X_i'$, $i = 1, \dots, n$, are independent and it is clear that, if $S = X'X$, then

$$\mathcal{L}(S) = \mathcal{L}\left(\sum_1^n S_i\right).$$

In other words, the noncentral Wishart distribution with n degrees of freedom can be represented as the convolution of n noncentral Wishart distributions each with one degree of freedom. This argument shows that, if $\mathcal{L}(S_i) = W(\Sigma, p, n_i; \Delta_i)$ for $i = 1, 2$ and if S_1 and S_2 are independent, then $\mathcal{L}(S_1 + S_2) = W(\Sigma, p, n_1 + n_2, \Delta_1 + \Delta_2)$. Since

$$\mathcal{E} X_i X_i' = \Sigma + \mu_i \mu_i',$$

it follows that

$$\mathcal{E}S = n\Sigma + \Delta$$

when $\mathcal{L}(S) = W(\Sigma, p, n; \Delta)$. Also,

$$\operatorname{Cov}(S) = \sum_1^n \operatorname{Cov}(S_i)$$

but an explicit expression for $\operatorname{Cov}(S_i)$ is not needed here. As with the central Wishart distribution, it is not difficult to prove that, when $\mathcal{L}(S) = W(\Sigma, p, n; \Delta)$, then S is positive definite with probability one iff $n \geqslant p$ and $\Sigma > 0$. Further, it is clear that if $\mathcal{L}(S) = W(\Sigma, p, n; \Delta)$ and A is an $r \times p$ matrix, then $\mathcal{L}(ASA') = W(A\Sigma A', r, n; A\Delta A')$. The next result provides an expression for the density function of S in a special case.

Proposition 8.12. Suppose $\mathcal{L}(S) = W(\Sigma, p, n; \Delta)$ where $n \geqslant p$ and $\Sigma > 0$, and assume that Δ has rank one, say $\Delta = \eta\eta'$ with $\eta \in R^p$. The density of S with respect to $\nu(dS) = dS/|S|^{(p+1)/2}$ is given by

$$p_1(S|\Sigma, \Delta) = p(S|\Sigma)\exp\left[-\tfrac{1}{2}\eta'\Sigma^{-1}\eta\right] H\left(\left(\eta'\Sigma^{-1}S\Sigma^{-1}\eta\right)^{1/2}\right)$$

where $p(S|\Sigma)$ is the density of a $W(\Sigma, p, n)$ distribution given in Proposition 8.2 and the function H is defined in Example 7.13.

Proof. Consider $X \in \mathcal{L}_{p,n}$ with $\mathcal{L}(X) = N(\mu, I_n \otimes \Sigma)$ where $\mu \in \mathcal{L}_{p,n}$ and $\mu'\mu = \Delta$. Since $S = X'X$ is a maximal invariant under the action of $\mathcal{O}_n$ on $\mathcal{L}_{p,n}$, the results of Example 7.15 show that the density of S with respect to the measure $\nu_0(dS) = (\sqrt{2\pi})^{np}\omega(n, p)|S|^{(n-p-1)/2}\,dS$ is

$$h(S) = \int_{\mathcal{O}_n} f(\Gamma X)\mu_0(d\Gamma).$$

Here, f is the density of X and μ_0 is the unique invariant probability measure on $\mathcal{O}_n$. The density of X is

$$f(X) = \left(\sqrt{2\pi}\right)^{-np}|\Sigma|^{-n/2}\exp\left[-\tfrac{1}{2}\operatorname{tr}(X-\mu)\Sigma^{-1}(X-\mu)'\right].$$

Substituting this into the expression for $h(S)$ and doing a bit of algebra shows that the density $p_1(S|\Sigma, \Delta)$ with respect to ν is

$$p_1(S|\Sigma, \Delta) = p(S|\Sigma)\exp\left[-\tfrac{1}{2}\operatorname{tr}\Sigma^{-1}\Delta\right]\int_{\mathcal{O}_n}\exp\left[\operatorname{tr}\Gamma X\Sigma^{-1}\mu'\right]\mu_0(d\Gamma).$$

The problem is now to evaluate the integral over $\mathcal{O}_n$. It is here where we use the assumption that Δ has rank one. Since $\Delta = \mu'\mu$, μ must have rank one so $\mu = \xi\eta'$ where $\xi \in R^n$, $\|\xi\| = 1$, and $\eta \in R^p$, $\Delta = \eta\eta'$. Since $\|\xi\| = 1$, $\xi = \Gamma_1\varepsilon_1$ for some $\Gamma_1 \in \mathcal{O}_n$ where $\varepsilon_1 \in R^n$ is the first unit vector. Setting $u = (\eta'\Sigma^{-1}S\Sigma^{-1}\eta)^{1/2}$, $X\Sigma^{-1}\eta = u\Gamma_2\varepsilon_1$ for some $\Gamma_2 \in \mathcal{O}_n$ as $u\varepsilon_1$ and $X\Sigma^{-1}\eta$ have the same length. Therefore,

$$\int_{\mathcal{O}_n} \exp[\operatorname{tr} \Gamma X\Sigma^{-1}\mu']\mu_0(d\Gamma) = \int_{\mathcal{O}_n} \exp[\operatorname{tr} \Gamma X\Sigma^{-1}\eta\xi']\mu_0(d\Gamma)$$

$$= \int_{\mathcal{O}_n} \exp[\xi'\Gamma X\Sigma^{-1}\eta]\mu_0(d\Gamma) = \int_{\mathcal{O}_n} \exp[u\varepsilon_1'\Gamma_1'\Gamma\Gamma_2\varepsilon_1]\mu_0(d\Gamma)$$

$$= \int_{\mathcal{O}_n} \exp[u\varepsilon_1'\Gamma\varepsilon_1]\mu_0(d\Gamma) = \int_{\mathcal{O}_n} \exp[u\gamma_{11}]\mu_0(d\Gamma) \equiv H(u).$$

The right and left invariance of μ_0 was used in the third to the last equality and γ_{11} is the (1, 1) element of Γ. The function H was evaluated in Example 7.13. Therefore, when $\Delta = \eta\eta'$,

$$p_1(S|\Sigma, \Delta) = p(S|\Sigma)\exp\left[-\tfrac{1}{2}\eta'\Sigma^{-1}\eta\right] \times H\left(\left(\eta'\Sigma^{-1}S\Sigma^{-1}\eta\right)^{1/2}\right). \qquad \square$$

The final result of this section is the analog of Proposition 8.5 for the noncentral Wishart distribution.

Proposition 8.13. Consider $X \in \mathcal{L}_{p,n}$ where $\mathcal{L}(X) = N(\mu, Q \otimes \Sigma)$ and let $S = X'PX$ where $P \geqslant 0$ is $n \times n$. Write $P = A^2$ with $A \geqslant 0$. If $B \equiv AQA$ is a rank k orthogonal projection and if $AQP\mu = A\mu$, then

$$\mathcal{L}(S) = W(\Sigma, p, k; \mu'P\mu).$$

Proof. The proof of this result is quite similar to that of Proposition 8.5 and is left to the reader. $\square$

It should be noted that there is not an analog of Proposition 8.7 for the noncentral Wishart distribution, at least as far as I know. Certainly, Proposition 8.7 is false as stated when S is noncentral Wishart.

8.4. DISTRIBUTIONS RELATED TO LIKELIHOOD RATIO TESTS

In the next two chapters, statistics that are the ratio of determinants of Wishart matrices arise as tests statistics related to likelihood ratio tests.

Since the techniques for deriving the distributions of these statistics are intimately connected with properties of the Wishart distribution, we have chosen to treat this topic here rather than interrupt the flow of the succeeding chapters with such considerations.

Let $X \in \mathcal{L}_{p,m}$ and $S \in \mathcal{S}_p^+$ be independent and suppose that $\mathcal{L}(X) = N(\mu, I_m \otimes \Sigma)$ and $\mathcal{L}(S) = W(\Sigma, p, n)$ where $n \geq p$ and $\Sigma > 0$. We are interested in deriving the distribution of the random variable

$$U = \frac{|S|}{|S + X'X|}$$

for some special values of the mean matrix μ of X. The argument below shows that the distribution of U depends on (μ, Σ) only through $\Sigma^{-1/2}\mu'\mu\Sigma^{-1/2}$ where $\Sigma^{1/2}$ is the positive definite square root of Σ. Let $S = \Sigma^{1/2}S_1\Sigma^{1/2}$ and $Y = X\Sigma^{-1/2}$. Then S_1 and Y are independent, $\mathcal{L}(S_1) = W(I, p, n)$, and $\mathcal{L}(Y) = N(\mu\Sigma^{-1/2}, I_m \otimes I_p)$. Also,

$$U = \frac{|S|}{|S + X'X|} = \frac{|S_1|}{|S_1 + Y'Y|}.$$

However, the discussion of the previous section shows that $Y'Y$ has a noncentral Wishart distribution, say $\mathcal{L}(Y'Y) = W(I, p, n; \Delta)$ where $\Delta = \Sigma^{-1/2}\mu'\mu\Sigma^{-1/2}$. In the following discussion we take $\Sigma = I_p$ and denote the distribution of U by

$$\mathcal{L}(U) = U(n, m, p; \Delta)$$

where $\Delta = \mu'\mu$. When $\mu = 0$, the notation

$$\mathcal{L}(U) = U(n, m, p)$$

is used. In the case that $p = 1$,

$$U = \frac{S}{S + X'X}$$

where $\mathcal{L}(S) = \chi_n^2$. Since $\mathcal{L}(X) = N(\mu, I_m)$, $\mathcal{L}(X'X) = \chi_m^2(\Delta)$ where $\Delta = \mu'\mu \geq 0$. Thus

$$U = \frac{1}{1 + \chi_m^2(\Delta)/\chi_n^2}.$$

When $\chi_m^2(\Delta)$ and χ_n^2 are independent, the distribution of the ratio

$$F(m, n; \Delta) \equiv \frac{\chi_m^2(\Delta)}{\chi_n^2}$$

is called a noncentral F distribution with parameters m, n, and Δ. When $\Delta = 0$, the distribution of $F(m, n; 0)$ is denoted by $F_{m,n}$ and is simply called an F distribution with (m, n) degrees of freedom. It should be noted that this usage is not standard as the above ratio has not been normalized by the constant n/m. At times, the relationship between the F distribution and the beta distribution is useful. It is not difficult to show that, when χ_m^2 and χ_n^2 are independent, the random variable

$$V = \frac{\chi_n^2}{\chi_n^2 + \chi_m^2}$$

has a beta distribution with parameters $n/2$ and $m/2$, and this is written as $\mathcal{L}(V) = \mathcal{B}(n/2, m/2)$. In other words, V has a density on $(0, 1)$ given by

$$p(v) = \frac{\Gamma(\alpha + \beta)}{\Gamma(\alpha)\Gamma(\beta)} v^{\alpha - 1}(1 - v)^{\beta - 1}$$

where $\alpha = n/2$ and $\beta = m/2$. More generally, the distribution of the random variable

$$V(\Delta) = \frac{\chi_n^2}{\chi_n^2 + \chi_m^2(\Delta)}$$

is called a noncentral beta distribution and the notation $\mathcal{L}(V(\Delta)) = \mathcal{B}(n/2, m/2; \Delta)$ is used. In summary, when $p = 1$,

$$\mathcal{L}(U) = \mathcal{B}\left(\frac{n}{2}, \frac{m}{2}; \Delta\right)$$

where $\Delta = \mu'\mu \geqslant 0$.

Now, we consider the distribution of U when $m = 1$. In this case, $\mathcal{L}(X') = N(\mu', I_p)$ where $X' \in R^p$ and

$$U = \frac{|S|}{|S + X'X|} = |I_p + S^{-1}X'X|^{-1} = (1 + XS^{-1}X')^{-1}.$$

The last equality follows from Proposition 1.35.

Proposition 8.14. When $m = 1$,

$$\mathcal{L}(U) = \mathcal{B}\left(\frac{n-p+1}{2}, \frac{p}{2}; \delta\right)$$

where $\delta = \mu\mu' \geqslant 0$.

Proof. It must be shown that

$$\mathcal{L}(XS^{-1}X') = F(p, n-p+1, \delta).$$

For X fixed, $X \neq 0$, Proposition 8.10 shows that

$$\mathcal{L}\left(\frac{XX'}{XS^{-1}X'}\right) = \chi^2_{n-p+1}$$

when $\mathcal{L}(S) = W(I, p, n)$. Since this distribution does not depend on X, we have that $(XX')/XS^{-1}X'$ and XX' are independent. Further,

$$\mathcal{L}(XX') = \chi^2_p(\delta)$$

since $\mathcal{L}(X') = N(\mu', I_p)$. Thus

$$\mathcal{L}(XS^{-1}X') = \mathcal{L}\left(\frac{XS^{-1}X'}{XX'}XX'\right) = F(n-p+1, p; \delta).$$ □

The next step in studying $\mathcal{L}(U)$ is the case when $m > 1$, $p > 1$, but $\mu = 0$.

Proposition 8.15. Suppose X and S are independent where $\mathcal{L}(S) = W(I, p, n)$ and $\mathcal{L}(X) = N(0, I_m \otimes I_p)$. Then

$$\mathcal{L}(U) = \mathcal{L}\left(\prod_1^m U_i\right)$$

where $U_1, \ldots, U_m$ are independent and $\mathcal{L}(U_i) = \mathcal{B}((n-p+i)/2, p/2)$.

Proof. The proof is by induction on m and, when $m = 1$, we know

$$\mathcal{L}(U) = \mathcal{B}((n-p+1)/2, p/2).$$

Since $X'X = \Sigma_1^m X_i X_i'$ where X has rows $X_1', \ldots, X_m'$,

$$U = \frac{|S|}{|S + X'X|} = \frac{|S|}{|S + X_1 X_1'|} \times \frac{|S + X_1 X_1'|}{|S + X_1 X_1' + \Sigma_2^m X_i X_i'|}.$$

The first claim is that

$$U_1 \equiv \frac{|S|}{|S + X_1 X_1'|}$$

and

$$W = \frac{|S + X_1 X_1'|}{|S + X_1 X_1' + \Sigma_2^m X_i X_i'|}$$

are independent random variables. Since $X_1, \ldots, X_m$ are independent and independent of S, to show U_1 and W are independent, it suffices to show that U_1 and $S + X_1 X_1'$ are independent. To do this, Proposition 7.19 is applicable. The group Gl_p acts on (S, X_1) by

$$A(S, X_1) = (ASA', AX_1)$$

and the induced group action on $T = S + X_1 X_1'$ sends T into ATA'. The induced group action is clearly transitive. Obviously, T is an equivariant function and also U_1 is an invariant function under the group action on (S, X_1). That T is a sufficient statistic for the parametric family generated by Gl_p and the fixed joint distribution of (S, X_1) is easily checked via the factorization criterion. By Proposition 7.19, U_1 and $S + X_1 X_1'$ are independent. Therefore,

$$\mathcal{L}(U) = \mathcal{L}(U_1 W)$$

where U_1 and W are independent and

$$\mathcal{L}(U_1) = \mathcal{B}\left(\frac{n - p + 1}{2}, \frac{p}{2}\right).$$

However, $\mathcal{L}(S + X_1 X_1') = W(I, p, n + 1)$ and the induction hypothesis applied to W yields

$$\mathcal{L}(W) = \mathcal{L}\left(\prod_{i=1}^{m-1} W_i\right)$$

where $W_1, \ldots, W_{m-1}$ are independent with

$$\mathcal{L}(W_i) = \mathcal{L}\left(\frac{n+1-p+i}{2}, \frac{p}{2}\right).$$

Setting $U_i = W_{i-1}$, $i = 2, \ldots, m$, we have

$$\mathcal{L}(U) = \mathcal{L}\left(\prod_{i=1}^{m} U_i\right)$$

where $U_1, \ldots, U_m$ are independent and

$$\mathcal{L}(U_i) = \mathcal{B}\left(\frac{n-p+i}{2}, \frac{p}{2}\right).$$ □

The above proof shows that U_i's are given by

$$U_i = \frac{|S + \Sigma_{j=1}^{i-1} X_j X_j'|}{|S + \Sigma_{j=1}^{i} X_j X_j|}, \qquad i = 1, \ldots, m$$

and that these random variables are independent. Since $\mathcal{L}(S + \Sigma_1^{i-1} X_j X_j') = W(I, p, n+i-1)$, Proposition 8.14 yields

$$\mathcal{L}(U_i) = \mathcal{B}\left(\frac{n-p+i}{2}, \frac{p}{2}\right).$$

In the special case that Δ has rank one, the distribution of U can be derived by an argument similar to that in the proof of Proposition 8.15.

Proposition 8.16. Suppose X and S are independent where $\mathcal{L}(S) = W(I, p, n)$ and $\mathcal{L}(X) = N(\mu, I_m \otimes I_p)$. Assume that $\mu = \xi\eta'$ with $\xi \in R^m$, $\|\xi\| = 1$, and $\eta \in R^p$. Then

$$\mathcal{L}(U) = \mathcal{L}\left(\prod_1^m U_i\right)$$

where $U_1, \ldots, U_m$ are independent,

$$\mathcal{L}(U_i) = \mathcal{B}\left(\frac{n-p+i}{2}, \frac{p}{2}\right), \qquad i = 1, \ldots, m-1$$

and

$$\mathcal{L}(U_m) = \mathcal{B}\left(\frac{n-p+m}{2}, \frac{p}{2}; \eta'\eta\right).$$

Proof. Let ε_m be the mth standard unit in R^m. Then $\Gamma\xi = \varepsilon_m$ for some $\Gamma \in \mathcal{O}_m$ as $\|\xi\| = \|\varepsilon_m\|$. Since

$$U = \frac{|S|}{|S + X'X|} = \frac{|S|}{|S + X'\Gamma'\Gamma X|}$$

and $\mathcal{L}(\Gamma X) = N(\varepsilon_m \eta', I_m \otimes I_p)$, we can take $\xi = \varepsilon_m$ without loss of generality. As in the proof of Proposition 8.15, $X'X = \sum_1^m X_i X_i'$ where $X_1, \dots, X_m$ are independent. Obviously, $\mathcal{L}(X_i) = N(0, I_p)$, $i = 1, \dots, m-1$, and $\mathcal{L}(X_m) = N(\eta, I_p)$. Now, write $U = \prod_1^m U_i$ where

$$U_i = \frac{|S + \sum_{j=1}^{i-1} X_j X_j'|}{|S + \sum_{j=1}^{i} X_j X_j'|}, \qquad i = 1, \dots, m.$$

The argument given in the proof of Proposition 8.15 shows that

$$U_1 = \frac{|S|}{|S + X_1 X_1'|}$$

and $\{S + X_1 X_1', X_2, \dots, X_m\}$ are independent. The assumption that X_1 has mean zero is essential here in order to verify the sufficiency condition necessary to apply Proposition 7.19. Since $U_2, \dots, U_m$ are functions of $\{S + X_1 X_1', X_2, \dots, X_m\}$, U_1 is independent of $\{U_2, \dots, U_m\}$. Now, we simply repeat this argument $m - 1$ times to conclude that $U_1, \dots, U_m$ are independent, keeping in mind that $X_1, \dots, X_{m-1}$ all have mean zero, but X_m need not have mean zero. As noted earlier,

$$\mathcal{L}(U_i) = \mathcal{B}\left(\frac{n-p+i}{2}, \frac{p}{2}\right); \qquad i = 1, \dots, m-1.$$

By Proposition 8.14,

$$\mathcal{L}(U_m) = \mathcal{L}\left(\frac{|S + \sum_1^{m-1} X_i X_i'|}{|S + \sum_1^m X_i X_i'|}\right) = \mathcal{B}\left(\frac{n-p+m}{2}, \frac{p}{2}; \eta'\eta\right). \qquad \square$$

Now, we return to the case when $\mu = 0$. In terms of the notation $\mathcal{L}(U) = U(n, m, p)$, Proposition 8.14 asserts that

$$U(n, 1, p) = \mathcal{B}\left(\frac{n-p+1}{2}, \frac{p}{2}\right).$$

Further, Proposition 8.15 can be written

$$U(n, m, p) = \prod_{i=1}^{m} U(n + i - 1, 1, p)$$

where this equation means that the distribution $U(n, m, p)$ can be represented as the distribution of the product of m independent random variables with distribution $U(n + i - 1, 1, p)$ for $i = 1, \ldots, m$. An alternative representation of $U(n, m, p)$ in terms of p independent random variables when $m \geqslant p$ follows. If $m \geqslant p$ and

$$U = \frac{|S|}{|S + X'X|}$$

with $\mathcal{L}(S) = W(I, p, n)$ and $\mathcal{L}(X) = N(0, I_m \otimes I_p)$, the matrix $T = X'X$ has a nonsingular Wishart distribution, $\mathcal{L}(T) = W(I, p, m)$. The following technical result provides the basic step for decomposing $U(n, m, p)$ into a product of p independent factors.

Proposition 8.17. Partition S into S_{ij} where S_{ij} is $p_i \times p_j$, $i, j = 1, 2$, and $p_1 + p_2 = p$. Partition T similarly and let

$$Z = S'_{12}S_{11}^{-1}S_{12} + T'_{12}T_{11}^{-1}T_{12} - (S_{12} + T_{12})'(S_{11} + T_{11})^{-1}(S_{12} + T_{12}).$$

Then the five random vectors S_{11}, T_{11}, $S_{22\cdot 1}$, $T_{22\cdot 1}$, and Z are mutually independent. Further,

$$\mathcal{L}(Z) = W(I, p_2, p_1).$$

Proof. Since S and T are independent by assumption, $(S_{11}, S_{12}, S_{22\cdot 1})$ and $(T_{11}, T_{12}, T_{22\cdot 1})$ are independent. Also, Proposition 8.8 shows that (S_{11}, S_{12}) and $S_{22\cdot 1}$ are independent with

$$\mathcal{L}(S_{22\cdot 1}) = W(I, p_2, n - p_1),$$

$$\mathcal{L}(S_{12}|S_{11}) = N(0, S_{11} \otimes I_{p_2}),$$

and

$$\mathcal{L}(S_{11}) = W(I, p_1, n).$$

Similar remarks hold for (T_{11}, T_{12}) and $T_{22\cdot 1}$ with n replaced by m. Thus the four random vectors (S_{11}, S_{12}), $S_{22\cdot 1}$, (T_{11}, T_{12}), and $T_{22\cdot 1}$ are mutually independent. Since Z is a function of (S_{11}, S_{12}) and (T_{11}, T_{12}), the proposition follows if we show that Z is independent of the vector (S_{11}, T_{11}). Conditional on (S_{11}, T_{11}),

$$\mathcal{L}\left(\begin{pmatrix} S_{12} \\ T_{12} \end{pmatrix} \middle| (S_{11}, T_{11}) \right) = N\left(0, \begin{pmatrix} S_{11} & 0 \\ 0 & T_{11} \end{pmatrix} \otimes I_{p_2}\right).$$

Let $A(B)$ be the positive definite square root of $S_{11}(T_{11})$. With $V = A^{-1}S_{12}$ and $W = B^{-1}T_{12}$,

$$\mathcal{L}\left(\begin{bmatrix} V \\ W \end{bmatrix} \middle| (S_{11}, T_{11}) \right) = N(0, I_{2p_1} \otimes I_{p_2}).$$

Also,

$$Z = S_{12}'S_{11}^{-1}S_{12} + T_{12}'T_{11}^{-1}T_{12} - (S_{12} + T_{12})'(S_{11} + T_{11})^{-1}(S_{12} + T_{12})$$
$$= \begin{bmatrix} V \\ W \end{bmatrix}'\begin{bmatrix} V \\ W \end{bmatrix} - \begin{bmatrix} V \\ W \end{bmatrix}'\begin{bmatrix} A \\ B \end{bmatrix}(A^2 + B^2)^{-1}\begin{bmatrix} A \\ B \end{bmatrix}'\begin{bmatrix} V \\ W \end{bmatrix} = \begin{bmatrix} V \\ W \end{bmatrix}'Q\begin{bmatrix} V \\ W \end{bmatrix}$$

where

$$Q = I_{2p_1} - \begin{bmatrix} A \\ B \end{bmatrix}(A^2 + B^2)^{-1}\begin{bmatrix} A \\ B \end{bmatrix}'.$$

However, Q is easily shown to be an orthogonal projection of rank p_1. By Proposition 8.5,

$$\mathcal{L}(Z|(S_{11}, T_{11})) = W(I, p_2, p_1)$$

for each value of (S_{11}, T_{11}). Therefore, Z is independent of (S_{11}, T_{11}) and the proof is complete. □

Proposition 8.18. If $m \geqslant p$, then

$$U(n, m, p) = \prod_{i=1}^{p} U(n - p + i, m, 1).$$

Proof. By definition,

$$U(n, m, p) = \mathcal{L}\left(\frac{|S|}{|S + T|}\right)$$

where S and T are independent, $\mathcal{L}(T) = W(I, p, m)$ and $\mathcal{L}(S) = W(I, p, n)$

with $n \geqslant p$. In the notation of Proposition 8.17, partition S and T with $p_1 = 1$ and $p_2 = p - 1$. Then S_{11}, T_{11}, $S_{22\cdot 1}$, $T_{22\cdot 1}$, and

$$Z = S_{12}'S_{22}^{-1}S_{12} + T_{12}'T_{11}^{-1}T_{12} - (S_{12} + T_{12})'(S_{11} + T_{11})^{-1}(S_{12} + T_{12})$$

are mutually independent. However,

$$|S| = |S_{11}||S_{22\cdot 1}|$$

and

$$|S + T| = |S_{11} + T_{11}||(S + T)_{22\cdot 1}| = |S_{11} + T_{11}||S_{22\cdot 1} + T_{22\cdot 1} + Z|.$$

Thus

$$\frac{|S|}{|S + T|} = \frac{|S_{11}|}{|S_{11} + T_{11}|} \times \frac{|S_{22\cdot 1}|}{|S_{22\cdot 1} + T_{22\cdot 1} + Z|}$$

and the two factors on the right side of this equality are independent by Proposition 8.17. Obviously,

$$\mathcal{L}\left(\frac{|S_{11}|}{|S_{11} + T_{11}|}\right) = U(n, m, 1).$$

Since $\mathcal{L}(T_{22\cdot 1}) = W(I, p - 1, m - 1)$, $\mathcal{L}(Z) = W(I, p - 1, 1)$, and $T_{22\cdot 1}$ and Z are independent, it follows that

$$\mathcal{L}(T_{22\cdot 1} + Z) = W(I, p - 1, m).$$

Therefore,

$$\mathcal{L}\left(\frac{|S_{22\cdot 1}|}{|S_{22\cdot 1} + T_{22\cdot 1} + Z|}\right) = U(n - 1, m, p - 1),$$

which implies the relation

$$U(n, m, p) = U(n, m, 1)U(n - 1, m, p - 1).$$

Now, an easy induction argument establishes

$$U(n, m, p) = \prod_{i=1}^{p} U(n - i + 1, m, 1),$$

which implies that

$$U(n, m, p) = \prod_{i=1}^{p} U(n - p + i, m, 1)$$

and this completes the proof. □

Combining Propositions 8.15 and 8.18 leads to the following.

Proposition 8.19. If $m \geqslant p$, then

$$U(n, m, p) = U(n - p + m, p, m).$$

Proof. For arbitrary m, Proposition 8.15 yields

$$U(n, m, p) = \prod_{i=1}^{m} \mathfrak{B}\left(\frac{n - p + i}{2}, \frac{p}{2}\right)$$

where this notation means that the distribution $U(n, m, p)$ can be represented as the product of m independent beta-random variables with the factors in the product having a $\mathfrak{B}((n - p + i)/2, p/2)$ distribution. Since

$$U(n - p + i, m, 1) = \mathfrak{B}\left(\frac{n - p + i}{2}, \frac{m}{2}\right),$$

Proposition 8.18 implies that

$$U(n, m, p) = \prod_{i=1}^{p} U(n - p + i, m, 1) = \prod_{i=1}^{p} \mathfrak{B}\left(\frac{n - p + i}{2}, \frac{m}{2}\right).$$

Applying Proposition 8.15 to $U(n - p + m, p, m)$ yields

$$\begin{aligned} U(n - p + m, p, m) &= \prod_{i=1}^{p} \mathfrak{B}\left(\frac{n - p + m - m + i}{2}, \frac{m}{2}\right) \\ &= \prod_{i=1}^{p} \mathfrak{B}\left(\frac{n - p + i}{2}, \frac{m}{2}\right), \end{aligned}$$

which is the distribution $U(n, m, p)$. □

In practice, the relationship $U(n, m, p) = U(n - p + m, p, m)$ shows that it is sufficient to deal with the case that $m \leqslant p$ when tabulating the distribution $U(n, m, p)$. Rather accurate approximations to the percentage points of the distribution $U(n, m, p)$ are available and these are discussed in detail in Anderson (1958, Chapter 8). This topic is not pursued further here.

PROBLEMS

1. Suppose S is $W(\Sigma, 2, n)$, $n \geqslant 2$, $\Sigma > 0$. Show that the density of $r = s_{12}/\sqrt{s_{11}s_{22}}$ can be written as

$$p(r|\rho) = \Gamma^2\left(\frac{n}{2}\right)2^n\omega(2, n)(1-\rho^2)^{n/2}(1-r^2)^{(n-1)/2}\psi(\rho r)$$

where $\rho = \sigma_{12}/\sqrt{\sigma_{11}\sigma_{22}}$ and ψ is defined as follows. Let X_1 and X_2 be independent chi-square random variables each with n degrees of freedom. Then $\psi(t) = \mathcal{E}\exp[t(X_1X_2)^{1/2}]$ for $|t| \leqslant 1$. Using this representation, prove that $p(r|\rho)$ has a monotone likelihood ratio.

2. The gamma distribution with parameters $\alpha > 0$ and $\lambda > 0$, denoted by $G(\alpha, \lambda)$, has the density

$$p(x|\alpha, \lambda) = \frac{x^{\alpha-1}}{\lambda^\alpha\Gamma(\alpha)}\exp\left[\frac{-x}{\lambda}\right], \qquad x > 0$$

with respect to Lebesgue measure on $(0, \infty)$.

(i) Show the characteristic function of this distribution is $(1 - i\lambda t)^{-\alpha}$.

(ii) Show that a $G(n/2, 2)$ distribution is that of a χ^2_n distribution.

3. The above problem suggests that it is natural to view the gamma family as an extension of the chi-squared family by allowing nonintegral degrees of freedom. Since the $W(\Sigma, p, n)$ distribution is a generalization of the chi-squared distribution, it is reasonable to ask if we can define a Wishart distribution for nonintegral degrees of freedom. One way to pose this question is to ask for what values of α is $\phi_\alpha(A) = |I_p - 2iA|^\alpha$, $A \in \mathbb{S}_p$, a characteristic function. (We have taken $\Sigma = I_p$ for convenience).

(i) Using Proposition 8.3 and Problem 7.1, show that ϕ_α is a characteristic function for $\alpha = 1/2, \ldots, (p-1)/2$ and all real $\alpha > (p-1)/2$. Give the density that corresponds to ϕ_α for $\alpha > (p-1)/2$. $W(I_p, p, 2\alpha)$ denotes such a distribution.

(ii) For any $\Sigma \geqslant 0$ and the values of α given in (i), show that $\phi_\alpha(\Sigma A)$, $A \in \mathbb{S}_p$, is a characteristic function.

4. Let S be a random element of the inner product space $(\mathbb{S}_p, \langle \cdot , \cdot \rangle)$ where $\langle \cdot , \cdot \rangle$ is the usual trace inner product on $\mathbb{S}_p$. Say that S has an $\mathcal{O}_p$-invariant distribution if $\mathcal{L}(S) = \mathcal{L}(\Gamma S \Gamma')$ for each $\Gamma \in \mathcal{O}_p$. Assume S has an $\mathcal{O}_p$-invariant distribution.

(i) Assuming $\mathcal{E}S$ exists, show that $\mathcal{E}S = cI_p$ where $c = \mathcal{E}s_{11}$ and s_{ij} is the i, j element of S.

(ii) Let $D \in \mathbb{S}_p$ be diagonal with diagonal elements $d_1, \ldots, d_p$. Show that $\operatorname{var}(\langle D, S \rangle) = (\gamma - \beta)\Sigma d_i^2 + \beta(\Sigma_1^p d_i)^2$ where $\gamma = \operatorname{var}(s_{11})$ and $\beta = \operatorname{cov}\{s_{11}, s_{22}\}$.

(iii) For $A \in \mathbb{S}_p$, show that $\operatorname{var}(\langle A, S \rangle) = (\gamma - \beta)\langle A, A \rangle + \beta(I_p, A)^2$. From this conclude that $\operatorname{Cov}(S) = (\gamma - \beta) I_p \otimes I_p + \beta I_p \square I_p$.

5. Suppose $S \in \mathbb{S}_p^+$ has a density f with respect to Lebesgue measure dS restricted to $\mathbb{S}_p^+$. For each $n \geqslant p$, show there exists a random matrix $X \in \mathcal{L}_{p,n}$ that has a density with respect to Lebesgue measure on $\mathcal{L}_{p,n}$ and $\mathcal{L}(X'X) = \mathcal{L}(S)$.

6. Show that Proposition 8.4 holds for all n_1, n_2 equal to $1, 2, \ldots, p - 1$ or any real number greater than $p - 1$.

7. (The inverse Wishart distribution.) Say that a positive definite $S \in \mathbb{S}_p^+$ has an inverse Wishart distribution with parameters Λ, p, and ν if $\mathcal{L}(S^{-1}) = W(\Lambda^{-1}, p, \nu + p - 1)$. Here $\Lambda \in \mathbb{S}_p^+$ and ν is a positive integer. The notation $\mathcal{L}(S) = IW(\Lambda, p, \nu)$ signifies that $\mathcal{L}(S^{-1}) = W(\Lambda^{-1}, p, \nu + p - 1)$.

(i) If $\mathcal{L}(S) = IW(\Lambda, p, \nu)$ and A is $r \times p$ of rank r, show that $\mathcal{L}(ASA') = IW(A\Lambda A', r, \nu)$.

(ii) If $\mathcal{L}(S) = IW(I_p, p, \nu)$ and $\Gamma \in \mathcal{O}_p$, show that $\mathcal{L}(\Gamma S \Gamma') = \mathcal{L}(S)$.

(iii) If $\mathcal{L}(S) = IW(\Lambda, p, \nu)$, show that $\mathcal{E}(S) = (\nu - 2)^{-1}\Lambda$. Show that $\operatorname{Cov}(S)$ has the form $c_1 \Lambda \otimes \Lambda + c_2 \Lambda \square \Lambda$—what are c_1 and c_2?

(iv) Now, partition S into $S_{11}: q \times q$, $S_{12}: q \times r$, and $S_{22}: r \times r$ with S as in (iii). Show that $\mathcal{L}(S_{11}) = IW(\Lambda_{11}, q, \nu)$. Also show that $\mathcal{L}(S_{22 \cdot 1}) = IW(\Lambda_{22 \cdot 1}, r, \nu + q)$.

8. (The matric t distribution.) Suppose X is $N(0, I_r \otimes I_p)$ and S is $W(I_p, p, m)$, $m \geqslant p$. Let $S^{-1/2}$ denote the inverse of the positive definite square root of S. When S and X are independent, the matrix $T = XS^{-1/2}$ is said to have a matric t distribution and is denoted by $\mathcal{L}(T) = T(m - p + 1, I_r, I_p)$.

(i) Show that the density of T with respect to Lebesgue measure on $\mathcal{L}_{p,r}$ is given by

$$p(T) = \frac{\omega(m, p)}{(\sqrt{2\pi})^{rp}\omega(m+r, p)} \frac{1}{|I_p + T'T|^{(m+r)/2}}.$$

Also, show that $\mathcal{L}(T) = \mathcal{L}(\Gamma T \Delta')$ for $\Gamma \in \mathcal{O}_r$ and $\Delta \in \mathcal{O}_p$. Using this, show $\mathcal{E}T = 0$ and $\text{Cov}(T) = c_1 I_r \otimes I_p$ when these exist. Here, c_1 is a constant equal to the variance of any element of T.

(ii) Suppose V is $IW(I_p, p, \nu)$ and that T given V is $N(0, I_r \otimes V)$. Show that the unconditional distribution of T is $T(\nu, I_r, I_p)$.

(iii) Using Problem 7 and (ii), show that if T is $T(\nu, I_r, I_p)$, and T_{11} is the $k \times q$ upper left-hand corner of T, then T_{11} is $T(\nu, I_k, I_q)$.

9. (Multivariate F distribution.) Suppose S_1 is $W(I_p, p, m)$ (for $m = 1, 2, \ldots$) and is independent of S_2, which is $W(I_p, p, \nu + p - 1)$ (for $\nu = 1, 2, \ldots$). The matrix $F = S_2^{-1/2} S_1 S_2^{-1/2}$ has a matric F distribution that is denoted by $F(m, \nu, I_p)$.

(i) If S is $IW(I_p, p, \nu)$ and V given S is $W(S, p, m)$, show that the unconditional distribution of V is $F(m, \nu, I_p)$.

(ii) Suppose T is $T(\nu, I_r, I_p)$. Show that $T'T$ is $F(r, \nu, I_p)$.

(iii) When $r \geqslant p$, show that the $F(r, \nu, I_p)$ distribution has a density with respect to $dF/|F|^{(p+1)/2}$ given by

$$p(F) = \frac{\omega(r, p)\omega(\nu + p - 1, p)}{\omega(r + \nu + p - 1, p)} \frac{|F|^{r/2}}{|I_p + F|^{(\nu+p+r-1)/2}}.$$

(iv) For $r \geqslant p$, show that, if F is $F(r, \nu, I_p)$, then F^{-1} is $F(\nu + p - 1, r - p + 1, I_p)$.

(v) If F is $F(r, \nu, I_p)$ and F_{11} is the $q \times q$ upper left block of F, use (ii) to show that F_{11} is $F(r, \nu, I_q)$.

(vi) Suppose X is $N(0, I_r \otimes I_p)$ with $r \leqslant p$ and S is $W(I_p, p, m)$ with $m \geqslant p$, X and S independent. Show that $XS^{-1}X'$ is $F(p, m - p + 1, I_r)$.

10. (Multivariate beta distribution.) Let S_1 and S_2 be independent and suppose $\mathcal{L}(S_i) = W(I_p, p, m_i)$, $i = 1, 2$, with $m_1 + m_2 \geqslant p$. The random matrix $B = (S_1 + S_2)^{-1/2} S_1 (S_1 + S_2)^{-1/2}$ has a p-dimensional multivariate beta distribution with parameters m_1 and m_2. This is written $\mathcal{L}(B) = B(m_1, m_2, I_p)$ (when $p = 1$, this is the univariate beta distribution with parameters $m_1/2$ and $m_2/2$).

(i) If B is $B(m_1, m_2, I_p)$ show that $\mathcal{L}(\Gamma B \Gamma') = \mathcal{L}(B)$ for all $\Gamma \in \mathcal{O}_p$. Use Example 7.16 to conclude that $\mathcal{L}(B) = \mathcal{L}(\Psi D \Psi')$ where $\Psi \in \mathcal{O}_p$ is uniform and is independent of the diagonal matrix D with elements $\lambda_1 \geqslant \cdots \geqslant \lambda_p \geqslant 0$. The distribution of D is determined by specifying the distribution of $\lambda_1, \ldots, \lambda_p$ and this is the distribution of the ordered roots of $(S_1 + S_2)^{-1/2} S_1 (S_1 + S_2)^{-1/2}$.

(ii) With S_1 and S_2 as in the definition of B, show that $S_1^{1/2}(S_1 + S_2)^{-1} S_1^{1/2}$ is $B(m_1, m_2, I_p)$.

(iii) Suppose F is $F(m, \nu, I_p)$. Use (i) and (ii) to show that $(I + F)^{-1}$ is $B(p + \nu - 1, m, I_p)$ and $F(I + F)^{-1}$ is $B(m, p + \nu - 1, I_p)$.

(iv) Suppose that X is $N(0, I_r \otimes I_p)$ and that it is independent of S, which is $W(I_p, p, m)$. When $r \leqslant p$ and $m \geqslant p$, show that $X(S + X'X)^{-1}X'$ is $B(p, r + m - p, I_r)$.

(v) If B is $B(m_1, m_2, I_p)$ and $m_1 \geqslant p$, show that $\det(B)$ is distributed as $U(m_1, m_2, p)$ in the notation of Section 7.4.

NOTES AND REFERENCES

1. The Wishart distribution was first derived in Wishart (1928).

2. For some alternative discussions of the Wishart distribution, see Anderson (1958), Dempster (1969), Rao (1973), and Muirhead (1982).

3. The density function of the noncentral Wishart distribution in the general case is obtained by "evaluating"

$$(8.1) \qquad \int_{\mathcal{O}_n} \exp[\operatorname{tr} \Gamma X \Sigma^{-1} \mu'] \mu_0(d\Gamma)$$

(see the proof of Proposition 8.12). The problem of evaluating

$$\psi(A) \equiv \int_{\mathcal{O}_n} \exp[\operatorname{tr} \Gamma A] \mu_0(d\Gamma)$$

for $A \in \mathcal{L}_{n,n}$ has received much attention since the paper of James (1954). Anderson (1946) first gave the noncentral Wishart density when

μ has rank 1 or rank 2. Much of the theory surrounding the evaluation of ψ and series expansions for ψ can be found in Muirhead (1982).

4. Wilks (1932) first proved Proposition 8.15 by calculating all the moments of U and showing these matched the moments of $\prod U_i$. Anderson (1958) also uses the moment method to find the distribution of U. This method was used by Box (1949) to provide asymptotic expansions for the distribution of U (see Anderson, 1958, Chapter 8).

CHAPTER 9

Inference for Means in Multivariate Linear Models

Essentially, this chapter consists of a number of examples of estimation and testing problems for means when an observation vector has a normal distribution. Invariance is used throughout to describe the structure of the models considered and to suggest possible testing procedures. Because of space limitations, maximum likelihood estimators are the only type of estimators discussed. Further, likelihood ratio tests are calculated for most of the examples considered.

Before turning to the concrete examples, it is useful to have a general model within which we can view the results of this chapter. Consider an n-dimensional inner product space $(V, (\cdot, \cdot))$ and suppose that X is a random vector in V. To describe the type of parametric models we consider for X, let f be a decreasing function on $[0, \infty)$ to $[0, \infty)$ such that $f[(x, x)]$ is a density with respect to Lebesgue measure on $(V, (\cdot, \cdot))$. For convenience, it is assumed that f has been normalized so that, if $Z \in V$ has density f, then $\text{Cov}(Z) = I$. Obviously, such a Z has mean zero. Now, let M be a subspace of V and let γ be a set of positive definite linear transformations on V to V such that $I \in \gamma$. The pair (M, γ) serves as the parameter space for a model for X. For $\mu \in M$ and $\Sigma \in \gamma$,

$$p(x|\mu, \Sigma) \equiv |\Sigma|^{-n/2} f\left[\left(x - \mu, \Sigma^{-1}(x - \mu)\right)\right]$$

is a density on V. The family

$$\{p(\cdot|\mu, \Sigma)|\mu \in M, \Sigma \in \gamma\}$$

determines a parametric model for X. It is clear that if $p(\cdot|\mu, \Sigma)$ is the density of X, then $\mathcal{E}X = \mu$ and $\text{Cov}(X) = \Sigma$. In particular, when

$$f(u) = (\sqrt{2\pi})^{-n}\exp[-\tfrac{1}{2}u], \qquad u \geqslant 0,$$

then X has a normal distribution with mean $\mu \in M$ and covariance $\Sigma \in \gamma$. The parametric model for X is in fact a linear model for X with parameter set (M, γ). Now, assume that $\Sigma(M) = M$ for all $\Sigma \in \gamma$. Since $I \in \gamma$, the least-squares and Gauss–Markov estimator of μ are equal to PX where P is the orthogonal projection onto M. Further, $\hat{\mu} \equiv PX$ is also the maximum likelihood estimator of μ. To see this, first note that $P\Sigma = \Sigma P$ for $\Sigma \in \gamma$ since M is invariant under $\Sigma \in \gamma$. With $Q = I - P$, we have

$$\begin{aligned}(x - \mu, \Sigma^{-1}(x - \mu)) &= (P(x - \mu) + Qx, \Sigma^{-1}(P(x - \mu) + Qx)) \\ &= (Px - \mu, \Sigma^{-1}(Px - \mu)) + (Qx, \Sigma^{-1}Qx).\end{aligned}$$

The last equality is a consequence of

$$(Qx, \Sigma^{-1}P(x - \mu)) = (x, Q\Sigma^{-1}P(x - \mu)) = (x, QP\Sigma^{-1}(x - \mu)) = 0$$

as $QP = 0$ and $\Sigma^{-1}P = P\Sigma^{-1}$. Therefore, for each $\Sigma \in \gamma$,

$$(x - \mu, \Sigma^{-1}(x - \mu)) \geqslant (Qx, \Sigma^{-1}Qx)$$

with equality iff $\mu = Px$. Since the function f was assumed to be decreasing, it follows that $\hat{\mu} = PX$ is the maximum likelihood estimator of μ, and $\hat{\mu}$ is unique if f is strictly decreasing. Thus under the assumptions made so far, $\hat{\mu} = PX$ is the maximum likelihood estimator for μ. These assumptions hold for most of the examples considered in this chapter. To find the maximum likelihood estimator of Σ, it is necessary to compute

$$\sup_{\Sigma \in \gamma} |\Sigma|^{-n/2} f[(Qx, \Sigma^{-1}Qx)]$$

and find the point $\hat{\Sigma} \in \gamma$ where the supremum is achieved, assuming it exists. The solution to this problem depends crucially on the set γ and this is what generates the infinite variety of possible models, even with the assumption that $\Sigma M = M$ for $\Sigma \in \gamma$. The examples of this chapter are generated by simply choosing some γ's for which $\hat{\Sigma}$ can be calculated explicitly.

We end this rather lengthy introduction with a few general comments about testing problems. In the notation of the previous paragraph, consider a parameter set (M, γ) with $I \in \gamma$ and assume $\Sigma M = M$ for $\Sigma \in \gamma$. Also, let

$M_0 \subset M$ be a subspace of V and assume that $\Sigma M_0 = M_0$ for $\Sigma \in \gamma$. Consider the problem of testing the null hypothesis that $\mu \in M_0$ versus the alternative that $\mu \in (M - M_0)$. Under the null hypothesis, the maximum likelihood estimator for μ is $\hat{\mu}_0 = P_0 X$ where P_0 is the orthogonal projection onto M_0. Thus the likelihood ratio test rejects the null hypothesis for small values of

$$\Lambda(x) = \frac{\sup_{\Sigma \in \gamma} |\Sigma|^{-n/2} f\left[(Q_0 x, \Sigma^{-1} Q_0 x)\right]}{\sup_{\Sigma \in \gamma} |\Sigma|^{-n/2} f\left[(Qx, \Sigma^{-1} Qx)\right]}$$

where $Q_0 = I - P_0$. Again, the set γ is the major determinant with regard to the distribution, invariance, and other properties of $\Lambda(x)$. The examples in this chapter illustrate some of the properties of γ that lead to tractable solutions to the estimation problem for Σ and the testing problem described above.

9.1. THE MANOVA MODEL

The multivariate general linear model introduced in Example 4.4, also known as the multivariate analysis of variance model (the MANOVA model), is the subject of this section. The vector space under consideration is $\mathcal{L}_{p,n}$ with the usual inner product $\langle \cdot , \cdot \rangle$ and the subspace M of $\mathcal{L}_{p,n}$ is

$$M = \{x | x = Z\beta, \beta \in \mathcal{L}_{p,k}\}$$

where Z is a fixed $n \times k$ matrix of rank k. Consider an observation vector $X \in \mathcal{L}_{p,n}$ and assume that

$$\mathcal{L}(X) = N(\mu, I_n \otimes \Sigma)$$

where $\mu \in M$ and Σ is an unknown $p \times p$ positive definite matrix. Thus the set of covariances for X is

$$\gamma = \{I_n \otimes \Sigma | \Sigma \in \mathcal{S}_p^+\}$$

and (M, γ) is the parameter set of the linear model for X. It was verified in Example 4.4 that M is invariant under each element of γ. Also, the orthogonal projection onto M is $P = P_z \otimes I_p$ where

$$P_z = Z(Z'Z)^{-1}Z'.$$

Further, $Q = I - P = Q_z \otimes I_p$ is the orthogonal projection onto $M^\perp$ where

$Q_z = I_n - P_z$. Thus

$$\hat{\mu} = PX = (P_z \otimes I_p)X = P_z X$$

is the maximum likelihood estimator of $\mu \in M$ and, from Example 7.10,

$$\hat{\Sigma} = \frac{1}{n} X' Q_z X$$

is the maximum likelihood estimator of Σ when $n - k \geqslant p$, which we assume throughout this discussion. Thus for the MANOVA model, the maximum likelihood estimators have been derived. The reader should check that the MANOVA model is a special case of the linear model described at the beginning of this chapter.

We now turn to a discussion of the classical MANOVA testing problem. Let K be a fixed $r \times k$ matrix of rank r and consider the problem of testing

$$H_0 : K\beta = 0 \quad \text{versus} \quad H_1 : K\beta \neq 0$$

where $\mu = Z\beta$ is the mean of X. It is not obvious that this testing problem is of the general type described in the introduction to this chapter. However, before proceeding further, it is convenient to transform this problem into what is called the canonical form of the MANOVA testing problem. The essence of the argument below is that it suffices to take

$$Z = Z_0 \equiv \begin{pmatrix} I_k \\ 0 \end{pmatrix}, \qquad K = K_0 \equiv (I_r \ \ 0)$$

in the above problem. In other words, a transformation of the original problem results in a problem where $Z = Z_0$ and $K = K_0$. We now proceed with the details. The parametric model for $X \in \mathcal{L}_{p,n}$ is

$$\mathcal{L}(X) = N(Z\beta, I_n \otimes \Sigma)$$

and the statistical problem is to test $H_0 : K\beta = 0$ versus $H_1 : K\beta \neq 0$. Since Z has rank k, $Z = \Psi U$ for some linear isometry $\Psi : n \times k$ and some $k \times k$ matrix $U \in G_U^+$. The k columns of Ψ are the first k columns of some $\Gamma \in \mathcal{O}_n$ so

$$\Psi = \Gamma \begin{pmatrix} I_k \\ 0 \end{pmatrix} = \Gamma Z_0.$$

Setting $\tilde{X} = \Gamma' X$, $\tilde{\beta} = U\beta$, and $\tilde{K} = KU^{-1}$, we have

$$\mathcal{L}(\tilde{X}) = N(Z_0 \tilde{\beta}, I_n \otimes \Sigma)$$

and the testing problem is $H_0: \tilde{K}\tilde{\beta} = 0$ versus $H_1: \tilde{K}\tilde{\beta} \neq 0$. Applying the same argument to $\tilde{K}'$ as we did to Z,

$$\tilde{K}' = \Delta \begin{pmatrix} I_r \\ 0 \end{pmatrix} U_1$$

for some $\Delta \in \mathcal{O}_k$ and some $r \times r$ matrix U_1 in G_U^+. Let

$$\Gamma_1 = \begin{pmatrix} \Delta' & 0 \\ 0 & I_{n-k} \end{pmatrix} \in \mathcal{O}_n$$

and set $Y = \Gamma_1 X$, $B = \Delta'\tilde{\beta}$. Since

$$\Gamma_1 Z_0 \tilde{\beta} = \begin{pmatrix} \Delta' & 0 \\ 0 & I_{n-k} \end{pmatrix} \begin{pmatrix} \tilde{\beta} \\ 0 \end{pmatrix} = \begin{pmatrix} B \\ 0 \end{pmatrix} = Z_0 B,$$

it follows that

$$\mathcal{L}(Y) = N(Z_0 B, I_n \otimes \Sigma)$$

and the testing problem is $H_0: K_0 B = 0$ versus $H_1: K_0 B \neq 0$. Thus after two transformations, the original problem has been transformed into a problem with $Z = Z_0$ and $K = K_0$. Since $K_0 = (I_r\ 0)$, the null hypothesis is that the first r rows of B are zero. Partition B into

$$B = \begin{pmatrix} B_1 \\ B_2 \end{pmatrix}; \qquad B_1: r \times p, \qquad B_2: (k-r) \times p$$

and partition Y into

$$Y = \begin{pmatrix} Y_1 \\ Y_2 \\ Y_3 \end{pmatrix}; \qquad Y_1: r \times p, \qquad Y_2: (k-r) \times p, \qquad Y_3: (n-k) \times p.$$

Because $\text{Cov}(Y) = I_n \otimes \Sigma$, Y_1, Y_2, and Y_3 are mutually independent and it is clear that

$$\mathcal{L}(Y_1) = N(B_1, I_r \otimes \Sigma)$$

$$\mathcal{L}(Y_2) = N(B_2, I_{(k-r)} \otimes \Sigma)$$

$$\mathcal{L}(Y_3) = N(0, I_{(n-k)} \otimes \Sigma).$$

Also, the testing problem is $H_0: B_1 = 0$ versus $H_1: B_1 \neq 0$. It is this form of the problem that is called the canonical MANOVA testing problem. The only reason for transforming from the original problem to the canonical problem is that certain expressions become simpler and the invariance of the MANOVA testing problem is more easily articulated when the problem is expressed in canonical form.

We now proceed to analyze the canonical MANOVA testing problem. To simplify some later formulas, the notation is changed a bit. Let Y_1, Y_2, and Y_3 be independent random matrices that satisfy

$$\mathcal{L}(Y_1) = N(B_1, I_r \otimes \Sigma)$$

$$\mathcal{L}(Y_2) = N(B_2, I_s \otimes \Sigma)$$

$$\mathcal{L}(Y_3) = N(0, I_m \otimes \Sigma)$$

so B_1 is $r \times p$ and B_2 is $s \times p$. As usual Σ is a $p \times p$ unknown positive definite matrix. To insure the existence of a maximum likelihood estimator for Σ, it is assumed that $m \geqslant p$ and the sample space for Y_3 is taken to be the set of all $m \times p$ real matrices of rank p. A set of Lebesgue measure zero has been deleted from the natural sample space $\mathcal{L}_{p,m}$ of Y_3. The testing problem is

$$H_0: B_1 = 0 \quad \text{versus} \quad H_1: B_1 \neq 0.$$

Setting $n = r + s + m$ and

$$Y = \begin{pmatrix} Y_1 \\ Y_2 \\ Y_3 \end{pmatrix} \in \mathcal{L}_{p,n},$$

$\mathcal{L}(Y) = N(\mu, I_n \otimes \Sigma)$ where μ is an element of the subspace

$$M = \left\{ \mu | \mu = \begin{pmatrix} B_1 \\ B_2 \\ 0 \end{pmatrix}; B_1 \in \mathcal{L}_{p,r}, B_2 \in \mathcal{L}_{p,s} \right\}.$$

In this notation, the null hypothesis is that $\mu \in M_0 \subset M$ where

$$M_0 = \left\{ \mu | \mu = \begin{pmatrix} 0 \\ B_2 \\ 0 \end{pmatrix}; B_2 \in \mathcal{L}_{p,s} \right\}.$$

Since M and M_0 are both invariant under $I_n \otimes \Sigma$ for all $\Sigma > 0$, the testing problem under consideration is of the type described in general terms earlier, and

$$\gamma = \{I_n \otimes \Sigma | \Sigma > 0\}.$$

When the model for Y is (M, γ), the density of Y is

$$\begin{aligned} p(Y|B_1, B_2, \Sigma) = (\sqrt{2\pi})^{-n} |\Sigma|^{-n/2} \\ \times \exp\left[-\tfrac{1}{2}\operatorname{tr}(Y_1 - B_1)\Sigma^{-1}(Y_1 - B_1)' \right. \\ \left. -\tfrac{1}{2}\operatorname{tr}(Y_2 - B_2)\Sigma^{-1}(Y_2 - B_2)' - \tfrac{1}{2}\operatorname{tr} Y_3\Sigma^{-1}Y_3'\right]. \end{aligned}$$

In this case, the maximum likelihood estimators of B_1, B_2, and Σ are easily seen to be

$$\hat{B}_1 = Y_1, \qquad \hat{B}_2 = Y_2, \qquad \hat{\Sigma} = \frac{1}{n} Y_3'Y_3.$$

When the model for Y is (M_0, γ), the density of Y is $p(Y|0, B_2, \Sigma)$ and the maximum likelihood estimators of B_2 and Σ are

$$\tilde{B}_2 = Y_2, \qquad \tilde{\Sigma} = \frac{1}{n}(Y_3'Y_3 + Y_1'Y_1).$$

Therefore, the likelihood ratio test rejects for small values of

$$\Lambda(Y) = \frac{p(Y|0, \tilde{B}_2, \tilde{\Sigma})}{p(Y|\hat{B}_1, \hat{B}_2, \hat{\Sigma})} = \frac{|\tilde{\Sigma}|^{-n/2}}{|\hat{\Sigma}|^{-n/2}} = \frac{|Y_3'Y_3|^{n/2}}{|Y_3'Y_3 + Y_1'Y_1|^{n/2}}.$$

Summarizing this, we have the following result.

Proposition 9.1. For the canonical MANOVA testing problem, the likelihood ratio test rejects the null hypothesis for small values of the statistic

$$U = \frac{|Y_3'Y_3|}{|Y_3'Y_3 + Y_1'Y_1|}.$$

Under H_0, $\mathcal{L}(U) = U(m, r, p)$ where the distribution $U(m, r, p)$ is given in Proposition 8.15.

Proof. The first assertion is clear. Under H_0, $\mathcal{L}(Y_1) = N(0, I_r \otimes \Sigma)$ and $\mathcal{L}(Y_3) = N(0, I_m \otimes \Sigma)$. Therefore, $\mathcal{L}(Y_1'Y_1) = W(\Sigma, p, r)$ and $\mathcal{L}(Y_3'Y_3) = W(\Sigma, p, m)$. Since $m \geqslant p$, Proposition 8.18 implies the result. □

Before attempting to interpret the likelihood ratio test, it is useful to see first what implications can be obtained from invariance considerations in the canonical MANOVA problem. In the notation of the previous paragraph, (M, γ) is the parameter set for the model for Y and under the null hypothesis, (M_0, γ) is the parameter set for Y. In order that the testing problem be invariant under a group of transformations, both of the parameter sets (M, γ) and (M_0, γ) must be invariant. With this in mind, consider the group G defined by

$$G = \{g | g = (\Gamma_1, \Gamma_2, \Gamma_3, \xi, A); \Gamma_1 \in \mathcal{O}_r, \Gamma_2 \in \mathcal{O}_s,$$

$$\Gamma_3 \in \mathcal{O}_m, \xi \in \mathcal{L}_{p,s}, A \in Gl_p\}$$

where the group action on the sample space is given by

$$(\Gamma_1, \Gamma_2, \Gamma_3, \xi, A)\begin{pmatrix} Y_1 \\ Y_2 \\ Y_3 \end{pmatrix} = \begin{pmatrix} \Gamma_1 Y_1 A' \\ \Gamma_2 Y_2 A' + \xi \\ \Gamma_3 Y_3 A' \end{pmatrix}.$$

The group composition, defined so that the above action is a left action on the sample space, is

$$(\Gamma_1, \Gamma_2, \Gamma_3, \xi, A)(\Delta_1, \Delta_2, \Delta_3, \eta, C) = (\Gamma_1\Delta_1, \Gamma_2\Delta_2, \Gamma_3\Delta_3, \Gamma_2\eta A' + \xi, AC).$$

Further, the induced group action on the parameter set (M, γ) is

$$(\Gamma_1, \Gamma_2, \Gamma_3, \xi, A)(B_1, B_2, \Sigma) = (\Gamma_1 B_1 A', \Gamma_2 B_2 A' + \xi, A\Sigma A'),$$

where the point

$$\begin{pmatrix} B_1 \\ B_2 \\ 0 \end{pmatrix} \in M, \qquad (I_n \otimes \Sigma) \in \gamma$$

has been represented simply by (B_1, B_2, Σ). Now it is routine to check that when Y has a normal distribution with $\mathcal{E}Y \in M (\mathcal{E}Y \in M_0)$ and $\text{Cov}(Y) \in \gamma$, then $\mathcal{E}gY \in M (\mathcal{E}gY \in M_0)$ and $\text{Cov}(gY) \in \gamma$, for $g \in G$. Thus the hypothesis testing problem is G-invariant and the likelihood ratio test is a G-invariant function of Y. To describe the invariant tests, a maximal invariant under the action of G on the sample space needs to be computed. The following result provides one form of a maximal invariant.

Proposition 9.2. Let $t = \min\{r, p\}$, and define $h(Y_1, Y_2, Y_3)$ to be the t-dimensional vector $(\lambda_1, \dots, \lambda_t)'$ where $\lambda_1 \geqslant \cdots \geqslant \lambda_t$ are the t largest eigenvalues of $Y_1'Y_1(Y_3'Y_3)^{-1}$. Then h is a maximal invariant under the action of G on the sample space of Y.

Proof. Note that $Y_1'Y_1(Y_3'Y_3)^{-1}$ has at most t nonzero eigenvalues, and these t eigenvalues are nonnegative. First, consider the case when $r \leqslant p$ so $t = r$. By Proposition 1.39, the nonzero eigenvalues of $Y_1'Y_1(Y_3'Y_3)^{-1}$ are the same as the nonzero eigenvalues of $Y_1(Y_3'Y_3)^{-1}Y_1'$, and these eigenvalues are obviously invariant under the action of g on Y. To show that h is maximal invariant for this case, a reduction argument similar to that in Example 7.4 is used. Given

$$Y = \begin{pmatrix} Y_1 \\ Y_2 \\ Y_3 \end{pmatrix},$$

we claim that there exists a $g_0 \in G$ such that

$$g_0(Y) = \begin{pmatrix} (D0) \\ 0 \\ \begin{pmatrix} I_p \\ 0 \end{pmatrix} \end{pmatrix} \in \begin{pmatrix} \mathcal{L}_{p,r} \\ \mathcal{L}_{p,s} \\ \mathcal{L}_{p,m} \end{pmatrix}$$

where D is $r \times r$ and diagonal and has diagonal elements $\sqrt{\lambda}_1, \dots, \sqrt{\lambda}_r$. For $g = (\Gamma_1, \Gamma_2, \Gamma_3, \xi, A)$,

$$gY = \begin{pmatrix} \Gamma_1 Y_1 A' \\ \Gamma_2 Y_2 A' + \xi \\ \Gamma_3 Y_3 A' \end{pmatrix}.$$

By Proposition 5.2, $Y_3 = \Psi_3 U_3$ where $\Psi_3 \in \mathcal{F}_{p,m}$ and $U_3 \in G_U^+$ is $p \times p$. Choose $A' = U_3^{-1}\Delta$ where $\Delta \in \mathcal{O}_p$ is, as yet, unspecified. Then

$$\Gamma_1 Y_1 A' = \Gamma_1 Y_1 U_3^{-1}\Delta$$

and, by the singular value decomposition theorem for matrices, there exists a $\Gamma_1 \in \mathcal{O}_r$ and a $\Delta \in \mathcal{O}_p$ such that

$$\Gamma_1 Y_1 U_3^{-1}\Delta = (D0)$$

where D is an $r \times r$ diagonal matrix whose diagonal elements are the square roots of the eigenvalues of $Y_1(U_3U_3')^{-1}Y_1' = Y_1(Y_3Y_3')^{-1}Y_1'$. With this choice for $\Delta \in \mathcal{O}_r$ it is clear that $Y_3A' = Y_3U_3^{-1}\Delta \in \mathcal{F}_{p,m}$ so there exists a $\Gamma_3 \in \mathcal{O}_m$ such that

$$\Gamma_3 Y_3 U_3^{-1}\Delta = \begin{pmatrix} I_p \\ 0 \end{pmatrix}.$$

Choosing $\Gamma_2 = I_s$, $\xi = -Y_2A'$, and setting

$$g_0 = \left(\Gamma_1, I_s, \Gamma_3, -Y_2U_3^{-1}\Delta, \left(U_3^{-1}\Delta\right)'\right),$$

g_0Y has the representation claimed. To show h is maximal invariant, suppose $h(Y_1, Y_2, Y_3) = h(Z_1, Z_2, Z_3)$. Let D be the $r \times r$ diagonal matrix, the squares of whose diagonal elements are the eigenvalues of $Y_1(Y_3Y_3')^{-1}Y_1'$ and $Z_1(Z_3Z_3')^{-1}Z_1'$. Then there exist g_0 and $g_1 \in G$ such that

$$g_0Y = \begin{pmatrix} (D0) \\ 0 \\ \begin{pmatrix} I_p \\ 0 \end{pmatrix} \end{pmatrix} = g_1Z$$

so $Y = g_0^{-1}g_1Z$. Thus Y and Z are in the same orbit and h is a maximal invariant function.

When $r > p$, basically the same argument establishes that h is a maximal invariant. To show h is invariant, if $g = (\Gamma_1, \Gamma_2, \Gamma_3, \xi, A)$, then the matrix $Y_1'Y_1(Y_3'Y_3)^{-1}$ gets transformed into $AY_1'Y_1(Y_3'Y_3)^{-1}A^{-1}$ when Y is transformed to gY. By Proposition 1.39, the eigenvalues of $AY_1'Y_1(Y_3'Y_3)^{-1}A^{-1}$ are the same as the eigenvalues of $Y_1'Y_1(Y_3'Y_3)^{-1}$, so h is invariant. To show h is maximal invariant, first show that, for each Y, there exists a $g_0 \in G$ such that

$$g_0Y = \begin{pmatrix} \begin{pmatrix} D \\ 0 \end{pmatrix} \\ 0 \\ \begin{pmatrix} I_p \\ 0 \end{pmatrix} \end{pmatrix} \in \begin{pmatrix} \mathcal{L}_{p,r} \\ \mathcal{L}_{p,s} \\ \mathcal{L}_{p,m} \end{pmatrix}$$

where D is the $p \times p$ diagonal matrix of square roots of eigenvalues

$(Y_1'Y_1)(Y_3'Y_3)^{-1}$. The argument for this is similar to that given previously and is left to the reader. Now, by mimicking the proof for the case $r \leqslant p$, it follows that h is maximal invariant. □

Proposition 9.3. The distribution of the maximal invariant $h(Y_1, Y_2, Y_3)$ depends on the parameters (B_1, B_2, Σ) only through the vector of the t largest eigenvalues of $B_1'B_1\Sigma^{-1}$.

Proof. Since h is a G-invariant function, the distribution of h depends on (B_1, B_2, Σ) only through a maximal invariant parameter under the induced action of G on the parameter space. This action, given earlier, is

$$(\Gamma_1, \Gamma_2, \Gamma_3, \xi, A)(B_1, B_2, \Sigma) = (\Gamma_1 B_1 A', \Gamma_2 B_2 A' + \xi, A\Sigma A').$$

However, an argument similar to that used to prove Proposition 9.2 shows that the vector of the t largest eigenvalues of $B_1'B_1\Sigma^{-1}$ is maximal invariant in the parameter space. □

An alternative form of the maximal invariant is sometimes useful.

Proposition 9.4. Let $t = \min\{r, p\}$ and define $h_1(Y_1, Y_2, Y_3)$ to be the t-dimensional vector $(\theta_1, \dots, \theta_t)'$ where $\theta_1 \leqslant \cdots \leqslant \theta_t$ are the t smallest eigenvalues of $Y_3'Y_3(Y_3'Y_3 + Y_1'Y_1)^{-1}$. Then $\theta_i = 1/(1 + \lambda_i)$, $i = 1, \dots, t$, where λ_i's are defined in Proposition 9.2. Further, $h_1(Y_1, Y_2, Y_3)$ is a maximal invariant.

Proof. For $\lambda \in [0, \infty)$, let $\theta = 1/(1 + \lambda)$. If λ satisfies the equation

$$\left| Y_1'Y_1(Y_3'Y_3)^{-1} - \lambda I_p \right| = 0,$$

then a bit of algebra shows that θ satisfies the equation

$$\left| Y_3'Y_3(Y_3'Y_3 + Y_1'Y_1)^{-1} - \theta I_p \right| = 0,$$

and conversely. Thus $\theta_i = 1/(1 + \lambda_i)$, $i = 1, \dots, t$, are the t smallest eigenvalues of $Y_3'Y_3(Y_3'Y_3 + Y_1'Y_1)^{-1}$. Since $h_1(Y_1, Y_2, Y_3)$ is a one-to-one function of $h(Y_1, Y_2, Y_3)$, it is clear that $h_1(Y_1, Y_2, Y_3)$ is a maximal invariant. □

Since every G-invariant test is a function of a maximal invariant, the problem of choosing a reasonable invariant test boils down to studying tests based on a maximal invariant. When $t \equiv \min\{p, r\} = 1$, the following result shows that there is only one sensible choice for an invariant test.

Proposition 9.5. If $t = 1$ in the MANOVA problem, then the test that rejects for large values of λ_1 is uniformly most powerful within the class of G-invariant tests. Further, this test is equivalent to the likelihood ratio test.

Proof. First consider the case when $p = 1$. Then $Y_1'Y_1(Y_3'Y_3)^{-1}$ is a nonnegative scalar and

$$\lambda_1 = \frac{Y_1'Y_1}{Y_3'Y_3}.$$

Also, $\mathcal{L}(Y_1) = N(B_1, \sigma^2 I_r)$ and $\mathcal{L}(Y_3) = N(0, \sigma^2 I_m)$ where Σ has been set equal to σ^2 to conform to classical notation when $p = 1$. By Proposition 8.14,

$$\mathcal{L}(\lambda_1) = F(r, m; \delta)$$

where $\delta = B_1'B_1/\sigma^2$ and the null hypothesis is that $\delta = 0$. Since the noncentral F distribution has a monotone likelihood ratio, it follows that the test that rejects for large values of λ_1 is uniformly most powerful for testing $\delta = 0$ versus $\delta > 0$. As every invariant test is a function of λ_1, the case for $p = 1$ follows.

Now, suppose $r = 1$. Then the only nonzero eigenvalue of $Y_1'Y_1(Y_3'Y_3)^{-1}$ is $Y_1(Y_3'Y_3)^{-1}Y_1'$ by Proposition 1.39. Thus

$$\lambda_1 = Y_1(Y_3'Y_3)^{-1}Y_1'$$

and, by Proposition 8.14,

$$\mathcal{L}(\lambda_1) = F(p, m - p + 1; \delta)$$

where $\delta = B_1\Sigma^{-1}B_1' \geqslant 0$. The problem is to test $\delta = 0$ versus $\delta > 0$. Again, the noncentral F distribution has a monotone likelihood ratio and the test that rejects for large values of λ_1 is uniformly most powerful among tests based on λ_1.

The likelihood ratio test rejects H_0 for small values of

$$\Lambda = \frac{|Y_3'Y_3|}{|Y_3'Y_3 + Y_1'Y_1|} = \frac{1}{|I_p + Y_1'Y_1(Y_3'Y_3)^{-1}|}.$$

If $p = 1$, then $\Lambda = (1 + \lambda_1)^{-1}$ and rejecting for small values of Λ is equivalent to rejecting for large values of λ_1. When $r = 1$, then

$$|I_p + Y_1'Y_1(Y_3'Y_3)^{-1}| = 1 + Y_1(Y_3'Y_3)^{-1}Y_1' = 1 + \lambda_1$$

so again $\Lambda = (1 + \lambda_1)^{-1}$. □

When $t > 1$, the situation is not so simple. In terms of the eigenvalues $\lambda_1, \ldots, \lambda_t$, the likelihood ratio criterion rejects H_0 for small values of

$$\Lambda = \frac{|Y_3'Y_3|}{|Y_3'Y_3 + Y_1'Y_1|} = \frac{1}{|I_p + Y_1'Y_1(Y_3'Y_3)^{-1}|} = \prod_{i=1}^{t} \frac{1}{1 + \lambda_i}.$$

However, there are no compelling reasons to believe that other tests based on $\lambda_1, \ldots, \lambda_t$ would not be reasonable. Before discussing possible alternatives to the likelihood ratio test, it is helpful to write the maximal invariant statistic in terms of the original variables that led to the canonical MANOVA problem. In the original MANOVA problem, we had an observation vector $X \in \mathcal{L}_{p,n}$ such that

$$\mathcal{L}(X) = N(Z\beta, I_n \otimes \Sigma)$$

and the problem was to test $K\beta = 0$. We know that

$$\hat{\beta} = (Z'Z)^{-1}ZX$$

and

$$\hat{\Sigma} = \frac{1}{n}X'Q_zX \equiv \frac{1}{n}S$$

are the maximum likelihood estimators of β and Σ.

Proposition 9.6. Let $t = \min\{p, r\}$. Suppose the original MANOVA problem is reduced to a canonical MANOVA problem. Then a maximal invariant in the canonical problem expressed in terms of the original variables is the vector $(\lambda_1, \ldots, \lambda_t)'$, $\lambda_1 \geqslant \cdots \geqslant \lambda_t$, of the t largest eigenvalues of

$$V \equiv \left[(K\hat{\beta})'\left(K(Z'Z)^{-1}K'\right)^{-1}K\hat{\beta}\right]S^{-1}.$$

Proof. The transformations that reduced the original problem to canonical form led to the three matrices Y_1, Y_2, and Y_3 where Y_1 is $r \times p$, Y_2 is $(k - r) \times p$, and Y_3 is $(n - k) \times p$. Expressing Y_1 and Y_3 in terms of X, Z, and K, it is not too difficult (but certainly tedious) to show that

$$Y_1'Y_1(Y_3'Y_3)^{-1} = V.$$

By Proposition 9.2, the vector $(\lambda_1, \ldots, \lambda_t)'$ of the t largest eigenvalues of

$Y_1'Y_1(Y_3'Y_3)^{-1}$ is a maximal invariant. Thus the vector of the t largest eigenvalues of V is maximal invariant. □

In terms of X, Z, and K, the likelihood ratio test rejects the null hypothesis if

$$\Lambda = \frac{|S|}{|\hat{\beta}'K'\big(K(Z'Z)^{-1}K'\big)^{-1}K\hat{\beta} + S|}$$

is too small. Also, the distribution of Λ under H_0 is given in Proposition 9.1 as $U(n-k, r, p)$. The distribution of Λ when $K\beta \neq 0$ is quite complicated when $t > 1$ except in the case when β has rank one. In this case, the distribution of Λ is given in Proposition 8.16.

We now turn to the question of possible alternatives to the likelihood ratio test. For notational convenience, the canonical form of the MANOVA problem is treated. However, the reader can express statistics in terms of the original variables by applying Proposition 9.6. Since our interest is in invariant tests, consider Y_1 and Y_3, which are independent, and satisfy

$$\mathcal{L}(Y_1) = N(B_1, I_n \otimes \Sigma)$$

$$\mathcal{L}(Y_3) = N(0, I_m \otimes \Sigma).$$

The random vector Y_2 need not be considered as invariant tests do not involve Y_2. Intuitively, the null hypothesis $H_0: B_1 = 0$ should be rejected, on the basis of an invariant test, if the nonzero eigenvalues $\lambda_1 \geqslant \cdots \geqslant \lambda_t$ of $Y_1'Y_1(Y_3'Y_3)^{-1}$ are too large in some sense. Since $\mathcal{L}(Y_1) = N(B_1, I_r \otimes \Sigma)$,

$$\mathcal{E}Y_1'Y_1 = B_1'B_1 + r\Sigma.$$

Also, it is not difficult to verify that (see the problems at the end of this chapter)

$$\mathcal{E}(Y_3'Y_3)^{-1} = \frac{1}{m-p-1}\Sigma^{-1}$$

when $m - p - 1 > 0$. Since Y_1 and Y_3 are independent,

$$\mathcal{E}Y_1'Y_1(Y_3'Y_3)^{-1} = \frac{r}{m-p-1}I_p + \frac{1}{m-p-1}B_1'B_1\Sigma^{-1}.$$

Therefore, the further B_1 is away from zero, the larger we expect the eigenvalues of $B_1'B_1\Sigma^{-1}$ to be, and hence the larger we expect the eigenvalues of $Y_1'Y_1(Y_3'Y_3)^{-1}$ to be. In particular,

$$\mathcal{E}\,\mathrm{tr}\,Y_1'Y_1(Y_3'Y_3)^{-1} = \frac{rp}{m-p-1} + \frac{1}{m-p-1}\,\mathrm{tr}\,B_1B_1'\Sigma^{-1}$$

and $\mathrm{tr}\,B_1'B_1\Sigma^{-1}$ is just the sum of the eigenvalues of $B_1'B_1\Sigma^{-1}$. The test that rejects for large values of the statistic

$$\sum_1^t \lambda_i = \mathrm{tr}\,Y_1'Y_1(Y_3'Y_3)^{-1}$$

is called the Lawley–Hotelling trace test and is one possible alternative to the likelihood ratio test. Also, the test that rejects for large values of

$$\sum_1^t \frac{\lambda_i}{1+\lambda_i} = \mathrm{tr}\,Y_1'Y_1(Y_3'Y_3 + Y_1'Y_1)^{-1}$$

was introduced by Pillai as a competitor to the likelihood ratio test. A third competitor is based on the following considerations. The null hypothesis $H_0: B_1 = 0$ is equivalent to the intersection over $u \in R^r$, $\|u\| = 1$, of the null hypotheses $H_u: u'B_1 = 0$. Combining Propositions 9.5 and 9.6, it follows that the test that accepts H_u iff

$$u'Y_1(Y_3'Y_3)^{-1}Y_1'u \leqslant c$$

is a uniformly most powerful test within the class of tests that are invariant under the group of transformations preserving H_u. Here, c is a constant. Under H_u,

$$\mathcal{L}\left(u'Y_1(Y_3'Y_3)^{-1}Y_1'u\right) = F_{p,m-p+1}$$

so it seems reasonable to require that c not depend on u. Since H_0 is equivalent to $\cap\{H_u \mid \|u\| = 1, u \in R^r\}$, H_0 should be accepted iff all the H_u are accepted—that is, H_0 should be accepted iff

$$\sup_{\|u\|=1} u'Y_1(Y_3'Y_3)^{-1}Y_1'u \leqslant c.$$

However, this supremum is just the largest eigenvalue of $Y_1(Y_3'Y_3)^{-1}Y_1'$, which is λ_1. Thus the proposed test is to accept H_0 iff $\lambda_1 \leqslant c$ or equivalently, to reject H_0 for large values of λ_1. This test is called Roy's maximum root test.

Unfortunately, there is very little known about the comparative behavior of the tests described above. A few numerical studies have been done for small values of r, m, and p but no single test stands out as dominating the others over a substantial portion of the set of alternatives. Since very accurate approximations are available for the null distribution of the likelihood ratio test, this test is easier to apply than the above competitors. Further, there is an interesting decomposition of the test statistic

$$\Lambda = \frac{|Y_3'Y_3|}{|Y_3'Y_3 + Y_1'Y_1|},$$

which has some applications in practice. Let $S = Y_3'Y_3$ so $\mathcal{L}(S) = W(\Sigma, p, m)$ and let $X_1', \ldots, X_r'$ denote the rows of Y_1. Under $H_0: B_1 = 0$, $X_1, \ldots, X_r$ are independent and $\mathcal{L}(X_i) = N(0, \Sigma)$. Further,

$$\Lambda = \frac{|S|}{\left|S + \sum_1^r X_i X_i'\right|} = \prod_{i=1}^r \Lambda_i$$

where

$$\Lambda_1 = \frac{|S|}{|S + X_1 X_1'|}$$

and

$$\Lambda_i = \frac{\left|S + \sum_1^{i-1} X_i X_i'\right|}{\left|S + \sum_1^i X_i X_i'\right|}, \qquad i = 2, \ldots, r.$$

Proposition 8.15 gives the distribution of Λ_i under H_0 and shows that $\Lambda_1, \ldots, \Lambda_r$ are independent under H_0. Let $\beta_1', \ldots, \beta_r'$ denote the rows of B_1 and consider the r testing problems given by the null hypotheses

$$H_i: \{(\beta_1, \ldots, \beta_r) | \beta_1 = \beta_2 = \cdots = \beta_i = 0\}$$

and the alternatives

$$\overline{H}_i: \{(\beta_1, \ldots, \beta_r) | \beta_1 = \beta_2 = \cdots = \beta_{i-1} = 0\}$$

for $i = 1, \ldots, r$. Obviously, $H_0 = \cap_1^r H_i$ and the likelihood ratio test for testing H_i against $\overline{H}_i$ rejects H_i iff Λ_i is too small. Thus the likelihood ratio test for H_0 can be thought of as one possible way of combining the r independent test statistics into an overall test of $\cap_1^r H_i$.

9.2. MANOVA PROBLEMS WITH BLOCK DIAGONAL COVARIANCE STRUCTURE

The parameter set of the MANOVA model considered in the previous section consisted of a subspace $M = \{\mu|\mu = ZB, B \in \mathcal{L}_{p,k}\} \subseteq \mathcal{L}_{p,n}$ and a set of covariance matrices

$$\gamma = \{I_n \otimes \Sigma | \Sigma \in \mathcal{S}_p^+\}.$$

It was assumed that the matrix Σ was completely unknown. In this section, we consider estimation and testing problems when certain things are known about Σ. For example, if $\Sigma = \sigma^2 I_p$ with σ^2 unknown and positive, then we have the linear model discussed in Section 3.1. In this case, the linear model with parameter set $\{M, \gamma\}$ is just a univariate linear model in the sense that $I_n \otimes \Sigma = \sigma^2 I_n \otimes I_p$ and $I_n \otimes I_p$ is the identity linear transformation on the vector space $\mathcal{L}_{p,n}$. This model is just the linear model of Section 9.1 when $p = 1$ and np plays the role of n. Of course, when $\Sigma = \sigma^2 I_p$, the subspace M need not have the structure above in order for Proposition 4.6 to hold.

In what follows, we consider another assumption concerning Σ and treat certain estimation and testing problems. For the models treated, it is shown that these models are actually "products" of the MANOVA models discussed in Section 9.1.

Suppose $Y \in \mathcal{L}_{p,n}$ is a random vector with $\mathcal{E}Y \in M$ where

$$M = \{\mu|\mu = ZB, B \in \mathcal{L}_{p,k}\}$$

and Z is a known $n \times k$ matrix of rank k. Write $p = p_1 + p_2$, $p_i \geqslant 1$, for $i = 1, 2$. The covariance of Y is assumed to be an element of

$$\gamma_0 = \left\{I_n \otimes \Sigma | \Sigma = \begin{pmatrix} \Sigma_{11} & 0 \\ 0 & \Sigma_{22} \end{pmatrix}, \Sigma_{ii} \in \mathcal{S}_{p_i}^+, i = 1, 2\right\}.$$

Thus the rows of Y, say $Y_1', \ldots, Y_n'$, are uncorrelated. Further, if Y_i is partitioned into $X_i \in R^{p_1}$ and $W_i \in R^{p_2}$, $Y_i' = (X_i', W_i')$, then X_i and W_i are also uncorrelated, since

$$\operatorname{Cov}(Y_i) = \operatorname{Cov}\left\{\begin{pmatrix} X_i \\ W_i \end{pmatrix}\right\} = \begin{pmatrix} \Sigma_{11} & 0 \\ 0 & \Sigma_{22} \end{pmatrix}.$$

Thus the interpretation of the assumed structure of γ_0 is that the rows of Y are uncorrelated and within each row, the first p_1 coordinates are uncorrelated with the last p_2 coordinates. This suggests that we decompose Y into $X \in \mathcal{L}_{p_1, n}$ and $W \in \mathcal{L}_{p_2, n}$ where

$$Y = (X, W) \in \mathcal{L}_{p, n}.$$

Obviously, the rows of $X(W)$ are $X_1', \ldots, X_n'(W_1', \ldots, W_n')$. Also, partition $B \in \mathcal{L}_{p,k}$ into $B_1 \in \mathcal{L}_{p_1,k}$ and $B_2 \in \mathcal{L}_{p_2,k}$. It is clear that

$$\mathcal{E}X \in M_1 \equiv \{\mu_1 | \mu_1 = ZB_1, B_1 \in \mathcal{L}_{p_1,k}\}$$

and

$$\mathcal{E}W \in M_2 \equiv \{\mu_2 | \mu_2 = ZB_2, B_2 \in \mathcal{L}_{p_2,k}\}.$$

Further,

$$\operatorname{Cov}(X) \in \gamma_1 \equiv \{I_n \otimes \Sigma_{11} | \Sigma_{11} \in \mathcal{S}_{p_1}^+\}$$

and

$$\operatorname{Cov}(W) \in \gamma_2 \equiv \{I_n \otimes \Sigma_{22} | \Sigma_{22} \in \mathcal{S}_{p_2}^+\}.$$

Since X and W are uncorrelated, if Y has a normal distribution, then X and W are independent and normal and we have a MANOVA model of Section 9.1 for both X and W (with parameter sets (M_1, γ_1) and (M_2, γ_2)). In summary, when Y has a normal distribution, Y can be partitioned into X and W, which are independent. Therefore, the density of Y is

$$f(Y|\mu, \Sigma) = f_1(X|\mu_1, \Sigma_{11}) f_2(W|\mu_2, \Sigma_{22})$$

where f, f_1, and f_2 are normal densities on the appropriate spaces. Since we have MANOVA models for both X and W, the maximum likelihood estimators of μ_1, μ_2, Σ_{11}, and Σ_{22} follow from the result of the first section. For testing the null hypothesis $H_0: KB = 0$, $K: r \times k$ of rank r, a similar decomposition occurs. As $B = (B_1 B_2)$, $H_0: KB = 0$ is equivalent to the two null hypotheses $H_0^1: KB_1 = 0$ and $H_0^2: KB_2 = 0$.

Proposition 9.7. Assume that $n - k \geqslant \max\{p_1, p_2\}$. For testing $H_0: KB = 0$, the likelihood ratio test rejects for small values of $\Lambda = \Lambda_1 \Lambda_2$ where

$$\Lambda_1 = \frac{|X'Q_z X|}{\left|X'Q_z X + (K\hat{B}_1)'\left(K(Z'Z)^{-1}K'\right)^{-1} K\hat{B}_1\right|}$$

and

$$\Lambda_2 = \frac{|W'Q_zW|}{|W'Q_zW + (K\hat{B}_2)'(K(Z'Z)^{-1}K')^{-1}K\hat{B}_2|}.$$

Here, $Q_z = I - P_z$ where $P_z = Z(Z'Z)^{-1}Z'$ and

$$\hat{B}_1 = (Z'Z)^{-1}Z'X, \qquad \hat{B}_2 = (Z'Z)^{-1}Z'W.$$

Proof. We need to calculate

$$\Psi(Y) \equiv \frac{\sup_{(\mu,\Sigma)\in H_0} f(Y|\mu,\Sigma)}{\sup_{(\mu,\Sigma)\in\mathfrak{M}} f(Y|\mu,\Sigma)}$$

where $\mathfrak{M}$ is the set of (μ, Σ) such that $\mu \in M$ and $I_n \otimes \Sigma \in \gamma_0$. As noted earlier,

$$f(Y|\mu,\Sigma) = f_1(X|\mu_1,\Sigma_{11})f_2(W|\mu_2,\Sigma_{22}).$$

Also, $(\mu, \Sigma) \in H_0$ iff $(\mu_1, \Sigma_{11}) \in H_0^1$ and $(\mu_2, \Sigma_{22}) \in H_0^2$. Further, $(\mu, \Sigma) \in \mathfrak{M}$ iff $(\mu_i, \Sigma_{ii}) \in \mathfrak{M}_i$ where $\mathfrak{M}_i$ is the set of (μ_i, Σ_{ii}) such that $\mu_i \in M_i$ and $I_n \otimes \Sigma_{ii} \in \gamma_i$ for $i = 1, 2$. From these remarks, it follows that

$$\Psi(Y) = \Psi_1(X)\Psi_2(W)$$

where

$$\Psi_1(X) = \frac{\sup_{(\mu_1,\Sigma_{11})\in H_0^1} f_1(X|\mu_1,\Sigma_{11})}{\sup_{(\mu_1,\Sigma_{11})\in\mathfrak{M}_1} f_1(X|\mu_1,\Sigma_{11})}$$

and

$$\Psi_2(W) = \frac{\sup_{(\mu_2,\Sigma_{22})\in H_0^2} f_2(W|\mu_2,\Sigma_{22})}{\sup_{(\mu_2,\Sigma_{22})\in\mathfrak{M}_2} f_2(W|\mu_2,\Sigma_{22})}.$$

However, $\Psi_1(X)$ is simply the likelihood ratio statistic for testing H_0^1 in the

MANOVA model for X. The results of Propositions 9.6 and 9.1 show that $\Psi_1(X) = (\Lambda_1)^{n/2}$. Similarly, $\Psi_2(W) = (\Lambda_2)^{n/2}$. Thus $\Psi(Y) = (\Lambda_1\Lambda_2)^{n/2}$ so the test that rejects for small values of $\Lambda = \Lambda_1\Lambda_2$ is equivalent to the likelihood ratio test. □

Since X and W are independent, the statistics Λ_1 and Λ_2 are independent. The distribution of Λ_i when H_0^i is true is $U(n - p_i, r, p_i)$ for $i = 1, 2$. Therefore, when H_0 is true, $\Lambda_1\Lambda_2$ is distributed as a product of independent beta random variables and the results in Anderson (1958) provide an approximation to the null distribution of $\Lambda_1\Lambda_2$.

We now turn to a discussion of the invariance aspects of testing $H_0 : KB = 0$ on the basis of the observation vector Y. The argument used to reduce the MANOVA model of Section 9.1 to canonical form is valid here, and this leads to a group of transformations G_1, which preserve the testing problem H_0^1 for the MANOVA model for X. Similarly, there is a group G_2 that preserves the testing problem H_0^2 for the MANOVA model for W. Since $Y = (X, W)$, we can define the product group $G_1 \times G_2$ acting on Y by

$$(g_1, g_2)Y \equiv (g_1X, g_2W)$$

and the testing problem H_0 is clearly invariant under this group action. A maximal invariant is derived as follows. Let $t_i = \min\{r, p_i\}$ for $i = 1, 2$, and in the notation of Proposition 9.7, let

$$V_1 = \left[(K\hat{B}_1)'\left(K(Z'Z)^{-1}K'\right)^{-1}K\hat{B}_1\right](X'Q_zX)^{-1}$$

and

$$V_2 = \left[(K\hat{B}_2)'\left(K(Z'Z)^{-1}K'\right)^{-1}K\hat{B}_2\right](W'Q_zW)^{-1}.$$

Let $\eta_1 \geqslant \cdots \geqslant \eta_{t_1}$ be the t_1 largest eigenvalues of V_1 and $\theta_1 \geqslant \cdots \geqslant \theta_{t_2}$ be the t_2 largest eigenvalues of V_2.

Proposition 9.8. A maximal invariant under the action of $G_1 \times G_2$ on Y is the $(t_1 + t_2)$-dimensional vector $(\eta_1, \ldots, \eta_{t_1}; \theta_1, \ldots, \theta_{t_2}) = h(Y) \equiv (h_1(X); h_2(W))$. Here, $h_1(X) = (\eta_1, \ldots, \eta_{t_1})$ and $h_2(W) = (\theta_1, \ldots, \theta_{t_2})$.

Proof. By Proposition 9.6, $h_1(X)(h_2(W))$ is maximal invariant under the action of $G_1(G_2)$ on $X(W)$. Thus h is G-invariant. If $h(Y_1) = h(Y_2)$ where $Y_1 = (X_1, W_1)$ and $Y_2 = (X_2, W_2)$, then $h_1(X_1) = h_1(X_2)$ and $h_2(W_1) = h_2(W_2)$. Thus there exists $g_1 \in G_1(g_2 \in G_2)$ such that $g_1X_1 = X_2(g_2W_1 =$

W_2). Therefore,

$$(g_1, g_2)Y_1 = (g_1 X_1, g_2 W_1) = (X_2, W_2) = Y_2$$

so h is maximal invariant. □

As a function of $h(Y)$, the likelihood ratio test rejects H_0 if

$$\Lambda = \Lambda_1 \Lambda_2 = \prod_1^{t_1}\left(\frac{1}{1+\eta_i}\right)\prod_1^{t_2}\left(\frac{1}{1+\theta_i}\right)$$

is too small. Since $t_1 + t_2 > 1$, the maximal invariant $h(Y)$ is always of dimension greater than one. Thus the situation described in Proposition 9.5 cannot arise in the present context. In no case will there exist a uniformly most powerful invariant test of $H_0: KB = 0$ even if K has rank 1. This completes our discussion of the present linear model.

It should be clear by now that the results described above can be easily extended to the case when Σ has the form

$$\Sigma = \begin{pmatrix} \Sigma_{11} & & & \\ & \Sigma_{22} & & \\ & & \ddots & \\ & & & \Sigma_{ss} \end{pmatrix}$$

where the off-diagonal blocks of Σ are zero. Here $\Sigma \in \mathbb{S}_p^+$ and $\Sigma_{ii} \in \mathbb{S}_{p_i}^+$, $\Sigma_1^s p_i = p$. In this case, the set of covariances for $Y \in \mathcal{L}_{p,n}$ is the set γ_0, which consists of all $I_n \otimes \Sigma$ where Σ has the above form and each Σ_{ii} is unknown. The mean space for Y is M as before. For this model, Y can be decomposed into s independent pieces and we have a MANOVA model in $\mathcal{L}_{p_i,n}$ for each piece. Also, the matrix $B(\mathcal{E}Y = ZB)$ decomposes into $B_1, \ldots, B_s$, $B_i \in \mathcal{L}_{p_i,k}$ and a null hypothesis $H_0: KB = 0$ is equivalent to the intersection of the s null hypotheses $H_0^i: KB_i = 0$, $i = 1, \ldots, s$. The likelihood ratio test of H_0 is now based on a product of s independent statistics, say $\Lambda \equiv \prod_1^s \Lambda_i$, where $\mathcal{L}(\Lambda_i) = U(n - p_i, r, p_i)$ and thus Λ is distributed as a product of independent beta random variables when H_0 is true. Further, invariance considerations lead to an s-fold product group that preserves the testing problem and a maximal invariant is of dimension $t_1 + \cdots + t_s$ where $t_i = \min\{r, p_i\}$, $i = 1, \ldots, s$. The details of all this, which are mainly notational, are left to the reader.

In this section, it has been shown that the linear model with a block diagonal covariance matrix can be decomposed into independent component models, each of which is a MANOVA model of the type treated in Section 9.1. This decomposition technique also appears in the next two sections in which we treat linear models with different types of covariance structure.

9.3. INTRACLASS COVARIANCE STRUCTURE

In some instances, it is natural to assume that the covariance matrix of a random vector possesses certain symmetry properties that are suggested by the sampling situation. For example, if n measurements are taken under the same experimental conditions, it may be reasonable to suppose that the order in which the observations are taken is immaterial. In other words, if $X_1, \ldots, X_p$ denote the observations and $X' = (X_1, \ldots, X_p)$ is the observation vector, then X and any permutation of X have the same distribution. Symbolically, this means that $\mathcal{L}(X) = \mathcal{L}(gX)$ where g is a permutation matrix. If $\Sigma \equiv \text{Cov}(X)$ exists, this implies that $\Sigma = g\Sigma g'$ for $g \in \mathcal{P}_p$ where $\mathcal{P}_p$ denotes the group of $p \times p$ permutation matrices. Our first task is to characterize those covariance matrices that are invariant under $\mathcal{P}_p$—that is, those covariance matrices that satisfy $\Sigma = g\Sigma g'$ for all $g \in \mathcal{P}_p$. Let $e \in R^p$ be the vector of ones and set $P_e = (1/p)ee'$ so P_e is the orthogonal projection onto span$\{e\}$. Also, let $Q_e = I_p - P_e$.

Proposition 9.9. Let Σ be a positive definite $p \times p$ matrix. The following are equivalent:

(i) $\Sigma = g\Sigma g'$ for $g \in \mathcal{P}_p$.

(ii) $\Sigma = \alpha P_e + \beta Q_e$ for $\alpha > 0$ and $\beta > 0$.

(iii) $\Sigma = \sigma^2 A(\rho)$ where $\sigma^2 > 0$, $-1/(p-1) < \rho < 1$, and $A(\rho)$ is a $p \times p$ matrix with elements $a_{ii} = 1$, $i = 1, \ldots, p$, and $a_{ij}(\rho) = \rho$ for $i \neq j$.

Proof. Since

$$A(\rho) = (1-\rho)I_p + \rho ee' = (1-\rho)I_p + p\rho P_e$$

$$= (1-\rho)Q_e + (1 + (p-1)\rho)P_{e'}$$

the equivalence of (ii) and (iii) follows by taking $\alpha = \sigma^2(1 + (p-1)\rho)$ and $\beta = \sigma^2(1-\rho)$. Since $ge = e$ for $g \in \mathcal{P}_p$, $gP_e = P_e g$. Thus if (ii) holds, then

$$g\Sigma g' = \alpha g P_e g' + \beta g Q_e g' = \alpha P_e + \beta Q_e = \Sigma$$

so (i) holds. To show (i) implies (ii), let $X \in R^p$ be a random vector with $\operatorname{Cov}(X) = \Sigma$. Then (i) implies that $\operatorname{Cov}(X) = \operatorname{Cov}(gX)$ for $g \in \mathcal{P}_p$. Therefore,

$$\operatorname{var}(X_i) = \operatorname{var}(X_j), \qquad i, j = 1, \dots, p$$

and

$$\operatorname{cov}(X_i, X_j) = \operatorname{cov}(X_{i'}, X_{j'}); \qquad i \neq j, i' \neq j'.$$

Let $\gamma = \operatorname{var}(X_1)$ and $\delta = \operatorname{cov}(X_1, X_2)$. Then

$$\begin{aligned}\Sigma &= \delta ee' + (\gamma - \delta) I_p = p\delta P_e + (\gamma - \delta)(P_e + Q_e) \\ &= (\gamma + (p-1)\delta) P_e + (\gamma - \delta) Q_e = \alpha P_e + \beta Q_e\end{aligned}$$

where $\alpha = \gamma + (p-1)\delta$ and $\beta = \gamma - \delta$. The positivity of α and β follows from the assumption that Σ is positive definite. □

A covariance matrix Σ that satisfies one of the conditions of Proposition 9.9 is called an *intraclass covariance matrix* and is said to have intraclass covariance structure. Now that intraclass covariance matrices have been described, suppose that $X \in \mathcal{L}_{p,n}$ has a normal distribution with $\mu \equiv \mathcal{E}X \in M$ and $\operatorname{Cov}(X) \in \gamma$ where M is a linear subspace of $\mathcal{L}_{p,n}$ and

$$\gamma = \{I_n \otimes \Sigma | \Sigma \in \mathcal{S}_p^+, \Sigma = \alpha P_e + \beta Q_e, \alpha > 0, \beta > 0\}.$$

The covariance structure assumed for X means that the rows of X are independent and each row of X has the same intraclass covariance structure. In terms of invariance, if $\Gamma \otimes g \in \mathcal{O}_n \otimes \mathcal{P}_p$, and $I_n \otimes \Sigma \in \gamma$, it is clear that

$$\operatorname{Cov}((\Gamma \otimes g)X) = \operatorname{Cov}(X)$$

since

$$(\Gamma \otimes g)(I_n \otimes \Sigma)(\Gamma \otimes g)' = (\Gamma I_n \Gamma') \otimes (g \Sigma g') = I_n \otimes \Sigma.$$

Conversely, if T is a positive definite linear transformation on $\mathcal{L}_{p,n}$ that satisfies

$$(\Gamma \otimes g)T(\Gamma \otimes g)' = T \quad \text{for } \Gamma \otimes g \in \mathcal{O}_n \otimes \mathcal{P}_p,$$

it is not difficult to show that $T \in \gamma$. The proof of this is left to the reader.

Since the identity linear transformation is an element of γ, in order that the least-squares estimator of $\mu \in M$ be the maximum likelihood estimator, it is sufficient that

$$(I_n \otimes \Sigma)M \subseteq M \quad \text{for } I_n \otimes \Sigma \in \gamma.$$

Our next task is to describe a class of linear subspaces M that satisfy the above condition.

Proposition 9.10. Let C be an $r \times p$ real matrix of rank r with rows $c_1', \dots, c_r'$. If $u_1, \dots, u_r$ is any basis for $N \equiv \text{span}\{c_1, \dots, c_r\}$ and U is an $r \times p$ matrix with rows $u_1', \dots, u_r'$, then there exists an $r \times r$ nonsingular matrix A such that $AU = C$.

Proof. Since $u_1, \dots, u_r$ is a basis for N,

$$c_i = \sum_{k=1}^{r} a_{ik} u_k, \qquad i = 1, \dots, r$$

for some real numbers a_{ik}. Setting $A = \{a_{ik}\}$, $AU = C$ follows. As the basis $\{u_1, \dots, u_r\}$ is mapped onto the basis $\{c_1, \dots, c_r\}$ by the linear transformation defined by A, the matrix A is nonsingular. □

Given positive integers n and p, let k and r be positive integers that satisfy $k < n$ and $r \leqslant p$. Define a subspace $M \subseteq \mathcal{L}_{p,n}$ by

$$M = \{\mu | \mu = Z_1 B Z_2;\ B \in \mathcal{L}_{r,k}\}$$

where Z_1 is $n \times k$ of rank k, Z_2 is $r \times p$ of rank r, and assume that $e \in R^p$ is an element of the subspace spanned by rows of Z_2, say $e \in N = \text{span}\{z_1, \dots, z_r\}$ and the rows of Z_2 are $z_1', \dots, z_r'$. At this point, it is convenient to relabel things a bit. Let $u_1 = e/\sqrt{p}$, $u_2, \dots, u_r$, be an orthonormal basis for N and let $U: r \times p$ have rows $u_1', \dots, u_r'$. By Proposition 9.10, $Z_2 = AU$ for some $r \times r$ nonsingular matrix A so

$$M = \{\mu | \mu = Z_1 B U, B \in \mathcal{L}_{r,k}\}.$$

Summarizing, $X \in \mathcal{L}_{p,n}$ is assumed to have a normal distribution with $\mathcal{E}X \in M$ and $\text{Cov}(X) \in \gamma$ where M and γ are given above. To decompose this model for X into the product of two simple univariate linear models, let $\Gamma \in \mathcal{O}_p$ have $u_1', \dots, u_r'$ as its first r rows. With $Y = (I_n \otimes \Gamma)X$,

$$\mathcal{E}Y = \mathcal{E}X\Gamma' = Z_1 B U \Gamma'$$

and

$$\begin{aligned}\operatorname{Cov}(Y) &= (I_n \otimes \Gamma)\operatorname{Cov}(X)(I_n \otimes \Gamma)' \\ &= (I_n \otimes \Gamma)(I_n \otimes (\alpha P_e + \beta Q_e))(I_n \otimes \Gamma)' \\ &= I_n \otimes (\alpha \Gamma P_e \Gamma' + \beta \Gamma Q_e \Gamma').\end{aligned}$$

However,

$$U\Gamma' = (I_r\, 0) \in \mathcal{L}_{p,r},$$

$$\Gamma P_e \Gamma' = \varepsilon_1 \varepsilon_1'$$

and

$$\Gamma Q_e \Gamma' = I_p - \varepsilon_1 \varepsilon_1'$$

where $\varepsilon_1' = (1, 0, \ldots, 0)$. Therefore, the matrix $D \equiv \alpha\Gamma P_e \Gamma' + \beta \Gamma Q_e \Gamma'$ is diagonal with diagonal elements $d_1, \ldots, d_p$ given by $d_1 = \alpha$ and $d_2 = \cdots = d_p = \beta$. Let $Y_1, \ldots, Y_p$ be the columns of Y and let $b_1, \ldots, b_r$ be the columns of B. Then it is clear that $Y_1, \ldots, Y_p$ are independent,

$$\mathcal{L}(Y_1) = N(Z_1 b_1, \alpha I_n)$$

$$\mathcal{L}(Y_i) = N(Z_1 b_i, \beta I_n), \qquad i = 2, \ldots, r,$$

and

$$\mathcal{L}(Y_i) = N(0, \beta I_n), \qquad i = r+1, \ldots, p.$$

To piece things back together, set $m = n(p-1)$ and let $V \in R^m$ be given by $V' = (Y_2', Y_3', \ldots, Y_p')$. Then

$$\mathcal{L}(V) = N(\tilde{Z}\delta, \beta I_m)$$

where $\delta \in R^{(r-1)p}$, $\delta' = (b_2', \ldots, b_r')$, and

$$\tilde{Z} = \begin{pmatrix} Z_1 & & & 0 \\ & Z_1 & & \\ & & \ddots & \\ 0 & & & Z_1 \\ \hdashline & & 0 & \end{pmatrix} : m \times ((r-1)p).$$

Thus X has been decomposed into the two independent random vectors Y_1 and V and the linear models for Y_1 and V are given by the parameter sets (M_1, γ_1) and (M_2, γ_2) where

$$M_1 = \{\mu_1 | \mu_1 = Z_1 b_1;\ b_1 \in R^k\}$$

$$\gamma_1 = \{\alpha I_n | \alpha > 0\}$$

$$M_2 = \{\mu_2 | \mu_2 = \tilde{Z}\delta,\ \delta \in R^{(r-1)p}\}$$

and

$$\gamma_2 = \{\beta I_m | \beta > 0\}.$$

Both of these linear models are univariate in the sense that γ_1 and γ_2 consist of a constant times an identity matrix.

It is obvious that the general theory developed in Section 9.1 for the MANOVA model applies directly to the above two linear models individually. In particular, the maximum likelihood estimators of b_1, α, δ, and β can simply be written down. Also, linear hypotheses about b_1 or δ can be tested separately, and uniformly most powerful invariant tests will exist for such testing problems when the two linear models are treated separately. However, an interesting phenomenon occurs when we test a joint hypothesis about b_1 and δ. For example, suppose the null hypothesis H_0 is that $b_1 = 0$ and $\delta = 0$ and the alternative is that $b_1 \neq 0$ or $\delta \neq 0$. This null hypothesis is equivalent to the hypothesis that $B = 0$ in the original model for X. By simply writing down the densities of Y_1 and V and substituting in the maximum likelihood estimators of the parameters, the likelihood ratio test for H_0 rejects if

$$\Lambda \equiv \left(\frac{\|Y_1 - Z_1\hat{b}_1\|^2}{\|Y_1\|^2}\right)^{n/2} \left(\frac{\|V - \tilde{Z}\hat{\delta}\|^2}{\|V\|^2}\right)^{m/2}$$

is too small. Here, $\|\cdot\|$ denotes the standard norm on the coordinate Euclidean space under consideration. Let

$$W_1 = \frac{\|Y_1 - Z_1\hat{b}_1\|^2}{\|Y_1\|^2}$$

and

$$W_2 = \frac{\|V - \tilde{Z}\hat{\delta}\|^2}{\|V\|^2}$$

so W_1 and W_2 are independent and each has a beta distribution. When $p \geq 3$, then $m = n(p-1) > n$ and it follows that $\Lambda^{2/n} = W_1 W_2^{m/n}$ is not in general distributed as a product of independent beta random variables. This is in contrast to the situation treated in Section 9.2 of this chapter.

We end this section with a brief description of what might be called multivariate intraclass covariance matrices. If $X \in R^p$ and $\text{Cov}(X) = \Sigma$, then Σ is an intraclass covariance matrix iff $\text{Cov}(gX) = \text{Cov}(X)$ for all $g \in \mathcal{P}_p$. When the random vector X is replaced by the random matrix $Y: p \times q$, then the expression $gY = (g \otimes I_q)Y$ still makes sense for $g \in \mathcal{P}_p$, and it is natural to seek a characterization of $\text{Cov}(Y)$ when $\text{Cov}(Y) = \text{Cov}((g \otimes I_q)Y)$ for all $g \in \mathcal{P}_p$. For $g \in \mathcal{P}_p$, the linear transformation $g \otimes I_q$ just permutes the rows of Y and, to characterize $T = \text{Cov}(Y)$, we must describe how permutations of the rows of Y affect T. The condition that $\text{Cov}(Y) = \text{Cov}((g \otimes I_q)Y)$ is equivalent to the condition

$$T = (g \otimes I_q)T(g \otimes I_q)', \qquad g \in \mathcal{P}_p.$$

For A and B in $\mathcal{S}_q^+$, consider

$$T_0 \equiv P_e \otimes A + Q_e \otimes B.$$

Then T_0 is a self-adjoint and positive definite linear transformation on $\mathcal{L}_{q,p}$ to $\mathcal{L}_{q,p}$. It is readily verified that

$$T_0 = (g \otimes I_q)T_0(g \otimes I_q)', \qquad g \in \mathcal{P}_p.$$

That T_0 is a possible generalization of an intraclass covariance matrix is fairly clear—the positive scalars α and β of Proposition 9.9 have become the positive definite matrices A and B. The following result shows that if T is $(\mathcal{P}_p \otimes I_q)$-invariant—that is, if T satisfies $T = (g \otimes I_q)T(g \otimes I_q)'$—then T must be a T_0 for some positive definite A and B.

Proposition 9.11. If T is positive definite and $(\mathcal{P}_p \otimes I_q)$-invariant, then there exist $q \times q$ positive definite matrices A and B such that

$$T = P_e \otimes A + Q_e \otimes B.$$

Proof. The proof of this is left to the reader. □

Unfortunately, space limitations prevent a detailed description of linear models that have covariances of the form $I_n \otimes T$ where T is given in

Proposition 9.11. However, the analysis of these models proceeds along the lines of that given for intraclass covariance models and, as usual, these models can be decomposed into independent pieces, each of which is a MANOVA model.

9.4. SYMMETRY MODELS: AN EXAMPLE

The covariance structures studied thus far in this chapter are special cases of a class of covariance models called symmetry models. To describe these, let $(V, (\cdot, \cdot))$ be an inner product space and let G be a compact subgroup of $\mathcal{O}(V)$. Define the class of positive definite transformations γ_G by

$$\gamma_G = \{\Sigma | \Sigma \in \mathcal{L}(V, V), \Sigma > 0, g\Sigma g' = \Sigma \quad \text{for all } g \in G\}.$$

Thus γ_G is the set of positive definite covariances that are invariant under G in the sense that $\Sigma = g\Sigma g'$ for $g \in G$. To justify the term symmetry model for γ_G, first observe that the notion of "symmetry" is most often expressed in terms of a group acting on a set. Further, if X is a random vector in V with $\text{Cov}(X) = \Sigma$, then $\text{Cov}(gX) = g\Sigma g'$. Thus the condition that $\Sigma = g\Sigma g'$ is precisely the condition that X and gX have the same covariance—hence, the term symmetry model.

Most of the covariance sets considered in this book have been symmetry models for a particular choice of $(V, (\cdot, \cdot))$ and G. For example, if $G = \mathcal{O}(V)$, then

$$\gamma_G = \{\Sigma | \Sigma = \sigma^2 I, \sigma^2 > 0\},$$

as Proposition 2.13 shows. Hence $\mathcal{O}(V)$ generates the weakly spherical symmetry model. The result of Proposition 2.19 establishes that when $(V, (\cdot, \cdot)) = (\mathcal{L}_{p,n}, \langle \cdot, \cdot \rangle)$ and

$$G = \{g | g = \Gamma \otimes I_p, \Gamma \in \mathcal{O}_n\},$$

then

$$\gamma_G = \{\Sigma | \Sigma = I_n \otimes A, A \in \mathcal{S}_p^+\}.$$

Of course, this symmetry model has occurred throughout this book. Using techniques similar to that in Proposition 2.19, the covariance models considered in Section 9.2 are easily shown to be symmetry models for an appropriate group. Moreover, Propositions 9.9 and 9.11 describe sets of covariances (the intraclass covariances and their multivariate extensions) in exactly the manner in which the set γ_G was defined. Thus symmetry models are not unfamiliar objects.

Now, given a closed group $G \subseteq \mathcal{O}(V)$, how can we explicitly describe the model γ_G? Unfortunately, there is no one method or approach that is appropriate for all groups G. For example, the results of Proposition 2.19 and Proposition 9.9 were proved by quite different means. However, there is a general structure theory known for the models γ_G (see Andersson, 1975), but we do not discuss that here. The general theory tells us what γ_G should look like, but does not tell us how to derive the particular form of γ_G.

The remainder of this section is devoted to an example where the methods are a bit different from those encountered thus far. To motivate the considerations below, consider observations $X_1, \dots, X_p$, which are taken at p equally spaced points on a circle and are numbered sequentially around the circle. For example, the observations might be temperatures at a fixed cross section on a cylindrical rod when a heat source is present at the center of the rod. Impurities in the rod and the interaction of adjacent measuring devices may make an exchangeability assumption concerning the joint distribution of $X_1, \dots, X_p$ unreasonable. However, it may be quite reasonable to assume that the covariance between X_j and X_k depends only on how far apart X_j and X_k are on the circle—that is, $\text{cov}(X_j, X_{j+1})$ does not depend on j, $j = 1, \dots, p$, where $X_{p+1} \equiv X_1$; $\text{cov}(X_j, X_{j+2})$ does not depend on j, $j = 1, \dots, p$, where $X_{p+2} \equiv X_2$, and so on. Assuming that $\text{cov}(X_j, X_j)$ does not depend on j, these assumptions can be succinctly expressed as follows. Let $X \in R^p$ have coordinates $X_1, \dots, X_p$ and let C be a $p \times p$ matrix with

$$c_{p1} = c_{j(j+1)} = 1, \qquad j = 1, \dots, p-1$$

and the remaining elements of C zero. A bit of reflection will convince the reader that the conditions assumed on the covariances is equivalent to the condition that $\text{Cov}(CX) = \text{Cov}(X)$. The matrix C is called a cyclic permutation matrix since, if $x \in R^p$ has coordinates $x_1, \dots, x_p$, then Cx has coordinates $x_2, x_3, \dots, x_p, x_1$. In the case that $p = 5$, a direct calculation shows that

$$\Sigma = \text{Cov}(X) = \text{Cov}(CX) = C\Sigma C'$$

iff Σ has the form

$$\Sigma = \sigma^2 \begin{pmatrix} 1 & \rho_1 & \rho_2 & \rho_2 & \rho_1 \\ & 1 & \rho_1 & \rho_2 & \rho_2 \\ & & 1 & \rho_1 & \rho_2 \\ & & & 1 & \rho_1 \\ & & & & 1 \end{pmatrix}$$

where $\sigma^2 > 0$. The conditions on ρ_1 and ρ_2 so that Σ is positive definite are given later. Covariances that satisfy the condition $\Sigma = C\Sigma C'$ are called *cyclic covariances*. Some further motivation for the study of cyclic covariances can be found in Olkin and Press (1969).

To begin the formal treatment of cyclic covariances, first observe that $C^p = I_p$ so the group generated by C is

$$G_0 = \{I_p, C, C^2, \ldots, C^{p-1}\}.$$

Since C generates G_0, it is clear that $C\Sigma C' = \Sigma$ iff $g\Sigma g' = \Sigma$ for all $g \in G_0$. In what follows, only the case of $p = 2q + 1$, $q \geqslant 1$, is treated. When p is even, slightly different expressions are obtained but the analyses are similar. Rather than characterize the covariance set γ_{G_0} directly, it is useful and instructive to first describe the set

$$\mathcal{A}_{G_0} = \{B | BC = CB, B \in \mathcal{C}_p\}.$$

Recall that $\mathbb{C}^p$ is the complex vector space of p-dimensional coordinate complex vectors and $\mathcal{C}_p$ is the set of all $p \times p$ complex matrices. Consider the complex number $r \equiv \exp[2\pi i/p]$ and define complex column vectors $w_k \in \mathbb{C}^p$ with jth coordinate given by

$$w_{kj} = p^{-1/2}\exp\left[\frac{2\pi i(j-1)(k-1)}{p}\right]; \quad j = 1, \ldots, p$$

for $k = 1, \ldots, p$. A direct calculation shows that

$$w_k^* w_l = \delta_{kl}, \qquad k, l = 1, \ldots, p$$

so $w_1, \ldots, w_p$ is an orthonormal basis for $\mathbb{C}^p$. For future reference note that

$$w_1 = p^{-1/2}e, \qquad \overline{w}_k = w_{p-k+2}, \qquad k = 2, \ldots, q+1$$

where $p = 2q + 1$, $q \geqslant 1$. Here, the bar over w_k denotes complex conjugate, and e is the vector of ones in $\mathbb{C}^p$. The basic relation

$$Cw_k = r^{k-1}w_k, \qquad k = 1, \ldots, p$$

shows that

$$C = \sum_{k=1}^{p} r^{k-1} w_k w_k^*. \tag{9.1}$$

As usual, * denotes conjugate transpose. Obviously, $1, r, \ldots, r^{p-1}$ are eigenvalues of C with corresponding eigenvectors $w_1, \ldots, w_p$. Let $D_0 \in \mathcal{C}_p$ be diagonal with $d_{kk} = r^{k-1}$, $k = 1, \ldots, p$ and let $U \in \mathcal{C}_p$ have columns $w_1, \ldots, w_k$. The relation (9.1) can be written $C = UD_0U^*$. Since $UU^* = I_p$, U is a unitary complex matrix.

Proposition 9.12. The set $\mathcal{Q}_{G_0}$ consists of those $B \in \mathcal{C}_p$ that have the form

$$B = \sum_1^p \beta_k w_k w_k^*. \tag{9.2}$$

where $\beta_1, \ldots, \beta_p$ are arbitrary complex numbers.

Proof. If B has the form (9.2), the identity $BC = CB$ follows easily from (9.1). Conversely, suppose $BC = CB$. Then

$$BUD_0U^* = UD_0U^*B$$

so

$$U^*BUD_0 = D_0U^*BU$$

since $U^*U = I_p$. In other words, U^*BU commutes with D_0. But D_0 is a diagonal matrix with distinct nonzero diagonal elements. This implies that U^*BU must be diagonal, say D, with diagonal elements $\beta_1, \ldots, \beta_p$. Thus $U^*BU = D$ so $B = UDU^*$. Then B has the form (9.2). □

The next step is to identify those elements of $\mathcal{Q}_{G_0}$ that are real and symmetric. Consider $B \in \mathcal{Q}_{G_0}$ so

$$B = \sum_1^p \beta_k w_k w_k^*.$$

Now, suppose that B is real and symmetric. Then the eigenvalues of B, namely $\beta_1, \ldots, \beta_p$, are real. Since $\beta_1, \ldots, \beta_p$ are real and B is real, we have

$$\sum_1^p \beta_k w_k w_k^* = B = \overline{B} = \sum_1^p \beta_k \overline{w}_k \overline{w}_k^*.$$

The relationship $\overline{w}_k = w_{p-k+2}$, $k = 2, \ldots, q+1$, implies that $\beta_k = \beta_{p-k+2}$,

$k = 2, \ldots, q + 1$, so

$$B = \beta_1 w_1 w_1^* + \sum_{k=2}^{q+1} \beta_k \left(w_k w_k^* + \overline{w}_k \overline{w}_k^* \right). \tag{9.3}$$

But any B given by (9.3) is real, symmetric, and commutes with C and conversely. We now show that (9.3) yields a spectral form for the real symmetric elements of $\mathcal{Q}_{G_0}$. Write $w_k = x_k + iy_k$ with $x_k, y_k \in R^p$, and define $u_k \in R^p$ by

$$u_k = x_k + y_k, \qquad k = 1, \ldots, p.$$

The two identities

$$w_k^* w_l = \delta_{kl}, \qquad k, l = 1, \ldots, p$$

$$\overline{w}_k = w_{p-k+2}, \qquad k = 2, \ldots, p$$

and the reality of w_1 yield the identities

$$u_k' u_l = \delta_{kl}, \qquad k, l = 1, \ldots, p$$

$$w_k w_k^* + \overline{w}_k \overline{w}_k^* = u_k u_k' + u_{p-k+2} u_{p-k+2}', \qquad k = 2, \ldots, p.$$

Thus $u_1, \ldots, u_p$ is an orthonormal basis for R^p. Hence any B of the form (9.3) can be written

$$B = \beta_1 u_1 u_1' + \sum_{2}^{q+1} \beta_k \left(u_k u_k' + u_{p-k+2} u_{p-k+2}' \right)$$

and this is a spectral form for B. Such a B is positive definite iff $\beta_k > 0$ for $k = 1, \ldots, q + 1$. This discussion yields the following.

Proposition 9.13. The symmetry model γ_{G_0} consists of those covariances Σ that have the form

$$\Sigma = \alpha_1 u_1 u_1' + \sum_{k=2}^{q+1} \alpha_k \left(u_k u_k' + u_{p-k+2} u_{p-k+2}' \right) \tag{9.4}$$

where $\alpha_k > 0$ for $k = 1, \ldots, q + 1$.

Let Γ have rows $u_1', \ldots, u_p'$. Then Γ is a $p \times p$ symmetric orthogonal matrix with elements

$$\gamma_{jk} = \cos\left[\frac{2\pi}{p}(j-1)(k-1)\right] + \sin\left[\frac{2\pi}{p}(j-1)(k-1)\right]$$

for $j, k = 1, \ldots, p$. Further, any Σ given by (9.4) will be diagonalized by Γ —that is, $\Gamma\Sigma\Gamma$ is diagonal, say D, with diagonal elements

$$d_k = \alpha_k, \qquad k = 1, \ldots, q+1; \qquad d_{p-k+2} = \alpha_k, \qquad k = 2, \ldots, q+1.$$

Since Γ simultaneously diagonalizes all the elements of γ_{G_0}, Γ can sometimes be used to simplify the analysis of certain models with covariances in γ_{G_0}. This is done in the following example.

As an application of the foregoing analysis, suppose $Y_1, \ldots, Y_n$ are independent with $Y_j \in R^p$, $p = 2q+1$, and $\mathcal{L}(Y_j) = N(\mu, \Sigma)$, $j = 1, \ldots, n$. It is assumed that Σ is a cyclic covariance so $\Sigma \in \gamma_{G_0}$. In what follows, we derive the likelihood ratio test for testing H_0, the null hypothesis that the coordinates of μ are all equal, versus H_1, the alternative that μ is completely unknown. As usual, form the matrix $Y: n \times p$ with rows Y_j', $j = 1, \ldots, n$, so

$$\mathcal{L}(Y) = N(e\mu', I_n \otimes \Sigma)$$

where $\mu \in R^p$ and $\Sigma \in \gamma_{G_0}$. Consider the new random vector $Z = (I_n \otimes \Gamma)Y$ where Γ is defined in the previous paragraph. Setting $\nu = \Gamma\mu$, we have

$$\mathcal{L}(Z) = N(e\nu', I_n \otimes D)$$

where $D = \Gamma\Sigma\Gamma$. As noted earlier, D is diagonal with diagonal elements

$$d_k = \alpha_k, \qquad k = 1, \ldots, q+1; \qquad d_{p-k+2} = \alpha_k, \qquad k = 2, \ldots, q+1.$$

Since Σ was assumed to be a completely unknown element of γ_{G_0}, the diagonal elements of D are unknown parameters subject only to the restriction that $\alpha_j > 0$, $j = 1, \ldots, q+1$. In terms of $\nu = \Gamma\mu$, the null hypothesis is $H_0: \nu_2 = \cdots = \nu_p = 0$. Because of the structure of D, it is convenient to relabel things once more. Denote the columns of Z by $Z_1, \ldots, Z_p$ and consider $W_1, \ldots, W_{q+1}$ defined by

$$W_1 = Z_1, \qquad W_j = (Z_j Z_{p-j+2}), \qquad j = 2, \ldots, q+1.$$

Thus $W_1 \in R^n$ and $W_j \in \mathcal{L}_{2,n}$ for $j = 2, \ldots, q+1$. Define vectors $\xi_j \in R^2$

by

$$\xi_j = \begin{pmatrix} \nu_j \\ \nu_{p-j+2} \end{pmatrix}, \qquad j = 2, \dots, q+1.$$

Now, it is clear that $W_1, \dots, W_{q+1}$ are independent and

$$\mathcal{L}(W_1) = N(\nu_1 e, \alpha_1 I_n), \qquad \mathcal{L}(W_j) = N(e\xi_j', \alpha_j I_n \otimes I_2),$$
$$j = 2, \dots, q+1.$$

Further, the null hypothesis is $H_0 : \xi_j = 0, j = 2, \dots, q+1$, and the alternative is that $\xi_j \neq 0$ for some $j = 2, \dots, q+1$. With the model written in this form, a derivation of the likelihood ratio test is routine. Let $P_e = ee'/n$ and let $\|\cdot\|$ denote the usual norm on $\mathcal{L}_{2,n}$. Then the likelihood ratio test rejects H_0 for small values of

$$\Lambda \equiv \prod_{j=2}^{q+1} \frac{\|W_j - P_e W_j\|^2}{\|W_j\|^2}.$$

Of course, the likelihood ratio test of $H_0^{(j)} : \xi_j = 0$ versus $H_1^{(j)} : \xi_j \neq 0$ rejects for small values of

$$\Lambda_j = \frac{\|W_j - P_e W_j\|^2}{\|W_j\|^2}, \qquad j = 2, \dots, q+1.$$

The random variables $\Lambda_2, \dots, \Lambda_{q+1}$ are independent, and under $H_0^{(j)}$,

$$\mathcal{L}(\Lambda_j) = \mathcal{B}(n-1, 1).$$

Thus under H_0, Λ is distributed as a product of the independent beta random variables, each with parameters $n-1$ and 1.

We end this section with a discussion that leads to a new type of structured covariance—namely, the complex covariance structure that is discussed more fully in the next section. This covariance structure arises when we search for an analog of Proposition 9.11 for the cyclic group G_0. To keep things simple, assume $p = 3$ (i.e., $q = 1$) so G_0 has three elements and is a subgroup of the permutation group $\mathcal{P}_3$, which has six elements. Since $p = 3$, Propositions 9.9 and 9.13 yield that $\gamma_{\mathcal{P}_3} = \gamma_{G_0}$ and these symmetry models consist of those covariances of the form

$$\Sigma = \alpha P_e + \beta Q_e, \qquad \alpha > 0, \beta > 0$$

where $P_e = \frac{1}{3}ee'$ and $Q_e = I_3 - P_e$.

Now, consider the two groups $\mathcal{P}_3 \otimes I_r$ and $G_0 \otimes I_r$ acting on $\mathcal{L}_{r,3}$ by

$$(g \otimes I_r)(x) = gx, \qquad g \in \mathcal{P}_3, \qquad x \in \mathcal{L}_{r,3}.$$

Proposition 9.11 states that a covariance T on $\mathcal{L}_{r,3}$ is $\mathcal{P}_3 \otimes I_r$ invariant iff

$$T = P_e \otimes A + Q_e \otimes B \tag{9.5}$$

for some $r \times r$ positive definite A and B. We now claim that for $r > 1$, there are covariances on $\mathcal{L}_{r,3}$ that cannot be written in the form (9.5), but that are $G_0 \otimes I_r$ invariant.

To establish the above claim, recall that the vectors u_1, u_2, and u_3 defined earlier are an orthonormal basis for R^3 and

$$P_e = u_1 u_1', \qquad Q_e = u_2 u_2' + u_3 u_3'.$$

These vectors were defined from the vectors $w_k = x_k + iy_k$, $k = 1, 2, 3$, by $u_k = x_k + y_k$, $k = 1, 2, 3$. Define the matrix J by

$$J = i[w_2 w_2^* - w_3 w_3^*].$$

By Proposition 9.12, J commutes with C. Consider vectors v_2 and v_3 given by

$$v_2 = \frac{1}{\sqrt{2}}(u_2 + u_3), \qquad v_3 = \frac{1}{\sqrt{2}}(u_2 - u_3)$$

so $\{v_2, v_3\}$ is an orthonormal basis for span $\{u_2, u_3\}$. Since $w_3 = \bar{w}_2$, we have $u_3 = x_2 - y_2$, which implies that $v_2 = \sqrt{2}\,x_2$ and $v_3 = \sqrt{2}\,y_2$. This readily implies that

$$J = v_2 v_3' - v_3 v_2'$$

so J is skew-symmetric, nonzero, and $Ju_1 = 0$. Now, consider the linear transformation T_0 on $\mathcal{L}_{r,3}$ to $\mathcal{L}_{r,3}$ given by

$$T_0 = P_e \otimes A + Q_e \otimes B + J \otimes F$$

where A and B are $r \times r$ and positive definite and F is skew-symmetric. It is now a routine matter to show that $(C \otimes I_r)T_0 = T_0(C \otimes I_r)$ since $CP_e = P_e C$, $CQ_e = Q_e C$, and $JC = CJ$. Thus T_0 commutes with each element of $G_0 \otimes I_r$ and T_0 is symmetric as both J and F are skew-symmetric. We now make two claims: first, that a nonzero F exists such that T_0 is positive definite, and second, that such a T_0 cannot be written in the form (9.5). Since $P_e \otimes A + Q_e \otimes B$ is positive definite, it follows that for all skew-symmetric F's that are sufficiently small,

$$P_e \otimes A + Q_e \otimes B + J \otimes F$$

is positive definite. Thus there exists a nonzero skew-symmetric F so that T_0 is positive definite. To establish the second claim, we have the following.

Proposition 9.14. Suppose that

$$P_e \otimes A_1 + Q_e \otimes B_1 + J \otimes F_1 = P_e \otimes A_2 + Q_e \otimes B_2 + J \otimes F_2$$

where A_j and B_j, $j = 1, 2$, are symmetric and F_j, $j = 1, 2$, is skew-symmetric. This implies that $A_1 = A_2$, $B_1 = B_2$, and $F_1 = F_2$.

Proof. Recall that $\{u_1, v_2, v_3\}$ is an orthonormal basis for R^3. The relation $Q_e u_1 = Ju_1 = 0$ implies that for $x \in R^r$

$$(P_e \otimes A_j + Q_e \otimes B_j + J \otimes F_j)(u_1 \Box x) = u_1 \Box (A_j x)$$

for $j = 1, 2$ so $u_1 \Box (A_1 x) = u_1 \Box (A_2 x)$. With $\langle \cdot, \cdot \rangle$ denoting the natural inner product on $\mathfrak{L}_{r,3}$, we have

$$x' A_1 x = \langle u_1 \Box x, u_1 \Box (A_1 x) \rangle = \langle u_1 \Box x, u_1 \Box (A_2 x) \rangle = x' A_2 x$$

for all $x \in R^r$. The symmetry of A_1 and A_2 yield $A_1 = A_2$. Since $P_e v_2 = 0$, $Q_e v_2 = v_2$, and $J v_2 = -v_3$, we have

$$\begin{aligned}(P_e \otimes A_1 + Q_e \otimes B_1 + J \otimes F_1)(v_2 \Box x) &= v_2 \Box (B_1 x) - v_3 \Box (F_1 x) \\ &= v_2 \Box (B_2 x) - v_3 \Box (F_2 x)\end{aligned}$$

for all $x \in R^r$. Thus

$$x' B_1 x = \langle v_2 \Box x, v_2 \Box B_1 x - v_3 \Box (F_1 x) \rangle = x' B_2 x,$$

which implies that $B_1 = B_2$. Further,

$$-y' F_1 x = \langle v_3 \Box y, v_2 \Box (B_1 x) - v_3 \Box F_1 x \rangle = -y' F_2 x$$

for all $x, y \in R^r$. Thus $F_1 = F_2$. □

In summary, we have produced a covariance

$$T_0 = P_e \otimes A + Q_e \otimes B + J \otimes F$$

that is $(G_0 \otimes I_r)$-invariant but is not $(\mathcal{P}_3 \otimes I_r)$-invariant when $r > 1$. Of course, when $r = 1$, the two symmetry models $\gamma_{\mathcal{P}_3}$ and γ_{G_0} are the same. At this point, it is instructive to write out the matrix of T_0 in a special ordered basis for $\mathcal{L}_{r,3}$. Let $\varepsilon_1, \ldots, \varepsilon_n$ be the standard basis for R^r so

$$\{u_1 \square \varepsilon_1, \ldots, u_1 \square \varepsilon_r, v_2 \square \varepsilon_1, \ldots, v_2 \square \varepsilon_r, v_3 \square \varepsilon_1, \ldots, v_3 \square \varepsilon_r\}$$

is an orthonormal basis for $(\mathcal{L}_{r,3}, \langle \cdot, \cdot \rangle)$. A straightforward calculation shows that the matrix of T_0 in this basis is

$$[T_0] = \begin{pmatrix} A & 0 & 0 \\ 0 & B & F \\ 0 & -F & B \end{pmatrix}.$$

Since $[T_0]$ is symmetric and positive definite, the $2r \times 2r$ matrix

$$\Sigma = \begin{pmatrix} B & F \\ -F & B \end{pmatrix}$$

has these properties also. In other words, for each positive definite B, there is a nonzero skew-symmetric F (in fact, there exist infinitely many such skew-symmetric F's) such that Σ is positive definite. This special type of structured covariance has not arisen heretofore. However, it arises again in a very natural way in the next section where we discuss the complex normal distribution. It is not proved here, but the symmetry model of $G_0 \otimes I_r$ when $p = 3$ consists of all covariances of the form

$$T_0 = P_e \otimes A + Q_e \otimes B + J \otimes F$$

where A and B are positive definite and F is skew-symmetric.

9.5. COMPLEX COVARIANCE STRUCTURES

This section contains an introduction to complex covariance structures. One situation where this type of covariance structure arises was described at the end of the last section. To provide further motivation for the study of such models, we begin this section with a brief discussion of the complex normal distribution. The complex normal distribution arises in a variety of contexts and it seems appropriate to include the definition and the elementary properties of this distribution.

The notation introduced in Section 1.6 is used here. In particular, $\mathbb{C}$ is the field of complex numbers, $\mathbb{C}^n$ is the n-dimensional complex vector space of n-tuples (columns) of complex numbers, and $\mathcal{C}_n$ is the set of all $n \times n$ complex matrices. For $x, y \in \mathbb{C}^n$, the *inner product* between x and y is

$$(x, y) \equiv \sum_{j=1}^{n} \bar{x}_j y_j = x^* y.$$

where x^* denotes the conjugate transpose of x. Each $x \in \mathbb{C}^n$ has the unique representation $x = u + iv$ with $u, v \in R^n$. Of course, u is the *real part* of x, v is the *imaginary part* of x, and $i = \sqrt{-1}$ is the imaginary unit. This representation of x defines a real vector space isomorphism between $\mathbb{C}^n$ and R^{2n}. More precisely, for $x \in \mathbb{C}^n$, let

$$[x] = \begin{pmatrix} u \\ v \end{pmatrix} \in R^{2n}$$

where $x = u + iv$. Then $[ax + by] = a[x] + b[y]$ for $x, y \in \mathbb{C}^n$, $a, b \in R$, and obviously, $[\cdot]$ is a one-to-one onto function. In particular, this shows that $\mathbb{C}^n$ is a $2n$-dimensional real vector space. If $C \in \mathcal{C}_n$, then $C = A + iB$ where A and B are $n \times n$ real matrices. Thus for $x = u + iv \in \mathbb{C}^n$,

$$Cx = (A + iB)(u + iv) = Au - Bv + i(Av + Bu)$$

so

$$[Cx] = \begin{pmatrix} Au - Bv \\ Av + Bu \end{pmatrix} = \begin{pmatrix} A & -B \\ B & A \end{pmatrix} \begin{pmatrix} u \\ v \end{pmatrix}.$$

This suggests that we let $\{C\}$ be the $(2n) \times (2n)$ partitioned matrix given by

$$\{C\} = \begin{pmatrix} A & -B \\ B & A \end{pmatrix} : (2n) \times (2n).$$

With this definition, $[Cx] = \{C\}[x]$. The whole point is that the matrix $C \in \mathcal{C}_n$ applied to $x \in \mathbb{C}^n$ can be represented by applying the real matrix $\{C\}$ to the real vector $[x] \in R^{2n}$.

A complex matrix $C \in \mathcal{C}_n$ is called Hermitian if $C = C^*$. Writing $C = A + iB$ with A and B both real, C is Hermitian iff

$$A + iB = A' - iB',$$

which is equivalent to the two conditions

$$A = A', \qquad B = -B'.$$

Thus C is Hermitian iff $\{C\}$ is a symmetric real matrix. A Hermitian matrix C is positive definite if $x^*Cx > 0$ for all $x \in \mathbb{C}^n$, $x \neq 0$. However, for Hermitian C,

$$x^*Cx = [x]'\{C\}[x]$$

so C is positive definite iff $\{C\}$ is a positive definite real matrix. Of course, a Hermitian matrix C is positive semidefinite if $x^*Cx \geqslant 0$ for $x \in \mathbb{C}^n$ and C is positive semidefinite iff $\{C\}$ is positive semidefinite.

Now consider a random variable X with values in $\mathbb{C}$. Then $X = U + iV$ where U and V are real random variables. It is clear that the mean value of X must be defined by

$$\mathcal{E}X = \mathcal{E}U + i\mathcal{E}V$$

assuming $\mathcal{E}U$ and $\mathcal{E}V$ both exist. The variance of X, assuming it exists, is defined by

$$\operatorname{var}(X) = \mathcal{E}\left[(X - \mathcal{E}(X))\left(\overline{X - \mathcal{E}(X)}\right)\right]$$

where the bar denotes complex conjugate. Since X is a complex random variable, the complex conjugate is necessary if we want the variance of X to be a nonnegative real number. In terms of U and V,

$$\operatorname{var}(X) = \operatorname{var}(U) + \operatorname{var}(V).$$

It also follows that

$$\operatorname{var}(aX + b) = a\bar{a}\operatorname{var}(X)$$

for $a, b \in \mathbb{C}$. For two random variables X_1 and X_2 in $\mathbb{C}$, define the covariance between X_1 and X_2 (in that order) to be

$$\operatorname{cov}\{X_1, X_2\} \equiv \mathcal{E}\left[(X_1 - \mathcal{E}(X_1))\left(\overline{X_2 - \mathcal{E}(X_2)}\right)\right],$$

assuming the expectations in question exist. With this definition it is clear that $\operatorname{cov}\{X_1, X_1\} = \operatorname{var}(X_1)$, $\operatorname{cov}\{X_2, X_1\} = \overline{\operatorname{cov}\{X_1, X_2\}}$, and

$$\operatorname{cov}\{X_1, X_2 + X_3\} = \operatorname{cov}\{X_1, X_2\} + \operatorname{cov}\{X_1, X_3\}.$$

Further,

$$\operatorname{cov}\{a_1X_1 + b_1, a_2X_2 + b_2\} = a_1\bar{a}_2 \operatorname{cov}\{X_1, X_2\}$$

for $a_1, a_2, b_1, b_2 \in \mathbb{C}$.

We now turn to the problem of defining a normal distribution on $\mathbb{C}^n$. Basically, the procedure is the same as defining a normal distribution on R^n. Step one is to define a normal distribution with mean zero and variance one on $\mathbb{C}$, then define an arbitrary normal distribution on $\mathbb{C}$ by an affine transformation of the distribution defined in step one, and finally we say that $Z \in \mathbb{C}^n$ has a complex normal distribution if $(a, Z) = a^*Z$ has a normal distribution in $\mathbb{C}$ for each $a \in \mathbb{C}^n$. However it is not entirely obvious how to carry out step one. Consider $X \in \mathbb{C}$ and let $\mathbb{C}N(0, 1)$ denote the distribution, yet to be defined, called the complex normal distribution with mean zero and variance one. Writing $X = U + iV$, we have

$$[X] = \begin{pmatrix} U \\ V \end{pmatrix} \in R^2$$

so the distribution of X on $\mathbb{C}$ determines the joint distribution of U and V on R^2 and, conversely, as $[\cdot]$ is one-to-one and onto. If $\mathcal{L}(X) = \mathbb{C}N(0, 1)$, then the following two conditions should hold:

(i) $\mathcal{L}(aX) = \mathbb{C}N(0, 1)$ for $a \in \mathbb{C}$ with $a\bar{a} = 1$.

(ii) $[X]$ has a bivariate normal distribution on R^2.

When $a\bar{a} = 1$ and X has mean zero and variance one, then aX has mean zero and variance one so condition (i) simply says that a scalar multiple of a complex normal is again complex normal. Condition (ii) is the requirement that a normal distribution on $\mathbb{C}$ be transformed into a normal distribution on R^2 under the real linear mapping $[\cdot]$. It can now be shown that conditions (i) and (ii) uniquely define the distribution of $[X]$ and hence provide us with the definition of a $\mathbb{C}N(0, 1)$ distribution. Since $\mathcal{E}X = 0$, we have $\mathcal{E}[X] = 0$. Condition (i) implies that

$$\mathcal{L}([X]) = \mathcal{L}([aX]), \qquad a\bar{a} = 1.$$

However, writing $a = \alpha + i\beta$,

$$[aX] = \begin{pmatrix} \alpha & -\beta \\ \beta & \alpha \end{pmatrix}[X] \equiv \Gamma[X]$$

where Γ is a 2×2 orthogonal matrix with determinant equal to one since $a\bar{a} = \alpha^2 + \beta^2 = 1$. Therefore,

$$\mathcal{L}([X]) = \mathcal{L}(\Gamma[X])$$

for all such orthogonal matrices. Using this together with the fact that $1 = \text{var}(X) = \text{var}(U) + \text{var}(V)$ implies that

$$\text{Cov}([X]) = \tfrac{1}{2}I_2.$$

Hence

$$\mathcal{L}([X]) = N_2(0, \tfrac{1}{2}I_2)$$

so the real and imaginary parts of X are independent normals with mean zero and variance one half.

Definition 9.1. A random variable $X = U + iV \in \mathbb{C}$ has a complex normal distribution with mean zero and variance one, written $\mathcal{L}(X) = \mathbb{C}N(0,1)$, if

$$\mathcal{L}\left(\begin{pmatrix} U \\ V \end{pmatrix}\right) = N_2(0, \tfrac{1}{2}I_2).$$

With this definition, it is clear that when $\mathcal{L}(X) = \mathbb{C}N(0,1)$, the density of X on $\mathbb{C}$ with respect to two-dimensional Lebesgue measure on $\mathbb{C}$ is

$$p(x) = \frac{1}{\pi}\exp[-x\bar{x}], \qquad x \in \mathbb{C}.$$

Given $\mu \in \mathbb{C}$ and σ^2, $\sigma > 0$, a random variable $X_1 \in \mathbb{C}$ has a complex normal distribution with mean μ and variance σ^2 if $\mathcal{L}(X_1) = \mathcal{L}(\sigma X + \mu)$ where $\mathcal{L}(X) = \mathbb{C}N(0,1)$. In such a case, we write $\mathcal{L}(X_1) = \mathbb{C}N(\mu, \sigma^2)$. It is clear that $X_1 = U_1 + iV_1$ has a $\mathbb{C}N(\mu, \sigma^2)$ distribution iff U_1 and V_1 are independent and normal with variance $\frac{1}{2}\sigma^2$ and means $\mathcal{E}U_1 = \mu_1$, $\mathcal{E}V_1 = \mu_2$, where $\mu = \mu_1 + i\mu_2$. As in the real case, a basic result is the following.

Proposition 9.15. Suppose $X_1, \dots, X_m$ are independent random variables in $\mathbb{C}$ with $\mathcal{L}(X_j) = \mathbb{C}N(\mu_j, \sigma_j^2)$, $j = 1, \dots, m$. Then

$$\mathcal{L}\left(\sum_{j=1}^{m}(a_jX_j + b_j)\right) = \mathbb{C}N\left(\sum_{j=1}^{m}(a_j\mu_j + b_j), \sum_{j=1}^{m} a_j\bar{a}_j\sigma_j^2\right)$$

for $a_j, b_j \in \mathbb{C}$, $j = 1, \dots, m$.

Proof. This is proved by considering the real and imaginary parts of each X_j. The details are left to the reader. □

Suppose Y is a random vector in $\mathbb{C}^n$ with coordinates $Y_1,\dots, Y_n$ and that $\text{var}(Y_j) < +\infty$ for $j = 1,\dots, n$. Define a complex matrix H with elements h_{jk} given by

$$h_{jk} \equiv \text{cov}\{Y_j, Y_k\}.$$

Since $h_{jk} = \overline{h_{kj}}$, H is a Hermitian matrix. For $a, b \in \mathbb{C}^n$, a bit of algebra shows that

$$\text{cov}\{a^*Y, b^*Y\} = a^*Hb = (a, Hb).$$

As in the real case, H is the *covariance matrix* of Y and is denoted by $\text{Cov}(Y) \equiv H$. Since $a^*Ha = \text{var}(a^*Y) \geqslant 0$, H is positive semidefinite. If $H = \text{Cov}(Y)$ and $A \in \mathcal{C}_n$, it is readily verified that $\text{Cov}(AY) = AHA^*$.

We now turn to the definition of a complex normal distribution on the n-dimensional complex vector space $\mathbb{C}^n$.

Definition 9.2. A random vector $X \in \mathbb{C}^n$ has a complex normal distribution if, for each $a \in \mathbb{C}^n$, $(a, X) = a^*X$ has a complex normal distribution on $\mathbb{C}$.

If $X \in \mathbb{C}_n$ has a complex normal distribution and if $A \in \mathcal{C}_n$, it is clear that AX also has a complex normal distribution since $(a, AX) = (A^*a, X)$. In order to describe all the complex normal distributions on $\mathbb{C}^n$, we proceed as in the real case. Let $X_1,\dots, X_n$ be independent with $\mathcal{L}(X_j) = \mathbb{C}N(0, 1)$ on $\mathbb{C}$ and let $X \in \mathbb{C}^n$ have coordinates $X_1,\dots, X_n$. Since

$$a^*X = \sum_{j=1}^{n} \bar{a}_j X_j,$$

Proposition 9.15 shows that $\mathcal{L}(a^*X) = \mathbb{C}N(0, \Sigma\bar{a}_j a_j)$. Thus X has a complex normal distribution. Further, $\mathcal{E}X = 0$ and

$$\text{cov}\{X_j, X_k\} = \delta_{jk}$$

so $\text{Cov}(X) = I$. For $A \in \mathcal{C}_n$ and $\mu \in \mathbb{C}^n$, it follows that $Y = AX + \mu$ has a complex normal distribution and

$$\mathcal{E}Y = \mu, \qquad \text{Cov}(Y) = AA^* \equiv H.$$

Since every nonnegative definite Hermitian matrix can be written as AA^* for some $A \in \mathcal{C}_n$, we have produced a complex normal distribution on $\mathbb{C}^n$ with an arbitrary mean vector $\mu \in \mathbb{C}^n$ and an arbitrary nonnegative definite Hermitian covariance matrix. However, it still must be shown that, if X and $\tilde{X}$ in $\mathbb{C}^n$ are complex normal with $\mathcal{E}X = \mathcal{E}\tilde{X}$ and $\text{Cov}(X) = \text{Cov}(\tilde{X})$, then $\mathcal{L}(X) = \mathcal{L}(\tilde{X})$. The proof of this assertion is left to the reader. Given this fact, it makes sense to speak of the complex normal distribution on $\mathbb{C}^n$ with mean vector μ and covariance matrix H as this specifies a unique probability distribution. If X has such a distribution, the notation

$$\mathcal{L}(X) = \mathbb{C}N(\mu, H)$$

is used. Writing $X = U + iV$, it is useful to describe the joint distribution of U and V when $\mathcal{L}(X) = \mathbb{C}N(\mu, H)$ on $\mathbb{C}^n$. First, consider $\tilde{X} = \tilde{U} + i\tilde{V}$ where $\mathcal{L}(\tilde{X}) = \mathbb{C}N(\mu, I)$. Then the coordinates of $\tilde{X}$ are independent and it follows that

$$\mathcal{L}\begin{pmatrix} \tilde{U} \\ \tilde{V} \end{pmatrix} = N\left(\begin{pmatrix} \mu_1 \\ \mu_2 \end{pmatrix}, \tfrac{1}{2}I_{2n}\right)$$

where $\mu = \mu_1 + i\mu_2$. For a general nonnegative definite Hermitian matrix H, write $H = AA^*$ with $A \in \mathcal{C}_n$. Then

$$\mathcal{L}(X) = \mathcal{L}(A\tilde{X} + \mu).$$

Since

$$[X] = \begin{pmatrix} U \\ V \end{pmatrix}$$

and

$$[AX + \mu] = \{A\}[\tilde{X}] + [\mu] = \begin{pmatrix} B & -C \\ C & B \end{pmatrix}\begin{pmatrix} \tilde{U} \\ \tilde{V} \end{pmatrix} + \begin{pmatrix} \mu_1 \\ \mu_2 \end{pmatrix}$$

where $A = B + iC$, it follows that

$$\mathcal{L}([X]) = \mathcal{L}(\{A\}[\tilde{X}] + [\mu]).$$

But $H = \Sigma + iF$ where Σ is positive semidefinite, F is skew-symmetric, and the real matrix

$$\{H\} = \begin{pmatrix} \Sigma & -F \\ F & \Sigma \end{pmatrix}$$

is positive semidefinite. Since $H = AA^*$, $\{H\} = \{A\}\{A\}'$, and therefore,

$$\mathcal{L}([X]) = \mathcal{L}(\{A\}[\tilde{X}] + [\mu]) = N\left(\begin{pmatrix} \mu_1 \\ \mu_2 \end{pmatrix}, \tfrac{1}{2}\{H\}\right)$$

$$= N\left(\begin{pmatrix} \mu_1 \\ \mu_2 \end{pmatrix}, \tfrac{1}{2}\begin{pmatrix} \Sigma & -F \\ F & \Sigma \end{pmatrix}\right).$$

In summary, we have the following result.

Proposition 9.16. Suppose $\mathcal{L}(X) = \mathbb{C}N(\mu, H)$ and write $X = U + iV$, $\mu = \mu_1 + i\mu_2$, and $H = \Sigma + iF$. Then

$$\mathcal{L}\begin{pmatrix} U \\ V \end{pmatrix} = N\left(\begin{pmatrix} \mu_1 \\ \mu_2 \end{pmatrix}, \tfrac{1}{2}\begin{pmatrix} \Sigma & -F \\ F & \Sigma \end{pmatrix}\right).$$

Conversely, with U and V jointly distributed as above, set $X = U + iV$, $\mu = \mu_1 + i\mu_2$, and $H = \Sigma + iF$. Then $\mathcal{L}(X) = \mathbb{C}N(\mu, H)$.

The above proposition establishes a one-to-one correspondence between n-dimensional complex normal distributions, say $\mathbb{C}N(\mu, H)$, and $2n$-dimensional real normal distributions with a special covariance structure given by

$$\tfrac{1}{2}\{H\} = \tfrac{1}{2}\begin{pmatrix} \Sigma & -F \\ F & \Sigma \end{pmatrix}$$

where $H = \Sigma + iF$. Given a sample of independent complex normal random vectors, the above correspondence provides us with the option of either analyzing the sample in the complex domain or representing everything in the real domain and performing the analysis there. Of course, the advantage of the real domain analysis is that we have developed a large body of theory that can be applied to this problem. However, this advantage is a bit illusory. As it turns out, many results for the complex normal distribution are clumsy to prove and difficult to understand when expressed in the real domain. From the point of view of understanding, the proper approach is simply to develop a theory of the complex normal distribution that parallels the development already given for the real normal distribution. Because of space limitations, this theory is not given in detail. Rather, we provide a brief list of results for the complex normal with the hope that the reader can see the parallel development. The proofs of many of these results are minor modifications of the corresponding real results.

Consider $X \in \mathbb{C}^p$ such that $\mathcal{L}(X) = \mathbb{C}N(\mu, H)$ where H is nonsingular. Then the density of X with respect to Lebesgue measure on $\mathbb{C}^p$ is

$$f(x) = \pi^{-p}(\det H)^{-1}\exp\left[-(x - \mu)^* H^{-1}(x - \mu)\right].$$

When $H = I$, then

$$\mathcal{L}(X^*X) = \tfrac{1}{2}\chi^2_{2p}(\mu^*\mu).$$

With this result and the spectral theorem for Hermitian matrices (see Halmos, 1958, Section 79), the distribution of quadratic forms, say X^*AX for a Hermitian, can be described in terms of linear combinations of independent noncentral chi-square random variables.

As in the real case, independence of jointly complex normal random vectors is equivalent to the absense of correlation. More precisely, if $\mathcal{L}(X) = \mathbb{C}N(\mu, H)$ and if $A : q \times p$ and $B : r \times p$ are complex matrices, then AX and BX are independent iff $AHB^* = 0$. In particular, if X is partitioned as

$$X = \begin{pmatrix} X_1 \\ X_2 \end{pmatrix}, \qquad X_j \in \mathbb{C}^{p_j}, j = 1, 2$$

and H is partitioned similarly as

$$H = \begin{pmatrix} H_{11} & H_{12} \\ H_{21} & H_{22} \end{pmatrix}$$

where H_{jk} is $p_j \times p_k$, then X_1 and X_2 are independent iff $H_{12} = 0$. When H_{22} is nonsingular, this implies that $X_1 - H_{12}H_{22}^{-1}X_2$ and X_2 are independent. This result yields the conditional distribution of X_1 given X_2, namely,

$$\mathcal{L}(X_1|X_2) = \mathbb{C}N\big(\mu_1 + H_{12}H_{22}^{-1}(X_2 - \mu_2),\ H_{11\cdot 2}\big)$$

where $H_{11\cdot 2} = H_{11} - H_{12}H_{22}^{-1}H_{21}$ and $\mu_j = \mathcal{E}X_j, j = 1, 2$.

The complex Wishart distribution arises in a natural way, just as the real Wishart distribution did.

Definition 9.3. A $p \times p$ random Hermitian matrix S has a complex Wishart distribution with parameters H, p, and n if

$$\mathcal{L}(S) = \mathcal{L}\left(\sum_{j=1}^{n} X_j X_j^*\right)$$

where $X_1, \ldots, X_n \in \mathbb{C}^p$ are independent with

$$\mathcal{L}(X_j) = \mathbb{C}N(0, H).$$

In such a case, we write

$$\mathcal{L}(S) = \mathbb{C}W(H, p, n).$$

In this definition, p is the dimension, n is the degrees of freedom and H is a $p \times p$ nonnegative definite Hermitian matrix. It is clear that S is always nonnegative definite and, as in the real case, S is positive definite with probability one iff H is positive definite and $n \geqslant p$. When $p = 1$ and $H = 1$, it is clear that

$$\mathbb{C}W(1, 1, n) = \tfrac{1}{2}\chi^2_{2n}.$$

Further, complex analogues of Proposition 8.8, 8.9, and 8.13 show that if $\mathcal{L}(S) = W(H, p, n)$ with $n \geqslant p$ and H positive definite, and if $\mathcal{L}(X) = N(0, H)$ with X and S independent, then

$$\mathcal{L}(X^*S^{-1}X) = F_{2p, 2(n-p+1)}.$$

We now turn to a brief discussion of one special case of the complex MANOVA problem. Suppose $X_1, \dots, X_n \in \mathbb{C}^p$ are independent with

$$\mathcal{L}(X_j) = \mathbb{C}N(\mu, H)$$

and assume that $H > 0$—that is, H is positive definite. The joint density of $X_1, \dots, X_n$ with respect to $2np$-dimensional Lebesgue measure is

$$\begin{aligned} p(X|\mu, H) &= \prod_{j=1}^{n} \pi^{-p}|H|^{-1}\exp\left[-(X_j - \mu)^*H^{-1}(X_j - \mu)\right] \\ &= \pi^{-np}|H|^{-n}\exp\left[-\sum_{j=1}^{n}(X_j - \mu)^*H^{-1}(X_j - \mu)\right] \\ &= \pi^{-np}|H|^{-n}\exp\left[-n(\bar{X} - \mu)^*H^{-1}(\bar{X} - \mu)\right. \\ &\qquad \left. - \operatorname{tr}\left(\sum_{j=1}^{n}(X_j - \bar{X})(X_j - \bar{X})^*\right)H^{-1}\right] \end{aligned}$$

where $\bar{X} = n^{-1}\Sigma X_j$ and tr denote the trace. Here, X is the np-dimensional vector in $\mathbb{C}^{np}$ consisting of $X_1, X_2, \ldots, X_n$. Setting

$$S = \sum_{j=1}^{n} (X_j - \overline{X})(X_j - \overline{X})^*,$$

we have

$$p(X|\mu, H) = \pi^{-np}|H|^{-n}\exp\left[-n(\overline{X}-\mu)^*H^{-1}(\overline{X}-\mu) - \operatorname{tr} SH^{-1}\right].$$

It follows that $(\overline{X}, S)$ is a sufficient statistic for this parametric family and $\hat{\mu} \equiv \overline{X}$ is the maximum likelihood estimator of μ. Thus

$$p(X|\hat{\mu}, H) = \pi^{-np}|H|^{-n}\exp[-\operatorname{tr} SH^{-1}].$$

A minor modification of the argument given in Example 7.10 shows that when $S > 0$, $p(X|\hat{\mu}, H)$ is maximized uniquely, over all positive definite H, at $\hat{H} = n^{-1}S$. When $n \geqslant p + 1$, then S is positive definite with probability one so in this case, the maximum likelihood estimator of H is $\hat{H} = n^{-1}S$. If $\mu = 0$, then

$$\begin{aligned} p(X|0, H) &= \pi^{-np}|H|^{-n}\exp\left[-\sum_{j=1}^{n} X_j^*H^{-1}X_j\right] \\ &= \pi^{-np}|H|^{-n}\exp[-\operatorname{tr} \tilde{S}H^{-1}] \end{aligned}$$

where

$$\tilde{S} = \sum_{j=1}^{n} X_j X_j^* = S + n\overline{X}\,\overline{X}^*.$$

Obviously, $p(X|0, H)$ is maximized at $\tilde{H} = n^{-1}\tilde{S}$. Thus the likelihood ratio test for testing $\mu = 0$ versus $\mu \neq 0$ rejects for small values of

$$\Lambda = \frac{p(X|0, \tilde{H})}{p(X|\hat{\mu}, \hat{H})} = \frac{|\tilde{S}|^{-n}}{|S|^{-n}} = \frac{|S|^n}{|S + n\overline{X}\,\overline{X}^*|^n}.$$

As in the real case, $\overline{X}$ and S are independent,

$$\mathcal{L}(S) = \mathbb{C}W(H, p, n-1)$$

and

$$\mathcal{L}(\sqrt{n}\,\overline{X}) = \mathbb{C}N(\sqrt{n}\,\mu, H).$$

Setting $Y = \sqrt{n}\,\overline{X}$,

$$\Lambda^{1/n} = \frac{|S|}{|S + YY^*|} = \frac{1}{1 + Y^*S^{-1}Y}$$

so the likelihood ratio test rejects for large values of $Y^*S^{-1}Y \equiv T^2$. Arguments paralleling those in the real case can be used to show that

$$\mathcal{L}(T^2) = F(2p, 2(n-p), \delta)$$

where $\delta = n\mu^*H^{-1}\mu$ is the noncentrality parameter in the F distribution. Further, the monotone likelihood ratio property of the F- distribution can be used to show that the likelihood ratio test is uniformly most powerful among tests that are invariant under the group of complex linear transformations that preserve the above testing problem.

In the preceeding discussion, we have outlined one possible analysis of the one-sample problem for the complex normal distribution. A theory for the complex MANOVA problem similar to that given in Section 9.1 for the real MANOVA problem would require complex analogues of many results given in the first eight chapters of this book. Of course, it is possible to represent everything in terms of real random vectors. This representation consists of an $n \times 2p$ random matrix $Y \in \mathcal{L}_{2p,n}$ where

$$\mathcal{L}(Y) = N(ZB, I_n \otimes \Psi).$$

As usual, Z is $n \times r$ of rank r and $B: r \times 2p$ is a real matrix of unknown parameters. The distinguishing feature of the model is that Ψ is assumed to have the form

$$\Psi = \begin{pmatrix} \Sigma & -F \\ F & \Sigma \end{pmatrix}$$

where $\Sigma: p \times p$ is positive definite and $F: p \times p$ is skew-symmetric. For reasons that should be obvious by now, Ψ's of the above form are said to have complex covariance structure. This model can now be analyzed using the results developed for the real normal linear model. However, as stated earlier, certain results are clumsy to prove and more difficult to understand when expressed in the real domain rather than the complex domain. Although not at all obvious, these models are not equivalent to a product of real MANOVA models of the type discussed in Section 9.1.

9.6. ADDITIONAL EXAMPLES OF LINEAR MODELS

The examples of this section have been chosen to illustrate how conditioning can sometimes be helpful in finding maximum likelihood estimators and also to further illustrate the use of invariance in analyzing linear models. The linear models considered now are not products of MANOVA models and the regression subspaces are not invariant under the covariance transformations of the model. Thus finding the maximum likelihood estimator of mean vector is not just a matter of computing the orthogonal projection onto the regression subspace. For the models below, we derive maximum likelihood estimators and likelihood ratio tests and then discuss the problem of finding a good invariant test.

The first model we consider consists of a variation on the one-sample problem. Suppose $X_1, \ldots, X_n$ are independent with $\mathcal{L}(X_i) = N(\mu, \Sigma)$ where $X_i \in R^p$, $i = 1, \ldots, n$. As usual, form the $n \times p$ matrix X whose rows are X_i', $i = 1, \ldots, n$. Then

$$\mathcal{L}(X) = N(e\mu', I_n \otimes \Sigma)$$

where $e \in R^n$ is the vector of ones. When μ and Σ are unknown, the linear model for X is a MANOVA model and the results in Section 9.1 apply directly. To transform this model to canonical form, let Γ be an $n \times n$ orthogonal matrix with first row $e'/\sqrt{n}$. Setting $Y = \Gamma X$ and $\beta = \sqrt{n}\,\mu'$,

$$\mathcal{L}(Y) = N(\varepsilon_1 \beta, I_n \otimes \Sigma)$$

where ε_1 is the first unit vector in R^n and $\beta \in \mathcal{L}_{p,1}$. Partition Y as

$$Y = \begin{pmatrix} Y_1 \\ Y_2 \end{pmatrix}$$

where $Y_1 \in \mathcal{L}_{p,1}$, $Y_2 \in \mathcal{L}_{p,m}$, and $m = n - 1$. Then

$$\mathcal{L}(Y_1) = N(\beta, \Sigma)$$

and

$$\mathcal{L}(Y_2) = N(0, I_m \otimes \Sigma).$$

For testing $H_0: \beta = 0$, the results of Section 9.1 show that the test that rejects for large values of $Y_1(Y_2'Y_2)^{-1}Y_1'$ (assuming $m \geqslant p$) is equivalent to the likelihood ratio test and this test is most powerful within the class of invariant tests.

We now turn to a testing problem to which the MANOVA results do not apply. With the above discussion in mind, consider $U \in \mathcal{L}_{p,1}$ and $Z \in \mathcal{L}_{p,m}$ where U and Z are independent with

$$\mathcal{L}(U) = N(\beta, \Sigma)$$

and

$$\mathcal{L}(Z) = N(0, I_m \otimes \Sigma).$$

Here, $\beta \in \mathcal{L}_{p,1}$ and $\Sigma > 0$ is a completely unknown $p \times p$ covariance matrix. Partition β into β_1 and β_1 where

$$\beta_i \in \mathcal{L}_{p_i,1}, \qquad i = 1,2, \quad p_1 + p_2 = p.$$

Consider the problem of testing the null hypothesis $H_0 : \beta_1 = 0$ versus $H_1 : \beta_1 \neq 0$ where β_2 and Σ are unknown. Under H_0, the regression subspace of the random matrix

$$\begin{pmatrix} U \\ Z \end{pmatrix} \in \mathcal{L}_{p,m+1}$$

is

$$M_0 = \left\{ \mu | \mu = \begin{pmatrix} 0 & \beta_2 \\ 0 & 0 \end{pmatrix} \in \mathcal{L}_{p,m+1}, \beta_2 \in \mathcal{L}_{p_2,1} \right\},$$

and the set of covariances is

$$\gamma = \{ I_{m+1} \otimes \Sigma | \Sigma \in \mathcal{S}_p^+ \}.$$

It is easy to verify that M_0 is not invariant under all the elements of γ so the maximum likelihood estimator of β_2 under H_0 cannot be found by least-squares (ignoring Σ). To calculate the likelihood ratio test for H_0 versus H_1, it is convenient to partition U and Z as

$$U = (U_1, U_2), \qquad U_i \in \mathcal{L}_{p_i,1}, \qquad i = 1,2$$

$$Z = (Z_1, Z_2), \qquad Z_i \in \mathcal{L}_{p_i,m}, \qquad i = 1,2$$

and then condition on U_1 and Z_1. Since U and Z are independent, the joint distribution of U and Z is specified by the two conditional distributions, $\mathcal{L}(U_2|U_1)$ and $\mathcal{L}(Z_2|Z_1)$, together with the two marginal distributions, $\mathcal{L}(U_1)$ and $\mathcal{L}(Z_1)$. Our results for the normal distribution show that these distributions are

$$\mathcal{L}(U_2|U_1) = N\big(\beta_2 + (U_1 - \beta_1)\Sigma_{11}^{-1}\Sigma_{12}, \Sigma_{22\cdot 1}\big)$$

$$\mathcal{L}(U_1) = N(\beta_1, \Sigma_{11})$$

$$\mathcal{L}(Z_2|Z_1) = N\big(Z_1\Sigma_{11}^{-1}\Sigma_{12}, I_m \otimes \Sigma_{22\cdot 1}\big)$$

$$\mathcal{L}(Z_1) = N(0, I_m \otimes \Sigma_{11})$$

where Σ is partitioned as

$$\Sigma = \begin{pmatrix} \Sigma_{11} & \Sigma_{12} \\ \Sigma_{21} & \Sigma_{22} \end{pmatrix}$$

with Σ_{ij} being $p_i \times p_j$, $i, j = 1, 2$. As usual, $\Sigma_{22\cdot 1} = \Sigma_{22} - \Sigma_{21}\Sigma_{11}^{-1}\Sigma_{12}$. By Proposition 5.8, the reparameterization defined by $\Psi_{11} = \Sigma_{11}$, $\Psi_{12} = \Sigma_{11}^{-1}\Sigma_{12}$, and $\Psi_{22} = \Sigma_{22\cdot 1}$ is one-to-one and onto. To calculate the likelihood ratio test for H_0 versus H_1, we need to find the maximum likelihood estimators under H_0 and H_1.

Proposition 9.17. The likelihood ratio test of $H_0 : \beta_1 = 0$ versus $H_1 : \beta_1 \neq 0$ rejects H_0 if the statistic

$$\Lambda = U_1 S_{11}^{-1} U_1'$$

is too large. Here, $S = Z'Z$ and

$$S = \begin{pmatrix} S_{11} & S_{12} \\ S_{21} & S_{22} \end{pmatrix}$$

where S_{ij} is $p_i \times p_j$.

Proof. Let $f_1(U_1|\beta_1, \Psi_{11})$ be the density of $\mathcal{L}(U_1)$, let $f_2(U_2|U_1, \beta_1, \beta_2, \Psi_{12}, \Psi_{22})$ be the conditional density of $\mathcal{L}(U_2|U_1)$, let $f_3(Z_1|\Psi_{11})$ be the density of $\mathcal{L}(Z_1)$, and let $f_4(Z_2|Z_1, \Psi_{12}, \Psi_{22})$ be the density of $\mathcal{L}(Z_2|Z_1)$. Under H_0, $\beta_1 = 0$ and the unique value of β_2 that maximizes $f_2(U_2|U_1, 0, \beta_2, \Psi_{12}, \Psi_{22})$ is

$$\hat{\beta}_2 = U_2 - U_1\Psi_{12},$$

for Ψ_{12} fixed. It is clear that

$$f_2\left(U_2|U_1, 0, \hat{\beta}_2, \Psi_{12}, \Psi_{22}\right) \,\dot{\alpha}\, |\Psi_{22}|^{-1/2}$$

where the symbol $\dot{\alpha}$ means "is proportional to." We now maximize with respect to Ψ_{12}. With $\beta_2 = \hat{\beta}_2$, Ψ_{12} occurs only in the density of Z_2 given Z_1. Since $\mathcal{L}(Z_2|Z_1) = N(Z_1\Psi_{12}, I_m \otimes \Psi_{22})$, it follows from our treatment of the MANOVA problem that

$$\hat{\Psi}_{12} = (Z_1'Z_1)^{-1}Z_1'Z_2 = S_{11}^{-1}S_{12}$$

and

$$f_4(Z_2|Z_1, \hat{\Psi}_{12}, \Psi_{22}) \mathrel{\dot{\propto}} |\Psi_{22}|^{-m/2}\exp[-\tfrac{1}{2}\operatorname{tr} S_{22\cdot 1}\Psi_{22}^{-1}].$$

Since $\beta_1 = 0$, it is now clear that

$$\hat{\Psi}_{11} = \frac{1}{m+1}[Z_1'Z_1 + U_1'U_1] = \frac{1}{m+1}[S_{11} + U_1'U_1]$$

and

$$\hat{\Psi}_{22} = \frac{1}{m+1}S_{22\cdot 1}.$$

Substituting these values into the product of the four densities shows that the maximum under H_0 is proportional to

$$\Lambda_0 = |S_{22\cdot 1}|^{-(m+1)/2}|S_{11} + U_1'U_1|^{-(m+1)/2}$$

Under the alternative H_1, we again maximize the likelihood function by first noting that

$$\tilde{\beta}_2 \equiv U_2 - (U_1 - \beta_1)\Psi_{12}$$

maximizes the density of U_2 given U_1. Also,

$$f_2(U_2|U_1, \beta_1, \tilde{\beta}_2, \Psi_{12}, \Psi_{22}) \mathrel{\dot{\propto}} |\Psi_{22}|^{-1/2}.$$

With this choice of $\tilde{\beta}_2$, β_1 occurs only in the density of U_1 so $\tilde{\beta}_1 = U_1$ maximizes the density of U_1 and

$$f_1(U_1|\tilde{\beta}_1, \Psi_{11}) \mathrel{\dot{\propto}} |\Psi_{11}|^{-1/2}.$$

It now follows easily that the maximum likelihood estimators of Ψ_{12}, Ψ_{11}, and Ψ_{22} are

$$\tilde{\Psi}_{12} = S_{11}^{-1}S_{12}$$

$$\tilde{\Psi}_{11} = \frac{1}{m+1}Z_1Z_1' = \frac{1}{m+1}S_{11}$$

$$\tilde{\Psi}_{22} = \frac{1}{m+1}S_{22\cdot 1}.$$

Substituting these into the product of the four densities shows that the maximum under H_1 is proportional to

$$\Lambda_1 = |S_{22\cdot 1}|^{-(m+1)/2}|S_{11}|^{-(m+1)/2}.$$

Hence the likelihood ratio test will reject H_0 for small values of

$$\Lambda_2 = \frac{\Lambda_0}{\Lambda_1} = \frac{|S_{11}|^{(m+1)/2}}{|S_{11} + U_1'U_1|^{(m+1)/2}} = \frac{1}{\left(1 + U_1S_{11}^{-1}U_1'\right)^{(m+1)/2}}.$$

Thus the likelihood ratio test rejects for large values of

$$\Lambda = U_1S_{11}^{-1}U_1'$$

and the proof is complete. □

We now want to show that the test derived above is a uniformly most powerful invariant test under a suitable group of affine transformations. Recall that U and Z are independent and

$$\mathcal{L}(U) = N(\beta, \Sigma), \qquad \mathcal{L}(Z) = N(0, I_m \otimes \Sigma).$$

The problem is to test $H_0: \beta_1 = 0$ where $\beta = (\beta_1, \beta_2)$ with $\beta_i \in \mathcal{L}_{p_i,1}$, $i = 1,2$. Consider the group G with elements $g = (\Gamma, A, (0, a))$ where

$$\Gamma \in \mathcal{O}_m, \qquad (0, a) \in \mathcal{L}_{p,1}, \qquad a \in \mathcal{L}_{p_2,1}$$

and

$$A = \begin{pmatrix} A_{11} & 0 \\ A_{21} & A_{22} \end{pmatrix}$$

where A_{ij} is $p_i \times p_j$ and A_{ii} is nonsingular for $i = 1,2$. The action of $g = (\Gamma, A, (0, a))$ is

$$g\begin{pmatrix} U \\ Z \end{pmatrix} \equiv \begin{pmatrix} UA' + (0, a) \\ \Gamma Z A' \end{pmatrix}.$$

The group operation, defined so G acts on the left of the sample space, is

$$(\Gamma_1, A_1, (0, a_1))(\Gamma_2, A_2, (0, a_2)) = (\Gamma_1\Gamma_2, A_1A_2, (0, a_2)A_1' + (0, a_1)).$$

It is routine to verify that the testing problem is invariant under G. Further, it is clear that the induced action of G on the parameter space is

$$(\Gamma, A, (0, a))(\beta, \Sigma) = (\beta A' + (0, a), A\Sigma A').$$

To characterize the invariant tests for the testing problem, a maximal invariant under the action of G on the sample space is needed.

Proposition 9.18. In the notation of Proposition 9.17, a maximal invariant is

$$\Lambda = U_1 S_{11}^{-1} U_1'.$$

Proof. As usual, the proof consists of showing that $\Lambda = U_1 S_{11}^{-1} U_1'$ is an orbit index. Since $m \geqslant p$, we deal with those Z's that have rank p, a set of probability one. The first claim is that for a given $U \in \mathcal{L}_{p,1}$ and $Z \in \mathcal{L}_{p,m}$ of rank p, there exists a $g \in G$ such that

$$g\begin{pmatrix} U \\ Z \end{pmatrix} = \begin{pmatrix} \Lambda^{1/2}\varepsilon_1' \\ Z_0 \end{pmatrix}$$

where $\varepsilon_1' = (1, 0, \dots, 0) \in \mathcal{L}_{p,1}$ and

$$Z_0 = \begin{pmatrix} I_p \\ 0 \end{pmatrix} \in \mathcal{L}_{p,m}.$$

Write $Z = \Psi V$ where $\Psi \in \mathcal{F}_{p,m}$ and V is a $p \times p$ upper triangular matrix so $S = Z'Z = V'V$. Then consider

$$A = \begin{pmatrix} \xi_1 & 0 \\ 0 & \xi_2 \end{pmatrix}(V')^{-1}$$

where $\xi_i \in \mathcal{O}_{p_i}$, $i = 1, 2$, and note that A is of the form

$$A = \begin{pmatrix} A_{11} & 0 \\ A_{21} & A_{22} \end{pmatrix}$$

since $(V')^{-1}$ is lower triangular. The values of ξ_i, $i = 1, 2$, are specified in a moment. With this choice of A,

$$ZA' = \Psi V V^{-1}\begin{pmatrix} \xi_1 & 0 \\ 0 & \xi_2 \end{pmatrix}' = \Psi\begin{pmatrix} \xi_1 & 0 \\ 0 & \xi_2 \end{pmatrix}'$$

which is in $\mathcal{F}_{p,m}$ for any choice of $\xi_i \in \mathcal{O}_{p_i}$, $i = 1, 2$. Hence there is a $\Gamma \in \mathcal{O}_m$ such that

$$\Gamma Z A' = Z_0.$$

Since V is upper triangular, write

$$V^{-1} = \begin{pmatrix} V^{11} & V^{12} \\ 0 & V^{22} \end{pmatrix}$$

with V^{ij} being $p_i \times p_j$. Then

$$UA' = UV^{-1}\begin{pmatrix} \xi_1' & 0 \\ 0 & \xi_2' \end{pmatrix} = (U_1, U_2)\begin{pmatrix} V^{11} & V^{12} \\ 0 & V^{22} \end{pmatrix}\begin{pmatrix} \xi_1' & 0 \\ 0 & \xi_2' \end{pmatrix}$$

$$= \left(U_1 V^{11}\xi_1', U_1 V^{12}\xi'_2 + U_2 V^{22}\xi_2'\right).$$

As $S = V'V$ and $V \in G_U^+$, it follows that $S_{11}^{-1} = V^{11}(V^{11})'$ so the vector $U_1 V^{11}$ has squared length $\Lambda = U_1 V^{11}(V^{11})' U_1' = U_1 S_{11}^{-1} U_1'$. Thus there exists $\xi_1' \in \mathcal{O}_{p_1}$ such that

$$U_1 V^{11} \xi_1' = \Lambda^{1/2} \tilde{\varepsilon}_1'$$

where $\tilde{\varepsilon}_1' = (1, 0, \dots, 0) \in \mathcal{L}_{p_1,1}$. Now choose $a \in \mathcal{L}_{p_2,1}$ to be

$$a = U_1 V^{12} \xi_2' - U_2 V^{22} \xi_2'$$

so

$$UA' + (0, a) = \Lambda^{1/2} \varepsilon_1'.$$

The above choices for A, ξ_1, Γ, and a yield $g = (\Gamma, A, (0, a))$, which satisfies

$$g\begin{pmatrix} U \\ Z \end{pmatrix} = \begin{pmatrix} \Lambda^{1/2}\varepsilon_1' \\ Z_0 \end{pmatrix}$$

and this establishes the claim. To show that

$$\Lambda = U_1 S_{11}^{-1} U_1'$$

is maximal invariant, first notice that Λ is invariant. Further, if

$$\begin{pmatrix} U_1 \\ Z_1 \end{pmatrix} \quad \text{and} \quad \begin{pmatrix} U_2 \\ Z_2 \end{pmatrix}$$

both yield the same value of Λ, then there exists $g_i \in G$ such that

$$g_i \begin{pmatrix} U_i \\ Z_i \end{pmatrix} = \begin{pmatrix} \Lambda^{1/2}\varepsilon_1' \\ Z_0 \end{pmatrix}, \qquad i = 1, 2.$$

Therefore,

$$g_2^{-1} g_1 \begin{pmatrix} U_1 \\ Z_1 \end{pmatrix} = \begin{pmatrix} U_2 \\ Z_2 \end{pmatrix}$$

and Λ is maximal invariant.

To show that a uniformly most powerful G-invariant test exists, the distribution of $\Lambda = U_1 S_{11}^{-1} U_1'$ is needed. However,

$$\mathcal{L}(U_1) = N(\beta_1, \Sigma_{11})$$

$$\mathcal{L}(S_{11}) = W(\Sigma_{11}, p_1, m)$$

and U_1 and S_{11} are independent. From Proposition 8.14, we see that

$$\mathcal{L}(\Lambda) = F(p_1, m - p_1 + 1, \delta)$$

where $\delta = \beta_1 \Sigma_{11}^{-1} \beta_1'$ and the null hypothesis is $H_0 : \delta = 0$. Since the noncentral F distribution has a monotone likelihood ratio, the test that rejects for large values of Λ is uniformly most powerful within the class of tests based on Λ. Since all G-invariant tests are functions of Λ, we conclude that the likelihood ratio test is uniformly most powerful invariant. □

The final problem to be considered in this chapter is a variation of the problem just solved. Again, the testing problem of interest is $H_0 : \beta_1 = 0$ versus $H_1 : \beta_1 \neq 0$, but it is assumed that the value of β_2 is known to be zero under both H_0 and H_1. Thus our model for U and Z is that U and Z are independent with

$$\mathcal{L}(U) = N((\beta_1, 0), \Sigma)$$

$$\mathcal{L}(Z) = N(0, I_m \otimes \Sigma)$$

where $U \in \mathcal{L}_{p,1}$, $\beta_1 \in \mathcal{L}_{p_1,1}$, $Z \in \mathcal{L}_{p,m}$, and $m \geqslant p$. In what follows, the likelihood ratio test of H_0 versus H_1 is derived and an invariance argument shows that there is no uniformly most powerful invariant test under a natural group that leaves the problem invariant. As usual, we partition U

into U_1 and U_2, $U_i \in \mathcal{L}_{p_i,1}$, and Z is partitioned into $Z_1 \in \mathcal{L}_{p_1,m}$ and $Z_2 \in \mathcal{L}_{p_2,m}$ so

$$U = (U_1, U_2), \qquad Z = (Z_1, Z_2).$$

Also

$$S = Z'Z = \begin{pmatrix} Z_1'Z_1 & Z_1'Z_2 \\ Z_2'Z_1 & Z_2'Z_2 \end{pmatrix} \equiv \begin{pmatrix} S_{11} & S_{12} \\ S_{21} & S_{22} \end{pmatrix}$$

and $S_{11\cdot 2} = S_{11} - S_{12}S_{22}^{-1}S_{21}$.

Proposition 9.19. The likelihood ratio test of H_0 versus H_1 rejects for large values of the statistic

$$\Lambda = \frac{(U_1 - U_2S_{22}^{-1}S_{21})S_{11\cdot 2}^{-1}(U_1 - U_2S_{22}^{-1}S_{21})'}{1 + U_2S_{22}^{-1}U_2'}.$$

Proof. Under H_0,

$$\mathcal{L}(U) = N(0, \Sigma)$$
$$\mathcal{L}(Z) = N(0, I_m \otimes \Sigma)$$

so the maximum likelihood estimator of Σ is

$$\hat{\Sigma} = \frac{1}{m+1}(Z'Z + U'U) = \frac{1}{m+1}(S + U'U).$$

The value of the maximized likelihood function is proportional to

$$\Lambda_0 \equiv |\hat{\Sigma}|^{-(m+1)/2}.$$

Under H_1, the situation is a bit more complicated and it is helpful to consider conditional distributions. Under H_1,

$$\mathcal{L}(U_1|U_2) = N(\beta_1 + U_2\Sigma_{22}^{-1}\Sigma_{21}, \Sigma_{11\cdot 2})$$
$$\mathcal{L}(U_2) = N(0, \Sigma_{22})$$
$$\mathcal{L}(Z_1|Z_2) = N(Z_2\Sigma_{22}^{-1}\Sigma_{21}, I_m \otimes \Sigma_{11\cdot 2})$$

and

$$\mathcal{L}(Z_2) = N(0, I_m \otimes \Sigma_{22}).$$

The reparameterization defined by $\Psi_{11} = \Sigma_{11\cdot 2}$, $\Psi_{21} = \Sigma_{22}^{-1}\Sigma_{21}$, and $\Psi_{22} = \Sigma_{22}$ is one-to-one and onto. Let $f_1(U_1|U_2, \beta_1, \Psi_{21}, \Psi_{11})$, $f_2(U_2|\Psi_{22})$, $f_3(Z_1|Z_2, \Psi_{21}, \Psi_{11})$, and $f_4(Z_2|\Psi_{22})$ be the density functions with respect to Lebesgue measure $dU_1\,dU_2\,dZ_1\,dZ_2$ of the four distributions above. It is clear that

$$\tilde{\beta}_1 = U_1 - U_2\Psi_{21}$$

maximizes $f_1(U_1|U_2, \beta_1, \Psi_{21}, \Psi_{11})$ and $f_1(U_1|U_2, \tilde{\beta}_1, \Psi_{21}, \Psi_{11}) \,\dot{\alpha}\, |\Psi_{11}|^{-1/2}$. With $\tilde{\beta}_2$ substituted into f_1, the parameter Ψ_{21} only occurs in the density $f_3(Z_1|Z_2, \Psi_{21}, \Psi_{11})$. Since

$$\mathcal{L}(Z_1|Z_2) = N(Z_2\Psi_{21}, I_m \otimes \Psi_{11}),$$

our results for the MANOVA model show that

$$\tilde{\Psi}_{21} = (Z_2'Z_2)^{-1}Z_2'Z_1 = S_{22}^{-1}S_{21}$$

maximizes $f_3(Z_1|Z_2, \Psi_{21}, \Psi_{11})$ for each value of Ψ_{11}. When $\tilde{\Psi}_{21}$ is substituted into f_3, an inspection of the resulting four density functions shows that the maximum likelihood estimators of Ψ_{11} and Ψ_{22} are

$$\tilde{\Psi}_{11} = \frac{1}{m+1}S_{11\cdot 2}$$

and

$$\tilde{\Psi}_{22} = \frac{1}{m+1}(Z_2'Z_2 + U_2'U_2) = \frac{1}{m+1}(S_{22} + U_2'U_2).$$

Under H_1, this yields a maximized likelihood function proportional to

$$\Lambda_1 = |\tilde{\Psi}_{11}|^{-(m+1)/2}|\tilde{\Psi}_{22}|^{-(m+1)/2}.$$

Therefore the likelihood ratio test rejects H_0 for small values of

$$\Lambda_3 \equiv \frac{\Lambda_0}{\Lambda_1} = \left[\frac{|S_{22} + U_2'U_2||S_{11\cdot 2}|}{|S + U'U|}\right]^{(m+1)/2}.$$

However,

$$|S_{22} + U_2'U_2| = |S_{22}|\left(1 + U_2 S_{22}^{-1} U_2'\right)$$

and

$$|S| = |S_{22}||S_{11\cdot 2}|.$$

Thus

$$[\Lambda_3]^{2/(m+1)} = \frac{|S|\left(1 + U_2 S_{22}^{-1} U'\right)}{|S + U'U|} = \frac{1 + U_2 S_{22}^{-1} U_2'}{1 + US^{-1}U}.$$

Now, the identity

$$US^{-1}U' = \left(U_1 - U_2 S_{22}^{-1} S_{21}\right) S_{11\cdot 2}^{-1} \left(U_1 - U_2 S_{22}^{-1} S_{21}\right)' + U_2 S_{22}^{-1} U_2'$$

follows from the problems in Chapter 5. Hence rejecting for small values of

$$[\Lambda_3]^{2/(m+1)} = \frac{1}{1 + \Lambda},$$

where Λ is given in the statement of this proposition, is equivalent to rejecting for large values of Λ. □

The above testing problem is now analyzed via invariance. The group G consists of elements $g = (\Gamma, A)$ where $\Gamma \in \mathcal{O}_m$ and

$$A = \begin{pmatrix} A_{11} & A_{12} \\ 0 & A_{22} \end{pmatrix}, \qquad A_{ii} \in Gl_{p_i}, \quad i = 1, 2.$$

The group action is

$$(\Gamma, A)\begin{pmatrix} U \\ Z \end{pmatrix} = \begin{pmatrix} UA' \\ \Gamma Z A' \end{pmatrix}$$

and group composition is

$$(\Gamma_1, A_1)(\Gamma_2, A_2) = (\Gamma_1 \Gamma_2, A_1 A_2).$$

The action of the group on the parameter space is

$$(\Gamma, A)(\beta_1, \Sigma) = (\beta_1 A_{11}', A \Sigma A').$$

It is clear that the testing problem is invariant under the group G.

Proposition 9.20. Under the action of G on the sample space, a maximal invariant is the pair (W_1, W_2) where

$$W_1 = \frac{(U_1 - U_2 S_{22}^{-1} S_{21}) S_{11\cdot 2}^{-1} (U_1 - U_2 S_{22}^{-1} S_{21})'}{1 + U_2 S_{22}^{-1} U_2'}$$

and

$$W_2 = U_2 S_{22}^{-1} U_2'.$$

A maximal invariant in the parameter space is

$$\delta = \beta_1 \Sigma_{11\cdot 2}^{-1} \beta_1'.$$

Proof. As usual, the method of proof is a reduction argument that provides a convenient index for the orbits in the sample space. Since $m \geqslant p$, a set of measure zero can be deleted from the sample space so that Z has rank p on the complement of this set. Let

$$Z_0 = \begin{pmatrix} I_p \\ 0 \end{pmatrix} \in \mathcal{L}_{p,m}$$

and set $u_1 = \varepsilon_1' \in \mathcal{L}_{p,1}$ and $u_2 = (0, \ldots, 0, 1, 0, \ldots, 0) \in \mathcal{L}_{p,1}$ where the one occurs in the $(p_1 + 1)$ coordinate of u_2. Now, given U and Z, we claim that there exists a $g = (\Gamma, A) \in G$ such that

$$g\begin{pmatrix} U \\ Z \end{pmatrix} = \begin{pmatrix} X_1 u_1 + X_2 u_2 \\ Z_0 \end{pmatrix}, \qquad X_i \geqslant 0,$$

where

$$X_1^2 = (U_1 - U_2 S_{22}^{-1} S_{21}) S_{11\cdot 2} (U_1 - U_2 S_{22}^{-1} S_{21})'$$

and

$$X_2^2 = U_2 S_{22}^{-1} U_2'.$$

To establish this claim, write $Z = \Psi T$ where $\Psi \in \mathcal{F}_{p,m}$ and $T \in G_T^+$ is a $p \times p$ lower triangular matrix. A modification of the proof of Proposition 5.2 establishes this representation for Z. Consider

$$A = \xi (T^{-1})' = \begin{pmatrix} \xi_1 & 0 \\ 0 & \xi_2 \end{pmatrix} (T^{-1})'$$

where $\xi_i \in \mathcal{O}_{p_i}$, $i = 1, 2$, so $\xi \in \mathcal{O}_p$ and

$$ZA' = \Psi TA' = \Psi\xi' \in \mathcal{F}_{p,m}.$$

Thus for any such ξ and $\Gamma \in \mathcal{O}_m$, $(\Gamma, A) \in G$. Also, Γ can be chosen so that

$$\Gamma ZA' = Z_0 \in \mathcal{F}_{p,m}.$$

Now,

$$UA' = (U_1, U_2)T^{-1}\xi' = (U_1, U_2)\begin{pmatrix} T^{11} & 0 \\ T^{21} & T^{22} \end{pmatrix}\begin{pmatrix} \xi_1' & 0 \\ 0 & \xi_2' \end{pmatrix}$$

$$= \left((U_1T^{11} + U_2T^{21})\xi_1', U_2T^{22}\xi_2'\right)$$

where T^{ij} is $p_i \times p_j$ and

$$T^{-1} \equiv \begin{pmatrix} T^{11} & 0 \\ T^{21} & T^{22} \end{pmatrix}.$$

Since

$$S = Z'Z = T'T,$$

a bit of algebra shows that

$$\begin{aligned}(U_1T^{11} + U_2T^{21})&(U_1T^{11} + U_2T^{21})' \\ &= (U_1 - U_2S_{22}^{-1}S_{21})S_{11\cdot 2}^{-1}(U_1 - U_2S_{22}^{-1}S_{21})' = X_1^2\end{aligned}$$

and

$$(U_2T^{22})(U_2T^{22})' = U_2S_{22}^{-1}U_2' = X_2^2.$$

Let $\tilde{\varepsilon}_1 = (1, 0, \ldots, 0) \in \mathcal{L}_{p_1,1}$ and $\tilde{\varepsilon}_2 = (1, 0, \ldots, 0) \in \mathcal{L}_{p_2,1}$. Since the vectors $X_1\tilde{\varepsilon}_1$ and $U_1T^{11} + U_2T^{21}$ have the same length, there exists $\xi_1' \in \mathcal{O}_{p_1}$ such that

$$(U_1T^{11} + U_2T^{22})\xi_1' = X_1\tilde{\varepsilon}_1.$$

For similar reasons, there exists a $\xi_2' \in \mathcal{O}_{p_2}$ such that

$$U_2T_{22}\xi_2' = X_2\tilde{\varepsilon}_2.$$

With these choices for ξ_1 and ξ_2,

$$\left((U_1T^{11} + U_2T^{21})\xi_1', U_2T^{22}\xi_2'\right) = (X_1u_1 + X_2u_2).$$

Thus there is a $g = (\Gamma, A) \in G$ such that

$$g\begin{pmatrix} U \\ Z \end{pmatrix} = \begin{pmatrix} X_1u_1 + X_2u_2 \\ Z_0 \end{pmatrix}.$$

This establishes the claim. It is now routine to show that $(X_1, X_2) = (X_1(U, Z), X_2(U, Z))$ is an invariant function. To show that (X_1, X_2) is maximal invariant, suppose (U, Z) and $(\tilde{U}, \tilde{Z})$ yield the same (X_1, X_2) values. Then there exist g and $\tilde{g}$ in G such that

$$g\begin{pmatrix} U \\ Z \end{pmatrix} = \begin{pmatrix} X_1u_1 + X_2u_2 \\ Z_0 \end{pmatrix} = \tilde{g}\begin{pmatrix} \tilde{U} \\ \tilde{Z} \end{pmatrix}$$

so

$$g^{-1}\tilde{g}\begin{pmatrix} \tilde{U} \\ \tilde{Z} \end{pmatrix} = \begin{pmatrix} U \\ Z \end{pmatrix}.$$

This shows that (X_1, X_2) is maximal invariant. Since the pair (W_1, W_2) is a one-to-one function of (X_1, X_2), it follows that (W_1, W_2) is maximal invariant. The proof that δ is a maximal invariant in the parameter space is similar and is left to the reader. □

In order to suggest an invariant test for $H_0: \beta_1 = 0$ based on (W_1, W_2), the distribution of (W_1, W_2) is needed. Since

$$\mathcal{L}((U_1, U_2)) = N((\beta_1, 0), \Sigma)$$

and

$$\mathcal{L}(S) = W(\Sigma, p, m)$$

with S and U independent,

$$\mathcal{L}(W_2) = \mathcal{L}\left(U_2 S_{22}^{-1} U_2'\right) = F_{p_2, m-p_2+1}.$$

Therefore, W_2 is an ancillary statistic as its distribution does not depend on any parameters under H_0 or H_1. We now compute the conditional distribution of W_1 given W_2. Proposition 8.7 shows that

$$\mathcal{L}(S_{11\cdot2}) = W(\Sigma_{11\cdot2}, p_1, m - p_2)$$

$$\mathcal{L}\left(S_{22}^{-1}S_{21} | S_{22}\right) = N\left(\Sigma_{22}^{-1}\Sigma_{21}, S_{22}^{-1} \otimes \Sigma_{11\cdot2}\right)$$

and

$$\mathcal{L}(S_{22}) = W(\Sigma_{22}, p_2, m)$$

where $S_{11\cdot 2}$ is independent of (S_{21}, S_{22}). Thus

$$\mathcal{L}\left(U_2 S_{22}^{-1} S_{21} | S_{22}, U_2\right) = N\left(U_2 \Sigma_{22}^{-1} \Sigma_{21}, \left(U_2 S_{22}^{-1} U_2'\right) \Sigma_{11\cdot 2}\right)$$

and conditional on (S_{22}, U_2),

$$\mathcal{L}(U_1 | S_{22}, U_2) = N\left(\beta_1 + U_2 \Sigma_{22}^{-1} \Sigma_{21}, \Sigma_{11\cdot 2}\right).$$

Further, U_1 and $U_2 S_{22}^{-1} S_{21}$ are conditionally independent—given (S_{22}, U_2). Therefore,

$$\mathcal{L}\left(U_1 - U_2 S_{22}^{-1} S_{21} | S_{22}, U_2\right) = N\left(\beta_1, \left(1 + U_2 S_{22}^{-1} U_2'\right) \Sigma_{11\cdot 2}\right)$$

so

$$\mathcal{L}\left(\frac{U_1 - U_2 S_{22}^{-1} S_{21}}{\sqrt{1 + W_2}} \Big| S_{22}, U_2\right) = N\left(\frac{\beta_1}{\sqrt{1 + W_2}}, \Sigma_{11\cdot 2}\right).$$

Since $S_{11\cdot 2}$ is independent of all other variables under consideration, and since

$$W_1 = \frac{\left(U_1 - U_2 S_{22}^{-1} S_{21}\right) S_{11\cdot 2} \left(U_1 - U_2 S_{22}^{-1} S_{21}\right)'}{1 + W_2},$$

it follows from Proposition 8.14 that

$$\mathcal{L}(W_1 | S_{22}, U_2) = F\left(p_1, m - p + 1; \frac{\delta}{1 + W_2}\right)$$

where $\delta = \beta_1 \Sigma_{11\cdot 2}^{-1} \beta_1'$. However, the conditional distribution of W_1 given (S_{22}, U_2) depends on (S_{22}, U_2) only through the function $W_2 = U_2 S_{22}^{-1} U_2'$. Thus

$$\mathcal{L}(W_1 | W_2) = F\left(p_1, m - p + 1; \frac{\delta}{1 + W_2}\right),$$

and

$$\mathcal{L}(W_2) = F_{p_2, m - p_2 + 1}.$$

Further, the null hypothesis is $H_0 : \delta = 0$ versus the alternative $H_1 : \delta > 0$. Under H_0, it is clear that W_1 and W_2 are independent. The likelihood ratio test rejects H_0 for large values of W_1 and ignores W_2. Of course, the level of this test is computed from a standard F-table, but the power of the test involves the marginal distribution of W_1 when $\delta > 0$. This marginal distribution, obtained by averaging the conditional distribution $\mathcal{L}(W_1|W_2)$ with respect to the distribution of W_2, is rather complicated.

To show that a uniformly most powerful test of H_0 versus H_1 does not exist, consider a particular alternative $\delta = \delta_0 > 0$. Let $f_1(w_1|w_2, \delta)$ denote the conditional density function of W_1 given W_2 and let $f_2(w_2)$ denote the density of W_2. For testing $H_0 : \delta = 0$ versus $H_1 : \delta = \delta_0$, the Neyman–Pearson Lemma asserts that the most powerful test of level α is to reject if

$$\frac{f_1(w_1|w_2, \delta_0)}{f_1(w_1|w_2, 0)} > c(\alpha)$$

where $c(\alpha)$ is chosen to make the test have level α. However, the rejection region for this test depends on the particular alternative δ_0 so a uniformly most powerful test cannot exist. Since W_2 is ancillary, we can argue that the test of H_0 should be carried out conditional on W_2, that is, the level and the power of tests should be compared only for the conditional distribution of W_1 given W_2. In this case, for w_2 fixed, the ratio

$$\frac{f_1(w_1|w_2, \delta_0)}{f_1(w_1|w_2, 0)}$$

is an increasing function of w_1 so rejecting for large values of the ratio (w_2 fixed) is equivalent to rejecting for $W_1 > k$. If k is chosen to make the test have level α, this argument leads to the level α likelihood ratio test.

PROBLEMS

1. Consider independent random vectors X_{ij} with $\mathcal{L}(X_{ij}) = N(\mu_i, \Sigma)$ for $j = 1, \ldots, n_i$ and $i = 1, \ldots, k$. For scalars $a_1, \ldots, a_k$ consider testing $H_0 : \Sigma a_i \mu_i = 0$ versus $H_1 : \Sigma a_i \mu_i \neq 0$. With $\tau^2 = \Sigma a_i^2 n_i^{-1}$, let $b_i = \tau^{-1} a_i$, set $\overline{X}_i = n_i^{-1} \Sigma_j X_{ij}$ and let $S_i = \Sigma_j (X_{ij} - \overline{X}_i)(X_{ij} - \overline{X}_i)'$. Write this problem in the canonical form of Section 9.1 and prove that the test that rejects for large values of $\Lambda = (\Sigma_i b_i \overline{X}_i)' S^{-1} (\Sigma_i b_i \overline{X}_i)$ is UMP invariant. Here $S = \Sigma_i S_i$. What is the distribution of Λ under H_0?

2. Given $Y \in \mathcal{L}_{p,n}$ and $X \in \mathcal{L}_{k,n}$ of rank k, the least-squares estimate $\hat{B} = (X'X)^{-1} X'Y$ of B can be characterized as the B that mini-

mizes $\operatorname{tr}(Y - XB)'(Y - XB)$ over all $k \times p$ matrices.

(i) Show that for any $k \times p$ matrix B,

$$(Y - XB)'(Y - XB) = (Y - X\hat{B})'(Y - X\hat{B}) + \big(X(B - \hat{B})\big)'\big(X(B - \hat{B})\big).$$

(ii) A real-valued function ϕ defined for $p \times p$ nonnegative definite matrices is *nondecreasing* if $\phi(S_1) \leqslant \phi(S_1 + S_2)$ for any S_1 and S_2 ($S_i \geqslant 0$, $i = 1, 2$). Using (i), show that, if ϕ is nondecreasing, then $\phi((Y - XB)'(Y - XB))$ is minimized by $B = \hat{B}$.

(iii) For A that is $p \times p$ and nonnegative definite, show that $\phi(S) = \operatorname{tr} AS$ is nondecreasing. Also, show that $\phi(S) = \det(A + S)$ is nondecreasing.

(iv) Suppose $\phi(S) = \phi(\Gamma S \Gamma')$ for $S \geqslant 0$ and $\Gamma \in \mathcal{O}_p$ so $\phi(S)$ can be written as $\phi(S) = \psi(\lambda(S))$ where $\lambda(S)$ is the vector of ordered characteristic roots of S. Show that, if ψ is nondecreasing in each argument, then ϕ is nondecreasing.

3. (The MANOVA model under non-normality.) Let E be a random $n \times p$ matrix that satisfies $\mathcal{L}(\Gamma E \psi') = \mathcal{L}(E)$ for all $\Gamma \in \mathcal{O}_n$ and $\psi \in \mathcal{O}_p$. Assume that $\operatorname{Cov}(E) = I_n \otimes I_p$ and consider a linear model for $Y \in \mathcal{L}_{p,n}$ generated by $Y = ZB + EA'$ where Z is a fixed $n \times k$ matrix of rank k, B is a $k \times p$ matrix of unknown parameters, and A is an element of Gl_p.

(i) Show that the distribution of Y depends on (B, A) only through (B, AA').

(ii) Let $M = \{\mu | \mu = ZB, B \in \mathcal{L}_{p,k}\}$ and $\gamma = \{I_n \otimes \Sigma | \Sigma > 0, \Sigma \text{ is } p \times p\}$. Show that $\{M, \gamma\}$ serves as a parameter space for the linear model (the distribution of E is assumed fixed).

(iii) Consider the problem of testing $H_0 : RB = 0$ where R is $r \times k$ of rank r. Show that the reduction to canonical form given in Section 9.1 can be used here to give a model of the form

$$\begin{pmatrix} \tilde{Y}_1 \\ \tilde{Y}_2 \\ \tilde{Y}_3 \end{pmatrix} = \begin{pmatrix} \tilde{B}_1 \\ \tilde{B}_2 \\ 0 \end{pmatrix} + \tilde{E}A' \tag{9.6}$$

where $\tilde{Y}_1$ is $r \times p$, $\tilde{Y}_2$ is $(k - r) \times p$, $\tilde{Y}_3$ is $(n - k) \times p$, $\tilde{B}_1$ is $r \times p$, $\tilde{B}_2$ is $(k - r) \times p$, $\tilde{E}$ is $n \times p$, and A is as in the original model. Further, E and $\tilde{E}$ have the same distribution and the null hypothesis is $H_0: \tilde{B}_1 = 0$.

(iv) Now, assume the form of the model in (9.6) and drop the tildas. Using the invariance argument given in Section 9.1, the testing problem is invariant and any invariant test is a function of the t largest eigenvalues of $Y_1(Y_3'Y_3)^{-1}Y_1'$ where $t = \min\{r, p\}$. Assume $n - k \geqslant p$ and partition E as Y is partitioned. Under H_0, show that

$$W \equiv Y_1(Y_3'Y_3)^{-1}Y_1' = E_1(E_3'E_3)^{-1}E_1'.$$

(v) Using Proposition 7.3 show that W has the same distribution no matter what the distribution of E as long as $\mathcal{L}(\Gamma E) = \mathcal{L}(E)$ for all $\Gamma \in \mathcal{O}_n$ and E_3 has rank p with probability one. This distribution of W is the distribution obtained by assuming the elements of E are i.i.d. $N(0, 1)$. In particular, any invariant test of H_0 has the same distribution under H_0 as when E is $N(0, I_n \otimes I_p)$.

4. When Y_1 is $N(B_1, I_r \otimes \Sigma)$ and Y_3 is $N(0, I_m \otimes \Sigma)$ with $m \geqslant p + 2$, verify the claim that

$$\mathcal{E} Y_1'Y_1(Y_3'Y_3)^{-1} = \frac{r}{m - p - 1} I_p + \frac{1}{m - p - 1} B_1'B_1\Sigma^{-1}.$$

5. Consider a data matrix $Y: n \times 2$ and assume $\mathcal{L}(Y) = N(ZB, I_n \otimes \Sigma)$ where Z is $n \times 2$ of rank two so B is 2×2. In some situations, it is reasonable to assume that $\sigma_{11} = \sigma_{22}$—that is, the diagonal elements of Σ are the same. Under this assumption, use the results of Section 9.2 to derive the likelihood ratio test for $H_0: b_{11} = b_{12}$, $b_{21} = b_{22}$ versus $H_1: b_{11} \neq b_{12}$ or $b_{21} \neq b_{22}$. Is this test UMP invariant?

6. Consider a "two-way layout" situation with observations Y_{ij}, $i = 1, \ldots, m$ and $j = 1, \ldots, r$. Assume $Y_{ij} = \mu + \alpha_i + \beta_j + e_{ij}$ where μ, α_i, and β_j are constants that satisfy $\Sigma\alpha_i = \Sigma\beta_j = 0$. The e_{ij} are random errors with mean zero (but not necessarily uncorrelated). Let Y be the $m \times n$ matrix of Y_{ij}'s, u_1 be the vector of ones in R^m, u_2 be the vector of ones in R^n, $\alpha \in R^m$ be the vector with coordinates α_i, and $\beta \in R^n$ be the vector with coordinates β_j. Let E be the matrix of e_{ij}'s.

(i) Show the model is $Y = \mu u_1 u_2' + \alpha u_2' + u_1\beta' + E$ in the vector space $\mathcal{L}_{n,m}$. Here, $\alpha \in R^m$ with $\alpha' u_1 = 0$ and $\beta \in R^n$ with $\beta' u_2$

$= 0$. Let

$$M_1 = \{x | x \in \mathcal{L}_{n,m}, x = \mu u_1 u_2', \mu \in R^1\}$$

$$M_2 = \{x | x = \alpha u_2', \alpha \in R^m, \alpha' u_1 = 0\}$$

$$M_3 = \{x | x = u_1 \beta', \beta \in R^n, \beta' u_2 = 0\}.$$

Also, let $\langle \cdot, \cdot \rangle$ be the usual inner product on $\mathcal{L}_{n,m}$.

(ii) Show $M_1 \perp M_2 \perp M_3 \perp M_1$ in $(\mathcal{L}_{n,m}, \langle \cdot, \cdot \rangle)$.

Now, assume $\text{Cov}(E) = I_m \otimes A$ where $A = \gamma P + \delta Q$ with $P = n^{-1} u_2 u_2'$, $Q = I - P$, and $\gamma > 0$ and $\delta > 0$ are unknown parameters.

(iii) Show the regression subspace $M = M_1 \oplus M_2 \oplus M_3$ is invariant under each $I_m \otimes A$. Find the Gauss–Markov estimates for μ, α, and β.

(iv) Now, assume E is $N(0, I_m \otimes A)$. Use an invariance argument to show that for testing $H_0 : \alpha = 0$ versus $H_1 : \alpha \neq 0$, the test that rejects for large values of $W = \|P_{M_2} Y\|^2 / \|Q_M Y\|^2$ is a UMP invariant test. Here, $Q_M = I - P_M$. What is the distribution of W?

7. The regression subspace for the MANOVA model was described as $M = \{\mu | \mu = ZB, B \in \mathcal{L}_{p,k}\} \subseteq \mathcal{L}_{p,n}$ where Z is $n \times k$ of rank k. The subspace of M associated with the null hypothesis $H_0 : RB = 0$ (R is $r \times r$ of rank r) is $\omega = \{\mu | \mu = ZB, B \in \mathcal{L}_{p,k}, RB = 0\}$. We know that $P_M = P_Z \otimes I_p$ where $P_Z = Z(Z'Z)^{-1}Z'$ (P_M is the orthogonal projection onto M in $(\mathcal{L}_{p,n}, \langle \cdot, \cdot \rangle)$). This problem gives one form for P_ω. Let $W = Z(Z'Z)^{-1}R'$.

(i) Show that W has rank r.

Let $P_W = W(W'W)^{-1}W'$ so $P_W \otimes I_p$ is an orthogonal projection.

(ii) Show that $\mathcal{R}(P_W \otimes I_p) \subseteq M - \omega$ where $M - \omega = M \cap \omega^\perp$. Also, show $\dim(\mathcal{R}(P_W \otimes I_p)) = rp$.

(iii) Show that $\dim \omega = (k - r)p$.

(iv) Now, show that $P_W \otimes I_p$ is the orthogonal projection onto $M - \omega$ so $P_Z \otimes I_p - P_W \otimes I_p$ is the orthogonal projection onto ω.

8. Assume $X_1, \ldots, X_n$ are i.i.d. from a five-dimensional $N(0, \Sigma)$ where Σ is a cyclic covariance matrix (Σ is written out explicitly at the beginning of Section 9.4). Find the maximum likelihood estimators of σ^2, ρ_1, ρ_2.

9. Suppose $X_1, \ldots, X_n$ are i.i.d. $N(0, \Psi)$ of dimension $2p$ and assume Ψ has the complex form

$$\Psi = \begin{pmatrix} \Sigma & F \\ -F & \Sigma \end{pmatrix}.$$

Let $S = \Sigma_1^n X_i X_i'$ and partition S as Ψ is partitioned. show that $\hat{\Sigma} = (2n)^{-1}(S_{11} + S_{22})$ and $\hat{F} = (2n)^{-1}(S_{12} - S_{21})$ are the maximum likelihood estimates of Σ and F.

10. Let $X_1, \ldots, X_n$ be i.i.d. $N(\mu, \Sigma)$ p-dimensional random vectors where μ and Σ are unknown, $\Sigma > 0$. Suppose R is $r \times p$ of rank r and consider testing $H_0 : R\mu = 0$ versus $H_1 : R\mu \neq 0$. Let $\overline{X} = (1/n)\Sigma_1^n X_i$ and $S = \Sigma_1^n (X_i - \overline{X})(X_i - \overline{X})'$. Show that the test that rejects for large values of $T = (R\overline{X})'(RSR')^{-1}(R\overline{X})$ is equivalent to the likelihood ratio test. Also, show this test is UMP invariant under a suitable group of transformations. Apply this to the problem of testing $\mu_1 = \mu_2 = \cdots = \mu_p$ where $\mu_1, \ldots, \mu_p$ are the coordinates of μ.

11. Consider a linear model of the form $Y = ZB + E$ with $Z : n \times k$ of rank k, $B : k \times p$ unknown, and E a matrix of errors. Assume the first column of Z is the vector e of ones (the regression equation has the constant term in it). Assume $\text{Cov}(E) = A(\rho) \otimes \Sigma$ where $A(\rho)$ has ones on the diagonal and ρ off the diagonal $(-1/(n-1) < \rho < 1)$.

(i) Show that the GM and least-squares estimates of B are the same.

(ii) When $\mathcal{L}(E) = N(0, A(\rho) \otimes \Sigma)$ with Σ and ρ unknown, argue via invariance to construct tests for hypotheses of the form $\dot{R}\dot{B} = 0$ where $\dot{R}$ is $r \times k - 1$ of rank r and $\dot{B} : (k-1) \times p$ consists of the last $k - 1$ rows of B.

NOTES AND REFERENCES

1. The material in Section 9.1 is fairly standard and can be found in many texts on multivariate analysis although the treatment and emphasis may be different than here. The likelihood ratio test in the MANOVA setting was originally proposed by Wilks (1932). Various competitors to the likelihood ratio test were proposed in Lawley (1938), Hotelling (1947), Roy (1953), and Pillai (1955).

2. Arnold (1973) applied the theory of products of problems (which he had developed in his Ph.D. dissertation at Stanford) to situations involving patterned covariance matrices. This notion appears in both this chapter and Chapter 10.

3. Given the covariance structure assumed in Section 9.2, the regression subspaces considered there are not the most general for which the Gauss–Markov and least-squares estimators are the same. See Eaton (1970) for a discussion.

4. Andersson (1975) provides a complete description of all symmetry models.

5. Cyclic covariance models were first studied systematically in Olkin and Press (1969).

6. For early papers on the complex normal distribution, see Goodman (1963) and Giri (1965a). Also, see Andersson (1975).

7. Some of the material in Section 9.6 comes from Giri (1964, 1965b).

8. In Proposition 9.5, when $r = 1$, the statistic λ_1 is commonly known as Hotelling's T^2 (see Hotelling (1931)).

CHAPTER 10

Canonical Correlation Coefficients

This final chapter is concerned with the interpretation of canonical correlation coefficients and their relationship to affine dependence and independence between two random vectors. After using an invariance argument to show that population canonical correlations are a natural measure of affine dependence, these population coefficients are interpreted as cosines of the angles between subspaces (as defined in Chapter 1). Next, the sample canonical correlations are defined and interpreted as cosines of angles. The distribution theory associated with the sample coefficients is discussed briefly.

When two random vectors have a joint normal distribution, independence between the vectors is equivalent to the population canonical correlations all being zero. The problem of testing for independence is treated in the fourth section of this chapter. The relationship between the MANOVA testing problem and testing for independence is discussed in the fifth and final section of the chapter.

10.1. POPULATION CANONICAL CORRELATION COEFFICIENTS

There are a variety of ways to introduce canonical correlation coefficients and three of these are considered in this section. We begin our discussion with the notion of affine dependence between two random vectors. Let $X \in (V, (\cdot, \cdot)_1)$ and $Y \in (W, (\cdot, \cdot)_2)$ be two random vectors defined on the same probability space so the random vector $Z = \{X, Y\}$ takes values in the vector space $V \oplus W$. It is assumed that $\text{Cov}(X) = \Sigma_{11}$ and $\text{Cov}(Y) = \Sigma_{22}$ both exist and are nonsingular. Therefore, $\text{Cov}(Z)$ exists (see Proposition

2.15) and is given by

$$\Sigma = \text{Cov}(Z) = \begin{pmatrix} \Sigma_{11} & \Sigma_{12} \\ \Sigma'_{12} & \Sigma_{22} \end{pmatrix}.$$

Also, the mean vector of Z is

$$\mu = \mathcal{E}Z = \{\mathcal{E}X, \mathcal{E}Y\} = \{\mu_1, \mu_2\}.$$

Definition 10.1. Two random vectors U and $\tilde{U}$, in $(V, (\cdot, \cdot)_1)$ are *affinely equivalent* if $U = A\tilde{U} + a$ for some nonsingular linear transformation A and some vector $a \in V$.

It is clear that affine equivalence is an equivalence relation among random vectors defined on the same probability space and taking values in V.

We now consider measures of affine dependence between X and Y, which are functions of $\mu = \{\mathcal{E}X, \mathcal{E}Y\}$ and $\Sigma = \text{Cov}(Z)$ where $Z = \{X, Y\}$. Let $m(\mu, \Sigma)$ be some real-valued function of μ and Σ that is supposed to measure affine dependence. If instead of X we observe $\tilde{X}$, which is affinely equivalent to X, then the affine dependence between X and Y should be the same as the affine dependence between $\tilde{X}$ and Y. Similarly, if $\tilde{Y}$ is affinely equivalent to Y, then the affine dependence between X and Y should be the same as the affine dependence between X and $\tilde{Y}$. These remarks imply that $m(\mu, \Sigma)$ should be invariant under affine transformations of both X and Y. If (A, a) is an affine transformation on V, then $(A, a)v = Av + a$ where A is nonsingular on V to V. Recall that the group of all affine transformations on V to V is denoted by $Al(V)$ and the group operation is given by

$$(A_1, a_1)(A_2, a_2) = (A_1A_2, A_1a_2 + a_1).$$

Also, let $Al(W)$ be the affine group for W. The product group $Al(V) \times Al(W)$ acts on the vector space $V \oplus W$ in the obvious way:

$$((A, a), (B, b))\{v, w\} = \{Av + b, Bw + b\}.$$

The argument given above suggests that the affine dependence between X and Y should be the same as the affine dependence between $(A, a)X$ and $(B, b)Y$ for all $(A, a) \in Al(V)$ and $(B, b) \in Al(W)$. We now need to interpret this requirement as a condition on $m(\mu, \Sigma)$. The random vector

$$((A, a), (B, b))\{X, Y\} = \{AX + a, BY + b\}$$

has a mean vector given by

$$((A, a), (B, b))\{\mu_1, \mu_2\} = \{A\mu_1 + a, A\mu_2 + b\}$$

and a covariance given by

$$\begin{pmatrix} A\Sigma_{11}A' & A\Sigma_{12}B' \\ B\Sigma_{12}'A' & B\Sigma_{22}B' \end{pmatrix}$$

Therefore, the group $Al(V) \times Al(W)$ acts on the set

$$\Theta = \{(\mu, \Sigma) | \mu \in V \oplus W, \Sigma \geqslant 0, \Sigma_{ii} > 0, i = 1, 2\}.$$

For $g \equiv ((A, a), (B, b)) \in Al(V) \times Al(W)$, the group action is given by

$$(\mu, \Sigma) \to (g\mu, g(\Sigma))$$

where

$$g\mu = \{A\mu_1 + a, B\mu_2 + b\}$$

and

$$g(\Sigma) = \begin{pmatrix} A\Sigma_{11}A' & A\Sigma_{12}B' \\ B\Sigma_{12}'A' & B\Sigma_{22}B' \end{pmatrix}.$$

Requiring the affine dependence between X and Y to be equal to the affine dependence between $(A, a)X$ and $(B, b)Y$ simply means that the function m defined on Θ must be invariant under the group action given above. Therefore, m must be a function of a maximal invariant function under the action of $Al(V) \times Al(W)$ on Θ. The following proposition gives one form of a maximal invariant.

Proposition 10.1. Let $q = \dim V$, $r = \dim W$, and let $t = \min\{q, r\}$. Given

$$\Sigma = \begin{pmatrix} \Sigma_{11} & \Sigma_{12} \\ \Sigma_{12}' & \Sigma_{22} \end{pmatrix},$$

which is positive definite on $V \oplus W$, let $\lambda_1 \geqslant \cdots \geqslant \lambda_t \geqslant 0$ be the t largest eigenvalues of

$$\Lambda(\Sigma) \equiv \Sigma_{11}^{-1}\Sigma_{12}\Sigma_{22}^{-1}\Sigma_{21}$$

where $\Sigma_{21} \equiv \Sigma'_{12}$. Define a function h on Θ by

$$h(\mu, \Sigma) = (\lambda_1, \lambda_2, \dots, \lambda_t),$$

where $\lambda_1 \geqslant \cdots \geqslant \lambda_t$ are defined in terms of Σ as above. Then h is a maximal invariant function under the action of $G \equiv Al(V) \times Al(W)$ on Θ.

Proof. Let $\{v_1, \dots, v_t\}$ and $\{w_1, \dots, w_t\}$ be fixed orthonormal sets in V and W. For each Σ, define $Q_{12}(\Sigma)$ by

$$Q_{12}(\Sigma) = \sum_{i=1}^{t} \lambda_i^{1/2} v_i \square w_i$$

where $\lambda_1 \geqslant \cdots \geqslant \lambda_t$ are the t largest eigenvalues of $\Lambda(\Sigma)$. Given $(\mu, \Sigma) \in \Theta$, we first claim that there exists a $g \in G$ such that $g\mu = 0$ and

$$g(\Sigma) = \begin{pmatrix} I_V & Q_{12}(\Sigma) \\ (Q_{12}(\Sigma))' & I_W \end{pmatrix}.$$

The proof of this claim follows. For $g = ((A, a), (B, b))$, we have

$$g(\Sigma) = \begin{pmatrix} A\Sigma_{11}A' & A\Sigma_{12}B' \\ B\Sigma_{21}A' & B\Sigma_{22}B' \end{pmatrix}.$$

Choose $A = \Gamma\Sigma_{11}^{-1/2}$ and $B = \Delta\Sigma_{22}^{-1/2}$ where $\Gamma \in \mathcal{O}(V)$, $\Delta \in \mathcal{O}(W)$, and $\Sigma_{ii}^{-1/2}$ is the inverse of the positive definite square root of Σ_{ii}, $i = 1, 2$. For each Γ and Δ,

$$A\Sigma_{11}A' = \Gamma\Sigma_{11}^{-1/2}\Sigma_{11}\Sigma_{11}^{-1/2}\Gamma' = I_V$$

$$B\Sigma_{22}B' = \Delta\Sigma_{22}^{-1/2}\Sigma_{22}\Sigma_{22}^{-1/2}\Delta' = I_W$$

and

$$A\Sigma_{12}B' = \Gamma\Sigma_{11}^{-1/2}\Sigma_{12}\Sigma_{22}^{-1/2}\Delta'.$$

Using the singular value decomposition, write

$$\Lambda_{12} \equiv \Sigma_{11}^{-1/2}\Sigma_{12}\Sigma_{22}^{-1/2} = \sum_{i=1}^{t} \lambda_i^{1/2} x_i \square y_i$$

where $\{x_1, \ldots, x_t\}$ and $\{y_1, \ldots, y_t\}$ are orthonormal sets in V and W, respectively. This representation follows by noting that the rank of Λ_{12} is at most t and

$$\Lambda_{12}\Lambda'_{12} = \Sigma_{11}^{-1/2}\Sigma_{12}\Sigma_{22}^{-1}\Sigma_{21}\Sigma_{11}^{-1/2}$$

has the same eigenvalues as $\Lambda(\Sigma)$, which are $\lambda_1 \geqslant \cdots \geqslant \lambda_t \geqslant 0$. For A and B as above, it now follows that

$$A\Sigma_{12}B' = \sum_{i=1}^{t} \lambda_i^{1/2}(\Gamma x_i)\Box(\Delta y_i).$$

Choose Γ so that $\Gamma x_i = v_i$ and choose Δ so that $\Delta y_i = w_i$. Then we have

$$A\Sigma_{12}B' = Q_{12}(\Sigma)$$

so $g(\Sigma)$ has the form claimed. With these choices for A and B, now choose $a = -A\mu_1$ and $b = -B\mu_2$. Then

$$\begin{aligned} g\mu = g\{\mu_1, \mu_2\} &= ((A, a), (B, b))\{\mu_1, \mu_2\} \\ &= \{A\mu_1 + a, B\mu_2 + b\} = \{0, 0\} = 0. \end{aligned}$$

The proof of the claim is now complete. To finish the proof of Proposition 10.1, first note that Proposition 1.39 implies that h is a G-invariant function. For the maximality of h, suppose that $h(\mu, \Sigma) = h(\nu, \Psi)$. Thus

$$Q_{12}(\Sigma) = Q_{12}(\Psi),$$

which implies that there exists a g and $\tilde{g}$ such that

$$g\mu = 0, \qquad \tilde{g}\nu = 0,$$

and

$$g(\Sigma) = \begin{pmatrix} I_V & Q_{12}(\Sigma) \\ (Q_{12}(\Sigma))' & I_W \end{pmatrix} = \tilde{g}(\Psi).$$

Therefore,

$$g^{-1}\tilde{g}(\nu, \Psi) = (\mu, \Sigma)$$

so h is maximal invariant. $\Box$

The form of the singular value decomposition used in the proof of Proposition 10.1 is slightly different than that given in Theorem 1.3. For a linear transformation C of rank k defined on $(V, (\cdot,\cdot)_1)$ to $(W, (\cdot,\cdot)_2)$, Theorem 1.3 asserts that

$$C = \sum_i^k \mu_i w_i \square x_i$$

where $\mu_i > 0$, $\{x_1, \ldots, x_k\}$, and $\{w_1, \ldots, w_k\}$ are orthonormal sets in V and W. With $q = \dim V$, $r = \dim W$, and $t = \min\{q, r\}$, obviously $k \leqslant t$. When $k < t$, it is clear that the orthonormal sets above can be extended to $\{x_1, \ldots, x_t\}$ and $\{w_1, \ldots, w_t\}$, which are still orthonormal sets in V and W. Also, setting $\mu_i = 0$ for $i = k+1, \ldots, t$, we have

$$C = \sum_1^t \mu_i w_i \square x_i,$$

and $\mu_1^2 \geqslant \cdots \geqslant \mu_t^2$ are the t largest eigenvalues of both CC' and $C'C$. This form of the singular value decomposition is somewhat more convenient in this chapter since the rank of C is not explicitly mentioned. However, the rank of C is just the number of μ_i, which are strictly positive. The corresponding modification of Proposition 1.48 should now be clear.

Returning to our original problem of describing measures of affine dependence, say $m(\mu, \Sigma)$, Proposition 10.1 demonstrates that m is invariant under affine relabelings of X and Y iff m is a function of the t largest eigenvalues, $\lambda_1, \ldots, \lambda_t$, of $\Lambda(\Sigma)$. Since the rank of $\Lambda(\Sigma)$ is at most t, the remaining eigenvalues of $\Lambda(\Sigma)$, if there are any, must be zero. Before suggesting some particular measures $m(\mu, \Sigma)$, the canonical correlation coefficients are discussed.

Definition 10.2. In the notation of Proposition 10.1, let $\rho_i = \lambda_i^{1/2}$, $i = 1, \ldots, t$. The numbers $\rho_1 \geqslant \rho_2 \geqslant \cdots \geqslant \rho_t \geqslant 0$ are called the *population canonical correlation coefficients*.

Since ρ_i is a one-to-one function of λ_i, it follows that the vector $(\rho_1, \ldots, \rho_t)$ also determines a maximal invariant function under the action of G on Θ. In particular, any measure of affine dependence should be a function of the canonical correlation coefficients.

The canonical correlation coefficients have a natural interpretation as cosines of angles between subspaces in a vector space. Recall that $Z = \{X, Y\}$ takes values in the vector space $V \oplus W$ where $(V, (\cdot,\cdot)_1)$ and $(W, (\cdot,\cdot)_2)$

are inner product spaces. The covariance of Z, with respect to the natural inner product, say $(\cdot, \cdot)$, on $V \oplus W$, is

$$\Sigma = \begin{pmatrix} \Sigma_{11} & \Sigma_{12} \\ \Sigma_{21} & \Sigma_{22} \end{pmatrix}.$$

In the discussion that follows, it is assumed that Σ is positive definite. Let $(\cdot, \cdot)_\Sigma$ denote the inner product on $V \oplus W$ defined by

$$(z_1, z_2)_\Sigma = (z_1, \Sigma z_2) = \text{cov}[(z_1, Z), (z_2, Z)],$$

for $z_1, z_2 \in V \oplus W$. The vector space V can be thought of as a subspace of $V \oplus W$—namely, just identify V with $V \oplus \{0\} \subseteq V \oplus W$. Similarly, W is a subspace of $V \oplus W$. The next result interprets the canonical correlations as the cosines of angles between the subspaces V and W when the inner product on $V \oplus W$ is $(\cdot, \cdot)_\Sigma$.

Proposition 10.2. Given Σ, the canonical correlation coefficients $\rho_1 \geqslant \cdots \geqslant \rho_t$ are the cosines of the angles between V and W as subspaces in the inner product space $(V \oplus W, (\cdot, \cdot)_\Sigma)$.

Proof. Let P_1 and P_2 be the orthogonal projections (relative to $(\cdot, \cdot)_\Sigma$) onto $V \oplus \{0\}$ and $W \oplus \{0\}$, respectively. In view of Proposition 1.48 and Definition 1.28, it suffices to show that the t largest eigenvalues of $P_1 P_2 P_1$ are $\lambda_i = \rho_i^2$, $i = 1, \ldots, t$. We claim that

$$C_1 = \begin{pmatrix} I_V & \Sigma_{11}^{-1}\Sigma_{12} \\ 0 & 0 \end{pmatrix}$$

is the orthogonal projection onto $V \oplus \{0\}$. For $\{v, w\} \in V \oplus W$,

$$\begin{pmatrix} I_V & \Sigma_{11}^{-1}\Sigma_{12} \\ 0 & 0 \end{pmatrix}\{v, w\} = \{v + \Sigma_{11}^{-1}\Sigma_{12}w, 0\}$$

so the range of C_1 is $V \oplus \{0\}$ and C_1 is the identity on $V \oplus \{0\}$. That $C_1^2 = C_1$ is easily verified. Also, since

$$\Sigma = \begin{pmatrix} \Sigma_{11} & \Sigma_{12} \\ \Sigma_{21} & \Sigma_{22} \end{pmatrix},$$

the identity $C_1'\Sigma = \Sigma C_1$ holds. Here C_1' is the adjoint of C_1 relative to the inner product $(\cdot, \cdot)$—namely,

$$C_1' = \begin{pmatrix} I_V & 0 \\ \Sigma_{21}\Sigma_{11}^{-1} & 0 \end{pmatrix}.$$

This shows that C_1 is self-adjoint relative to the inner product $(\cdot, \cdot)_\Sigma$. Hence C_1 is the orthogonal projection onto $V \oplus \{0\}$ in $(V \oplus W, (\cdot, \cdot)_\Sigma)$. A similar argument yields

$$C_2 = \begin{pmatrix} 0 & 0 \\ \Sigma_{22}^{-1}\Sigma_{21} & I_W \end{pmatrix}$$

as the orthogonal projection onto $\{0\} \oplus W$ in $(V \oplus W, (\cdot, \cdot)_\Sigma)$. Therefore $P_i = C_i$, $i = 1, 2$, and a bit of algebra shows that

$$P_1 P_2 P_1 = \begin{pmatrix} \Lambda(\Sigma) & C \\ 0 & 0 \end{pmatrix}$$

where $\Lambda(\Sigma) = \Sigma_{11}^{-1}\Sigma_{12}\Sigma_{22}^{-1}\Sigma_{21}$ and

$$C = \Lambda(\Sigma)\Sigma_{11}^{-1}\Sigma_{12}.$$

Thus the characteristic polynomial of $P_1 P_2 P_1$ is given by

$$p(\alpha) = \det[P_1 P_2 P_1 - \alpha I] = (-\alpha)^r \det[\Lambda(\Sigma) - \alpha I_V]$$

where $r = \dim W$. Since $t = \min\{q, r\}$ where $q = \dim V$, it follows that the t largest eigenvalues of $P_1 P_2 P_1$ are the t largest eigenvalues of $\Lambda(\Sigma)$. These are $\rho_1^2 \geqslant \cdots \geqslant \rho_t^2$, so the proof is complete. □

Another interpretation of the canonical correlation coefficients can be given using Proposition 1.49 and the discussion following Definition 1.28. Using the notation adopted in the proof of Proposition 10.2, write

$$P_2 P_1 = \sum_{i=1}^{t} \rho_i \xi_i \Box \eta_i$$

where $\{\eta_1, \ldots, \eta_t\}$ is an orthonormal set in $V \oplus \{0\}$ and $\{\xi_1, \ldots, \xi_t\}$ is an orthonormal set in $\{0\} \oplus W$. Here orthonormal refers to the inner product $(\cdot, \cdot)_\Sigma$ on $V \oplus W$, as does the symbol $\Box$ in the expression for $P_2 P_1$—that is, for $z_1, z_2 \in V \oplus W$,

$$(z_1 \Box z_2) z = (z_2, z)_\Sigma z_1 = (z_2, \Sigma z) z_1.$$

The existence of this representation for P_2P_1 follows from Proposition 1.48, as does the relationship

$$(\eta_i, \xi_j)_\Sigma = \delta_{ij}\rho_j$$

for $i, j = 1, \ldots, t$. Define the sets D_{1i} and D_{2i}, $i = 1, \ldots, t$, as in Proposition 1.49 (with $M_1 = V \oplus \{0\}$ and $M_2 = \{0\} \oplus W$), so

$$\sup_{\eta \in D_{1i}} \sup_{\xi \in D_{2i}} (\eta, \xi)_\Sigma = (\eta_i, \xi_i)_\Sigma = \rho_i$$

for $i = 1, \ldots, t$. To interpret ρ_1, first consider the case $i = 1$. A vector η is in D_{11} iff

$$\eta = \{v, 0\}, \qquad v \in V$$

and

$$1 = (\eta, \Sigma\eta) = (v, \Sigma_{11}v)_1 = \text{var}(v, X)_1.$$

Similarly, $\xi \in D_{21}$ iff

$$\xi = \{0, w\}, \qquad w \in W$$

and

$$1 = (\xi, \Sigma\xi) = (w, \Sigma_{22}w)_2 = \text{var}(w, Y)_2.$$

However, for $\eta = \{v, 0\} \in D_{11}$ and $\xi = \{0, w\} \in D_{21}$,

$$(\eta, \xi)_\Sigma = (v, \Sigma_{12}w)_1 = \text{cov}\{(v, X)_1, (w, Y)_2\}.$$

This is just the ordinary correlation between $(v, X)_1$ and $(w, Y)_2$ as v and w have been normalized so that $1 = \text{var}(v, X)_1 = \text{var}(w, Y)_2$. Since $(\eta, \xi)_\Sigma \leqslant \rho_1$ for all $\eta \in D_{11}$ and $\xi \in D_{21}$, it follows that for every $x \in V$, $x \neq 0$, and $y \in W$, $y \neq 0$, the correlation between $(x, X)_1$ and $(y, Y)_2$ is no greater than ρ_1. Further, writing $\eta_1 = \{v_1, 0\}$ and $\xi_1 = \{0, w_1\}$, we have

$$\begin{aligned} \rho_1 &= (\eta_1, \xi_1)_\Sigma = (\eta_1, \Sigma\xi_1) \\ &= (v_1, \Sigma_{12}w_1)_1 = \text{cov}\{(v_1, X)_1, (w_1, Y)_2\}, \end{aligned}$$

which is the correlation between $(v_1, X)_1$ and $(w_1, Y)_2$. Therefore, ρ_1 is the maximum correlation between $(x, X)_1$ and $(y, Y)_2$ for all nonzero $x \in V$ and $y \in W$. Further, this maximum correlation is achieved by choosing $x = v_1$ and $y = w_1$.

The second largest canonical correlation coefficient, ρ_2, satisfies the equality

$$\sup_{\eta \in D_{12}} \sup_{\xi \in D_{22}} (\eta, \xi)_\Sigma = (\eta_2, \xi_2)_\Sigma = \rho_2.$$

A vector η is in D_{12} iff

$$\eta = \{v, 0\}, \qquad v \in V$$

$$1 = (\eta, \eta)_\Sigma = (v, \Sigma_{11} v)_1$$

and

$$0 = (\eta, \eta_1)_\Sigma = (v, \Sigma_{11} v_1)_1.$$

Also, a vector ξ is in D_{22} iff

$$\xi = \{0, w\}, \qquad w \in W$$

$$1 = (\xi, \xi)_\Sigma = (w, \Sigma_{22} w)_2$$

and

$$0 = (\xi, \xi_1)_\Sigma = (w, \Sigma_{22} w_1)_2.$$

These relationships provide the following interpretation of ρ_2. The maximum correlation between $(x, X)_1$ and $(y, Y)_2$ is ρ_1 and is

$$\rho_1 = \operatorname{cov}\{(v_1, X)_1, (w_1, Y)_2\}$$

since $1 = \operatorname{var}(v_1, X)_1 = \operatorname{var}(w_1, Y)_2$. Suppose we now want to find the maximum correlation between $(x, X)_1$ and $(y, Y)_2$ subject to the condition

$$\text{(i)} \quad \begin{cases} \operatorname{cov}\{(x, X)_1, (v_1, X)_1\} = 0 \\ \operatorname{cov}\{(y, Y)_2, (w_1, Y)_2\} = 0. \end{cases}$$

Clearly (i) is equivalent to

$$\text{(ii)} \quad \begin{cases} (x, \Sigma_{11} v_1)_1 = 0 \\ (y, \Sigma_{22} w_1)_2 = 0. \end{cases}$$

Since correlation is invariant under multiplication of the random variables by positive constants, to find the maximum correlation between $(x, X)_1$ and $(y, Y)_2$ subject to (ii), it suffices to maximize $\text{cov}\{(x, X)_1, (y, Y)_2\}$ over those x's and y's that satisfy

$$\text{(iii)} \quad \begin{cases} (x, \Sigma_{11}x)_1 = 1, (x, \Sigma_{11}v_1)_1 = 0 \\ (y, \Sigma_{22}y)_2 = 1, (y, \Sigma_{22}w_1)_2 = 0. \end{cases}$$

However, $x \in V$ satisfies (iii) iff $\eta = \{x, 0\}$ is in D_{12} and $y \in W$ satisfies (iii) iff $\xi = \{0, y\}$ is in D_{22}. Further, for such x, y, η, and ξ,

$$\text{cov}\{(x, X)_1, (y, Y)_2\} = (\eta, \xi)_\Sigma.$$

Thus maximizing this covariance subject to (iii) is the same as maximizing $(\eta, \xi)_\Sigma$ for $\eta \in D_{12}$ and $\xi \in D_{22}$. Of course, this maximum is ρ_2 and is achieved at $\eta_2 \in D_{12}$ and $\xi_2 \in D_{22}$. Writing $\eta_2 = \{v_2, 0\}$ and $\xi_2 = \{0, w_2\}$, it is clear that $v_2 \in V$ and $w_2 \in W$ satisfy (iii) and

$$\text{cov}\{(v_2, X)_1, (w_2, Y)_2\} = \rho_2.$$

Furthermore, Proposition 1.48 shows that

$$0 = (\eta_1, \xi_2)_\Sigma = (\eta_2, \xi_1)_\Sigma,$$

which implies that

$$0 = \text{cov}\{(v_1, X)_1, (w_2, Y)_2\} = \text{cov}\{(v_2, X)_1, (w_1, Y)_2\}.$$

Therefore, the problem of maximizing the correlation between $(x, X)_1$ and $(y, Y)_2$ (subject to the condition that the correlation between $(x, X)_1$ and $(v_1, X)_1$ be zero and the correlation between $(y, Y)_2$ and $(w_1, Y)_2$ be zero) has been solved.

It should now be fairly clear how to interpret the remaining canonical correlation coefficients. The easiest way to describe the coefficients is by induction. The coefficient ρ_1 is the largest possible correlation between $(x, X)_1$ and $(y, Y)_2$ for nonzero vectors $x \in V$ and $y \in W$. Further, there exist vectors $v_1 \in V$ and $w_1 \in W$ such that

$$\text{cov}\{(v_1, X)_1, (w_1, Y)_2\} = \rho_1$$

and

$$1 = \text{var}(v_1, X)_1 = \text{var}(w_1, Y)_2.$$

These vectors came from η_1 and ξ_1 in the representation

$$P_2 P_1 = \sum_{i=1}^{t} \rho_i \xi_i \Box \eta_i$$

given earlier. Since $\eta_i \in V \oplus \{0\}$, we can write $\eta_i = \{v_i, 0\}$, $i = 1, \ldots, t$. Similarly, $\xi_i = \{0, w_i\}$, $i = 1, \ldots, t$. Using Proposition 1.48, it is easy to check that

$$\operatorname{cov}\{(v_j, X)_1, (w_k, Y)_2\} = \rho_j \delta_{jk}$$

$$\operatorname{cov}\{(v_j, X)_1, (v_k, X)_1\} = \delta_{jk}$$

$$\operatorname{cov}\{(w_j, Y)_2, (w_k, Y)_2\} = \delta_{jk}$$

for $j, k = 1, \ldots, t$. Of course, these relationships are simply a restatement of the properties of $\xi_1, \ldots, \xi_t$ and $\eta_1, \ldots, \eta_t$. For example,

$$\operatorname{cov}\{(v_j, X)_1, (w_k, Y)_2\} = (v_j, \Sigma_{12} w_k)_1 = (\eta_j, \xi_k)_\Sigma = \rho_j \delta_{jk}.$$

However, as argued in the case of ρ_2, we can say more. Given $\rho_1, \ldots, \rho_t$ and the vectors $v_1, \ldots, v_{i-1}$ and $w_1, \ldots, w_{i-1}$ obtained from $\eta_1, \ldots, \eta_{i-1}$ and $\xi_1, \ldots, \xi_{i-1}$, consider the problem of maximizing the correlation between $(x, X)_1$ and $(y, Y)_2$ subject to the conditions that

$$\begin{cases} \operatorname{cov}\{(x, X)_1, (v_j, X)_1\} = 0, & j = 1, \ldots, i-1 \\ \operatorname{cov}\{(y, Y)_2, (w_j, Y)_2\} = 0, & j = 1, \ldots, i-1. \end{cases}$$

By simply unravelling the notation and using Proposition 1.49, this maximum correlation is ρ_i and is achieved for $x = v_i$ and $y = w_i$. This successive maximization of correlation is often a useful interpretation of the canonical correlation coefficients.

The vectors $v_1, \ldots, v_t$ and $w_1, \ldots, w_t$ lead to what are called the *canonical variates*. Recall that $q = \dim V$, $r = \dim W$ and $t = \min\{q, r\}$. For definiteness, assume that $q \leqslant r$ so $t = q$. Thus $\{v_1, \ldots, v_q\}$ is a basis for V and satisfies

$$(v_j, \Sigma_{11} v_k)_1 = \delta_{jk}$$

for $j, k = 1, \ldots, q$ so $\{v_1, \ldots, v_q\}$ is an orthonormal basis for V relative to the inner product determined by Σ_{11}. Further, the linearly independent set $\{w_1,\ldots, w_q\}$ satisfies

$$(w_j, \Sigma_{22}w_k)_2 = \delta_{jk}$$

so $\{w_1,\ldots, w_q\}$ is an orthonormal set relative to the inner product determined by Σ_{22}. Now, extend this set to $\{w_1,\ldots, w_r\}$ so that this is an orthonormal basis for W in the Σ_{22} inner product.

Definition 10.3. The real-valued random variables defined by

$$X_i = (v_i, X)_1, \qquad i = 1,\ldots, q$$

and

$$Y_i = (w_i, Y)_2, \qquad i = 1,\ldots, r$$

are called the *canonical variates* of X and Y, respectively.

Proposition 10.3. The canonical variates satisfy the relationships

(i) $\operatorname{var} X_j = \operatorname{var} Y_k = 1$.

(ii) $\operatorname{cov}\{X_j, Y_k\} = \rho_j\delta_{jk}$.

These relationships hold for $j = 1,\ldots, q$ and $k = 1,\ldots, r$. Here, $\rho_1,\ldots, \rho_q$ are the canonical correlation coefficients.

Proof. This is just a restatement of part of what we have established above. □

Let us briefly review what has been established thus far about the population canonical correlation coefficients $\rho_1,\ldots, \rho_t$. These coefficients were defined in terms of a maximal invariant under a group action and this group action arose quite naturally in an attempt to define measures of affine dependence. Using Proposition 1.48 and Definition 1.28, it was then shown that $\rho_1,\ldots, \rho_t$ are cosines of angles between subspaces with respect to an inner product defined by Σ. The statistical interpretation of the coefficients came from the detailed information given in Proposition 1.49 and this interpretation closely resembled the discussion following Definition 1.28. Given X in $(V, (\cdot, \cdot)_1)$ and Y in $(W, (\cdot, \cdot)_2)$ with a nonsingular covariance

$$\Sigma = \begin{pmatrix} \Sigma_{11} & \Sigma_{12} \\ \Sigma_{21} & \Sigma_{22} \end{pmatrix},$$

the existence of special bases $\{v_1,\ldots, v_q\}$ and $\{w_1,\ldots, w_r\}$ for V and W was established. In terms of the canonical variates

$$X_i = (v_i, X)_1, \qquad Y_j = (w_j, Y)_2,$$

the properties of these bases can be written

$$1 = \operatorname{var} X_i = \operatorname{var} Y_j$$

and

$$\operatorname{cov}\{X_i, Y_j\} = \rho_i \delta_{ij}$$

for $i = 1,\ldots, q$ and $j = 1,\ldots, r$. Here, the convention that $\rho_i = 0$ for $i > t = \min\{q, r\}$ has been used although ρ_i is not defined for $i > t$. When $q \leqslant r$, the covariance matrix of the variates $X_1,\ldots, X_q, Y_1,\ldots, Y_r$ (in that order) is

$$\Sigma_0 = \begin{pmatrix} I_q & (DO) \\ (DO)' & I_r \end{pmatrix}$$

where D is a $q \times q$ diagonal matrix with diagonal entries $\rho_1 \geqslant \cdots \geqslant \rho_q$ and O is a $q \times (r-q)$ block of zeroes. The reader should compare this matrix representation of Σ to the assertion of Proposition 5.7.

The final point of this section is to relate a prediction problem to that of suggesting a particular measure of affine dependence. Using the ideas developed in Chapter 4, a slight generalization of Proposition 2.22 is presented below. Again, consider $X \in (V, (\cdot,\cdot)_1)$ and $Y \in (W, (\cdot,\cdot)_2)$ with $\mathcal{E}X = \mu_1$, $\mathcal{E}Y = \mu_2$, and

$$\operatorname{Cov}\{X, Y\} = \begin{pmatrix} \Sigma_{11} & \Sigma_{12} \\ \Sigma_{21} & \Sigma_{22} \end{pmatrix}.$$

It is assumed that Σ_{11} and Σ_{22} are both nonsingular. Consider the problem of predicting X by an affine function of Y—say $CY + v_0$ where $C \in \mathcal{L}(W, V)$ and $v_0 \in V$. Let $[\cdot,\cdot]$ be any inner product on V and let $\|\cdot\|$ be the norm defined by $[\cdot,\cdot]$. The following result shows how to choose C and v_0 to minimize

$$\mathcal{E}\|X - (CY + v_0)\|^2.$$

Of course, the inner product $[\cdot,\cdot]$ on V is related to the inner product $(\cdot,\cdot)_1$

by

$$[v_1, v_2] = (v_1, A_0 v_2)_1$$

for some positive definite A_0.

Proposition 10.4. For any $C \in \mathcal{L}(W, V)$ and $v_0 \in V$, the inequality

$$\mathcal{E}\|X - (CY + v_0)\|^2 \geqslant \langle A_0, \Sigma_{11} - \Sigma_{12}\Sigma_{22}^{-1}\Sigma_{21}\rangle$$

holds. There is equality in this inequality iff

$$v_0 = \hat{v}_0 \equiv \mu_1 - \Sigma_{12}\Sigma_{22}^{-1}\mu_2$$

and

$$C = \hat{C} \equiv \Sigma_{12}\Sigma_{22}^{-1}.$$

Here, $\langle \cdot, \cdot \rangle$ is the natural inner product on $\mathcal{L}(V, V)$ inherited from $(V, (\cdot, \cdot)_1)$.

Proof. First, write

$$X - (CY + v_0) = U_1 + U_2$$

where

$$U_1 = X - (\hat{C}Y + \hat{v}_0) = X - \mu_1 - \Sigma_{12}\Sigma_{22}^{-1}(Y - \mu_2)$$

and

$$U_2 = (\hat{C} - C)Y + \hat{v}_0 - v_0.$$

Clearly, U_1 has mean zero. It follows from Proposition 2.17 that U_1 and U_2 are uncorrelated and

$$\text{Cov}(U_1) = \Sigma_{11} - \Sigma_{12}\Sigma_{22}^{-1}\Sigma_{21}.$$

Further, from Proposition 4.3 we have $\mathcal{E}[U_1, U_2] = 0$. Therefore,

$$\begin{aligned}
\mathcal{E}\|X - (CY + v_0)\|^2 &= \mathcal{E}\|U_1 + U_2\|^2 = \mathcal{E}\|U_1\|^2 + \mathcal{E}\|U_2\|^2 \\
&= \mathcal{E}(U_1, A_0 U_1) + \mathcal{E}\|U_2\|^2 = \mathcal{E}\langle A_0, U_1 \square U_1\rangle + \mathcal{E}\|U_2\|^2 \\
&= \langle A_0, \Sigma_{11} - \Sigma_{12}\Sigma_{22}^{-1}\Sigma_{21}\rangle + \mathcal{E}\|U_2\|^2,
\end{aligned}$$

where the last equality follows from the identity

$$\mathcal{E}U_1 \square U_1 = \Sigma_{11} - \Sigma_{12}\Sigma_{22}^{-1}\Sigma_{21}$$

established in Proposition 2.21. Thus the desired inequality holds and there is equality iff $\mathcal{E}\|U_2\|^2 = 0$. But $\mathcal{E}\|U_2\|^2$ is zero iff U_2 is zero with probability one. This holds iff $v_0 = \hat{v}_0$ and $C = \hat{C}$ since $\text{Cov}(Y) = \Sigma_{22}$ is positive definite. This completes the proof. □

Now, choose A_0 to be Σ_{11}^{-1} in Proposition 10.4. Then the mean squared error due to predicting X by $\hat{C}Y + \hat{v}_0$, measured relative to Σ_{11}^{-1}, is

$$\phi(\Sigma) \equiv \langle \Sigma_{11}^{-1}, \Sigma_{11} - \Sigma_{12}\Sigma_{22}^{-1}\Sigma_{21} \rangle = \mathcal{E}\|X - (\hat{C}Y + \hat{v}_0)\|^2.$$

Here, $\|\cdot\|$ is obtained from the inner product defined by

$$[v_1, v_2] = (v_1, \Sigma_{11}^{-1}v_2).$$

We now claim that ϕ is invariant under the group of transformations discussed in Proposition 10.1, and thus ϕ is a possible measure of affine dependence between X and Y. To see this, first recall that $\langle \cdot , \cdot \rangle$ is just the trace inner product for linear transformations. Using properties of the trace, we have

$$\begin{aligned}\phi(\Sigma) &= \langle I, I - \Sigma_{11}^{-1/2}\Sigma_{12}\Sigma_{22}^{-1}\Sigma_{21}\Sigma_{11}^{-1/2} \rangle \\ &= \text{tr}\left(I - \Sigma_{11}^{-1/2}\Sigma_{12}\Sigma_{22}^{-1}\Sigma_{21}\Sigma_{11}^{-1/2}\right) \\ &= \sum_{i=1}^{q} (1 - \lambda_i)\end{aligned}$$

where $\lambda_1 \geqslant \cdots \geqslant \lambda_q \geqslant 0$ are the eigenvalues of $\Sigma_{11}^{-1/2}\Sigma_{12}\Sigma_{22}^{-1}\Sigma_{21}\Sigma_{11}^{-1/2}$. However, at most $t = \min\{q, r\}$ of these eigenvalues are nonzero and, by definition, $\rho_i = \lambda_i^{1/2}$, $i = 1, \ldots, t$, are the canonical correlation coefficients. Thus

$$\phi(\Sigma) = \sum_{1}^{t}\left(1 - \rho_i^2\right) + (q - t)$$

is a function of $\rho_1, \ldots, \rho_t$ and hence is an invariant measure of affine dependence. Since the constant $q - t$ is irrelevant, it is customary to use

$$\phi_1(\Sigma) = \sum_{i=1}^{t} (1 - \rho_i^2)$$

rather than $\phi(\Sigma)$ as a measure of affine dependence.

10.2. SAMPLE CANONICAL CORRELATIONS

To introduce the sample canonical correlation coefficients, again consider inner product spaces $(V, (\cdot, \cdot)_1)$ and $(W, (\cdot, \cdot)_2)$ and let $(V \oplus W, (\cdot, \cdot))$ be the direct sum space with the natural inner product $(\cdot, \cdot)$. The observations consist of n random vectors $Z_i = \{X_i, Y_i\} \in V \oplus W$, $i = 1, \ldots, n$. It is assumed that these random vectors are uncorrelated with each other and $\mathcal{L}(Z_i) = \mathcal{L}(Z_j)$ for all i, j. Although these assumptions are not essential in much of what follows, it is difficult to interpret canonical correlations without these assumptions. Given $Z_1, \ldots, Z_n$, define the random vector Z by specifying that Z takes on the values Z_i with probability $1/n$. Obviously, the distribution of Z is discrete in $V \oplus W$ and places mass $1/n$ at Z_i for $i = 1, \ldots, n$. Unless otherwise specified, when we speak of the distribution of Z, we mean the conditional distribution of Z given $Z_1, \ldots, Z_n$ as described above. Since the distribution of Z is nothing but the sample probability measure of $Z_1, \ldots, Z_n$, we can think of Z as a sample approximation to a random vector whose distribution is $\mathcal{L}(Z_1)$. Now, write $Z = \{X, Y\}$ with $X \in V$ and $Y \in W$ so X is X_i with probability $1/n$ and Y is Y_i with probability $1/n$. Given $Z_1, \ldots, Z_n$, the mean vector of Z is

$$\mathcal{E}Z = \bar{Z} \equiv \frac{1}{n} \sum_{i=1}^{n} Z_i = \{\bar{X}, \bar{Y}\}$$

and the covariance of Z is

$$\operatorname{Cov} Z = S \equiv \frac{1}{n} \sum_{i=1}^{n} (Z_i - \bar{Z}) \,\square\, (Z_i - \bar{Z}).$$

This last assertion follows from Proposition 2.21 by noting that

$$\operatorname{Cov} Z = \mathcal{E}(Z - \bar{Z}) \,\square\, (Z - \bar{Z})$$

since the mean of Z is $\bar{Z}$. When $V = R^q$ and $W = R^r$ are the standard coordinate spaces with the usual inner products, then S is just the sample covariance matrix. Since S is a linear transformation on $V \oplus W$ to $V \oplus W$, S can be written as

$$S = \begin{pmatrix} S_{11} & S_{12} \\ S_{21} & S_{22} \end{pmatrix}.$$

It is routine to show that

$$S_{11} = \frac{1}{n}\sum_{i=1}^{n}(X_i - \bar{X})\Box(X_i - \bar{X})$$

$$S_{12} = \frac{1}{n}\sum_{i=1}^{n}(X_i - \bar{X})\Box(Y_i - \bar{Y})$$

$$S_{22} = \frac{1}{n}\sum_{i=1}^{n}(Y_i - \bar{Y})\Box(Y_i - \bar{Y})$$

and $S_{21} = S'_{12}$. The reader should note that the symbol $\Box$ appearing in the expressions for S_{11}, S_{12}, and S_{22} has a different meaning in each of the three expressions—namely, the outer product depends on the inner products on the spaces in question. Since it is clear which vectors are in which spaces, this multiple use of $\Box$ should cause no confusion.

Now, to define the sample canonical correlation coefficients, the results of Section 10.1 are applied to the random vector Z. For this reason, we assume that $S = \operatorname{Cov} Z$ is nonsingular. With $q = \dim V$, $r = \dim W$, and $t = \min\{q, r\}$, the canonical correlation coefficients are the square roots of the t largest eigenvalues of

$$\Lambda(S) = S_{11}^{-1}S_{12}S_{22}^{-1}S_{21}.$$

In the sampling situation under discussion, these roots are denoted by $r_1 \geqslant \cdots \geqslant r_t \geqslant 0$ and are called the *sample canonical correlation coefficients*. The justification for such nomenclature is that $r_1^2, \ldots, r_t^2$ are the t largest eigenvalues of $\Lambda(S)$ where S is the sample covariance based on $Z_1, \ldots, Z_n$. Of course, all of the discussion of the previous section applies directly to the situation at hand. In particular, the vector $(r_1, \ldots, r_t)$ is a maximal invariant under the group action described in Proposition 10.1. Also, $r_1, \ldots, r_t$ are the cosines of the angles between the subspaces $V \oplus \{0\}$ and $\{0\} \oplus W$ in the vector space $V \oplus W$ relative to the inner product determined by S.

Now, let $\{v_1, \ldots, v_q\}$ and $\{w_1, \ldots, w_r\}$ be the canonical bases for V and W. Then we have

$$\operatorname{cov}\{(v_i, X)_1, (w_j, Y)_2\} = r_i \delta_{ij}$$

for $i = 1, \ldots, q$ and $j = 1, \ldots, r$. The convention that $r_i \equiv 0$ for $i > t$ is being used. To interpret what this means in terms of the sample $Z_1, \ldots, Z_n$, consider r_1. For nonzero $x \in V$ and $y \in W$, the maximum correlation between $(x, X)_1$ and $(y, Y)_2$ is r_1 and is achieved for $x = v_1$ and $y = w_1$. However, given $Z_1, \ldots, Z_n$, we have

$$\operatorname{var}(x, X)_1 = \operatorname{var}(\{x, 0\}, Z) = (\{x, 0\}, S\{x, 0\})$$

$$= (x, S_{11}x)_1 = \frac{1}{n} \sum_{i=1}^{n} (x, X_i - \bar{X})_1^2$$

and, similarly,

$$\operatorname{var}(y, Y)_2 = \frac{1}{n} \sum_{i=1}^{n} (y, Y_i - \bar{Y})_2^2.$$

An analogous calculation shows that

$$\operatorname{cov}\{(x, X)_1, (y, Y)_2\} = \frac{1}{n} \sum_{i=1}^{n} (x, X_i - \bar{X})_1 (y, Y_i - \bar{Y})_2.$$

Thus $\operatorname{var}(x, X)_1$ is just the sample variance of the random variables $(x, X_i)_1$, $i = 1, \ldots, n$, and $\operatorname{var}(y, Y)_2$ is the sample variance of $(y, Y_i)_2$, $i = 1, \ldots, n$. Also, $\operatorname{cov}\{(x, X)_1, (y, Y)_2\}$ is the sample covariance of the random variables $(x, X_i)_1$, $(y, Y_i)_2$, $i = 1, \ldots, n$. Therefore, the correlation between $(x, X)_1$ and $(y, Y)_2$ is the ordinary sample correlation coefficient for the random variables $(x, X_i)_1$, $(y, Y_i)_2$, $i = 1, \ldots, n$. This observation implies that the maximum possible sample correlation coefficient for $(x, X_i)_1$, $(y, Y_i)_2$, $i = 1, \ldots, n$ is the largest sample canonical correlation coefficient, r_1, and this maximum is attained by choosing $x = v_1$ and $y = w_1$. The interpretation of $r_2, \ldots, r_t$ should now be fairly obvious. Given i, $2 \leqslant i \leqslant t$, and given $r_1, \ldots, r_{i-1}$, $v_1, \ldots, v_{i-1}$, and $w_1, \ldots, w_{i-1}$, consider the problem of maximizing the correlation between $(x, X)_1$ and $(y, Y)_2$ subject to the conditions

$$\operatorname{cov}\{(x, X)_1, (v_j, X)_1\} = 0, \qquad j = 1, \ldots, i-1$$

$$\operatorname{cov}\{(y, Y)_2, (w_j, Y)_2\} = 0, \qquad j = 1, \ldots, i-1.$$

These conditions are easily shown to be equivalent to the conditions that the sample correlation for

$$(x, X_k)_1, \qquad (v_j, X_k)_1, \qquad k = 1,\dots, n$$

be zero for $j = 1,\dots, i-1$ with a similar statement concerning the Y's. Further, the correlation between $(x, X)_1$ and $(y, Y)_2$ is the sample correlation for $(x, X_k)_1, (y, Y_k)_2, k = 1,\dots, n$. The maximum sample correlation is r_i and is attained by choosing $x = v_i$ and $y = w_i$. Thus the sample interpretation of $r_1,\dots, r_t$ is completely analogous to the population interpretation of the population canonical correlation coefficients.

For the remainder of this section, it is assumed that $V = R^q$ and $W = R^r$ are the standard coordinate spaces with the usual inner products, so $V \oplus W$ is just R^p where $p = q + r$. Thus our sample is $Z_1,\dots, Z_n$ with $Z_i \in R^p$ and we write

$$Z_i = \begin{pmatrix} X_i \\ Y_i \end{pmatrix} \in R^p$$

with $X_i \in R^q$ and $Y_i \in R^r$, $i = 1,\dots, n$. The sample covariance matrix, assumed to be nonsingular, is

$$S = \frac{1}{n}\sum_1^n (Z_i - \bar{Z})(Z_i - \bar{Z})' = \begin{pmatrix} S_{11} & S_{12} \\ S_{21} & S_{22} \end{pmatrix}$$

where

$$S_{11} = \frac{1}{n}\sum_1^n (X_i - \bar{X})(X_i - \bar{X})'$$

$$S_{22} = \frac{1}{n}\sum_1^n (Y_i - \bar{Y})(Y_i - \bar{Y})'$$

$$S_{12} = \frac{1}{n}\sum_1^n (X_i - \bar{X})(Y_i - \bar{Y})'$$

and $S_{21} = S'_{12}$. Now, form the random matrix $\tilde{Z}: n \times p$ whose rows are $(Z_i - \bar{Z})'$ and partition $\tilde{Z}$ into $U: n \times q$ and $V: n \times r$ so that

$$\tilde{Z} = (UV).$$

The rows of U are $(X_i - \bar{X})'$ and the rows of V are $(Y_i - \bar{Y})'$, $i = 1, \ldots, n$. Obviously, we have $nS = \tilde{Z}'\tilde{Z}$, $nS_{11} = U'U$, $nS_{22} = V'V$, and $nS_{12} = U'V$. The sample canonical correlation coefficients $r_1 \geqslant \cdots \geqslant r_t$ are the square roots of the t largest eigenvalues of

$$\Lambda(S) = S_{11}^{-1}S_{12}S_{22}^{-1}S_{21} = (U'U)^{-1}U'V(V'V)^{-1}V'U.$$

However, the t largest eigenvalues of $\Lambda(S)$ are the same as the t largest eigenvalues of $P_X P_Y$ where

$$P_X = U(U'U)^{-1}U'$$

and

$$P_Y = V(V'V)^{-1}V'.$$

Now, P_X is the orthogonal projection onto the q-dimensional subspace of R^n, say M_X, spanned by the columns of U. Also, P_Y is the orthogonal projection onto the r-dimensional subspace of R^n, say M_Y, spanned by the columns of V. It follows from Proposition 1.48 and Definition 1.28 that the sample canonical correlation coefficients $r_1, \ldots, r_t$ are the cosines of the angles between the two subspaces M_X and M_Y contained in R^n. Summarizing, we have the following proposition.

Proposition 10.5. Given random vectors

$$Z_i = \begin{pmatrix} X_i \\ Y_i \end{pmatrix} \in R^p, \qquad i = 1, \ldots, n$$

where $X_i \in R^q$ and $Y_i \in R^r$, form the matrices $U: n \times q$ and $V: n \times r$ as above. Let $M_X \subseteq R^n$ be the subspace spanned by the columns of U and let $M_Y \subseteq R^n$ be the subspace spanned by the columns of V. Assume that the sample covariance matrix

$$S = \frac{1}{n}\sum_1^n (Z_i - \bar{Z})(Z_i - \bar{Z})'$$

is nonsingular. Then the sample canonical correlation coefficients are the cosines of the angles between M_X and M_Y.

The sample coefficients $r_1, \ldots, r_t$ have been shown to be the cosines of angles between subspaces in two different vector spaces. In the first case, the interpretation followed from the material developed in Section 10.1 of this chapter: namely, $r_1, \ldots, r_t$ are the cosines of the angles between $R^q \oplus \{0\} \subseteq R^p$ and $\{0\} \oplus R^r \subseteq R^p$ when R^p has the inner product determined by the sample covariance matrix. In the second case, described in Proposition 10.5, $r_1, \ldots, r_t$ are the cosines of the angles between M_X and M_Y in R^n when R^n has the standard inner product. The subspace M_X is spanned by the columns of U where U has rows $(X_i - \overline{X})'$, $i = 1, \ldots, n$. Thus the coordinates of the jth column of U are $X_{ij} - \overline{X}_j$ for $i = 1, \ldots, n$ where X_{ij} is the jth coordinate of $X_i \in R^q$, and $\overline{X}_j$ is the jth coordinate of $\overline{X}$. This is the reason for the subscript X on the subspace M_X. Of course, similar remarks apply to M_Y.

The vector $(r_1, \ldots, r_t)$ can also be interpreted as a maximal invariant under a group action on the sample matrix. Given

$$Z_i = \begin{pmatrix} X_i \\ Y_i \end{pmatrix} \in R^p, \qquad i = 1, \ldots, n,$$

let $\tilde{X}: n \times q$ have rows X_i', $i = 1, \ldots, n$ and let $\tilde{Y}: n \times r$ have rows Y_i', $i = 1, \ldots, n$. Then the data matrix of the whole sample is

$$\tilde{Z} = (\tilde{X}\tilde{Y}) : n \times p,$$

which has rows Z_i', $i = 1, \ldots, n$. Let $e \in R^n$ be the vector of all ones. It is assumed that $\tilde{Z} \in \mathcal{Z} \subseteq \mathcal{L}_{p,n}$ where $\mathcal{Z}$ is the set of all $n \times p$ matrices such that the sample covariance mapping

$$s(\tilde{Z}) = (\tilde{Z} - e\overline{Z}')'(\tilde{Z} - e\overline{Z}')$$

has rank p. Assuming that $n \geqslant p + 1$, the complement of $\mathcal{Z}$ in $\mathcal{L}_{p,n}$ has Lebesgue measure zero. To describe the group action on $\mathcal{Z}$, let G be the set of elements $g = (\Gamma, c, C)$ where

$$\Gamma \in \mathcal{O}_n(e) = \{\Gamma | \Gamma \in \mathcal{O}_n, \Gamma e = e\}, \qquad c \in R^p$$

and

$$C = \begin{pmatrix} A & 0 \\ 0 & B \end{pmatrix}, \qquad A \in Gl_q, \quad B \in Gl_r.$$

For $g = (\Gamma, c, C)$, the value of g at $\tilde{Z}$ is

$$g\tilde{Z} = \Gamma \tilde{Z} C' + ec'.$$

Since

$$s(g\tilde{Z}) = Cs(\tilde{Z})C'.$$

it follows that each $g \in G$ is a one-to-one onto mapping of $\mathfrak{Z}$ to $\mathfrak{Z}$. The composition in G, defined so G acts on the left of $\mathfrak{Z}$, is

$$(\Gamma_1, c_1, C_1)(\Gamma_2, c_2, C_2) \equiv (\Gamma_1\Gamma_2, c_1 + C_1c_2, C_1C_2).$$

Proposition 10.6. Under the action of G on $\mathfrak{Z}$, a maximal invariant is the vector of canonical correlation coefficients $r_1, \ldots, r_t$ where $t = \min\{q, r\}$.

Proof. Let $\mathbb{S}_p^+$ be the space of $p \times p$ positive definite matrices so the sample covariance mapping $s: \mathfrak{Z} \to \mathbb{S}_p^+$ is onto. Given $S \in \mathbb{S}_p^+$, partition S as

$$S = \begin{pmatrix} S_{11} & S_{12} \\ S_{21} & S_{22} \end{pmatrix}$$

where S_{11} is $q \times q$, S_{22} is $r \times r$, and S_{12} is $q \times r$. Define h on $\mathbb{S}_p^+$ by letting $h(S)$ be the vector $(\lambda_1, \ldots, \lambda_t)'$ of the t largest eigenvalues of

$$\Lambda(S) = S_{11}^{-1}S_{12}S_{22}^{-1}S_{21}.$$

Since $r_i = \sqrt{\lambda_i}$, $i = 1, \ldots, t$, the proposition will be proved if it is shown that

$$\varphi(\tilde{Z}) \equiv h(s(\tilde{Z}))$$

is a maximal invariant function. This follows since $h(s(\tilde{Z})) = (\lambda_1, \ldots, \lambda_t)'$, which is a one-to-one function of $(r_1, \ldots, r_t)$. The proof that φ is maximal invariant proceeds as follows. Consider the two subgroups G_1 and G_2 of G defined by

$$G_1 = \{g | g = (\Gamma, c, I_p) \in G\}$$

and

$$G_2 = \{g | g = (I_n, 0, C) \in G\}.$$

Note that G_2 acts on the space $\mathbb{S}_p^+$ in the obvious way— namely, if $g_2 = (I_n, 0, C)$, then

$$g_2(S) \equiv CSC', \qquad S \in \mathbb{S}_p^+.$$

Further, since

$$(\Gamma, c, C) = (\Gamma, c, I_p)(I_n, 0, C),$$

it follows that each $g \in G$ can be written as $g = g_1 g_2$ where $g_i \in G_i$, $i = 1, 2$. Now, we make two claims:

(i) $s : \mathfrak{Z} \to \mathbb{S}_p^+$ is a maximal invariant under the action of G_1 on $\mathfrak{Z}$.
(ii) $h : \mathbb{S}_p^+ \to R^t$ is a maximal invariant under the action of G_2 on $\mathbb{S}_p^+$.

Assuming (i) and (ii), we now show that $\varphi(\tilde{Z}) = h(s(\tilde{Z}))$ is maximal invariant. For $g \in G$, write $g = g_1 g_2$ with $g_i \in G_i$, $i = 1, 2$. Since

$$s(g_1 \tilde{Z}) = s(\tilde{Z}), \qquad g_1 \in G_1$$

and

$$s(g_2 \tilde{Z}) = g_2(s(\tilde{Z})), \qquad g_2 \in G_2,$$

we have

$$\varphi(g\tilde{Z}) = h(s(g_1 g_2 \tilde{Z})) = h(s(g_2 \tilde{Z})) = h(g_2 s(\tilde{Z})) = h(s(\tilde{Z})).$$

It follows that φ is invariant. To show that φ is maximal invariant, assume $\varphi(\tilde{Z}_1) = \varphi(\tilde{Z}_2)$. A $g \in G$ must be found so that $g\tilde{Z}_1 = \tilde{Z}_2$. Since h is maximal invariant under G_2 and

$$h(s(\tilde{Z}_1)) = h(s(\tilde{Z}_2)),$$

there is a $g_2 \in G_2$ such that

$$g_2(s(\tilde{Z}_1)) = s(\tilde{Z}_2).$$

However,

$$g_2(s(\tilde{Z}_1)) = s(g_2 \tilde{Z}_1) = s(\tilde{Z}_2)$$

and s is maximal invariant under G_1 so there exists a g_1 such that

$$g_1 g_2 \tilde{Z}_1 = \tilde{Z}_2.$$

This completes the proof that φ, and hence $r_1, \ldots, r_t$, is a maximal invariant

—assuming claims (i) and (ii). The proof that $s: \mathfrak{Z} \to \mathfrak{S}_p^+$ is a maximal invariant is an easy application of Proposition 1.20 and is left to the reader. That $h: \mathfrak{S}_p^+ \to R^t$ is maximal invariant follows from an argument similar to that given in the proof of Proposition 10.1. □

The group action on $\mathfrak{Z}$ treated in Proposition 10.6 is suggested by the following considerations. Assuming that the observations $Z_1, \ldots, Z_n$ in R^p are uncorrelated random vectors and $\mathcal{L}(Z_i) = \mathcal{L}(Z_1)$ for $i = 1, \ldots, n$, it follows that

$$\mathcal{E}\tilde{Z} = e\mu'$$

and

$$\operatorname{Cov}\tilde{Z} = I_n \otimes \Sigma$$

where $\mu = \mathcal{E}Z_1$ and $\operatorname{Cov} Z_1 = \Sigma$. When $\tilde{Z}$ is transformed by $g = (\Gamma, c, C)$, we have

$$\mathcal{E}g\tilde{Z} = e(C\mu + c)'$$

and

$$\operatorname{Cov} g\tilde{Z} = I_n \otimes (C\Sigma C').$$

Thus the induced action of g on (μ, Σ) is exactly the group action considered in Proposition 10.1. The special structure of $\mathcal{E}\tilde{Z}$ and $\operatorname{Cov}\tilde{Z}$ is reflected by the fact that, for $g = (\Gamma, 0, I_p)$, we have $\mathcal{E}g\tilde{Z} = \mathcal{E}\tilde{Z}$ and $\operatorname{Cov} g\tilde{Z} = \operatorname{Cov}\tilde{Z}$.

10.3. SOME DISTRIBUTION THEORY

The distribution theory associated with the sample canonical correlation coefficients is, to say the least, rather complicated. Most of the results in this section are derived under the assumption of normality and the assumption that the population canonical correlations are zero. However, the distribution of the sample multiple correlation coefficient is given in the general case of a nonzero population multiple correlation coefficient.

Our first result is a generalization of Example 7.12. Let $Z_1, \ldots, Z_n$ be a random sample of vectors in R^p and partition Z_i as

$$Z_i = \begin{pmatrix} X_i \\ Y_i \end{pmatrix}, \qquad X_i \in R^q, \quad Y_i \in R^r.$$

Assume that Z_1 has a density on R^p given by

$$p(z|\mu, \Sigma) = |\Sigma|^{-1/2} f((z-\mu)'\Sigma^{-1}(z-\mu))$$

where f has been normalized so that

$$\int zz'f(z'z)\,dz = I_p.$$

Thus when the density of Z_1 is $p(\cdot|\mu, \Sigma)$, then

$$\mathcal{E}Z_1 = \mu, \qquad \operatorname{Cov} Z_1 = \Sigma.$$

Assuming that $n \geqslant p + 1$, the sample covariance matrix

$$S = \sum_1^n (Z_i - \bar{Z})(Z_i - \bar{Z})' = \begin{pmatrix} S_{11} & S_{12} \\ S_{21} & S_{22} \end{pmatrix}$$

is positive definite with probability one. Here S_{11} is $q \times q$, S_{22} is $r \times r$, and S_{12} is $q \times r$. Partitioning Σ as S is partitioned, we have

$$\Sigma = \begin{pmatrix} \Sigma_{11} & \Sigma_{12} \\ \Sigma_{21} & \Sigma_{22} \end{pmatrix}.$$

Thus the squared sample coefficients, $r_1^2 \geqslant \cdots \geqslant r_t^2$, are the t largest eigenvalues of $S_{11}^{-1}S_{12}S_{22}^{-1}S_{21}$ and the squared population coefficients, $\rho_1^2 \geqslant \cdots \geqslant \rho_t^2$, are the t largest eigenvalues of $\Sigma_{11}^{-1}\Sigma_{12}\Sigma_{22}^{-1}\Sigma_{21}$. In the present generality, an invariance argument is given to show that the joint distribution of $(r_1, \ldots, r_t)$ depends on (μ, Σ) only through $(\rho_1, \ldots, \rho_t)$. Consider the group G whose elements are $g = (C, c)$ where $c \in R^p$ and

$$C = \begin{pmatrix} A & 0 \\ 0 & B \end{pmatrix}, \qquad A \in Gl_q, \quad B \in Gl_r.$$

The action of G on R^p is

$$(C, c)z = Cz + c$$

and group composition is

$$(C_1, c_1)(C_2, c_2) = (C_1C_2, C_1c_2 + c_1).$$

The group action on the sample is

$$g(Z_1,\ldots, Z_n) = (gZ_1,\ldots, gZ_n).$$

With the induced group action on (μ, Σ) given by

$$g(\mu, \Sigma) = (g\mu, C\Sigma C')$$

where $g = (C, c)$, it is clear that the family of distributions of $(Z_1,\ldots, Z_n)$ that are indexed by elements of

$$\Theta = \{(\mu, \Sigma)|\mu \in R^p, \Sigma \in \mathbb{S}_p^+\}$$

is a G-invariant family of probability measures.

Proposition 10.7. The joint distribution of $(r_1,\ldots, r_t)$ depends on (μ, Σ) only through $(\rho_1,\ldots, \rho_t)$.

Proof. From Proposition 10.6, we know that $(r_1,\ldots, r_t)$ is a G-invariant function of $(Z_1,\ldots, Z_n)$. Thus the distribution of $(r_1,\ldots, r_t)$ will depend on the parameter $\theta = (\mu, \Sigma)$ only through a maximal invariant in the parameter space. However, Proposition 10.1 shows that $(\rho_1,\ldots, \rho_t)$ is a maximal invariant under the action of G on Θ. □

Before discussing the distribution of canonical correlation coefficients, even for $t = 1$, it is instructive to consider the bivariate correlation coefficient. Consider pairs of random variables (X_i, Y_i), $i = 1,\ldots, n$, and let $X \in R^n$ and $Y \in R^n$ have coordinates X_i and Y_i, $i = 1,\ldots, n$. With $e \in R^n$ being the vector of ones, $P_e = ee'/n$ and $Q_e = I - P_e$, the sample correlation coefficient is defined by

$$r = \left(\frac{Q_eY}{\|Q_eY\|}\right)'\frac{Q_eX}{\|Q_eX\|}.$$

The next result describes the distribution of r when (X_i, Y_i), $i = 1,\ldots, n$, is a random sample from a bivariate normal distribution.

Proposition 10.8. Suppose $(X_i, Y_i)' \in R^2$, $i = 1,\ldots, n$, are independent random vectors with

$$\mathcal{L}\begin{pmatrix} X_i \\ Y_i \end{pmatrix} = N(\mu, \Sigma), \qquad i = 1,\ldots, n$$

where $\mu \in R^2$ and

$$\Sigma = \begin{pmatrix} \sigma_{11} & \sigma_{12} \\ \sigma_{21} & \sigma_{22} \end{pmatrix}$$

is positive definite. Consider random variables (U_1, U_2, U_3) with:

(i) (U_1, U_2) independent of U_3.
(ii) $\mathcal{L}(U_3) = \chi^2_{n-2}$.
(iii) $\mathcal{L}(U_2) = \chi^2_{n-1}$.
(iv) $\mathcal{L}(U_1 | U_2) = N(\frac{\rho}{\sqrt{1-\rho^2}} U_2^{1/2}, 1)$.

where $\rho = \sigma_{12}/(\sigma_{11}\sigma_{22})^{1/2}$ is the correlation coefficient. Then we have

$$\mathcal{L}\left(\frac{r}{\sqrt{1-r^2}}\right) = \mathcal{L}\left(\frac{U_1}{U_3^{1/2}}\right).$$

Proof. The assumption of independence and normality implies that the matrix $(XY) \in \mathcal{L}_{2,n}$ has a distribution given by

$$\mathcal{L}(XY) = N(e\mu', I_n \otimes \Sigma).$$

It follows from Proposition 10.7 that we may assume, without loss of generality, that Σ has the form

$$\Sigma = \begin{pmatrix} 1 & \rho \\ \rho & 1 \end{pmatrix}.$$

When Σ has this form, the conditional distribution of X given Y is

$$\mathcal{L}(X|Y) = N\big((\mu_1 - \rho\mu_2)e + \rho Y, (1-\rho^2)I_n\big)$$

so

$$\mathcal{L}(Q_e X|Y) = N\big(\rho Q_e Y, (1-\rho^2)Q_e\big).$$

Now, let $v_1, \ldots, v_n$ be an orthonormal basis for R^n with $v_1 = e/\sqrt{n}$ and

$$v_2 = \frac{Q_e Y}{\|Q_e Y\|}.$$

Expressing $Q_e X$ in this basis leads to

$$Q_e X = \sum_2^n (v_i' Q_e X) v_i$$

since $Q_e e = 0$. Setting

$$\xi_i = \frac{v_i' Q_e X}{\sqrt{1 - \rho^2}}, \qquad i = 2, \dots, n,$$

it is easily seen that, conditional on Y, we have that $\xi_2, \dots, \xi_n$ are independent with

$$\mathcal{L}(\xi_2 | Y) = N\left(\rho(1 - \rho^2)^{-1/2} \|Q_e Y\|, 1\right)$$

and

$$\mathcal{L}(\xi_i | Y) = N(0, 1), \qquad i = 3, \dots, n.$$

Since

$$\|Q_e X\|^2 = \sum_2^n (v_i' Q_e X)^2 = (1 - \rho^2) \sum_2^n \xi_i^2,$$

the identity

$$r = \frac{\xi_2}{\sqrt{\xi_2^2 + \Sigma_3^n \xi_i^2}}$$

holds. This leads to

$$\frac{r}{\sqrt{1 - r^2}} = \frac{\xi_2}{\sqrt{\Sigma_3^n \xi_i^2}}.$$

Setting $U_1 = \xi_2$, $U_2 = \|Q_e Y\|^2$, and $U_3 = \Sigma_3^n \xi_i^2$ yields the assertion of the proposition. □

The result of this proposition has a couple of interesting consequences. When $\rho = 0$, then the statistic

$$W = \sqrt{n - 2}\, \frac{r}{\sqrt{1 - r^2}} = \sqrt{n - 2}\, \frac{U_1}{U_3^{1/2}}$$

has a Students t distribution with $n - 2$ degrees of freedom. In the general case, the distribution of W can be described by saying that: conditional on U_2, W has a noncentral t distribution with $n - 2$ degrees of freedom and noncentrality parameter

$$\delta = \frac{\rho}{\sqrt{1-\rho^2}} U_2^{1/2}$$

where $\mathcal{L}(U_2) = \chi^2_{n-1}$. Let $p_m(\cdot|\delta)$ denote the density function of a noncentral t distribution with m degrees of freedom and noncentrality parameter δ. The results in the Appendix show that $p_m(\cdot|\delta)$ has a monotone likelihood ratio. It is clear that the density of W is

$$h(w|\rho) = \int_0^\infty p_m\left(w|\left(\rho(1-\rho^2)^{-1/2}\right)u^{1/2}\right) f(u)\, du$$

where f is the density of U_2 and $m = n - 2$. From this representation and the results in the Appendix, it is not difficult to show that $h(\cdot|\rho)$ has a monotone likelihood ratio. The details of this are left to the reader.

In the case that the two random vectors X and Y in R^n are independent, the conditions under which W has a t_{n-2} distribution can be considerably weakened.

Proposition 10.9. Suppose X and Y in R^n are independent and both $\|Q_e X\|$ and $\|Q_e Y\|$ are positive with probability one. Also assume that, for some number $\mu_1 \in R$, the distribution of $X - \mu_1 e$ is orthogonally invariant. Under these assumptions, the distribution of

$$W = \sqrt{n-2}\,\frac{r}{\sqrt{1-r^2}}$$

where

$$r = \left(\frac{Q_e Y}{\|Q_e Y\|}\right)' \frac{Q_e X}{\|Q_e X\|}$$

is a t_{n-2} distribution.

Proof. The two random vectors $Q_e X$ and $Q_e Y$ take values in the $(n-1)$-dimensional subspace

$$M = \{x | x \in R^n, x'e = 0\}.$$

Fix Y so the vector

$$y \equiv \frac{Q_e Y}{\|Q_e Y\|} \in M$$

has length one. Since the distribution of $X - \mu_1 e$ is $\mathcal{O}_n$ invariant, it follows that the distribution of $Q_e X$ is invariant under the group

$$G = \{\Gamma | \Gamma \in \mathcal{O}_n, \Gamma e = e\},$$

which acts on M. Therefore, the distribution of $Q_e X / \|Q_e X\|$ is G-invariant on the set

$$\mathcal{X} = \{x | x \in M, \|x\| = 1\}.$$

But G is compact and acts transitively on $\mathcal{X}$ so there is a unique G-invariant distribution for $Q_e X / \|Q_e X\|$ in $\mathcal{X}$. From this uniqueness it follows that

$$\mathcal{L}\left(\frac{Q_e X}{\|Q_e X\|}\right) = \mathcal{L}\left(\frac{Q_e Z}{\|Q_e Z\|}\right)$$

where $\mathcal{L}(Z) = N(0, I_n)$ on R^n. Therefore, we have

$$\mathcal{L}(r) = \mathcal{L}\left(y' \frac{Q_e Z}{\|Q_e Z\|}\right),$$

and for each y, the claimed result follows from the argument given to prove Proposition 10.8. □

We now turn to the canonical correlation coefficients in the special case that $t = 1$. Consider random vectors X_i and Y_i with $X_i \in R^1$ and $Y_i \in R^r$, $i = 1, \ldots, n$. Let $X \in R^n$ have coordinates $X_1, \ldots, X_n$ and let $Y \in \mathcal{L}_{r,n}$ have rows $Y_1', \ldots, Y_n'$. Assume that $Q_e Y$ has rank r so

$$P \equiv Q_e Y[(Q_e Y)' Q_e Y]^{-1} (Q_e Y)'$$

is the orthogonal projection onto the subspace spanned by the columns of $Q_e Y$. Since $t = 1$, the canonical correlation coefficient is the square root of the largest, and only nonzero, eigenvalue of

$$\frac{(Q_e X)(Q_e X)'}{\|Q_e X\|^2} P,$$

which is

$$r_1^2 \equiv \frac{(Q_e X)'P(Q_e X)}{\|Q_e X\|^2} = \frac{\|PQ_e X\|^2}{\|Q_e X\|^2}.$$

For the case at hand, r_1 is commonly called the *multiple correlation coefficient*. The distribution of r_1^2 is described next under the assumption of normality.

Proposition 10.10. Assume that the distribution of $(XY) \in \mathcal{L}_{r+1,n}$ is given by

$$\mathcal{L}(XY) = N(e\mu', I_n \otimes \Sigma)$$

and partition Σ as

$$\Sigma = \begin{pmatrix} \sigma_{11} & \Sigma_{12} \\ \Sigma_{21} & \Sigma_{22} \end{pmatrix} : (r+1) \times (r+1)$$

where $\sigma_{11} > 0$, Σ_{12} is $1 \times r$, and Σ_{22} is $r \times r$. Consider random variables U_1, U_2, and U_3 whose joint distribution is specified by:

(i) (U_1, U_2) and U_3 are independent.
(ii) $\mathcal{L}(U_3) = \chi^2_{n-r-1}$.
(iii) $\mathcal{L}(U_2) = \chi^2_{n-1}$.
(iv) $\mathcal{L}(U_1|U_2) = \chi^2_r(\Delta)$, where $\Delta = \rho^2(1-\rho^2)^{-1}U_2$.

Here $\rho = (\Sigma_{12}\Sigma_{22}^{-1}\Sigma_{21}/\sigma_{11})^{1/2}$ is the population multiple correlation coefficient. Then we have

$$\mathcal{L}\left(\frac{r_1^2}{1-r_1^2}\right) = \mathcal{L}\left(\frac{U_1}{U_3}\right).$$

Proof. Combining the results of Proposition 10.1 and Proposition 5.7, without loss of generality, Σ can be assumed to have the form

$$\Sigma = \begin{pmatrix} 1 & \rho\varepsilon_1' \\ \rho\varepsilon_1 & I_r \end{pmatrix}$$

where $\varepsilon_1 \in R^r$ and $\varepsilon_1' = (1, 0, \dots, 0)$. When Σ has this form, the conditional distribution of X given Y is

$$\mathcal{L}(X|Y) = N\big((\mu_1 - \rho\mu_2'\varepsilon_1)e + \rho Y\varepsilon_1, (1-\rho^2)I_n\big)$$

where $\mathcal{E}X = \mu_1 e$ and $\mathcal{E}Y = e\mu_2'$. Since $Q_e e = 0$, we have

$$\mathcal{L}(Q_e X|Y) = N\big(\rho Q_e Y\varepsilon_1, (1-\rho^2)Q_e\big).$$

The subspace spanned by the columns of $Q_e Y$ is contained in the range of Q_e and this implies that $Q_e P = PQ_e = P$ so

$$\|Q_e X\|^2 = \|(Q_e - P)X\|^2 + \|PX\|^2 = \|(Q_e - P)Q_e X\|^2 + \|PQ_e X\|^2.$$

Since

$$r_1^2 = \frac{\|PQ_e X\|^2}{\|Q_e X\|^2},$$

it follows that

$$\frac{r_1^2}{1 - r_1^2} = \frac{\|PQ_e X\|^2}{\|(Q_e - P)Q_e X\|^2}.$$

Given Y, the conditional covariance of $Q_e X$ is $(1-\rho^2)Q_e$ and, therefore, the identity $PQ_e(Q_e - P) = 0$ implies that $PQ_e X$ and $(Q_e - P)Q_e X$ are conditionally independent. It is clear that

$$\mathcal{L}((Q_e - P)Q_e X|Y) = N\big(0, (1-\rho^2)(Q_e - P)\big),$$

so we have

$$\mathcal{L}\big(\|(Q_e - P)Q_e X\|^2|Y\big) = (1-\rho^2)\chi^2_{n-r-1}$$

since $Q_e - P$ is an orthogonal projection of rank $n - r - 1$. Again, conditioning on Y,

$$\mathcal{L}(PQ_e X|Y) = N\big(\rho Q_e Y\varepsilon_1, (1-\rho^2)P\big)$$

since $PQ_e = P$ and $Q_e Y\varepsilon_1$ is in the range of P. It follows from Proposition 3.8 that

$$\mathcal{L}\big(\|PQ_e X\|^2|Y\big) = (1-\rho^2)\chi_r^2(\Delta)$$

where the noncentrality parameter Δ is given by

$$\Delta = \frac{\rho^2}{1-\rho^2}\varepsilon_1' Y' Q_e Y \varepsilon_1.$$

That $U_2 \equiv \varepsilon_1' Y Q_e Y \varepsilon_1$ has a χ^2_{n-1} distribution is clear. Defining U_1 and U_3 by

$$U_1 = (1-\rho^2)^{-1}\|PQ_e X\|^2$$

and

$$U_3 = (1-\rho^2)^{-1}\|(Q_e - P)Q_e X\|^2,$$

the identity

$$\frac{r_1^2}{1-r_1^2} = \frac{U_1}{U_3}$$

holds. That U_3 is independent of (U_1, U_2) follows by conditioning on Y. Since

$$\mathcal{L}(U_1 | Y) = \chi_n(\Delta)$$

where

$$\Delta = \frac{\rho^2}{1-\rho^2}\varepsilon_1' Y' Q_e Y \varepsilon_1 = \frac{\rho^2}{1-\rho^2} U_2,$$

the conditional distribution of U_1 given Y is the same as the conditional distribution of U_1 given U_2. This completes the proof. □

When $\rho = 0$, Proposition 10.10 shows that

$$\mathcal{L}\left(\frac{r_1^2}{1-r_1^2}\right) = \mathcal{L}\left(\frac{\chi_r^2}{\chi_{n-r-1}^2}\right) = F_{r,\,n-r-1},$$

which is the unnormalized F-distribution on $(0, \infty)$. More generally,

$$\mathcal{L}\left(\frac{r_1^2}{1-r_1^2}\right) = F(r, n-r-1; \Delta)$$

where

$$\Delta = \frac{\rho^2}{1-\rho^2}\chi^2_{n-1}$$

is random. This means that, conditioning on $\Delta = \delta$,

$$\mathcal{L}\left(\frac{r_1^2}{1-r_1^2}\bigg|\delta\right) = F(r, n-r-1; \delta).$$

Let $f(\cdot|\delta)$ denote the density function of an $F(r, n-r-1; \delta)$ distribution, and let $h(\cdot)$ be the density of a χ^2_{n-1} distribution. Then the density of $r_1^2/(1-r_1^2)$ is

$$k(w|\rho) = \int_0^\infty f\left(w|\rho^2(1-\rho^2)^{-1}u\right)h(u)\,du.$$

From this representation, it can be shown, using the results in the Appendix, that $k(w|\rho)$ has a monotone likelihood ratio.

The final exact distributional result of this section concerns the function of the sample canonical correlations given by

$$W = \prod_{i=1}^{t}(1-r_i^2)$$

when the random sample $(X_i, Y_i)'$, $i = 1,\ldots, n$, is from a normal distribution and the population coefficients are all zero. This statistic arises in testing for independence, which is discussed in detail in the next section. To be precise, it is assumed that the random sample

$$Z_i = \begin{pmatrix} X_i \\ Y_i \end{pmatrix} \in R^p, \qquad i = 1,\ldots, n$$

satisfies

$$\mathcal{L}\begin{pmatrix} X_i \\ Y_i \end{pmatrix} = N(\mu, \Sigma).$$

As usual, $X_i \in R^q$, $Y_i \in R^r$, and the sample covariance matrix

$$S = \sum_1^n (Z_i - \bar{Z})(Z_i - \bar{Z})'$$

is partitioned as

$$S = \begin{pmatrix} S_{11} & S_{12} \\ S_{21} & S_{22} \end{pmatrix}$$

where S_{11} is $q \times q$ and S_{22} is $r \times r$. Under the assumptions made this far, S has a Wishart distribution—namely,

$$\mathcal{L}(S) = W(\Sigma, p, n-1).$$

Partitioning Σ, we have

$$\Sigma = \begin{pmatrix} \Sigma_{11} & \Sigma_{12} \\ \Sigma_{21} & \Sigma_{22} \end{pmatrix}$$

and the population canonical correlation coefficients, say $\rho_1 \geqslant \cdots \geqslant \rho_t$, are all zero iff $\Sigma_{12} = 0$.

Proposition 10.11. Assume $n - 1 \geqslant p$ and let $r_1 \geqslant \cdots \geqslant r_t$ be the sample canonical correlations. When $\Sigma_{12} = 0$, then

$$\mathcal{L}\left(\prod_1^t (1 - r_i^2)\right) = U(n - r - 1, r, q)$$

where the distribution $U(n - r - 1, r, q)$ is described in Proposition 8.14.

Proof. Since $r_1^2, \ldots, r_t^2$ are the t largest eigenvalues of

$$\Lambda(S) = S_{11}^{-1} S_{12} S_{22}^{-1} S_{21}$$

and the remaining $q - t$ eigenvalues of $\Lambda(S)$ are zero, it follows that

$$W \equiv \prod_{i-1}^t (1 - r_i^2) = |I_q - S_{11}^{-1} S_{12} S_{22}^{-1} S_{21}|.$$

Since W is a function of the sample canonical correlations and $\Sigma_{12} = 0$, Proposition 10.1 implies that we can take

$$\Sigma = \begin{pmatrix} I_q & 0 \\ 0 & I_r \end{pmatrix}$$

without loss of generality to find the distribution of W. Using properties of determinants, we have

$$W = |S_{11}^{-1}||S_{11} - S_{12}S_{22}^{-1}S_{21}| = \frac{|S_{11\cdot 2}|}{|S_{11\cdot 2} + S_{12}S_{22}^{-1}S_{21}|}.$$

Proposition 8.7 implies that

$$\mathcal{L}(S_{11\cdot 2}) = W(I_q, q, n - r - 1)$$

$$\mathcal{L}\left(S_{22}^{-1/2}S_{21}|S_{22}\right) = N(0, I_r \otimes I_q)$$

and $S_{11\cdot 2}$ and $S_{12}S_{22}^{-1}S_{21}$ are independent. Therefore,

$$\mathcal{L}\left(S_{12}S_{22}^{-1}S_{21}\right) = W(I_q, q, r)$$

and by definition, it follows that

$$\mathcal{L}(W) = U(n - r - 1, r, q). \qquad \square$$

Since

$$W = \frac{|S_{22\cdot 1}|}{|S_{22\cdot 1} + S_{21}S_{11}^{-1}S_{12}|},$$

the proof of Proposition 10.11 shows that $\mathcal{L}(W) = U(n - q - 1, q, r)$ so $U(n - q - 1, q, r) = U(n - r - 1, r, q)$ as long as $n - 1 \geqslant q + r$. Using the ideas in the proof of Proposition 8.15, the distribution of W can be derived when Σ_{12} has rank one—that is, when one population canonical correlation is positive and the rest are zero. The details of this are left to the reader.

We close this section with a discussion that provides some qualitative information about the distribution of $r_1 \geqslant \cdots \geqslant r_t$ when the data matrices $X \in \mathcal{L}_{q,n}$ and $Y \in \mathcal{L}_{r,n}$ are independent. As usual, let P_X and P_Y denote the orthogonal projections onto the column spaces of Q_eX and Q_eY. Then the sample canonical correlations are the t largest eigenvalues of P_YP_X—say

$$\varphi(P_YP_x) \equiv \begin{pmatrix} r_1 \\ \vdots \\ r_t \end{pmatrix} \in R^t.$$

It is assumed that Q_eX has rank q and Q_eY has rank r. Since the distribution of $\varphi(P_YP_X)$ is of interest, it is reasonable to investigate the distributional properties of the two random projections P_X and P_Y. Since X and Y are assumed to be independent, it suffices to focus our attention on P_X. First note that P_X is an orthogonal projection onto a q-dimensional subspace contained in

$$M = \{x | x \in R^n, x'e = 0\}.$$

Therefore, P_X is an element of

$$\mathcal{P}_{q,n}(e) = \left\{ P \middle| \begin{array}{c} P \text{ is an } n \times n \text{ rank } q \text{ orthogonal} \\ \text{projection, } Pe = 0 \end{array} \right\}$$

Furthermore, the space $\mathcal{P}_{q,n}(e)$ is a compact subset of R^{n^2} and is acted on by the compact group

$$\mathcal{O}_n(e) = \{\Gamma | \Gamma \in \mathcal{O}_n, \Gamma e = e\},$$

with the group action given by $P \to \Gamma P \Gamma'$. Since $\mathcal{O}_n(e)$ acts transitively on $\mathcal{P}_{q,n}(e)$, there is a unique $\mathcal{O}_n(e)$-invariant probability distribution on $\mathcal{P}_{q,n}(e)$. This is called the *uniform distribution* on $\mathcal{P}_{q,n}(e)$.

Proposition 10.12. If $\mathcal{L}(X) = \mathcal{L}(\Gamma X)$ for $\Gamma \in \mathcal{O}_n(e)$, then P_X has a uniform distribution on $\mathcal{P}_{q,n}(e)$.

Proof. It is readily verified that

$$P_{\Gamma X} = \Gamma P_X \Gamma', \qquad \Gamma \in \mathcal{O}_n(e).$$

Therefore, if $\mathcal{L}(\Gamma X) = \mathcal{L}(X)$, then

$$\mathcal{L}(P_X) = \mathcal{L}(\Gamma P_X \Gamma'),$$

which implies that the distribution $\mathcal{L}(P_X)$ on $\mathcal{P}_{q,n}(e)$ is $\mathcal{O}_n(e)$-invariant. The uniqueness of the uniform distribution on $\mathcal{P}_{q,n}(e)$ yields the result. □

When $\mathcal{L}(X) = N(e\mu_1', I_n \otimes \Sigma_{11})$, then $\mathcal{L}(X) = \mathcal{L}(\Gamma X)$ for $\Gamma \in \mathcal{O}_n(e)$, so Proposition 10.12 applies to this case. For any two $n \times n$ positive semidefinite matrices B_1 and B_2, define the function $\varphi(B_1 B_2)$ to be the vector of the t largest eigenvalues of $B_1 B_2$. In particular, $\varphi(P_Y P_X)$ is the vector of sample canonical correlations.

Proposition 10.13. Assume X and Y are independent, $\mathcal{L}(\Gamma X) = \mathcal{L}(X)$ for $\Gamma \in \mathcal{O}_n(e)$, $Q_e X$ has rank q, and $Q_e Y$ has rank r. Then

$$\mathcal{L}(\varphi(P_Y P_X)) = \mathcal{L}(\varphi(P_0 P_X))$$

where P_0 is any fixed rank r projection in $\mathcal{P}_{r,n}(e)$.

Proof. First note that

$$\varphi(P_Y \Gamma P_X \Gamma') = \varphi(\Gamma' P_Y \Gamma P_X)$$

since the eigenvalues of $P_Y \Gamma P_X \Gamma'$ are the same as the eigenvalues of $\Gamma' P_Y \Gamma P_X$. From Proposition 10.12, we have

$$\mathcal{L}(P_X) = \mathcal{L}(\Gamma P_X \Gamma'), \qquad \Gamma \in \mathcal{O}_n(e).$$

Conditioning on Y, the independence of X and Y implies that

$$\mathcal{L}(\varphi(P_Y P_X)|Y) = \mathcal{L}(\varphi(P_Y \Gamma P_X \Gamma')|Y) = \mathcal{L}(\varphi(\Gamma' P_Y \Gamma P_X)|Y)$$

for all $\Gamma \in \mathcal{O}_n(e)$. The group $\mathcal{O}_n(e)$ acts transitively on $\mathcal{P}_{r,n}(e)$, so for Y fixed, there exists a $\Gamma \in \mathcal{O}_n(e)$ such that $\Gamma' P_Y \Gamma = P_0$. Therefore, the equation

$$\mathcal{L}(\varphi(P_Y P_X)|Y) = \mathcal{L}(\varphi(P_0 P_X)|Y) = \mathcal{L}(\varphi(P_0 P_X))$$

holds for each Y since X and Y are independent. Averaging $\mathcal{L}(\varphi(P_Y P_X)|Y)$ over Y yields $\mathcal{L}(\varphi(P_Y P_X))$, which must then equal $\mathcal{L}(\varphi(P_0 P_X))$. This completes the proof. □

The preceeding result shows that $\mathcal{L}(\varphi(P_Y P_X))$ does not depend on the distribution of Y as long as X and Y are independent and $\mathcal{L}(X) = \mathcal{L}(\Gamma X)$ for $\Gamma \in \mathcal{O}_n(e)$. In this case, the distribution of $\varphi(P_Y P_X)$ can be derived under the assumption that $\mathcal{L}(X) = N(0, I_n \otimes I_q)$ and $\mathcal{L}(Y) = N(0, I_n \otimes I_r)$. Suppose that $q \leqslant r$ so $t = q$. Then $\mathcal{L}(\varphi(P_Y P_X))$ is the distribution of $r_1 \geqslant \cdots \geqslant r_q$ where $\lambda_i = r_i^2$, $i = 1, \ldots, q$, are the eigenvalues of $S_{11}^{-1} S_{12} S_{22}^{-1} S_{21}$ and

$$S = \begin{pmatrix} S_{11} & S_{12} \\ S_{21} & S_{22} \end{pmatrix}$$

is the sample covariance matrix. To find the distribution of $r_1, \ldots, r_q$, it would obviously suffice to find the distribution of $\gamma_i = 1 - \lambda_i$, $i = 1, \ldots, q$, which are the eigenvalues of

$$I_q - S_{11}^{-1}S_{12}S_{22}^{-1}S_{21} = (T_1 + T_2)^{-1}T_1$$

where

$$T_1 = S_{11} - S_{12}S_{22}^{-1}S_{21}; \qquad T_2 = S_{12}S_{22}^{-1}S_{21}.$$

It was shown in the proof of Proposition 10.11 that T_1 and T_2 are independent and

$$\mathcal{L}(T_1) = W(I_q, q, n - r - 1)$$

and

$$\mathcal{L}(T_2) = W(I_q, q, r).$$

Since the matrix

$$B = (T_1 + T_2)^{-1/2}T_1(T_1 + T_2)^{-1/2}$$

has the same eigenvalues as $(T_1 + T_2)^{-1}T_1$, it suffices to find the distribution of the eigenvalues of B. It is not too difficult to show (see the Problems at end of this chapter) that the density of B is

$$p(B) = \frac{\omega(n - r - 1, q)\omega(r, q)}{\omega(n - 1, q)}|B|^{(n-r-q-2)/2}|I_q - B|^{(r-q-1)/2}$$

with respect to Lebesgue measure dB restricted to the set

$$\mathfrak{X} = \{B | B \in \mathcal{S}_q^+, I_q - B \in \mathcal{S}_q^+\}.$$

Here, $\omega(\cdot, \cdot)$ is the Wishart constant defined in Example 5.1. Now, the ordered eigenvalues of B are a maximal invariant under the action of the group $\mathcal{O}_q$ on $\mathfrak{X}$ given by $B \to \Gamma B \Gamma'$, $\Gamma \in \mathcal{O}_q$. Let λ be the vector of ordered eigenvalues of B so $\lambda \in R^q$, $1 \geqslant \lambda_1 \geqslant \cdots \geqslant \lambda_q \geqslant 0$. Since $p(\Gamma B \Gamma') = p(B)$, $\Gamma \in \mathcal{O}_q$, it follows from Proposition 7.15 that the density of λ is $q(\lambda) = p(D_\lambda)$ where D_λ is a $q \times q$ diagonal matrix with diagonal entries $\lambda_1, \ldots, \lambda_q$. Of course, $q(\cdot)$ is the density of λ with respect to the measure $\nu(d\lambda)$ induced by the maximal invariant mapping. More precisely, let

$$\mathfrak{Z} = \{\lambda | \lambda \in R^q, 1 \geqslant \lambda_1 \geqslant \cdots \geqslant \lambda_q \geqslant 0\}$$

and consider the mapping φ on $\mathfrak{X}$ to $\mathfrak{Z}$ defined by $\varphi(B) = \lambda$ where λ is the vector of eigenvalues of B. For any Borel set $C \subseteq \mathfrak{Z}$, $\nu(C)$ is defined by

$$\nu(C) = \int_{\varphi^{-1}(C)} dB.$$

Since $q(\lambda)$ has been calculated, the only step left to determine the distribution of λ is to find the measure ν. However, it is rather nontrivial to find ν and the details are not given here. We have included the above argument to show that the only step in obtaining $\mathcal{L}(\lambda)$ that we have not solved is the calculation of ν. This completes our discussion of distributional problems associated with canonical correlations.

The measure ν above is just the restriction to $\mathfrak{Z}$ of the measure ν_2 discussed in Example 6.1. For one derivation of ν_2, see Muirhead (1982, p. 104).

10.4. TESTING FOR INDEPENDENCE

In this section, we consider the problem of testing for independence based on a random sample from a normal distribution. Again, let $Z_1, \ldots, Z_n$ be independent random vectors in R^p and partition Z_i as

$$Z_i = \begin{pmatrix} X_i \\ Y_i \end{pmatrix}, \qquad X_i \in R^q, \qquad Y_i \in R^r.$$

It is assumed that $\mathcal{L}(Z_i) = N(\mu, \Sigma)$, so

$$\text{Cov}(Z_i) = \Sigma = \begin{pmatrix} \Sigma_{11} & \Sigma_{12} \\ \Sigma_{21} & \Sigma_{22} \end{pmatrix} = \text{Cov}\begin{pmatrix} X_i \\ Y_i \end{pmatrix}$$

for $i = 1, \ldots, n$. The problem is to test the null hypothesis $H_0 : \Sigma_{12} = 0$ against the alternative $H_1 : \Sigma_{12} \neq 0$. As usual, let Z have rows Z_i', $i = 1, \ldots, n$ so $\mathcal{L}(Z) = N(e\mu', I_n \otimes \Sigma)$. Assuming $n \geqslant p + 1$, the set $\mathfrak{Z} \subseteq \mathcal{L}_{p,n}$ where

$$S \equiv (Z - e\bar{Z}')'(Z - e\bar{Z}') = \begin{pmatrix} S_{11} & S_{12} \\ S_{21} & S_{22} \end{pmatrix}$$

has rank p is a set of probability one and $\mathfrak{Z}$ is taken as the sample space for Z. The group G considered in Proposition 10.6 acts on $\mathfrak{Z}$ and a maximal invariant is the vector of canonical correlation coefficients $r_1, \ldots, r_t$ where $t = \min\{q, r\}$.

Proposition 10.14. The problem of testing $H_0 : \Sigma_{12} = 0$ versus $H_1 : \Sigma_{12} \neq 0$ is invariant under G. Every G-invariant test is a function of the sample canonical correlation coefficients $r_1, \dots, r_t$. When $t = 1$, the test that rejects for large values of r_1 is a uniformly most powerful invariant test.

Proof. That the testing problem is G-invariant is easily checked. From Proposition 10.6, the function mapping Z into $r_1, \dots, r_t$ is a maximal invariant so every invariant test is a function of $r_1, \dots, r_t$. When $t = 1$, the test that rejects for large values of r_1 is equivalent to the test that rejects for large values of $U \equiv r_1^2/(1 - r_1^2)$. It was argued in the last section (see Proposition 10.10) that the density of U, say $k(u|\rho)$, has a monotone likelihood ratio where ρ is the only nonzero population canonical correlation coefficient. Since the null hypothesis is that $\rho = 0$ and since every invariant test is a function of U, it follows that the test that rejects for large values of U is a uniformly most powerful invariant test.

When $t = 1$, the distribution of U is specified in Proposition 10.10, and this can be used to construct a test of level α for H_0. For example, if $q = 1$, then $\mathcal{L}(U) = F_{r, n-r-1}$ and a constant $c(\alpha)$ can be found from standard tables of the normalized $\mathcal{F}$-distribution such that, under H_0, $P\{U > c(\alpha)\} = \alpha$.

In the case that $t > 1$, there is no obvious function of $r_1, \dots, r_t$ that provides an optimum test of H_0 versus H_1. Intuitively, if some of the r_i's are "too big," there is reason to suspect that H_0 is not true. The likelihood ratio test provides one possible criterion for testing $\Sigma_{12} = 0$.

Proposition 10.15. The likelihood ratio test of H_0 versus H_1 rejects if the statistic

$$W = \prod_{i=1}^{t} \left(1 - r_i^2\right) = \frac{|S|}{|S_{11}||S_{22}|}$$

is too small. Under H_0, $\mathcal{L}(W) = U(n - r - 1, r, q)$, which is the distribution described in Proposition 8.14.

Proof. The density function of Z is

$$p(Z|\mu, \Sigma) = \left(\sqrt{2\pi}\right)^{-np} |\Sigma|^{-n/2} \exp\left[-\tfrac{1}{2} \operatorname{tr}(Z - e\mu')'(Z - e\mu')\Sigma^{-1}\right].$$

Under both H_0 and H_1, the maximum likelihood estimate of μ is $\hat{\mu} = \bar{Z}$. Under H_1, the maximum likelihood estimate of Σ is $\hat{\Sigma} = (1/n)S$. Partitioning S as Σ is partitioned, we have

$$S = \begin{pmatrix} S_{11} & S_{12} \\ S_{21} & S_{22} \end{pmatrix}$$

where S_{11} is $q \times q$, S_{12} is $q \times r$, and S_{22} is $r \times r$. Under H_0, Σ has the form

$$\Sigma = \begin{pmatrix} \Sigma_{11} & 0 \\ 0 & \Sigma_{22} \end{pmatrix}$$

so

$$\Sigma^{-1} = \begin{pmatrix} \Sigma_{11}^{-1} & 0 \\ 0 & \Sigma_{22}^{-1} \end{pmatrix}.$$

When Σ has this form,

$$\begin{aligned} p(Z|\hat{\mu}, \Sigma) &= (\sqrt{2\pi})^{-np}|\Sigma_{11}|^{-n/2}|\Sigma_{22}|^{-n/2}\exp\left[-\tfrac{1}{2}\operatorname{tr} S\Sigma^{-1}\right] \\ &= (\sqrt{2\pi})^{-np}|\Sigma_{11}|^{-n/2}\exp\left[-\tfrac{1}{2}\operatorname{tr} S_{11}\Sigma_{11}^{-1}\right]|\Sigma_{22}|^{-n/2} \\ &\quad \times \exp\left[-\tfrac{1}{2}\operatorname{tr} S_{22}\Sigma_{22}^{-1}\right]. \end{aligned}$$

From this it is clear that, under H_0, $\hat{\Sigma}_{11} = (1/n)S_{11}$ and $\hat{\Sigma}_{22} = (1/n)S_{22}$. Substituting these estimates into the densities under H_0 and H_1 leads to a likelihood ratio of

$$\Lambda(Z) = \left(\frac{|S_{11}||S_{22}|}{|S|}\right)^{-n/2}.$$

Rejecting H_0 for small values of $\Lambda(Z)$ is equivalent to rejecting for small values of

$$W = (\Lambda(Z))^{2/n} = \frac{|S|}{|S_{11}||S_{22}|}.$$

The identity $|S| = |S_{11}||S_{22} - S_{21}S_{11}^{-1}S_{12}|$ shows that

$$W = \frac{|S_{22} - S_{21}S_{11}^{-1}S_{21}|}{|S_{22}|} = |I_r - S_{22}^{-1}S_{21}S_{11}^{-1}S_{12}| = \prod_{i=1}^{t}(1 - r_i^2)$$

where $r_1^2, \ldots, r_t^2$ are the t largest eigenvalues of $S_{22}^{-1}S_{21}S_{11}^{-1}S_{12}$. Thus the likelihood ratio test is equivalent to the test that rejects for small values of W. That $\mathcal{L}(W) = U(n - r - 1, r, q)$ under H_0 follows from Proposition 10.11. □

The distribution of W under H_1 is quite complicated to describe except in the case that Σ_{12} has rank one. As mentioned in the last section, when Σ_{12} has rank one, the methods used in Proposition 8.15 yields a description of the distribution of W.

Rather than discuss possible alternatives to the likelihood test, in the next section we show that the testing problem above is a special case of the MANOVA testing problem considered in Chapter 9. Thus the alternatives to the likelihood ratio test for the MANOVA problem are also alternatives to the likelihood ratio test for independence.

We now turn to a slight generalization of the problem of testing that $\Sigma_{12} = 0$. Again suppose that $Z \in \mathfrak{L}$ satisfies $\mathcal{L}(Z) = N(e\mu', I_n \otimes \Sigma)$ where $\mu \in R^p$ and Σ are both unknown parameters and $n \geqslant p + 1$. Given an integer $k \geqslant 2$, let $p_1, \dots, p_k$ be positive integers such that $\Sigma_1^k p_i = p$. Partition Σ into blocks Σ_{ij} of dimension $p_i \times p_j$ for $i, j = 1, \dots, k$. We now discuss the likelihood ratio test for testing $H_0 : \Sigma_{ij} = 0$ for all i, j with $i \neq j$. For example, when $k = p$ and each $p_i = 1$, then the null hypothesis is that Σ is diagonal with unknown diagonal elements. By mimicking the proof of Proposition 10.15, it is not difficult to show that the likelihood ratio test for testing H_0 versus the alternative that Σ is completely unknown rejects for small values of

$$\Lambda = \frac{|S|}{\prod_{i=1}^{k} |S_{ii}|}.$$

Here, $S = (Z - e\bar{Z}')'(Z - e\bar{Z}')$ is partitioned into $S_{ij} : p_i \times p_j$ as Σ was partitioned. Further, for $i = 1, \dots, k$, define $S_{(ii)}$ by

$$S_{(ii)} = \begin{pmatrix} S_{ii} & S_{i(i+1)} & \cdots & S_{ik} \\ & & & \vdots \\ & & & S_{kk} \end{pmatrix}$$

so $S_{(ii)}$ is $(p_i + \cdots + p_k) \times (p_i + \cdots + p_k)$. Noting that $S_{(11)} = S$, we can write

$$\Lambda = \frac{|S|}{\prod_{i=1}^{k} |S_{ii}|} = \prod_{i=1}^{k-1} \frac{|S_{(ii)}|}{|S_{ii}||S_{(i+1, i+1)}|}.$$

Define W_i, $i = 1, \ldots, k-1$, by

$$W_i = \frac{|S_{(ii)}|}{|S_{ii}||S_{(i+1,i+1)}|}.$$

Under the null hypothesis, it follows from Proposition 10.11 that

$$\mathcal{L}(W_i) = U\left(n - 1 - \sum_{j=i+1}^{k} p_j, \sum_{j=i+1}^{k} p_j, p_i\right).$$

To derive the distribution of Λ under H_0, we now show that $W_1, \ldots, W_{k-1}$ are independent random variables under H_0. From this it follows that, under H_0,

$$\mathcal{L}(\Lambda) = \prod_{i=1}^{k-1} U\left(n - 1 - \sum_{j=i+1}^{k} p_j, \sum_{j=i+1}^{k} p_j, p_i\right)$$

so Λ is distributed as a product of independent beta random variables. The independence of $W_1, \ldots, W_{k-1}$ for a general k follows easily by induction once independence has been verified for $k = 3$.

For $k = 3$, we have

$$\Lambda = W_1 W_2 = \frac{|S|}{|S_{11}||S_{(22)}|} \frac{|S_{(22)}|}{|S_{22}||S_{33}|}$$

and, under H_0,

$$\mathcal{L}(S) = W(\Sigma, p, n-1)$$

where Σ has the form

$$\Sigma = \begin{pmatrix} \Sigma_{11} & 0 & 0 \\ 0 & \Sigma_{22} & 0 \\ 0 & 0 & \Sigma_{33} \end{pmatrix} = \begin{pmatrix} \Sigma_{11} & 0 \\ 0 & \Sigma_{(22)} \end{pmatrix}.$$

To show W_1 and W_2 are independent, Proposition 7.19 is applied. The sample space for S is $\mathbb{S}_p^+$—the space of $p \times p$ positive definite matrices. Fix Σ of the above form and let P_0 denote the probability measure of S so P_0 is the probability measure of a $W(\Sigma, p, n-1)$ distribution on $\mathbb{S}_p^+$. Consider the group G whose elements are (A, B) where $A \in Gl_{p_1}$ and $B \in Gl_{(p_2+p_3)}$

and the group composition is

$$(A_1, B_1)(A_2, B_2) = (A_1A_2, B_1B_2).$$

It is easy to show that the action

$$(A, B)[S] \equiv \begin{pmatrix} A & 0 \\ 0 & B \end{pmatrix} S \begin{pmatrix} A & 0 \\ 0 & B \end{pmatrix}'$$

defines a left action of G on $\mathbb{S}_p^+$. If $\mathcal{L}(S) = W(\Sigma, p, n-1)$, then

$$\mathcal{L}((A, B)[S]) = W((A, B)[\Sigma], p, n-1)$$

where

$$(A, B)[\Sigma] = \begin{pmatrix} A & 0 \\ 0 & B \end{pmatrix} \Sigma \begin{pmatrix} A & 0 \\ 0 & B \end{pmatrix}' = \begin{pmatrix} A\Sigma_{11}A_1 & 0 \\ 0 & B\Sigma_{(22)}B' \end{pmatrix}.$$

This last equality follows from the special form of Σ. The first thing to notice is that

$$W_1 = W_1(S) = \frac{|S|}{|S_{11}||S_{(22)}|}$$

is invariant under the action of G on $\mathbb{S}_p^+$. Also, because of the special form of Σ, the statistic

$$\tau(S) \equiv (S_{11}, S_{(22)}) \in \mathbb{S}_{p_1}^+ \times \mathbb{S}_{(p_2+p_3)}^+$$

is a sufficient statistic for the family of distributions $\{gP_0 | g \in G\}$. This follows from the factorization criterion applied to the family $\{gP_0 | g \in G\}$, which is the Wishart family

$$\left\{ W(\Sigma, p, n-1) | \Sigma = \begin{pmatrix} \gamma_{11} & 0 \\ 0 & \gamma_{22} \end{pmatrix}, \quad \gamma_{11} \in \mathbb{S}_{p_1}^+, \gamma_{22} \in \mathbb{S}_{(p_2+p_3)}^+ \right\}.$$

However, G acts transitively on $\mathbb{S}_{p_1}^+ \times \mathbb{S}_{(p_2+p_3)}^+$ in the obvious way:

$$(A, B)[S_1, S_2] \equiv (AS_1A', BS_2B')$$

for $[S_1, S_2] \in \mathbb{S}_{p_1}^+ \times \mathbb{S}_{(p_2+p_3)}^+$. Further, the sufficient statistic $\tau(S) \in \mathbb{S}_{p_1}^+ \times \mathbb{S}_{(p_2+p_3)}^+$ satisfies

$$\tau((A, B)[S]) = (A, B)[\tau(S)]$$

so $\tau(\cdot)$ is an equivariant function. It now follows from Proposition 7.19 that the invariant statistic $W_1(S)$ is independent of the sufficient statistic $\tau(S)$. But

$$W_2(S) = \frac{|S_{(22)}|}{|S_{22}||S_{33}|}$$

is a function of $S_{(22)}$ and so is a function of $\tau(S) = [S_{11}, S_{(22)}]$. Thus W_1 and W_2 are independent for each value of Σ in the null hypothesis. Summarizing, we have the following result.

Proposition 10.16. Assume $k = 3$ and Σ has the form specified under H_0. Then, under the action of the group G on both $\mathcal{S}_p^+$ and $\mathcal{S}_{p_1}^+ \times \mathcal{S}_{(p_2+p_3)}^+$, the invariant statistic

$$W_1(S) = \frac{|S|}{|S_{11}||S_{(22)}|}$$

and the equivariant statistic

$$\tau(S) = \left[S_{11}, S_{(22)}\right]$$

are independent. In particular, the statistic

$$W_2(S) = \frac{|S_{(22)}|}{|S_{22}||S_{33}|},$$

being a function of $\tau(S)$, is independent of W_1.

The application and interpretation of the previous paragraph for general k should be fairly clear. The details are briefly outlined. Under the null hypothesis that $\Sigma_{ij} = 0$ for $i, j = 1, \ldots, k$ and $i \neq j$, we want to describe the distribution of

$$\Lambda = \prod_{i=1}^{k-1} \frac{|S_{(ii)}|}{|S_{ii}||S_{(i+1,\,i+1)}|} = \prod_{i=1}^{k-1} W_i.$$

It was remarked earlier that each W_i is distributed as a product of independent beta random variables. To see that $W_1, \ldots, W_{k-1}$ are independent, Proposition 10.16 shows that

$$W_1 = \frac{|S|}{|S_{11}||S_{(22)}|}$$

and $S_{(22)}$ are independent. Since $(W_2, \dots, W_{k-1})$ is a function of $S_{(22)}$, W_1 and $(W_2, \dots, W_{k-1})$ are independent. Next, apply Proposition 10.16 to $S_{(22)}$ to conclude that

$$W_2 = \frac{|S_{(22)}|}{|S_{22}||S_{(33)}|}$$

and $S_{(33)}$ are independent. Since $(W_3, \dots, W_{k-1})$ is a function of $S_{(33)}$, W_2 and $(W_3, \dots, W_{k-1})$ are independent. The conclusion that $W_1, \dots, W_{k-1}$ are independent now follows easily. Thus the distribution of Λ under H_0 has been described.

To interpret the decomposition of Λ into the product $\prod_1^{k-1} W_i$, first consider the null hypothesis

$$H_0^{(1)} : \Sigma_{1j} = 0 \quad \text{for } j = 2, \dots, k.$$

An application of Proposition 10.15 shows that the likelihood ratio test of $H_0^{(1)}$ versus the alternative that Σ is unknown rejects for small values of

$$W_1 = \frac{|S|}{|S_{11}||S_{(22)}|}.$$

Assuming $H_0^{(1)}$ to be true, consider testing

$$H_0^{(2)} : \Sigma_{2j} = 0 \quad \text{for } j = 3, \dots, k$$

versus

$$H_1^{(2)} : \Sigma_{2j} \neq 0 \quad \text{for some } j = 3, \dots, k.$$

A minor variation of the proof of Proposition 10.15 yields a likelihood ratio test of $H_0^{(2)}$ versus $H_1^{(2)}$ (given $H_0^{(1)}$) that rejects for small values of

$$W_2 = \frac{|S_{(22)}|}{|S_{22}||S_{(33)}|}.$$

Proceeding by induction, assume null hypotheses $H_0^{(i)}$, $i = 1, \dots, m - 1$, to be true and consider testing

$$H_0^{(m)} : \Sigma_{mj} = 0, \qquad j = m + 1, \dots, k$$

versus

$$H_1^{(m)} : \Sigma_{mj} \neq 0 \quad \text{for some } j = m + 1, \dots, k.$$

Given the null hypotheses $H_0^{(i)}$, $i = 1, \ldots, m - 1$, the likelihood ratio test of $H_0^{(m)}$ versus $H_1^{(m)}$ rejects for small values of

$$W_m = \frac{|S_{(mm)}|}{|S_{mm}||S_{(m+1,\, m+1)}|}.$$

The overall likelihood ratio test is one possible way of combining the likelihood ratio tests of $H_0^{(m)}$ versus $H_1^{(m)}$, given that $H_0^{(i)}$, $i = 1, \ldots, m - 1$, is true.

10.5. MULTIVARIATE REGRESSION

The purpose of this section is to show that testing for independence can be viewed as a special case of the general MANOVA testing problem treated in Chapter 9. In fact, the results below extend those of the previous section by allowing a more general mean structure for the observations. In the notation of the previous section, consider a data matrix $Z: n \times p$ that is partitioned as $Z = (XY)$ where X is $n \times q$ and Y is $n \times r$ so $p = q + r$. It is assumed that

$$\mathcal{L}(Z) = N(TB, I_n \otimes \Sigma)$$

where T is an $n \times k$ known matrix of rank k and B is a $k \times p$ matrix of unknown parameters. As usual, Σ is a $p \times p$ positive definite matrix. This is precisely the linear model discussed in Section 9.1 and clearly includes the model of previous sections of this chapter as a special case.

To test that X and Y are independent, it is illuminating to first calculate the conditional distribution of Y given X. Partition the matrix B as $B = (B_1 B_2)$ where B_1 is $k \times q$ and B_2 is $k \times r$. In describing the conditional distribution of Y given X, say $\mathcal{L}(Y|X)$, the notation

$$\Sigma_{22\cdot1} \equiv \Sigma_{22} - \Sigma_{21}\Sigma_{11}^{-1}\Sigma_{12}$$

is used. Following Example 3.1, we have

$$\begin{aligned}\mathcal{L}(Y|X) &= N\big(TB_2 + \big(I_n \otimes \Sigma_{21}\Sigma_{11}^{-1}\big)(X - TB_1), I_n \otimes \Sigma_{22\cdot1}\big)\\ &= N\big(T\big(B_2 - B_1\Sigma_{11}^{-1}\Sigma_{12}\big) + X\Sigma_{11}^{-1}\Sigma_{12}, I_n \otimes \Sigma_{22\cdot1}\big)\end{aligned}$$

and the marginal distribution of X is

$$\mathcal{L}(X) = N(TB_1, I_n \otimes \Sigma_{11}).$$

Let W be the $n \times (q + k)$ matrix (XT) and let C be the $(q + k) \times r$ matrix of parameters

$$C = \begin{pmatrix} C_1 \\ C_2 \end{pmatrix} \equiv \begin{pmatrix} \Sigma_{11}^{-1}\Sigma_{12} \\ B_2 - B_1\Sigma_{11}^{-1}\Sigma_{12} \end{pmatrix}$$

so

$$X\Sigma_{11}^{-1}\Sigma_{12} + T(B_2 - B_1\Sigma_{11}^{-1}\Sigma_{12}) = (XT)\begin{pmatrix} C_1 \\ C_2 \end{pmatrix} = WC.$$

In this notation, we have

$$\mathcal{L}(Y|X) = N(WC, I_n \otimes \Sigma_{22\cdot 1})$$

and

$$\mathcal{L}(X) = N(TB_1, I_n \otimes \Sigma_{11}).$$

Assuming $n \geqslant p + k$, the matrix W has rank $q + k$ with probability one so the conditional model for Y is of the MANOVA type. Further, testing $H_0 : \Sigma_{12} = 0$ versus $H_1 : \Sigma_{12} \neq 0$ is equivalent to testing $\tilde{H}_0 : C_1 = 0$ versus $\tilde{H}_1 : C_1 \neq 0$. In other words, based on the model for Z,

$$\mathcal{L}(Z) = N(TB, I_n \otimes \Sigma),$$

the null hypothesis concerns the covariance matrix. But in terms of the conditional model, the null hypothesis concerns the matrix of regression parameters.

With the above discussion and models in mind, we now want to discuss various approaches to testing H_0 and $\tilde{H}_0$. In terms of the model

$$\mathcal{L}(Z) = N(TB, I_n \otimes \Sigma)$$

and assuming H_1, the maximum likelihood estimators of B and Σ are

$$\hat{B} = (T'T)^{-1}T'Z, \qquad \hat{\Sigma} = \frac{1}{n}S$$

where

$$S = (Z - T\hat{B})'(Z - T\hat{B}),$$

so

$$\mathcal{L}(S) = W(\Sigma, p, n-k).$$

Under H_0, the maximum likelihood estimator of B is still $\hat{B}$ as above and since

$$\Sigma = \begin{pmatrix} \Sigma_{11} & 0 \\ 0 & \Sigma_{22} \end{pmatrix},$$

it is readily verified that

$$\hat{\Sigma}_{ii} = \frac{1}{n} S_{ii}, \qquad i = 1, 2$$

where

$$S = \begin{pmatrix} S_{11} & S_{12} \\ S_{21} & S_{22} \end{pmatrix}.$$

Substituting these estimators into the density of Z under H_0 and H_1 demonstrates that the likelihood ratio test rejects for small values of

$$\Lambda(Z) = \frac{|S|}{|S_{11}||S_{22}|}.$$

Under H_0, the proof of Proposition 10.11 shows that the distribution of $\Lambda(Z)$ is $U(n-k-r, r, q)$ as described in Proposition 8.14. Of course, symmetry in r and q implies that $U(n-k-r, r, q) = U(n-k-q, q, r)$. An alternative derivation of this likelihood ratio test can be given using the conditional distribution of Y given X and the marginal distribution of X. This follows from two facts: (i) the density of Z is proportional to the conditional density of Y given X multiplied by the marginal density of X, and (ii) the relabeling of the parameters is one-to-one—namely, the mapping from (B, Σ) to $(C, B_1, \Sigma_{11}, \Sigma_{22\cdot 1})$ is a one-to-one onto mapping of $\mathcal{L}_{p,k} \times \mathcal{S}_p^+$ to $\mathcal{L}_{r,(q+k)} \times \mathcal{L}_{q,k} \times \mathcal{S}_q^+ \times \mathcal{S}_r^+$. We now turn to the likelihood ratio test of $\tilde{H}_0$ versus $\tilde{H}_1$ based on the conditional model

$$\mathcal{L}(Y|X) = N(WC, I_n \otimes \Sigma_{22\cdot 1})$$

where X is treated as fixed. With X fixed, testing $\tilde{H}_0$ versus $\tilde{H}_1$ is a special case of the MANOVA testing problem and the results in Chapter 9 are directly applicable. To express $\tilde{H}_0$ in the MANOVA testing problem form, let K be the $q \times (q + k)$ matrix $K = (I_q\ 0)$, so the null hypothesis $\tilde{H}_0$ is

$$\tilde{H}_0 : KC = 0.$$

Recall that

$$\hat{C} \equiv (W'W)^{-1}W'Y$$

is the maximum likelihood estimator of C under $\tilde{H}_1$. Let $P_W = W(W'W)^{-1}W'$ denote the orthogonal projection onto the column space of W, let $Q_W = I_n - P_W$, and define $V \in \mathbb{S}_r^+$ by

$$V \equiv Y'Q_WY = (Y - W\hat{C})'(Y - W\hat{C}).$$

As shown in Section 9.1, based on the model

$$\mathcal{L}(Y|X) = N(WC, I_n \otimes \Sigma_{22\cdot 1}),$$

the likelihood ratio test of $\tilde{H}_0 : KC = 0$ versus $\tilde{H}_1 : KC \neq 0$ rejects H_0 for small values of

$$\Lambda_1(Y) \equiv \frac{|V|}{|V + (K\hat{C})'\left(K(W'W)^{-1}K'\right)^{-1}(K\hat{C})|}.$$

For each fixed X, Proposition 9.1 shows that under H_0, the distribution of $\Lambda_1(Y)$ is $U(n - q - k, q, r)$, which is the distribution (unconditional) of $\Lambda(Z)$ under H_0. In fact, much more is true.

Proposition 10.17. In the notation above:

(i) $V = S_{22\cdot 1}$.
(ii) $(K\hat{C})'(K(W'W)^{-1}K')^{-1}(K\hat{C}) = S_{21}S_{11}^{-1}S_{12}$.
(iii) $\Lambda_1(Y) = \Lambda(Z)$.

Further, under H_0, the conditional (given X) and unconditional distribution of $\Lambda_1(Y)$ and $\Lambda(Z)$ are the same.

Proof. To establish (i), first write S as

$$S = (Z - T\hat{B})'(Z - T\hat{B}) = Z'(I - P_T)Z$$

where $P_T = T(T'T)^{-1}T'$ is the orthogonal projection onto the column space of T. Setting $Q_T = I - P_T$ and writing $Z = (XY)$, we have

$$S = Z'Q_TZ = \begin{pmatrix} X' \\ Y' \end{pmatrix} Q_T(XY) = \begin{pmatrix} X'Q_TX & X'Q_TY \\ Y'Q_TX & Y'Q_TY \end{pmatrix} = \begin{pmatrix} S_{11} & S_{12} \\ S_{21} & S_{22} \end{pmatrix}.$$

This yields the identity

$$S_{22\cdot 1} = Y'Q_TY - Y'Q_TX(X'Q_TX)^{-1}X'Q_TY = Y'(I - P_T)Y - Y'P_0Y$$

where $P_0 = Q_TX(X'Q_TX)^{-1}X'Q_T$ is the orthogonal projection onto the column space of Q_TX. However, a bit of reflection reveals that $P_0 = P_W - P_T$ so

$$S_{22\cdot 1} = Y'(I - P_T)Y - Y'(P_W - P_T)Y = Y'(I - P_W)Y = Y'Q_WY = V.$$

This establishes assertion (i). For (ii), we have

$$S_{21}S_{11}^{-1}S_{12} = Y'P_0Y$$

and

$$\begin{aligned}(K\hat{C})'\left(K(W'W)^{-1}K'\right)^{-1}K\hat{C} &= Y'W(W'W)^{-1}K'\left(K(W'W)^{-1}W'W(W'W)^{-1}K'\right)^{-1} \\ &\quad \times K(W'W)^{-1}W'Y \\ &= Y'U(U'U)^{-1}U'Y \equiv Y'P_UY\end{aligned}$$

where $U \equiv W(W'W)^{-1}K'$ and P_U is the orthogonal projection onto the column space of U. Thus it must be shown that $P_U = P_0$ or, equivalently, that the column space of U is the same as the column space of Q_TX. Since $W = (XT)$, the relationship

$$W'U = W'W(W'W)^{-1}K' = K' = \begin{pmatrix} I_q \\ 0 \end{pmatrix}$$

proves that the q columns of U are orthogonal to the k columns of T. Thus the columns of U span a q-dimensional subspace contained in the column space of W and orthogonal to the column space of T. But there is only one subspace with these properties. Since the column space of $Q_T X$ also has these properties, it follows that $P_U = P_0$ so (ii) holds. Relationship (iii) is a consequence of (i), (ii), and

$$\Lambda(Z) = \frac{|S|}{|S_{11}||S_{22}|} = \frac{|S_{22\cdot 1}|}{|S_{22\cdot 1} + S_{21}S_{11}^{-1}S_{12}|}.$$

The validity of the final assertion concerning the distribution of $\Lambda_1(Y)$ and $\Lambda(Z)$ was established earlier. □

The results of Proposition 10.17 establish the connection between testing for independence and the MANOVA testing problem. Further, under H_0, the conditional distribution of $\Lambda_1(Y)$ is $U(n-q-k, q, r)$ for each value of X, so the marginal distribution of X is irrelevant. In other words, as long as the conditional model for Y given X is valid, we can test $\tilde{H}_0$ using the likelihood ratio test and under H_0, the distribution of the test statistic does not depend on the value of X. Of course, this implies that the conditional (given X) distribution of $\Lambda(Z)$ is the same as the unconditional distribution of $\Lambda(Z)$ under H_0. However, under H_1, the conditional and unconditional distributions of $\Lambda(Z)$ are not the same.

PROBLEMS

1. Given positive integers t, q, and r with $t \leqslant q, r$, consider random vectors $U \in R^t$, $V \in R^q$, and $W \in R^r$ where $\text{Cov}(U) = I_t$ and U, V, and W are uncorrelated. For $A: q \times t$ and $B: r \times t$, construct $X = AU + V$ and $Y = BU + W$.
   (i) With $\Lambda_{11} = \text{Cov}(V)$ and $\Lambda_{22} = \text{Cov}(W)$, show that
   $$\text{Cov}(X) = AA' + \Lambda_{11}$$
   $$\text{Cov}(Y) = BB' + \Lambda_{22}$$
   and the cross covariance between X and Y is AB'. Conclude that the number of nonzero population canonical correlations between X and Y is at most t.
   (ii) Conversely, given $\tilde{X} \in R^q$ and $\tilde{Y} \in R^r$ with t nonzero population canonical correlations, construct U, V, W, A, and B as above so that $X = AU + V$ and $Y = BU + W$ have the same joint covariance as $\tilde{X}$ and $\tilde{Y}$.

2. Consider $X \in R^q$ and $Y \in R^r$ and assume that $\text{Cov}(X) = \Sigma_{11}$ and $\text{Cov}(Y) = \Sigma_{22}$ exist. Let Σ_{12} be the cross covariance of X with Y. Recall that $\mathcal{P}_n$ denotes the group of $n \times n$ permutation matrices.

(i) If $g\Sigma_{12}h = \Sigma_{12}$ for all $g \in \mathcal{P}_q$ and $h \in \mathcal{P}_r$, show that $\Sigma_{12} = \delta e_1 e_2'$ for some $\delta \in R^1$ where e_1 (e_2) is the vector of ones in R^q (R^r).

(ii) Under the assumptions in (i), show that there is at most one nonzero canonical correlation and it is $|\delta|(e_1'\Sigma_{11}^{-1}e_1)^{1/2}(e_2'\Sigma_{22}^{-1}e_2)^{1/2}$. What is a set of canonical coordinates?

3. Consider $X \in R^p$ with $\text{Cov}(X) = \Sigma > 0$ (for simplicity, assume $\mathcal{E}X = 0$). This problem has to do with the approximation of X by a lower dimensional random vector—say $Y = BX$ where B is a $t \times p$ matrix of rank t.

(i) In the notation of Proposition 10.4, suppose $A_0 : p \times p$ is used to define the inner product $[\cdot, \cdot]$ on R^n and prediction error is measured by $\mathcal{E}\|X - CY\|^2$ where $\|\cdot\|$ is defined by $[\cdot, \cdot]$ and C is $p \times t$. Show that the minimum prediction error (B fixed) is

$$\delta(B) = \text{tr}\, A_0\big(\Sigma - \Sigma B'(B\Sigma B')^{-1}B\Sigma\big)$$

and the minimum is achieved for $C = \hat{C} = \Sigma B(B\Sigma B')^{-1}$.

(ii) Let $A = \Sigma^{1/2}A_0\Sigma^{1/2}$ and write A in spectral form as $A = \Sigma_1^p \lambda_i a_i a_i'$ where $\lambda_1 \geqslant \cdots \geqslant \lambda_p > 0$ and $a_1, \ldots, a_p$ is an orthonormal basis for R^p. Show that $\delta(B) = \text{tr}\, A(I - Q(B))$ where $Q(B) = \Sigma^{1/2}B'(B\Sigma B')^{-1}B\Sigma^{1/2}$ is a rank t orthogonal projection. Using this, show that $\delta(B)$ is minimized by choosing $Q = \hat{Q} = \Sigma_1^t a_i a_i'$, and the minimum is $\Sigma_{t+1}^p \lambda_i$. What is a corresponding $\hat{B}$ and $\hat{X} = \hat{C}\hat{B}X$ that gives the minimum? Show that $\hat{X} = \hat{C}\hat{B}X = \Sigma^{1/2}\hat{Q}\Sigma^{-1/2}X$.

(iii) In the special case that $A_0 = I_p$, show that

$$\hat{X} = \sum_{i=1}^{t} (a_i'X)a_i$$

where $a_1, \ldots, a_p$ are the eigenvectors of Σ and $\Sigma a_i = \lambda_i a_i$ with $\lambda_1 \geqslant \cdots \geqslant \lambda_p$. (The random variables $a_i'X$ are often called the *principal components* of X, $i = 1, \ldots, p$. It is easily verified that $\text{cov}(a_i'X, a_j'X) = \delta_{ij}\lambda_i$.)

4. In R^p, consider a translated subspace $M + a_0$ where $a_0 \in R^p$—such a set in R^p is called a *flat* and the dimension of the flat is the dimension of M.

(i) Given any flat $M + a_0$, show that $M + a_0 = M + b_0$ for some *unique* $b_0 \in M^{\perp}$.

Consider a flat $M + a_0$, and define the orthogonal projection onto $M + a_0$ by $x \to P(x - a_0) + a_0$ where P is the orthogonal projection onto M. Given n points $x_1, \ldots, x_n$ in R^p, consider the problem of finding the "closest" k-dimensional flat $M + a_0$ to the n points. As a measure of distance of the n points from $M + a_0$, we use $\Delta(M, a_0) = \Sigma_1^n \|x_i - \hat{x}_i\|^2$ where $\|\cdot\|$ is the usual Euclidean norm and $\hat{x}_i = P(x_i - a_0) + a_0$ is the projection of x_i onto $M + a_0$. The problem is to find M and a_0 to minimize $\Delta(M, a_0)$ over all k-dimensional subspaces M and all a_0.

(ii) First, regard a_0 as fixed, and set $S(b) = \Sigma_1^n (x_i - b)(x_i - b)'$ for any $b \in R^p$. With $Q = I - P$, show that $\Delta(M, a_0) = \operatorname{tr} S(a_0)Q = \operatorname{tr} S(\bar{x})Q + n(a_0 - \bar{x})'Q(a_0 - \bar{x})$ where $\bar{x} = n^{-1}\Sigma_1^n x_i$.

(iii) Write $S(\bar{x}) = \Sigma_1^p \lambda_i v_i v_i'$ in spectral form where $\lambda_1 \geqslant \cdots \geqslant \lambda_p \geqslant 0$ and $v_1, \ldots, v_p$ is an orthonormal basis for R^p. Using (ii), show that $\Delta(M, a_0) \geqslant \Sigma_{k+1}^p \lambda_i$ with equality for $z_0 = \bar{x}$ and for $M = \operatorname{span}\{v_1, \ldots, v_k\}$.

5. Consider a sample covariance matrix

$$S = \begin{pmatrix} S_{11} & S_{12} \\ S_{21} & S_{22} \end{pmatrix}$$

with $S_{ii} > 0$ for $i = 1, 2$. With $t = \min\{\dim S_{ii}, i = 1, 2\}$, show that the t sample canonical correlations are the t largest solutions (in λ) to the equation $|S_{12}S_{22}^{-1}S_{21} - \lambda^2 S_{11}| = 0$, $\lambda \in [0, \infty)$.

6. (The Eckhart–Young Theorem, 1936.) Given a matrix $A : n \times p$ (say $n \geqslant p$), let $k \leqslant p$. The problem is to find a matrix $B : n \times p$ of rank no greater than k that is "closest" to A in the usual trace inner product on $\mathcal{L}_{p,n}$. Let $\mathcal{B}_k$ be all the $n \times p$ matrices of rank no larger than k, so the problem is to find

$$\inf_{B \in \mathcal{B}_k} \|A - B\|^2$$

where $\|M\|^2 = \operatorname{tr} MM'$ for $M \in \mathcal{L}_{p,n}$.

(i) Show that every $B \in \mathcal{B}_k$ can be written ψC where ψ is $n \times k$, $\psi'\psi = I_k$, and C is $k \times p$. Conversely, $\psi C \in \mathcal{B}_k$, for each such ψ and C.

(ii) Using the results of Example 4.4, show that, for A and ψ fixed,

$$\inf_{C \in \mathcal{L}_{p,k}} \|A - \psi C\|^2 = \|A - \psi\psi' A\|^2.$$

(iii) With $Q = I - \psi\psi'$, Q is a rank $n - k$ orthogonal projection. Show that, for each $B \in \mathcal{B}_k$,

$$\|A - B\|^2 \geqslant \inf_Q \|AQ\|^2 = \inf_Q \operatorname{tr} QAA' = \Sigma_{k+1}^p \lambda_i^2$$

where $\lambda_1 \geqslant \cdots \geqslant \lambda_p$ are the singular values of A. Here Q ranges over all rank $n - k$ orthogonal projections.

(iv) Write $A = \Sigma_1^p \lambda_i u_i v_i'$ as the singular value decomposition for A. Show that $\hat{B} = \Sigma_1^k \lambda_i u_i v_i'$ achieves the infimum of part (iii).

7. In the case of a random sample from a bivariate normal distribution $N(\mu, \Sigma)$, use Proposition 10.8 and Karlin's Lemma in the Appendix to show that the density of $W = \sqrt{n-2}\, r(1 - r^2)^{-1/2}$ (r is the sample correlation coefficient) has a monotone likelihood ratio in $\theta = \rho(1 - \rho^2)^{-1/2}$. Conclude that the density of r has a monotone likelihood ratio in ρ.

8. Let $f_{p,q}$ denote the density function on $(0, \infty)$ of an unnormalized $F_{p,q}$ random variable. Under the assumptions of Proposition 10.10, show that the distribution of $W = r_1^2(1 - r_1^2)^{-1}$ has a density given by

$$f(w|\rho) = \sum_{k=1}^{\infty} f_{r+2k,\, n-r-1}(w) h(k|\rho)$$

where

$$h(k|\rho) = \frac{(1-\rho^2)^{(n-1)/2}\Gamma((n-1)/2 + k)}{k!\Gamma((n-1)/2)}(\rho^2)^k,$$

$$k = 0, 1, \ldots .$$

Note that $h(\cdot|\rho)$ is the probability mass function of a negative binomial distribution, so $f(w|\rho)$ is a mixture of F distributions. Show that $f(\cdot|\rho)$ has a monotone likelihood ratio.

9. (A generalization of Proposition 10.12.) Consider the space R^n and an integer k with $1 \leqslant k < n$. Fix an orthogonal projection P of rank k, and for $s \leqslant n - k$, let $\mathcal{P}_s$ be the set of all $n \times n$ orthogonal projections R of rank s that satisfy $RP = 0$. Also, consider the group $\mathcal{O}(P) = \{\Gamma | \Gamma \in \mathcal{O}_n, \Gamma P = P\Gamma\}$.

(i) Show that the group $\mathcal{O}(P)$ acts transitively on $\mathcal{P}_s$ under the action $R \to \Gamma R \Gamma'$.

(ii) Argue that there is a unique $\mathcal{O}(P)$ invariant probability distribution on $\mathcal{P}_s$.

(iii) Let Δ have a uniform distribution on $\mathcal{O}(P)$ and fix $R_0 \in \mathcal{P}_s$. Show that $\Delta R_0 \Delta'$ has the unique $\mathcal{O}(P)$ invariant distribution on $\mathcal{P}_s$.

10. Suppose $Z \in \mathcal{L}_{p,n}$ has an $\mathcal{O}_n$-left invariant distribution and has rank p with probability one. Let Q be a rank $n-k$ orthogonal projection with $p+k \leqslant n$ and form $W = QZ$.

(i) Show that W has rank p with probability one.

(ii) Show that $R = W(W'W)^{-1}W$ has the uniform distribution on $\mathcal{P}_p$ (in the notation of Problem 9 above with $P = I - Q$ and $s = p$).

11. After the proof of Proposition 10.13, it was argued that, when $q \leqslant r$, to find the distribution of $r_1 \geqslant \cdots \geqslant r_q$, it suffices to find the distribution of the eigenvalues of the matrix $B = (T_1 + T_2)^{-1/2}T_1(T_1 + T_2)^{-1/2}$ where T_1 and T_2 are independent with $\mathcal{L}(T_1) = W(I_q, q, n-r-1)$ and $\mathcal{L}(T_2) = W(I_q, q, r)$. It is assumed that $q \leqslant n - r - 1$. Let $f(\cdot|m)$ denote the density function of the $W(I_q, q, m)$ distribution $(m \geqslant q)$ with respect to Lebesgue measure dS on $\mathcal{S}_q$. Thus $f(S|m) = \omega(m,q)|S|^{(m-q-1)/2}\exp[-\frac{1}{2}\operatorname{tr} S]I(S)$ where

$$I(S) = \begin{cases} 1 & \text{if } S > 0 \\ 0 & \text{otherwise} \end{cases}.$$

(i) With $W_1 = T_1$ and $W_2 = T_1 + T_2$, show that the joint density of W_1 and W_2 with respect to $dW_1\,dW_2$ is $f(W_1|n-r-1)f(W_2 - W_1|r)$.

(ii) On the set where $W_1 > 0$ and $W_2 > 0$, define $B = W_2^{-1/2}W_1W_2^{-1/2}$ and $W_2 = V$. Using Proposition 5.11, show that the Jacobian of this transformation is $|\det V|^{(q+1)/2}$. Show that the joint density of B and V on the set where $B > 0$ and $V > 0$ is given by

$$f\left(V^{1/2}BV^{1/2}|n-r-1\right)f\left(V^{1/2}(I-B)V^{1/2}|r\right)|\det V|^{(q+1)/2}.$$

(iii) Now, integrate out V to show that the density of B on the set $0 < B < I_q$ is

$$p(B) = \frac{\omega(n-r-1,q)\omega(r,q)}{\omega(n-1,q)} \times |B|^{(n-r-q-2)/2}|I_q - B|^{(r-q-1)/2}.$$

12. Suppose the random orthogonal transformation Γ has a uniform distribution on $\mathcal{O}_n$. Let Δ be the upper left-hand $k \times p$ block of Γ and assume $p \leqslant k$. Under the additional assumption that $p \leqslant n - k$, the following argument shows that Δ has a density with respect to Lebesgue measure on $\mathcal{L}_{p,k}$.

(i) Let $\psi: n \times p$ consist of the first p columns of Γ so $\Delta: k \times p$ has rows that are the first k rows of ψ. Show that ψ has a uniform distribution on $\mathcal{F}_{p,n}$. Conclude that ψ has the same distribution as $Z(Z'Z)^{-1/2}$ where $Z: n \times p$ is $N(0, I_n \otimes I_p)$.

(ii) Now partition Z as $Z = \begin{pmatrix} X \\ Y \end{pmatrix}$ where X is $k \times p$ and Y is $(n-k) \times p$. Show that $Z'Z = X'X + Y'Y$ and that Δ has the same distribution as $X(X'X + Y'Y)^{-1/2}$.

(iii) Using (ii) and Problem 11, show that $B = \Delta'\Delta$ has the density

$$p(B) = \frac{\omega(k, p)\omega(n-k, p)}{\omega(n, p)}|B|^{(k-p-1)/2}|I_p - B|^{(n-k-p-1)/2}$$

with respect to Lebesgue measure on the set $0 < B < I_p$.

(iv) Consider a random matrix $L: k \times p$ with a density with respect to Lebesgue measure given by

$$h(L) = c|I_p - L'L|^{(n-k-p-1)/2}\phi(L'L)$$

where for $B \in \mathcal{S}_p$,

$$\phi(B) = \begin{cases} 1 & \text{if } 0 < B < I_p \\ 0 & \text{otherwise} \end{cases}$$

and

$$c = \frac{\omega(n-k, p)}{(\sqrt{2\pi})^{kp}\omega(n, p)}.$$

Show that $B = L'L$ has the density $p(B)$ given in part (iii) (use Proposition 7.6).

(v) Now, to conclude that Δ has h as its density, first prove the following proposition: Suppose $\mathfrak{X}$ is acted on measurably by a compact group G and $\tau: \mathfrak{X} \to \mathfrak{Y}$ is a maximal invariant. If P_1 and P_2 are both G-invariant measures on $\mathfrak{X}$ such that $P_1(\tau^{-1}(C)) = P_2(\tau^{-1}(C))$ for all measurable $C \subseteq \mathfrak{Y}$, then $P_1 = P_2$.

(vi) Now, apply the proposition above with $\mathcal{X} = \mathcal{L}_{p,k}$, $G = \mathcal{O}_k$, $\tau(x) = x'x$, P_1 the distribution of Δ, and P_2 the distribution of L as given in (iv). This shows that Δ has density h.

13. Consider a random matrix $Z: n \times p$ with a density given by $f(Z|B, \Sigma) = |\Sigma|^{-n/2} h(\text{tr}(Z - TB)\Sigma^{-1}(Z - TB)')$ where $T: n \times k$ of rank k is known, $B: k \times p$ is a matrix of unknown parameters, and $\Sigma: p \times p$ is positive definite and unknown. Assume that $n \geqslant p + k$, that

$$\sup_{C \in \mathcal{S}_p^+} |C|^{n/2} h(\text{tr}(C)) < +\infty,$$

and that h is a nonincreasing function defined on $[0, \infty)$. Partition Z into $X: n \times q$ and $Y: n \times r$, $q + r = p$, so $Z = (XY)$. Also, partition Σ into Σ_{ij}, $i, j = 1, 2$, where Σ_{11} is $q \times q$, Σ_{22} is $r \times r$, and Σ_{12} is $q \times r$.

(i) Show that the maximum likelihood estimator of B is $\hat{B} = (T'T)^{-1}TZ$ and $f(Z|\hat{B}, \Sigma) = |\Sigma|^{-n/2}h(\text{tr}\, S\Sigma^{-1})$ where $S = Z'QZ$ with $Q = I - P$ and $P = T(T'T)^{-1}T'$.

(ii) Derive the likelihood ratio test of $H_0: \Sigma_{12} = 0$ versus $H_1: \Sigma_{12} \neq 0$. Show that the test rejects for small values of

$$\Lambda(Z) = \frac{|S|}{|S_{11}||S_{22}|}.$$

(iii) For $U: n \times q$ and $V: n \times r$, establish the identity $\text{tr}(UV)\Sigma^{-1}(UV)' = \text{tr}(V - U\Sigma_{11}^{-1}\Sigma_{12})\Sigma_{22\cdot 1}^{-1}(V - U\Sigma_{11}^{-1}\Sigma_{12}) + \text{tr}\, U\Sigma_{11}^{-1}U'$. Use this identity to derive the conditional distribution of Y given X in the above model. Using the notation of Section 10.5, show that the conditional density of Y given X is

$$f_1(Y|C, B_1, \Sigma_{11}, \Sigma_{22\cdot 1}, X)$$
$$= |\Sigma_{22\cdot 1}|^{-n/2}h\left(\text{tr}(Y - WC)\Sigma_{22\cdot 1}^{-1}(Y - WC)' + \eta\right)\phi(\eta)$$

where $\eta = \text{tr}(X - TB_1)\Sigma_{11}^{-1}(X - TB_1)$ and $(\phi(\eta))^{-1} = \int_{\mathcal{L}_{r,n}} h(\text{tr}\, uu' + \eta)\, du$.

(iv) The null hypothesis is now that $C_1 = 0$. Show that, for each fixed η, the likelihood ratio test (with C and $\Sigma_{22\cdot 1}$ as parameters) based on the conditional density rejects for large values of $\Lambda(Z)$. Verify (i), (ii), and (iii) of Proposition 10.17.

(v) Now, assume that

$$\sup_{\eta>0}\ \sup_{C\in \mathbb{S}_r^+} |C|^{n/2} h(\operatorname{tr} C + \eta)\phi(\eta) = k_2 < +\infty.$$

Show that the likelihood ratio test for $C_1 = 0$ (with C, $\Sigma_{22\cdot 1}$, B_1, and Σ_{11} as parameters) rejects for large values of $\Lambda(Z)$.

(vi) Show that, under H_0, the sample canonical correlations based on S_{11}, S_{12}, S_{22} (here $S = Z'QZ$) have the same distribution as when Z is $N(TB, I_n \otimes \Sigma)$. Conclude that under H_0, $\Lambda(Z)$ has the same distribution as when Z is $N(TB, I_n \otimes \Sigma)$.

NOTES AND REFERENCES

1. Canonical correlation analysis was first proposed in Hotelling (1935, 1936). There are as many approaches to canonical correlation analysis as there are books covering the subject. For a sample of these, see Anderson (1958), Dempster (1969), Kshirsagar (1972), Rao (1973), Mardia, Kent, and Bibby (1979), Srivastava and Khatri (1979), and Muirhead (1982).

2. See Eaton and Kariya (1981) for some material related to Proposition 10.13.

Appendix

We begin this appendix with a statement and proof of a result due to Basu (1955). Consider a measurable space $(\mathcal{X}, \mathcal{B})$ and a probability model $\{P_\theta | \theta \in \Theta\}$ defined on $(\mathcal{X}, \mathcal{B})$. Consider a statistic T defined on $(\mathcal{X}, \mathcal{B})$ to $(\mathcal{Y}, \mathcal{B}_1)$, and let $\mathcal{B}_T = \{T^{-1}(B) | B \in \mathcal{B}_1\}$. Thus $\mathcal{B}_T$ is a σ-algebra and $\mathcal{B}_T \subseteq \mathcal{B}$. Conditional expectation given $\mathcal{B}_T$ is denoted by $\mathcal{E}(\cdot | \mathcal{B}_T)$.

Definition A.1. The statistic T is a *sufficient statistic* for the family $\{P_\theta | \theta \in \Theta\}$ if, for each bounded $\mathcal{B}$ measurable function f, there exists a $\mathcal{B}_T$ measurable function $\hat{f}$ such that $\mathcal{E}_\theta(f | \mathcal{B}_T) = \hat{f}$ a.e. P_θ for all $\theta \in \Theta$.

Note that the null set where the above equality does not hold is allowed to depend on both θ and f. The usual intuitive description of sufficiency is that the conditional distribution of $X \in \mathcal{X}$ ($\mathcal{L}(X) = P_\theta$ for some $\theta \in \Theta$) given $T(X) = t$ does not depend on θ. Indeed, if $P(\cdot | t)$ is such a version of the conditional distribution of X given $T(X) = t$, then $\hat{f}$ defined by $\hat{f}(x) = h(T(x))$ where

$$h(t) = \int f(x) P(dx | t)$$

serves as a version of $\mathcal{E}_\theta(f | \mathcal{B}_T)$ for each $\theta \in \Theta$.

Now, consider a statistic U defined on $(\mathcal{X}, \mathcal{B})$ to $(\mathcal{Z}, \mathcal{B}_2)$.

Definition A.2. The statistic U is called an *ancillary statistic* for the family $\{P_\theta | \theta \in \Theta\}$ if the distribution of U on $(\mathcal{Z}, \mathcal{B}_2)$ does not depend on $\theta \in \Theta$—that is, if for all $B \in \mathcal{B}_2$,

$$P_\theta\{U^{-1}(B)\} = P_\eta\{U^{-1}(B)\}$$

for all $\theta, \eta \in \Theta$.

In many instances, ancillary statistics are functions of maximal invariant statistics in a situation where a group acts transitively on the family of probabilities in question—see Section 7.5 for a discussion.

Finally, given a statistic T on $(\mathfrak{X}, \mathfrak{B})$ to $(\mathfrak{Y}, \mathfrak{B}_1)$ and the parametric family $\{P_\theta | \theta \in \Theta\}$, let $\{Q_\theta | \theta \in \Theta\}$ be the induced family of distributions of T on $(\mathfrak{Y}, \mathfrak{B}_1)$—that is,

$$Q_\theta(B) = P_\theta(T^{-1}(B)), \qquad B \in \mathfrak{B}_1.$$

Definition A.3. The family $\{Q_\theta | \theta \in \Theta\}$ is called *boundedly complete* if the only bounded solution to the equation

$$\int h(y) Q_\theta(dy) = 0, \qquad \theta \in \Theta$$

is the function $h = 0$ a.e. Q_θ for all $\theta \in \Theta$.

At times, a statistic T is called boundedly complete—this means that the induced family of distributions of T is boundedly complete according to the above definition. If the family $\{Q_\theta | \theta \in \Theta\}$ is an exponential family on a Euclidean space and if Θ contains a nonempty open set, then $\{Q_\theta | \theta \in \Theta\}$ is boundedly complete—see Lehmann (1959, page 132).

Theorem (Basu, 1955). If T is a boundedly complete sufficient statistic and if U is an ancillary statistic, then, for each θ, $T(X)$ and $U(X)$ are independent.

Proof. It suffices to show that, for bounded measurable functions h and k on $\mathfrak{Y}$ and $\mathfrak{Z}$, we have

$$\text{(A.1)} \qquad \mathcal{E}_\theta h(T(X)) k(U(X)) = \mathcal{E}_\theta h(T(X)) \mathcal{E}_\theta k(U(X)).$$

Since U is ancillary, $a = \mathcal{E}_\theta k(U(X))$ does not depend on θ, so $\mathcal{E}_\theta(k(U) - a) = 0$ for all θ. Hence

$$\mathcal{E}_\theta\left[\mathcal{E}_\theta((k(U) - a) | \mathfrak{B}_T)\right] = 0 \quad \text{for all } \theta.$$

Since T is sufficient, there is a $\mathfrak{B}_T$ measurable function, say $\hat{f}$, such that $\mathcal{E}_\theta((k(U) - a) | \mathfrak{B}_T) = \hat{f}$ a.e. P_θ. But since $\hat{f}$ is $\mathfrak{B}_T$ measurable, we can write $\hat{f}(x) = \psi(T(x))$ (see Lehmann, 1959, Lemma 1, page 37). Also, since k is bounded, $\hat{f}$ can be taken to be bounded. Hence

$$\mathcal{E}_\theta \psi(T) = 0 \qquad \text{for all } \theta$$

and ψ is bounded. The bounded completeness of T implies that ψ is 0 a.e. Q_θ where $Q_\theta(B) = P_\theta(T^{-1}(B))$, $B \in \mathcal{B}_1$. Thus $h(T)\psi(T) = 0$ a.e. Q_θ, so

$$\begin{aligned} 0 = \mathcal{E}_\theta h(T)\psi(T) &= \mathcal{E}_\theta\left[h(T)\mathcal{E}_\theta((k(U) - a)|\mathcal{B}_T)\right] \\ &= \mathcal{E}_\theta\left[\mathcal{E}_\theta(h(T)(k(U) - a)|\mathcal{B}_T)\right] \\ &= \mathcal{E}_\theta h(T)(k(U) - a). \end{aligned}$$

Thus (A.1) holds. □

This Theorem can be used in many of the examples in the text where we have used Proposition 7.19.

The second topic in this Appendix concerns monotone likelihood ratio and its implications. Let $\mathcal{X}$ and $\mathcal{Y}$ be subsets of the real line.

Definition A.4. A nonnegative function k defined on $\mathcal{X} \times \mathcal{Y}$ is *totally positive of order 2* (TP-2) if, for $x_1 < x_2$ and $y_1 < y_2$, we have

$$k(x_1, y_1)k(x_2, y_2) \geqslant k(x_1, y_2)k(x_2, y_1). \tag{A.2}$$

In the case that $\mathcal{Y}$ is a parameter space and $k(\cdot, y)$ is a density with respect to some fixed measure, it is customary to say that k has a monotone likelihood ratio when k is TP-2. This nomenclature arises from the observation that, when k is TP-2 and $y_1 < y_2$, then the ratio $k(\cdot, y_2)/k(\cdot, y_1)$ is nondecreasing in x—assuming that $k(\cdot, y_1)$ does not vanish. Some obvious examples of TP-2 functions are: $\exp[xy]$, x^y for $x > 0$, y^x for $y > 0$. If $x = g(s)$ and $y = h(t)$ where g and h are both increasing or decreasing, then $k_1(s, t) \equiv k(g(s), h(t))$ is TP-2 whenever k is TP-2. Further, if $\psi_1(x) \geqslant 0$, $\psi_2(y) \geqslant 0$, and k is TP-2, then $k_1(x, y) = \psi_1(x)\psi_2(y)k(x, y)$ is also TP-2.

The following result due to Karlin (1956) is of use in verifying that some of the more complicated densities that arise in statistics are TP-2. Here is the setting. Let $\mathcal{X}$, $\mathcal{Y}$, and $\mathcal{Z}$ be Borel subsets of R^1 and let μ be a σ-finite measure on the Borel subsets of $\mathcal{Y}$.

Lemma (Karlin, 1956). Suppose g is TP-2 on $\mathcal{X} \times \mathcal{Y}$ and h is TP-2 on $\mathcal{Y} \times \mathcal{Z}$. If

$$k(x, z) = \int g(x, y)h(y, z)\mu(dy)$$

is finite for all $x \in \mathcal{X}$ and $z \in \mathcal{Z}$, then k is TP-2.

Proof. For $x_1 < x_2$ and $z_1 < z_2$, the difference

$$\Delta = k(x_1, z_1)k(x_2, z_2) - k(x_1, z_2)k(x_2, z_1)$$

can be written

$$\Delta = \int\int g(x_1, y_1)g(x_2, y_2)[h(y_1, z_1)h(y_2, z_2) - h(y_1, z_2)h(y_2, z_1)] \times \mu(dy_1)\mu(dy_2).$$

Now, write Δ as the double integral over the set $\{y_1 < y_2\}$ plus the double integral over the set $\{y_1 > y_2\}$. In the integral over the set $\{y_1 > y_2\}$, interchange y_1 and y_2 and then combine with the integral over $\{y_1 < y_2\}$. This yields

$$\Delta = \int\int_{\{y_1 < y_2\}} [g(x_1, y_1)g(x_2, y_2) - g(x_1, y_2)g(x_2, y_1)] \cdot [h(y_1, z_1)h(y_2, z_2) - h(y_1, z_2)(h(y_2, z_1)]\mu(dy_1)\mu(dy_2).$$

On the set $\{y_1 < y_2\}$, both of the bracketed expressions are non-negative as g and h are TP-2. Hence $\Delta \geqslant 0$ so k is TP-2. □

Here are some examples.

◆ **Example A.1.** With $\mathcal{X} = (0, \infty)$, let

$$f(x, m) = \frac{x^{(m/2)-1}\exp[-\frac{1}{2}x]}{2^{m/2}\Gamma(m/2)}$$

be the density of a chi-squared distribution with m degrees of freedom. Since $x^{m/2}$, $x \in \mathcal{X}$ and $m > 0$, is TP-2, $f(x, m)$ is TP-2. Recall that the density of a noncentral chi-squared distribution with

p degrees of freedom and noncentrality parameter $\lambda \geqslant 0$ is given by

$$h(x, \lambda) = \sum_{j=0}^{\infty} \frac{(\lambda/2)^j \exp[-\frac{1}{2}\lambda]}{j!} f(x, p + 2j).$$

Observe that $f(x, p + 2j)$ is TP-2 in x and j and $(\lambda/2)^j \exp[-\frac{1}{2}\lambda]/j!$ is TP-2 in j and λ. With $\mathcal{Y} = \{0, 1, \dots\}$ and μ as counting measure, Karlin's Lemma implies that $h(x, \lambda)$ is TP-2. ◆

◆ **Example A.2.** Recall that, if χ_m^2 and χ_n^2 are independent random variables, then $Y = \chi_m^2/\chi_n^2$ has a density given by

$$f(y|m, n) = \frac{\Gamma((m+n)/2)}{\Gamma(m/2)\Gamma(n/2)} \frac{y^{(m/2)-1}}{(1+y)^{(m+n)/2}}, \qquad y > 0.$$

If the random variable χ_m^2 is noncentral chi-squared, say $\chi_p^2(\lambda)$, rather than central chi-squared, then $Y = \chi_p^2(\lambda)/\chi_m^2$ has a density

$$h(y|\lambda) = \sum_{j=0}^{\infty} \frac{(\lambda/2)^j \exp[-\frac{1}{2}\lambda]}{j!} f(y|p + 2j, n).$$

Of course, Y has an unnormalized $F(p, m; \lambda)$ distribution according to our usage in Section 8.4. Since $f(y|p + 2j, n)$ is TP-2 in y and j, it follows as in Example A.1 that h is TP-2. ◆

The next result yields the useful fact that the noncentral Student's t distribution is TP-2.

Proposition A.1. Suppose $f \geqslant 0$ defined on $(0, \infty)$ satisfies

(i) $\int_0^\infty e^{ux} f(x)\, dx < +\infty$ for $u \in R^1$
(ii) $f(x/\eta)$ is TP-2 on $(0, \infty) \times (0, \infty)$.

For $\theta \in R^1$ and $t \in R^1$, define k by $k(t, \theta) = \int_0^\infty e^{\theta t x} f(x)\, dx$. Then k is TP-2.

Proof. First consider $t \in R^1$ and $\theta > 0$. Set $v = \theta x$ in the integral defining k to obtain

$$k(t, \theta) = \frac{1}{\theta} \int_0^\infty e^{tv} f\left(\frac{v}{\theta}\right) dv.$$

Now apply Karlin's Lemma to conclude that k is TP-2 on $R^1 \times (0, \infty)$. A similar argument shows that k is TP-2 on $R^1 \times (-\infty, 0)$. Since $k(t, 0)$ is a constant, it is now easy to show that k is TP-2 on $R^1 \times R^1$. □

◆ **Example A.3.** Suppose X is $N(\mu, 1)$ and Y is χ^2_n. The random variable $T = X/\sqrt{Y}$, which is, up to a factor of $\sqrt{n}$, a noncentral Student's t random variable, has a density that depends on μ—the noncentrality parameter. The density of T_1 (derived by writing down the joint density of X and Y, changing variables to T and $W = \sqrt{Y}$, and integrating out W) can be written

$$h(t, \mu) = \frac{2\exp\left[-\frac{1}{2}\mu^2\right]}{\Gamma(n/2)(1 + t^2)^{(n+1)/2}} \times \int_0^\infty \exp[\psi(t)\mu x]\exp[-x^2]x^{-n}\,dx$$

where

$$\psi(t) = \sqrt{2}\,t(1 + t^2)^{-1/2}$$

is an increasing function of t. Consider the function

$$k(v, \mu) = \int_0^\infty \exp[v\mu x]\exp[-x^2]x^{-n}\,dx.$$

With $f(x) = \exp[-x^2]x^{-n}$, Proposition A.1 shows k, and hence h, is TP-2. ◆

We conclude this appendix with a brief description of the role of TP-2 in one sided testing problems. Consider a TP-2 density $p(x|\theta)$ for $x \in \mathcal{X} \subseteq R^1$ and $\theta \in \Theta \subseteq R^1$. Suppose we want to test the null hypothesis $H_0: \theta \in (-\infty, \theta_0] \cap \Theta$ versus $H_1: \theta \in (\theta_0, \infty) \cap \Theta$. The following basic result is due to Karlin and Rubin (1956).

Proposition A.2. Given any test ϕ_0 of H_0 versus H_1, there exists a test ϕ of the form

$$\phi(x) = \begin{cases} 1 & \text{if } x > x_0 \\ \gamma & \text{if } x = x_0 \\ 0 & \text{if } x < x_0 \end{cases}$$

with $0 \leqslant \gamma \leqslant 1$ such that $\mathcal{E}_\theta\phi \leqslant \mathcal{E}_\theta\phi_0$ for $\theta \leqslant \theta_0$ and $\mathcal{E}_\theta\phi \geqslant \mathcal{E}_\theta\phi_0$ for $\theta > \theta_0$. For any such test ϕ, $\mathcal{E}_\theta\phi$ is nondecreasing in θ.

Comments on Selected Problems

CHAPTER 1

4. This problem gives the direct sum version of partitioned matrices. For (ii), identify V_1 with vectors of the form $\{v_1, 0\} \in V_1 \oplus V_2$ and restrict T to these. This restriction is a map from V_1 to $V_1 \oplus V_2$ so $T\{v_1, 0\} = \{z_1(v_1), z_2(v_1)\}$ where $z_1(v_1) \in V_1$ and $z_2(v_1) \in V_2$. Show that z_1 is a linear transformation on V_1 to V_1 and z_2 is a linear transformation on V_1 to V_2. This gives A_{11} and A_{21}. A similar argument gives A_{12} and A_{22}. Part (iii) is a routine computation.
5. If $x_{r+1} = \Sigma_1^r c_i x_i$, then $w_{r+1} = \Sigma_1^r c_i w_i$.
8. If $u \in R^k$ has coordinates $u_1, \dots, u_k$, then $Au = \Sigma_1^k u_i x_i$ and all such vectors are just span $\{x_1, \dots, x_k\}$. For (ii), $r(A) = r(A')$ so $\dim \mathcal{R}(A'A) = \dim \mathcal{R}(AA')$.
10. The algorithm of projecting $x_2, \dots, x_k$ onto $\{\text{span } x_1\}^{\perp}$ is known as Björk's algorithm (Björk, 1967) and is an alternative method of doing Gram–Schmidt. Once you see that $y_2, \dots, y_k$ are perpendicular to y_1, this problem is not hard.
11. The assumptions and linearity imply that $[Ax, w] = [Bx, w]$ for all $x \in V$ and $w \in W$. Thus $[(A - B)x, w] = 0$ for all w. Choose $w = (A - B)x$ so $(A - B)x = 0$.
12. Choose z such that $[y_1, z] \neq 0$. Then $[y_1, z]x_1 = [y_2, z]x_2$ so set $c = [y_2, z]/[y_1, z]$. Thus $cx_2 \square y_1 = x_2 \square y_2$ so $cy_1 \square x_2 = y_2 \square x_2$. Hence $c\|x_2\|^2 y_1 = \|x_2\|^2 y_2$ so $y_1 = c^{-1} y_2$.
13. This problem shows the topologies generated by inner products are all the same. We know $[x, y] = (x, Ay)$ for some $A > 0$. Let c_1 be the minimum eigenvalue of A, and let c_2 be the maximum eigenvalue of A.

14. This is just the Cauchy–Schwarz Inequality.
15. The classical two-way *ANOVA* table is a consequence of this problem. That A, B_1, B_2, and B_3 are orthogonal projections is a routine but useful calculation. Just keep the notation straight and verify that $P^2 = P = P'$, which characterizes orthogonal projections.
16. To show that $\Gamma(M^{\perp}) \subseteq M^{\perp}$, verify that $(u, \Gamma v) = 0$ for all $u \in M$ when $v \in M^{\perp}$. Use the fact that $\Gamma'\Gamma = I$ and $u = \Gamma u_1$ for some $u_1 \in M$ (since $\Gamma(M) \subseteq M$ and Γ is nonsingular).
17. Use Cauchy–Schwarz and the fact that $P_M x = x$ for $x \in M$.
18. This is Cauchy–Schwarz for the non-negative definite bilinear form $[C, D] = \operatorname{tr} ACBD'$.
20. Use Proposition 1.36 and the assumption that A is real.
21. The representation $\alpha P + \beta(I - P)$ is a spectral type representation—see Theorem 1.2a. If $M = \mathcal{R}(P)$, let $x_1, \ldots, x_r, x_{r+1}, \ldots, x_n$ be any orthonormal basis such that $M = \text{span}\{x_1, \ldots, x_r\}$. Then $Ax_i = \alpha x_i$, $i = 1, \ldots, r$, and $Ax_i = \beta x_i$, $i = r + 1, \ldots, n$. The characteristic polynomial of A must be $(\alpha - \lambda)^r(\beta - \lambda)^{n-r}$.
22. Since $\lambda_1 = \sup_{\|x\|=1}(x, Ax)$, $\mu_1 = \sup_{\|x\|=1}(x, Bx)$, and $(x, Ax) \geqslant (x, Bx)$, obviously $\lambda_1 \geqslant \mu_1$. Now, argue by contradiction—let j be the smallest index such that $\lambda_j < \mu_j$. Consider eigenvectors $x_1, \ldots, x_n$ and $y_1, \ldots, y_n$ with $Ax_i = \lambda_i x_i$ and $By_i = \mu_i y_i$, $i = 1, \ldots, n$. Let $M = \text{span}\{x_j, x_{j+1}, \ldots, x_n\}$ and let $N = \text{span}\{y_1, \ldots, y_j\}$. Since dim $M = n - j + 1$, $\dim M \cap N \geqslant 1$. Using the identities $\lambda_j = \sup_{x \in M, \|x\|=1}(x, Ax)$, $\mu_j = \inf_{x \in N, \|x\|=1}(x, Bx)$, for any $x \in M \cap N$, $\|x\| = 1$, we have $(x, Ax) \leqslant \lambda_j < \mu_j \leqslant (x, Bx)$, which is a contradiction.
23. Write $S = \Sigma_1^n \lambda_i x_i \Box x_i$ in spectral form where $\lambda_i > 0$, $i = 1, \ldots, n$. Then $0 = \langle S, T \rangle = \Sigma_1^n \lambda_i (x_i, Tx_i)$, which implies $(x_i, Tx_i) = 0$ for $i = 1, \ldots, n$ as $T \geqslant 0$. This implies $T = 0$.
24. Since $\operatorname{tr} A$ and $\langle A, I \rangle$ are both linear in A, it suffices to show equality for A's of the form $A = x \Box y$. But $\langle x \Box y, I \rangle = (x, y)$. However, that $\operatorname{tr} x \Box y = (x, y)$ is easily verified by choosing a coordinate system.
25. Parts (i) and (ii) are easy but (iii) is not. It is false that $A^2 \geqslant B^2$ and a 2×2 matrix counter example is not hard to construct. It is true that $A^{1/2} \geqslant B^{1/2}$. To see this, let $C = B^{1/2}A^{-1/2}$, so by hypothesis, $I \geqslant C'C$. Note that the eigenvalues of C are real and positive—being the same as those of $B^{1/4}A^{-1/2}B^{1/4}$ which is positive definite. If λ is any eigenvalue for C, there is a corresponding eigenvector—say x such that $\|x\| = 1$ and $Cx = \lambda x$. The relation $I \geqslant C'C$ implies $\lambda^2 \leqslant 1$, so $0 < \lambda \leqslant 1$ as λ is positive. Thus all the eigenvalues of C are in $(0, 1]$ so

the same is true of $A^{-1/4}B^{1/2}A^{-1/4}$. Hence $A^{-1/4}B^{1/2}A^{-1/4} \leqslant I$ so $B^{1/2} \leqslant A^{1/2}$.

26. Since P is an orthogonal projection, all its eigenvalues are zero or one and the multiplicity of one is the rank of P. But tr P is just the sum of the eigenvalues of P.

28. Since any $A \in \mathcal{L}(V, V)$ can be written as $(A + A')/2 + (A - A')/2$, it follows that $M + N = \mathcal{L}(V, V)$. If $A \in M \cap N$, then $A = A' = -A$, so $A = 0$. Thus $\mathcal{L}(V, V)$ is the direct sum of M and N so dim M + dim $N = n^2$. A direct calculation shows that $\{x_i \square x_j + x_j \square x_i | i \leqslant j\} \cup \{x_i \square x_j - x_j \square x_i | i < j\}$ is an orthogonal set of vectors, none of which is zero, and hence the set is linearly independent. Since the set has n^2 elements, it forms a basis for $\mathcal{L}(V, V)$. Because $x_i \square x_j + x_j \square x_i \in M$ and $x_i \square x_j - x_j \square x_i \in N$, dim $M \geqslant n(n + 1)/2$ and dim $N \geqslant n(n - 1)/2$. Assertions (i), (ii), and (iii) now follow. For (iv), just verify that the map $A \to (A + A')/2$ is idempotent and self-adjoint.

29. Part (i) is a consequence of $\sup_{\|v\|=1}\|Av\| = \sup_{\|v\|=1}[Av, Av]^{1/2} = \sup_{\|v\|=1}(v, A'Av)^{1/2}$ and the spectral theorem. The triangle inequality follows from $|||A + B||| = \sup_{\|v\|=1} \|Av + Bv\| \leqslant \sup_{\|v\|=1}(\|Av\| + \|Bv\|) \leqslant \sup_{\|v\|=1} \|Av\| + \sup_{\|v\|=1} \|Bv\|$.

30. This problem is easy, but it is worth some careful thought—it provides more evidence that $A \otimes B$ has been defined properly and $\langle \cdot , \cdot \rangle$ is an appropriate inner produce on $\mathcal{L}(W, V)$. Assertion (i) is easy since $(A \otimes B)(x_i \square w_j) = (Ax_i) \square (Bw_j) = (\lambda_i x_i) \square (\mu_j w_j) = \lambda_i \mu_j x_i \square w_j$. Obviously, $x_i \square w_j$ is an eigenvector of the eigenvalue $\lambda_i \mu_j$. Part (ii) follows since the two linear transformations agree on the basis $\{x_i \square w_j | i = 1, \ldots, m,\ j = 1, \ldots, n\}$ for $\mathcal{L}(W, V)$. For (iii), if the eigenvalues of A and B are positive, so are the eigenvalues of $A \otimes B$. Since the trace of a self-adjoint linear transformation in the sum of the eigenvalues (this is true even without self-adjointness, but the proof requires a bit more than we have established here), we have tr $A \otimes B = \Sigma_{i,\,j}\lambda_i \mu_j = (\Sigma_i \lambda_i)(\Sigma_j \mu_j) =$ (tr A)(tr B). Since the determinant is the product of the eigenvalues, $\det(A \otimes B) = \prod_{i,\,j}(\lambda_i \mu_j) = (\prod \lambda_i)^n (\prod \mu_j)^m = (\det A)^n (\det B)^m$.

31. Since $\psi'\psi = I_p$, ψ is a linearly isometry and its columns form an orthonormal set. Since $R(\psi) \subseteq M$ and the two subspaces have the same dimension, (i) follows. (ii) is immediate.

32. If C is $n \times k$ and D is $k \times n$, the set of nonzero eigenvalues of CD is the same as the set of nonzero eigenvalues of DC.

33. Apply Problem 32.

34. Orthogonal transformations preserve angles.

35. This problem requires that you have a facility in dealing with conditional expectation. If you do, the problem requires a bit of calculation but not much more. If you don't, proceed to Chapter 2.

CHAPTER 2

1. Write $x = \Sigma_1^n c_i x_i$ so $(x, X) = \Sigma c_i(x_i, X)$. Thus $\mathcal{E}|(x, X)| \leqslant \Sigma_1^n |c_i| \mathcal{E}|(x_i, X)|$ and $\mathcal{E}|(x_i, X)|$ is finite by assumption. To show that Cov(X) exists, it suffices to verify that var(x, X) exists for each $x \in V$. But $\text{var}(x, X) = \text{var}\{\Sigma c_i(x_i, X)\} = \Sigma\Sigma \text{ cov}\{c_i(x_i, X), c_j(x_j, X)\}$. Then $\text{var}\{c_i(x_i, X)\} = \mathcal{E}[c_i(x_i X)]^2 - [\mathcal{E}c_i(x_i, X)]^2$, which exists by assumption. The Cauchy–Schwarz Inequality shows that $[\text{cov}\{c_i(x_i, X), c_j(x_j, X)\}]^2 \leqslant \text{var}\{c_i(x_i, X)\}\,\text{var}\{c_j(x_j, X)\}$. But, $\text{var}\{c_i(x_i, X)\}$ exists by the above argument.
2. All inner products on a finite dimensional vector space are related via the positive definite quadratic forms. An easy calculation yields the result of this problem.
3. Let $(\cdot, \cdot)_i$ be an inner product on V_i, $i = 1, 2$. Since f_i is linear on V_i, $f_i(x) = (x_i, x)_i$ for $x_i \in V_i$, $i = 1, 2$. Thus if X_1 and X_2 are uncorrelated (the choice of inner product is irrelevant by Problem 2), (2.2) holds. Conversely, if (2.2) holds, then $\text{Cov}\{(x_1, X_1)_1, (x_2, X_2)_2\} = 0$ for $x_i \in V_i$, $i = 1, 2$ since $(x_1, \cdot)_1$ and $(x_2, \cdot)_2$ are linear functions.
4. Let $s = n - r$ and consider $\Gamma \in \mathcal{O}_r$ and a Borel set B_1 of R^r. Then

$$Pr\{\Gamma\dot{X} \in B_1\} = Pr\{\Gamma\dot{X} \in B_1, \ddot{X} \in R^s\}$$

$$= Pr\left\{\begin{pmatrix}\Gamma & 0\\ 0 & I_s\end{pmatrix}\begin{pmatrix}\dot{X}\\ \ddot{X}\end{pmatrix} \in B_1 \times R^s\right\}$$

$$= Pr\left\{\begin{pmatrix}\dot{X}\\ \ddot{X}\end{pmatrix} \in B_1 \times R^s\right\} = Pr\{\dot{X} \in B_1\}.$$

The third equality holds since the matrix

$$\begin{pmatrix}\Gamma & 0\\ 0 & I_s\end{pmatrix}$$

is in $\mathcal{O}_n$. Thus $\dot{X}$ has an $\mathcal{O}_r$-invariant distribution. That $\dot{X}$ given $\ddot{X}$ has an $\mathcal{O}_r$-invariant distribution is easy to prove when X has a density with respect to Lebesgue measure on R^n (the density has a version that satisfies $f(x) = f(\psi x)$ for $x \in R^n$, $\psi \in \mathcal{O}_n$). The general case requires some fiddling with conditional expections—this is left to the interested reader.

5. Let $A_i = \text{Cov}(X_i)$, $i = 1, \ldots, n$. It suffices to show that $\text{var}(x, \Sigma X_i) = \Sigma(x, A_i x)$. But (x, X_i), $i = 1, \ldots, n$, are uncorrelated, so $\text{var}[\Sigma(x, X_i)] = \Sigma\,\text{var}(x, X_i) = \Sigma(x, A_i x)$.
6. $\mathcal{E}U = \Sigma p_i \varepsilon_i = p$. Let U have coordinates $U_1, \ldots, U_k$. Then $\text{Cov}(U) = \mathcal{E}UU' - pp'$ and UU' is a $p \times p$ matrix with elements $U_i U_j$. For $i \neq j$, $U_i U_j = 0$ and for $i = j$, $U_i U_j = U_i$. Since $\mathcal{E}U_i = p_i$, $\mathcal{E}UU' = D_p$. When $0 < p_i < 1$, D_p has rank k and the rank of $\text{Cov}(U)$ is the rank of $I_k - D_p^{-1/2} pp' D_p^{-1/2}$. Let $u = D_p^{-1/2} p$, so $u \in R^k$ has length one. Thus $I_k - uu'$ is a rank $k - 1$ orthogonal projection. The null space of $\text{Cov}\, U$ is $\text{span}\{e\}$ where e is the vector of ones in R^k. The rest is easy.
7. The random variable X takes on $n!$ values—namely the $n!$ permutations of x—each with probability $1/n!$. A direct calculation gives $\mathcal{E}X = \bar{x}e$ where $\bar{x} = n^{-1}\Sigma_1^n x_i$. The distribution of X is permutation invariant, which implies that $\text{Cov}\, X$ has the form $\sigma^2 A$ where $a_{ii} = 1$ and $a_{ij} = \rho$ for $i \neq j$ where $-1/(n-1) \leqslant \rho \leqslant 1$. Since $\text{var}(e'X) = 0$, we see that $\rho = -1/(n-1)$. Thus $\sigma^2 = \text{var}(X_1) = n^{-1}[\Sigma_1^n (x_i - \bar{x})^2]$ where X_1 is the first coordinate of X.
8. Setting $D = -I$, $\mathcal{E}X = -\mathcal{E}X$ so $\mathcal{E}X = 0$. For $i \neq j$, $\text{cov}\{X_i, X_j\} = \text{cov}\{-X_i, X_j\} = -\text{cov}\{X_i, X_j\}$ so X_i and X_j are uncorrelated. The first equality is obtained by choosing D with $d_{ii} = -1$ and $d_{jj} = 1$ in the relation $\mathcal{L}(X) = \mathcal{L}(DX)$.
9. This is a direct calculation.
10. It suffices to verify the equality for $A = x \square y$ as both sides of the equality are linear in A. For $A = x \square y$, $\langle A, \Sigma \rangle = (x, \Sigma y)$ and $(\mu, A\mu) = (\mu, x)(\mu, y)$, so the equality is obvious.
11. To say $\text{Cov}(X) = I_n \otimes \Sigma$ is to say that $\text{cov}\{(\text{tr}\, AX'), (\text{tr}\, BX')\} = \text{tr}\, A\Sigma B'$. To show rows 1 and 2 are uncorrelated, pick $A = \varepsilon_1 v'$ and $B = \varepsilon_2 u'$ where $u, v \in R^p$. Let X_1' and X_2' be the first two rows of X. Then $\text{tr}\, AX' = v'X_1$, $\text{tr}\, BX' = u'X_2$, and $\text{tr}\, A\Sigma B = 0$. The desired equality is established by first showing that it is valid for $A = xy'$, $x, y \in R^n$, and using linearity. When $A = xy'$, a useful equality is $X'AX = \Sigma_i \Sigma_j x_i y_j X_i X_j'$ where the rows of X are $X_1', \ldots, X_n'$.
12. The equation $\Gamma A \Gamma' = A$ for $\Gamma \in \mathcal{O}_p$ implies that $A = cI_p$ for some c.
13. $\text{Cov}((\Gamma \otimes I)X) = \text{Cov}(X)$ implies $\text{Cov}(X) = I \otimes \Sigma$ for some Σ. $\text{Cov}((I \otimes \psi)X) = \text{Cov}(X)$ then implies $\psi \Sigma \psi' = \Sigma$, which necessitates $\Sigma = cI$ for some $c \geqslant 0$. Part (ii) is immediate since $\Gamma \otimes \psi$ is an orthogonal transformation on $(\mathcal{L}(V, W), \langle \cdot, \cdot \rangle)$.

14. This problem is a nasty calculation intended to inspire an appreciation for the equation $\mathrm{Cov}(X) = I_n \otimes \Sigma$.
15. Since $\mathcal{L}(X) = \mathcal{L}(-X)$, $\mathcal{E}X = 0$. Also, $\mathcal{L}(X) = \mathcal{L}(\Gamma X)$ implies $\mathrm{Cov}(X) = cI$ for some $c > 0$. But $\|X\|^2 = 1$ implies $c = 1/n$. Best affine predictor of X_1 given $\dot{X}$ is 0. I would predict X_1 by saying that X_1 is $\sqrt{1 - \dot{X}'\dot{X}}$ with probability $\frac{1}{2}$ and X_1 is $-\sqrt{1 - \dot{X}'\dot{X}}$ with probability $\frac{1}{2}$.
16. This is just the definition of $\square$.
17. For (i), just calculate. For (ii), $\mathrm{Cov}(S) = 2I_2 \otimes I_2$ by Proposition 2.23. The coordinate inner product on R^3 is not the inner product $\langle \cdot, \cdot \rangle$ on $\mathcal{S}_2$.

CHAPTER 3

2. Since $\mathrm{var}(X_1) = \mathrm{var}(Y_1) = 1$ and $\mathrm{cov}\{X_1, Y_1\} = \rho$, $|\rho| \leqslant 1$. Form $Z = (XY)$—an $n \times 2$ matrix. Then $\mathrm{Cov}(Z) = I_n \otimes A$ where
$$A = \begin{pmatrix} 1 & \rho \\ \rho & 1 \end{pmatrix}.$$
When $|\rho| < 1$, A is positive definite, so $I_n \otimes A$ is positive definite. Conditioning on Y, $\mathcal{L}(X|Y) = N(\rho Y, (1-\rho^2)I_n)$, so $\mathcal{L}(Q(Y)X|Y) = N(0, (1-\rho^2)Q(Y))$ as $Q(Y)Y = 0$ and $Q(Y)$ is an orthogonal projection. Now, apply Proposition 3.8 for Y fixed to get $\mathcal{L}(W) = (1-\rho^2)\chi^2_{n-1}$.
3. Just do the calculations.
4. Since $p(x)$ is zero in the second and fourth quadrants, X cannot be normal. Just find the marginal density of X_1 to show that X_1 is normal.
5. Write U in the form $X'AX$ where A is symmetric. Then apply Propositions 3.8 and 3.11.
6. Note that $\mathrm{Cov}(X \square X) = 2I \otimes I$ by Proposition 2.23. Since $(X, AX) = \langle X \square X, A \rangle$, and similarly for (X, BX), $0 = \mathrm{cov}\{(X, AX), (X, BX)\} = \mathrm{cov}\{\langle X \square X, A\rangle, \langle X \square X, B\rangle\} = \langle A, 2(I \otimes I)B\rangle = 2\,\mathrm{tr}\,AB$. Thus $0 = \mathrm{tr}\,A^{1/2}BA^{1/2}$ so $A^{1/2}BA^{1/2} = 0$, which shows $A^{1/2}B^{1/2} = 0$ and hence $AB = 0$.
7. Since $\mathcal{E}[\exp(itW_j)] = \exp\{it\mu_j - \sigma_j|t|\}$, $\mathcal{E}[\exp(it\Sigma a_j W_j)] = \exp[it\Sigma a_j\mu_j - (\Sigma|a_j|\sigma_j)|t|]$, so $\mathcal{L}(\Sigma a_j W_j) = C(\Sigma a_j\mu_j, \Sigma|a_j|\sigma_j)$. Part (ii) is immediate from (i).
8. For (i), use the independence of R and Z_0 to compute as follows: $P\{U \leqslant u\} = P\{Z_0 \leqslant u/R\} = \int_0^\infty P\{Z_0 \leqslant u/t\}G(dt) = \int_0^\infty \Phi(u/t)G(dt)$ where Φ is the distribution function of Z_0. Now, differentiate. Part (ii) is clear.

9. Let $\mathcal{B}_1$ be the sub σ-algebra induced by $T_1(X) = X_2$ and let $\mathcal{B}_2$ be the sub σ-algebra induced by $T_2(X) = X_2'X_2$. Since $\mathcal{B}_2 \subseteq \mathcal{B}_1$, for any bounded function $f(X)$, we have $\mathcal{E}(f(X)|\mathcal{B}_2) = \mathcal{E}(\mathcal{E}(f(X)|\mathcal{B}_1)|\mathcal{B}_2)$. But for $f(X) = h(X_2'X_1)$, the conditional expectation given $\mathcal{B}_1$ can be computed via the conditional distribution of $X_2'X_1$ given X_2, which is

$$(3.3) \qquad \mathcal{L}(X_2'X_1|X_2) = N\big(X_2'X_2\Sigma_{22}^{-1}\Sigma_{21},\ X_2'X_2 \otimes \Sigma_{11\cdot 2}\big).$$

Hence $\mathcal{E}(h(X_2'X_1)\mathcal{B}_1)$ is $\mathcal{B}_2$ measurable, so $\mathcal{E}(h(X_2'X_1)|\mathcal{B}_2) = \mathcal{E}(h(X_2'X_1)|\mathcal{B}_1)$. This implies that the conditional distribution (3.3) serves as a version of the conditional distribution of $X_2'X_1$ given $X_2'X_2$.

10. Show that $T^{-1}T_1 : R^n \to R^n$ is an orthogonal transformation so $l(C) = l((T^{-1}T_1)(C))$. Setting $B = T_1(C)$, we have $\nu_0(B) = \nu_1(B)$ for Borel B.

11. The measures ν_0 and ν_1 are equal up to a constant so all that needs to be calculated is $\nu_0(C)/\nu_1(C)$ for some set C with $0 < \nu_1(C) < +\infty$. Do the calculation for $C = \{v|[v, v] \leqslant 1\}$.

12. The inner product $\langle \cdot , \cdot \rangle$ on $\mathcal{S}_p$ is not the coordinate inner product. The "Lebesgue measure" on $(\mathcal{S}_p, \langle \cdot , \cdot \rangle)$ given by our construction is not $l(dS) = \prod_{i \leqslant j} ds_{ij}$, but is $\nu_0(dS) = (\sqrt{2})^{p(p-1)}l(dS)$.

13. Any matrix M of the form

$$M = a\begin{pmatrix} 1 & b & \cdots & b \\ b & 1 & & \vdots \\ \vdots & & \ddots & b \\ b & \cdots & b & 1 \end{pmatrix} : p \times p$$

can be written as $M = a[(p-1)b + 1]A + a(1-b)(I - A)$. This is a spectral decomposition for M so M has eigenvalues $a((p-1)b+1)$ and $a(1-b)$ (of multiplicity $p-1$). Setting $\alpha = a[(p-1)b+1]$ and $\beta = a(1-b)$ solves (i). Clearly, $M^{-1} = \alpha^{-1}A + \beta^{-1}(I - A)$ whenever α and β are not zero. To do part (ii), use the parameterization (μ, α, β) given above ($a = \sigma^2$ and $b = p$). Then use the factorization criterion on the likelihood function.

CHAPTER 4

1. Part (i) is clear since $Z\beta = \sum_1^k \beta_i z_i$ for $\beta \in R^k$. For (ii), use the singular value decomposition to write $Z = \sum_1^r \lambda_i x_i u_i'$ where r is the rank of Z, $\{x_1, \ldots, x_r\}$ is an orthonormal set in R^n, $\{u_1, \ldots, u_r\}$ is an orthonormal set in R^k, $M = \text{span}\{x_1, \ldots, x_r\}$, and $\mathcal{N}(Z) = (\text{span}\{u_1, \ldots, u_r\})^{\perp}$.

Thus $(Z'Z)^- = \Sigma_1^r \lambda_i^{-2} u_i u_i'$ and a direct calculation shows that $Z(Z'Z)^- Z' = \Sigma_1^r x_i x_i'$, which is the orthogonal projection onto M.

2. Since $\mathcal{L}(X_i) = \mathcal{L}(\beta + \varepsilon_i)$ where $\mathcal{E}\varepsilon_i = 0$ and $\text{var}(\varepsilon_i) = 1$, it follows that $\mathcal{L}(X) = \mathcal{L}(\beta e + \varepsilon)$ where $\mathcal{E}\varepsilon = 0$ and $\text{Cov}(\varepsilon) = I_n$. A direct application of least-squares yields $\hat{\beta} = \bar{X}$ for this linear model. For (iii), since the same β is added to each coordinate of ε, the vector of ordered X's has the same distribution as the $\beta e + \nu$ where ν is the vector of ordered ε's. Thus $\mathcal{L}(U) = \mathcal{L}(\beta e + \nu)$ so $\mathcal{E}U = \beta e + a_0$ and $\text{Cov}(U) = \text{Cov}(\nu) = \Sigma_0$. Hence $\mathcal{L}(U - a_0) = \mathcal{L}(\beta e + (\nu - a_0))$. Based on this model, the Gauss–Markov estimator for β is $\tilde{\beta} = (e'\Sigma_0^{-1}e)^{-1}e'\Sigma_0^{-1}(U - a_0)$. Since $\bar{X} = (1/n)e'(U - a_0)$ (show $e'a_0 = 0$ using the symmetry of f), it follows from the Gauss–Markov Theorem that $\text{var}(\tilde{\beta}) < \text{var}(\hat{\beta})$.

3. That $M - \omega = M \cap \omega^\perp$ is clear since $\omega \subseteq M$. The condition $(P_M - P_\omega)^2 = P_M - P_\omega$ follows from observing that $P_M P_\omega = P_\omega P_M = P_\omega$. Thus $P_M - P_\omega$ is an orthogonal projection onto its range. That $\mathcal{R}(P_M - P_\omega) = M - \omega$ is easily verified by writing $x \in V$ as $x = x_1 + x_2 + x_3$ where $x_1 \in \omega$, $x_2 \in M - \omega$, and $x_3 \in M^\perp$. Then $(P_M - P_\omega)(x_1 + x_2 + x_3) = x_1 + x_2 - x_1 = x_2$. Writing $P_M = P_M - P_\omega + P_\omega$ and noting that $(P_M - P_\omega)P_\omega = 0$ yields the final identity.

4. That $\mathcal{R}(A) = M_0$ is clear. To show $\mathcal{R}(B_1) = M_1 - M_0$, first consider the transformation C defined by $(Cy)_{ij} = \bar{y}_{i\cdot}$, $i = 1,\ldots, I$, $j = 1,\ldots, J$. Then $C^2 = C = C'$, and clearly, $\mathcal{R}(C) \subseteq M_1$. But if $y \in M_1$, then $Cy = y$ so C is the orthogonal projection onto M_1. From Problem 3 (with $M = M_1$ and $\omega = M_0$), we see that $C - A_0$ is the orthogonal projection onto $M_1 - M_0$. But $((C - A_0)y)_{ij} = \bar{y}_{i\cdot} - \bar{y}_{\cdot\cdot}$, which is just $(B_1 y)_{ij}$. Thus $B_1 = C - A_0$ so $\mathcal{R}(B_1) = M_1 - M_0$. A similar argument shows $\mathcal{R}(B_2) = M_2 - M_0$. For (ii), use the fact that $A_0 + B_1 + B_2 + B_3$ is the identity and the four orthogonal projections are perpendicular to each other. For (iii), first observe that $M = M_1 + M_2$ and $M_1 \cap M_2 = M_0$. If μ has the assumed representation, let ν be the vector with $\nu_{ij} = \alpha + \beta_i$ and let ξ be the vector with $\xi_{ij} = \gamma_j$. Then $\nu \in M_1$ and $\xi \in M_2$ so $\mu = \nu + \xi \in M_1 + M_2$.. Conversely, suppose $\mu \in M_0 \oplus (M_1 - M_0) \oplus (M_2 - M_0)$—say $\mu = \delta + \nu + \xi$. Since $\delta \in M_0$, $\delta_{ij} = \bar{\delta}_{\cdot\cdot}$ for all i, j, so set $\alpha = \bar{\delta}_{\cdot\cdot}$. Since $\nu \in M_1 - M_0$, $\nu_{ij} - \nu_{ik} = 0$ for all j, k for each fixed i and $\bar{\nu}_{\cdot\cdot} = 0$. Take $j = 1$ and set $\beta_i = \nu_{i1}$. Then $\nu_{ij} = \beta_i$ for $j = 1,\ldots, J$ and, since $\bar{\nu}_{\cdot\cdot} = 0$, $\Sigma\beta_i = 0$. Similarly, setting $\gamma_j = \xi_{1j}$, $\xi_{ij} = \gamma_j$ for all i, j and since $\bar{\xi}_{\cdot\cdot} = 0$, $\Sigma\gamma_j = 0$. Thus $\mu_{ij} = \alpha + \beta_i + \gamma_j$ where $\Sigma\beta_i = \Sigma\gamma_j = 0$.

5. With $n = \dim V$, the density of Y is (up to constants) $f(y|\mu, \sigma^2) = \sigma^{-n}\exp[-(1/2\sigma^2)\|y - \mu\|^2]$. Using the results and notation Problem

3, write $V = \omega \oplus (M - \omega) \oplus M^{\perp}$ so $(M - \omega) \oplus M^{\perp} = \omega^{\perp}$. Under H_0, $\mu \in \omega$ so $\hat{\mu}_0 = P_\omega y$ is the maximum likelihood estimator of μ and

$$(4.4) \qquad f(y|\mu_0, \sigma^2) = \sigma^{-n}\exp\left[-\frac{1}{2\sigma^2}\|Q_\omega y\|^2\right]$$

where $Q_\omega = I - P_\omega$. Maximizing (4.4) over σ^2 yields $\hat{\sigma}_0^2 = n^{-1}\|Q_\omega y\|^2$. A similar analysis under H_1 shows that the maximum likelihood estimator of μ is $\hat{\mu}_1 = P_M y$ and $\hat{\sigma}_1^2 = n^{-1}\|Q_M y\|^2$ is the maximum likelihood estimator of σ^2. Thus the likelihood ratio test rejects for small values of the ratio

$$\Lambda(y) = \frac{f(y|\hat{\mu}_0, \hat{\sigma}_0^2)}{f(y|\hat{\mu}_1, \hat{\sigma}_1^2)} = \frac{\hat{\sigma}_0^{-n}}{\hat{\sigma}_1^{-n}} = \left(\frac{\|Q_M y\|^2}{\|Q_\omega y\|^2}\right)^{n/2}.$$

But $Q_\omega = Q_M + P_{M-\omega}$ and $Q_M P_{M-\omega} = 0$, so $\|Q_\omega y\|^2 = \|Q_M y\|^2 + \|P_{M-\omega} Y\|^2$. But rejecting for small values of $\Lambda(y)$ is equivalent to rejecting for large values of $(\Lambda(y))^{-2/n} - 1 = \|P_{M-\omega} y\|^2/\|Q_M y\|^2$. Under H_0, $\mu \in \omega$ so $\mathcal{L}(P_{M-\omega} Y) = N(0, \sigma^2 P_{M-\omega})$ and $\mathcal{L}(Q_M Y) = N(0, \sigma^2 Q_M)$. Since $Q_M P_{M-\omega} = 0$, $Q_M Y$ and $P_{M-\omega} Y$ are independent and $\mathcal{L}(\|P_{M-\omega} Y\|) = \sigma^2 \chi_r^2$ where $r = \dim M - \dim \omega$. Also, $\mathcal{L}(\|Q_M Y\|^2) = \sigma^2 \chi_{n-k}^2$ where $k = \dim M$.

6. We use the notation of Problems 4 and 5. In the parameterization described in (iii) of Problem 4, $\beta_1 = \beta_2 = \cdots = \beta_I$ iff $\mu \in M_2$. Thus $\omega = M_2$ so $M - \omega = M_1 - M_0$. Since $M^{\perp}$ is the range of B_3 (Problem 1.15), $\|B_3 y\|^2 = \|Q_M y\|^2$, and it is clear that $\|B_3 y\|^2 = \Sigma\Sigma(y_{ij} - \bar{y}_{i\cdot} - \bar{y}_{\cdot j} + \bar{y}_{\cdot\cdot})^2$. Also, since $M - \omega = M_1 - M_0$, $P_{M-\omega} = P_{M_1} - P_{M_0}$ and $\|P_{M-\omega} y\|^2 = \|P_{M_1} y\|^2 - \|P_{M_0} y\|^2 = \Sigma_i \Sigma_j \bar{y}_{i\cdot}^2 - \Sigma_i \Sigma_j \bar{y}_{\cdot\cdot}^2 = J\Sigma_i(\bar{y}_{i\cdot} - \bar{y}_{\cdot\cdot})^2$.
7. Since $\mathfrak{R}(X') = \mathfrak{R}(X'X)$ and $X'y$ is in the range of X', there exists a $b \in R^k$ such that $X'Xb = X'y$. Now, suppose that b is any solution. First note that $P_M X = X$ since each column of X is in M. Since $X'Xb = X'y$, we have $X'[Xb - P_M y] = X'Xb - X'P_M y = X'Xb - (P_M X)'y = X'Xb - X'y = 0$. Thus the vector $v = Xb - P_M y$ is perpendicular to each column of X ($X'v = 0$) so $v \in M^{\perp}$. But $Xb \in M$, and obviously, $P_M y \in M$, so $v \in M$. Hence $v = 0$, so $Xb = P_M y$.
8. Since $I \in \gamma$, Gauss–Markov and least-squares agree iff

$$(4.5) \qquad (\alpha P_e + \beta Q_e)M \subseteq M, \qquad \text{for all } \alpha, \beta > 0.$$

But (4.5) is equivalent to the two conditions $P_e M \subseteq M$ and $Q_e M \subseteq M$.

But if $e \in M$, then $M = \text{span}\{e\} \oplus M_1$ where $M_1 \subseteq (\text{span}\{e\})^{\perp}$. Thus $P_e M = \text{span}\{e\} \subseteq M$ and $Q_e M = M_1 \subseteq M$, so Gauss–Markov equals least-squares. If $e \in M^{\perp}$, then $M \subseteq \{\text{span}\, e\}^{\perp}$, so $P_e M = \{0\}$ and $Q_e M = M$, so again Gauss–Markov equals least-squares. For (ii), if $e \notin M^{\perp}$ and $e \notin M$, then one of the two conditions $P_e M \subseteq M$ or $Q_e M \subseteq M$ is violated, so least-squares and Gauss–Markov cannot agree for all α and β. For (ii), since $M \subseteq (\text{span}\{e\})^{\perp}$ and $M \neq (\text{span}\{e\})^{\perp}$, we can write $R^n = \text{span}\{e\} \oplus M \oplus M_1$ where $M_1 = (\text{span}\{e\})^{\perp} - M$ and $M_1 \neq \{0\}$. Let P_1 be the orthogonal projection onto M_1. Then the exponent in the density for Y is (ignoring the factor $-\frac{1}{2}$) $(y-\mu)'(\alpha^{-1}P_e + \beta^{-1}Q_e)(y-\mu) = (P_e y + P_1 y + P_M(y-\mu))'(\alpha^{-1}P_e + \beta^{-1}Q_e)(P_e y + P_1 y + P_M(y-\mu)) = \alpha^{-1}y'P_e y + \beta^{-1}y'P_1 y + \beta^{-1}(y-\mu)'P_M(y-\mu)$ where we have used the fact that $Q_e = P_1 + P_M$ and $P_1 P_M = 0$. Since $\det(\alpha P_e + \beta Q_e) = \alpha\beta^{n-1}$, the usual arguments yields $\hat{\mu} = P_M y$, $\hat{\alpha} = y'P_e y$, and $\hat{\beta} = (n-1)^{-1}y'P_1 y$ as maximum likelihood estimators. When $M = \text{span}\{e\}$, then the maximum likelihood estimators for (α, μ) do not exist—other than the solution $\hat{\mu} = P_e y$ and $\hat{\alpha} = 0$ (which is outside the parameter space). The whole point is that when $e \in M$, you must have replications to estimate α when the covariance structure is $\alpha P_e + \beta Q_e$.

9. Define the inner product $(\cdot, \cdot)$ on R^n by $(x, y) = x'\Sigma_1^{-1}y$. In the inner product space $(R^n, (\cdot, \cdot))$, $\mathcal{E}Y = X\beta$ and $\text{Cov}(Y) = \sigma^2 I$. The transformation P defined by the matrix $X(X'\Sigma_1^{-1}X)^{-1}X'\Sigma_1^{-1}$ satisfies $P^2 = P$ and is self-adjoint in $(R^n, (\cdot, \cdot))$. Thus P is an orthogonal projection onto its range, which is easily shown to be the column space of X. The Gauss–Markov Theorem implies that $\hat{\mu} = PY$ as claimed. Since $\mu = X\beta$, $X'\mu = X'X\beta$ so $\beta = (X'X)^{-1}X'\mu$. Hence $\hat{\beta} = (X'X)^{-1}X'\hat{\mu}$, which is just the expression given.

10. For (i), each $\Gamma \in \mathcal{O}(V)$ is nonsingular so $\Gamma(M) \subseteq M$ is equivalent to $\Gamma(M) = M$—hence $\Gamma^{-1}(M) = M$ and $\Gamma^{-1} = \Gamma'$. Parts (ii) and (iii) are easy. To verify (iv), $t_0(c\Gamma Y + x_0) = P_M(c\Gamma Y + x_0) = cP_M\Gamma Y + x_0 = c\Gamma P_M Y + x_0 = c\Gamma t_0(Y) + x_0$. The identity $P_M\Gamma = \Gamma P_M$ for $\Gamma \in \mathcal{O}_M(V)$ was used to obtain the third equality. For (v), first set $\Gamma = I$ and $x_0 = -P_M y$ to obtain

$$t(y) = t(Q_M y) + P_M y. \tag{4.6}$$

Then to calculate t, we need only know t for vectors $u \in M^{\perp}$ as $Q_M y \in M^{\perp}$. Fix $u \in M^{\perp}$ and let $z = t(u)$ so $z \in M$ by assumption. Then there exists a $\Gamma \in \mathcal{O}_M(V)$ such that $\Gamma u = u$ and $\Gamma z = -z$. For this Γ, we have $z = t(u) = t(\Gamma u) = \Gamma t(u) = \Gamma z = -z$ so $z = 0$. Hence $t(u) = 0$ for all $u \in M^{\perp}$ and the result follows.

11. Part (i) follows by showing directly that the regression subspace M is invariant under each $I_n \otimes A$. For (ii), an element of M has the form $\mu = \{Z_1\beta_1, Z_2\beta_2\} \in \mathcal{L}_{2,n}$ for some $\beta_1 \in R^k$ and $\beta_2 \in R^k$. To obtain an example where M is not invariant under all $I_n \otimes \Sigma$, take $k = 1$, $Z_1 = \varepsilon_1$, and $Z_2 = \varepsilon_2$ so μ is

$$\mu = \begin{pmatrix} \beta_1 & 0 \\ 0 & \beta_2 \\ 0 & 0 \\ \vdots & \vdots \\ 0 & 0 \end{pmatrix}.$$

That the set of such μ's is not invariant under all $I_n \otimes \Sigma$ is easily verified. When $Z_1 = Z_2$, then $\mu = Z_1 B$ where B is $k \times 2$ with ith column β_i, $i = 1, 2$. Thus Example 4.4 applies. For (iii), first observe that Z_1 and Z_2 have the same column space (when they are of full rank) iff $Z_2 = Z_1 C$ where C is $k \times k$ and nonsingular. Now, apply part (ii) with β_2 replaced by $C\beta_2$, so M is the set of μ's of the form $\mu = Z_1 B$ where $B \in \mathcal{L}_{2,k}$.

CHAPTER 5

1. Let $a_1, \ldots, a_p$ be the columns of A and apply Gram–Schmidt to these vectors in the order $a_p, a_{p-1}, \ldots, a_1$. Now argue as in Proposition 5.2.
2. Follows easily from the uniqueness of $F(S)$.
3. Just modify the proof of Proposition 5.4.
4. Apply Proposition 5.7
5. That F is one-to-one and onto follows from Proposition 5.2. Given $A \in \mathcal{L}^0_{p,n}$, $F^{-1}(A) \in \mathcal{F}_{p,n} \times G_u^+$ is the pair (ψ, U) where $A = \psi U$. For (ii), $F(\Gamma\psi, UT') = \Gamma\psi UT' = (\Gamma \otimes T)(\psi U) = (\Gamma \otimes T)(F(\psi, U))$. If $F^{-1}(A) = (\psi, U)$, then $A = \psi U$ and ψ and U are unique. Then $(\Gamma \otimes T)A = \Gamma A T' = \Gamma\psi UT'$ and $\Gamma\psi \in \mathcal{F}_{p,n}$ and $UT' \in G_U^+$. Uniqueness implies that $F^{-1}(\Gamma\psi UT') = (\Gamma\psi, UT')$.
6. When $D_g(x_0)$ exists, it is the unique $n \times n$ matrix that satisfies

$$\lim_{x \to x_0} \frac{\|g(x) - g(x_0) - D_g(x_0)(x - x_0)\|}{\|x - x_0\|} = 0. \tag{5.3}$$

But by assumption, (5.3) is satisfied by A (for $D_g(x_0)$). By definition $J_g(x_0) = \det(D_g(x_0))$.

7. With t_{ii} denoting the ith diagonal element of T, the set $\{T|t_{ii} > 0\}$ is open since the function $T \to t_{ii}$ is continuous on V to R^1. But $G_T^+ = \cap_1^p\{T|t_{ii} > 0\}$, which is open. That g has the given representation is just a matter of doing a little algebra. To establish the fact that $\lim_{x\to 0}(\|R(x)\|/\|x\|) = 0$, we are free to use any norm we want on V and $\mathbb{S}_p^+$ (all norms defined by inner products define the same topology). Using the trace inner product on V and $\mathbb{S}_p^+$, $\|R(x)\|^2 = \|xx'\|^2 = \operatorname{tr} xx'xx'$ and $\|x\|^2 = \operatorname{tr} xx'$, $x \in V$. But for $S \geqslant 0$, $\operatorname{tr} S^2 \leqslant (\operatorname{tr} S)^2$ so $\|R(x)\|/\|x\| \leqslant \operatorname{tr} xx'$, which converges to zero as $x \to 0$. For (iii), write $S = L(x)$, string the S coordinates out as a column vector in the order $s_{11}, s_{21}, s_{22}, s_{31}, s_{32}, s_{33}, \ldots,$ and string the x coordinates out in the same order. Then the matrix of L is lower triangular and its determinant is easily computed by induction. Part (iv) is immediate from Problem 6.

8. Just write out the equations $SS^{-1} = I$ in terms of the blocks and solve.

9. That $P^2 = P$ is easily checked. Also, some algebra and Problem 8 show that $(Pu, v) = (u, Pv)$ so P is self-adjoint in the inner product $(\cdot, \cdot)$. Thus P is an orthogonal projection on $(R^p, (\cdot, \cdot))$. Obviously,

$$R(P) = \left\{x|x = \begin{pmatrix} y \\ z \end{pmatrix}, z = 0\right\}.$$

Since

$$Px = \begin{pmatrix} y - \Sigma_{12}\Sigma_{22}^{-1}z \\ 0 \end{pmatrix},$$

$$\|Px\|^2 = (Px, Px) = \begin{pmatrix} y - \Sigma_{12}\Sigma_{22}^{-1}z \\ 0 \end{pmatrix}' \Sigma^{-1} \begin{pmatrix} y - \Sigma_{12}\Sigma_{22}^{-1}z \\ 0 \end{pmatrix}$$

$$= (y - \Sigma_{12}\Sigma_{22}^{-1}z)'\Sigma^{11}(y - \Sigma_{12}\Sigma_{22}^{-1}z).$$

A similar calculation yields $\|(I - P)x\|^2 = z'\Sigma_{22}^{-1}z$. For (iii), the exponent in the density of X is $-\frac{1}{2}(x, x) = -\frac{1}{2}\|Px\|^2 - \frac{1}{2}\|(I - P)x\|^2$. Marginally, Z is $N(0, \Sigma_{22})$, so the exponent in Z's density is $-\frac{1}{2}\|(I - P)x\|^2$. Thus dividing shows that the exponent in the conditional density of Y given Z is $-\frac{1}{2}\|Px\|^2$, which corresponds to a normal distribution with mean $\Sigma_{12}\Sigma_{22}^{-1}Z$ and covariance $(\Sigma^{11})^{-1} = \Sigma_{11} - \Sigma_{12}\Sigma_{22}^{-1}\Sigma_{21}$.

10. On G_T^+, for $j < i$, t_{ij} ranges from $-\infty$ to $+\infty$ and each integral contributes $\sqrt{2\pi}$—there are $p(p-1)/2$ of these. For $j = i$, t_{ii} ranges

from 0 to ∞ and the change of variable $u_{ii} = t_{ii}^2/2$ shows that the integral over t_{ii} is $(\sqrt{2})^{r-i-1}\Gamma((r-i+1)/2)$. Hence the integral is equal to

$$\pi^{(p(p-1))/4} 2^{(p(p-1))/4} 2^{1/2\Sigma(r-i-1)} \prod_1^p \Gamma\left(\frac{r-i+1}{2}\right),$$

which is just $2^{-p}c(r,p)$.

CHAPTER 6

1. Each $g \in Gl(V)$ maps a linearly independent set into a linearly independent set. Thus $g(M) \subseteq M$ implies $g(M) = M$ as $g(M)$ and M have the same dimension. That $G(M)$ is a group is clear. For (ii),

$$\begin{pmatrix} g_{11} & g_{12} \\ g_{21} & g_{22} \end{pmatrix}\begin{pmatrix} y \\ 0 \end{pmatrix} \in M \qquad \text{for } y \in R^q$$

iff $g_{21}y = 0$ for $y \in R^q$ iff $g_{21} = 0$. But

$$\begin{pmatrix} g_{11} & g_{12} \\ 0 & g_{22} \end{pmatrix}$$

is nonsingular iff both g_{11} and g_{22} are nonsingular. That G_1 and G_2 are subgroups of $G(M)$ is obvious. To show G_2 is normal, consider $h \in G_2$ and $g \in G(M)$. Then

$$ghg^{-1} = \begin{pmatrix} g_{11} & g_{12} \\ 0 & g_{22} \end{pmatrix}\begin{pmatrix} h_{11} & h_{12} \\ 0 & I_r \end{pmatrix}\begin{pmatrix} g_{11}^{-1} & -g_{11}^{-1}g_{12}g_{22}^{-1} \\ 0 & g_{22}^{-1} \end{pmatrix}$$

has its 2, 2 element I_r, so is in G_2. For (iv), that $G_1 \cap G_2 = \{I\}$ is clear. Each $g \in G$ can be written as

$$g = \begin{pmatrix} g_{11} & g_{12} \\ 0 & g_{22} \end{pmatrix} = \begin{pmatrix} I_q & 0 \\ 0 & g_{22} \end{pmatrix}\begin{pmatrix} g_{11} & g_{12} \\ 0 & I_r \end{pmatrix},$$

which has the form $g = hk$ with $h \in G_1$ and $k \in G_2$. The representation is unique as $G_1 \cap G_2 = \{I\}$. Also, $g_1g_2 = h_1k_1h_2k_2 = h_1h_2h_2^{-1}k_1h_2k_2 = h_3k_3$ by the uniqueness of the representation.

2. $G(M)$ does not act transitively on $V - \{0\}$ since the vector $\binom{y}{0}$, $y \neq 0$ remains in M under the action of each $g \in G$. To show $G(M)$ is

transitive on $V \cap M^c$, consider

$$x_i = \begin{pmatrix} y_i \\ z_i \end{pmatrix}, \qquad i = 1, 2$$

with $z_1 \neq 0$ and $z_2 \neq 0$. It is easy to argue there is a $g \in G(M)$ such that $gx_1 = x_2$ (since $z_1 \neq 0$ and $z_2 \neq 0$).

3. Each $n \times n$ matrix $\Gamma \in \mathcal{O}_n$ can be regarded as an n^2-dimensional vector. A sequence $\{\Gamma_j\}$ converges to a point $x \in R^m$ iff each element of Γ_j converges to the corresponding element of x. It is clear that the limit of a sequence of orthogonal matrices is another orthogonal matrix. To show $\mathcal{O}_n$ is a topological group, it must be shown that the map $(\Gamma, \psi) \to \Gamma\psi'$ is continuous from $\mathcal{O}_n \times \mathcal{O}_n$ to $\mathcal{O}_n$—this is routine. To show $\chi(\Gamma) = 1$ for all Γ, first observe that $H = \{\chi(\Gamma) | \Gamma \in \mathcal{O}_n\}$ is a subgroup of the multiplicative group $(0, \infty)$ and H is compact as it is the continuous image of a compact set. Suppose $r \in H$ and $r \neq 1$. Then $r^j \in H$ for $j = 1, 2, \ldots$ as H is a group, but $\{r^j\}$ has no convergent subsequence—this contradicts the compactness of H. Hence $r = 1$.
4. Set $x = e^u$ and $\xi(u) = \log \chi(e^u)$, $u \in R^1$. Then $\xi(u_1 + u_2) = \xi(u_1) + \xi(u_2)$ so ξ is a continuous homomorphism on R^1 to R^1. It must be shown that $\xi(u) = \nu u$ for some fixed real ν. This follows from the solution to Problem 6 below in the special case that $V = R^1$.
5. This problem is easy, but the result is worth noting.
6. Part (i) is easy and for part (ii), all that needs to be shown is that ϕ is linear. First observe that

$$(6.6) \qquad \phi(v_1 + v_2) = \phi(v_1) + \phi(v_2)$$

so it remains to verify that $\phi(\lambda v) = \lambda\phi(v)$ for $\lambda \in R^1$. (6.6) implies $\phi(0) = 0$ and $\phi(nv) = n\phi(v)$ for $n = 1, 2, \ldots$. Also, $\phi(-v) = -\phi(v)$ follows from (6.6). Setting $w = nv$ and dividing by n, we have $\phi(w/n) = (1/n)\phi(w)$ for $n = 1, 2, \ldots$. Now $\phi((m/n)v) = m\phi((1/n)v) = (m/n)\phi(v)$ and by continuity, $\phi(\lambda v) = \lambda\phi(v)$ for $\lambda > 0$. The rest is easy.
7. Not hard with the outline given.
8. By the spectral theorem, every rank r orthogonal projection can be written $\Gamma x_0 \Gamma'$ for some $\Gamma \in \mathcal{O}_n$. Hence transitivity holds. The equation $\Gamma x_0 \Gamma' = x_0$ holds for $\Gamma \in \mathcal{O}_n$ iff Γ has the form

$$\Gamma = \begin{pmatrix} \Gamma_{11} & 0 \\ 0 & \Gamma_{22} \end{pmatrix} \in \mathcal{O}_n,$$

and this gives the isotropy subgroup of x_0. For $\Gamma \in \mathcal{O}_n$, $\Gamma x_0 \Gamma' = \Gamma x_0(\Gamma x_0)'$ and Γx_0 has the form $(\psi 0)$ where $\psi : n \times r$ has columns that are the first r columns of Γ. Thus $\Gamma x_0 \Gamma' = \psi\psi'$. Part (ii) follows by observing that $\psi_1\psi_1' = \psi_2\psi_2'$ if $\psi_1 = \psi_2\Delta$ for some $\Delta \in \mathcal{O}_r$.

9. The only difficulty here is (iii). The problem is to show that the only continuous homomorphisms χ on G_2 to (∞, ∞) are t_{pp}^{α} for some real α. Consider the subgroups G_3 and G_4 of G_2 given by

$$G_3 = \left\{ \begin{pmatrix} I_{p-1} & 0 \\ x & 1 \end{pmatrix} \middle| x' \in R^{p-1} \right\}, \qquad G_4 = \left\{ \begin{pmatrix} I_{p-1} & 0 \\ 0 & u \end{pmatrix} \middle| u \in (0, \infty) \right\}.$$

The group G_3 is isomorphic to R^{p-1} so the only homomorphisms are $x \to \exp[\Sigma_1^{p-1} a_i x_i]$ and G_4 is isomorphic to $(0, \infty)$ so the only homomorphisms are $u \to u^{\alpha}$ for some real α. For $k \in G_2$, write

$$k = \begin{pmatrix} I_{p-1} & 0 \\ x & u \end{pmatrix} = \begin{pmatrix} I_{p-1} & 0 \\ x & 1 \end{pmatrix} \begin{pmatrix} I_{p-1} & 0 \\ 0 & u \end{pmatrix}$$

so $\chi(k) = \exp[\Sigma a_i x_i] u^{\alpha}$. Now, use the condition $\chi(k_1 k_2) = \chi(k_1) \cdot \chi(k_2)$ to conclude $a_1 = a_2 = \cdots = a_{p-1} = 0$ so χ has the claimed form.

10. Use (6.4) to conclude that

$$I_{\gamma} = 2^p (\sqrt{2\pi})^{np} \omega(n, p) \int_{G_U^+} \prod_1^p U_{ii}^{2\gamma+n-i} \exp\left[-\tfrac{1}{2} \sum_{i \leqslant j}\sum U_{ij}^2 \right] dU$$

and then use Problem 5.10 to evaluate the integral over G_U^+. You will find that, for $2\gamma + n > p - 1$, the integral is finite and is $I_{\gamma} = (\sqrt{2\pi})^{np} \omega(n, p)/\omega(2\gamma + n, p)$. If $2\gamma + n \leqslant p - 1$, the integral diverges.

11. Examples 6.14 and 6.17 give Δ_r for $G(M)$ and all the continuous homomorphisms for $G(M)$. Pick $x_0 \in R^p \cap M^c$ to be

$$x_0 = \begin{pmatrix} 0 \\ z_0 \end{pmatrix}$$

where $z_0' = (1, 0, \ldots, 0)$, $z_0 \in R^r$. Then H_0 consists of those g's with the first column of g_{12} being 0 and the first column of g_{22} being z_0. To apply Theorem 6.3, all that remains is to calculate the right-hand modulus of H_0—say Δ_r^0. This is routine given the calculations of Examples 6.14 and 6.17. You will find that the only possible multipliers are $\chi(g) = |g_{11}||g_{33}|$ and Lebesgue measure on $R^p \cap M^c$ is the only (up to a positive constant) invariant measure.

12. Parts (i), (ii), (iii), and (iv) are routine. For (v), $J_1(f) = \int f(x)\mu(dx)$ and $J_2(f) = \int f(\tau^{-1}(y))\nu(dy)$ are both invariant integrals on $\mathcal{K}(\mathcal{X})$. By Theorem 6.3, $J_1 = kJ_2$ for some constant k. To find k, take $f(x) = (\sqrt{2\pi})^{-n}s^n(x)\exp[-\frac{1}{2}x'x]$ so $J_1(f) = 1$. Since $s(\tau^{-1}(y)) = v$ for $y = (u, v, w)$,

$$J_2(f) = (\sqrt{2\pi})^{-n}\int_y v^n \exp\left[-\tfrac{1}{2}v^2 - \tfrac{1}{2}nu^2\right] du\frac{dv}{v^2}\nu(dw)$$

$$= \frac{1}{2}\frac{\Gamma((n-1)/2)}{(\sqrt{\pi})^{n-1}} = \frac{1}{k}.$$

For (vi), the expected value of any function of $\bar{x}$ and $s(x)$, say $q(\bar{x}, s(x))$ is

$$\mathcal{E}q(\bar{x}, s(x)) = \int q(\bar{x}, s(x))f(x)s^n(x)\mu(dx)$$

$$= k\int q(u, v)f(\tau^{-1}(u, v, w))v^n\, du\frac{dv}{v^2}\nu(dw)$$

$$= k\int q(u, v)\frac{v^{n-2}}{\sigma^2}h\left(\frac{v^2}{\sigma^2} + \frac{n(u-\delta)^2}{\sigma^2}\right) du\, dv.$$

Thus the joint density of $\bar{x}$ and $s(x)$ is

$$p(u, v) = \frac{kv^{n-2}}{\sigma^n}h\left(\frac{v^2}{\sigma^2} + \frac{n(u-\delta)^2}{\sigma^2}\right) \qquad \text{(with respect to } du\, dv\text{)}.$$

13. We need to show that, with $Y(X) = X/\|X\|$, $P\{\|X\| \in B, Y \in C\} = P\{\|X\| \in B\}P\{Y \in C\}$. If $P\{\|X\| \in B\} = 0$, the above is obvious. If not, set $\nu(C) = P\{Y \in C, \|X\| \in B\}/P\{\|X\| \in B\}$ so ν is a probability measure on the Borel sets of $\{y \mid \|y\| = 1\} \subseteq R^n$. But the relation $\phi(\Gamma x) = \Gamma\phi(x)$ and the $\mathcal{O}_n$ invariance of $\mathcal{L}(X)$ implies that ν is an $\mathcal{O}_n$-invariant probability measure and hence is unique —(for all Borel B)—namely, ν is uniform probability measure on $\{y \mid \|y\| = 1\}$.

14. Each $x \in \mathcal{X}$ can be uniquely written as gy with $g \in \mathcal{P}_n$ and $y \in \mathcal{Y}$ (of course, y is the order statistic of x). Define $\mathcal{P}_n$ acting on $\mathcal{P}_n \times \mathcal{Y}$ by

$g(P, y) = (gP, y)$. Then $\phi^{-1}(gx) = g\phi^{-1}(x)$. Since $P(gx) = gP(x)$, the argument used in Problem 13 shows that $P(X)$ and $Y(X)$ are independent and $P(X)$ is uniform on $\mathcal{P}_n$.

CHAPTER 7

1. Apply Propositions 7.5 and 7.6.
2. Write $X = \psi U$ as in Proposition 7.3 so ψ and U are independent. Then $P(X) = \psi\psi'$ and $S(X) = U'U$ and the independence is obvious.
3. First, write Q in the form

$$Q = M'\begin{pmatrix} I_r & 0 \\ 0 & 0 \end{pmatrix} M$$

where M is $n \times n$ and nonsingular. Since M is nonsingular, it suffices to show that $(M^{-1}(A))^c$ has measure zero. Write $x = \binom{\dot{x}}{\ddot{x}}$ where $\dot{x}$ is $r \times p$. It then suffices to show that $B^c = \{x | x \in \mathcal{L}_{p,n}, \text{rank}(\dot{x}) = p\}^c$ has measure zero. For this, use the argument given in Proposition 7.1.
4. That the ϕ's are the only equivariant functions follows as in Example 7.6.
5. Part (i) is obvious. For (ii), just observe that knowledge of F_n allows you to write down the order statistic and conversely.
6. Parts (i) and (ii) are clear. For (iii), write $x = Px + Qx$. If t is equivariant $t(x + y) = t(x) + y$, $y \in M$. This implies that $t(Qx) = t(x) + Px$ (pick $y = Px$). Thus $t(x) = Px + t(Qx)$. Since $Q = I - P$, $Qx \in M^{\perp}$, so $BQx = Qx$ for any B with $(B, y) \in G$. Since $t(Qx) \in M$, pick B such that $Bx = -x$ for $x \in M$. The equivariance of t then gives $t(Qx) = t(BQx) = Bt(Qx) = -t(Qx)$, so $t(Qx) = 0$.
7. Part (i) is routine as is the first part of (ii) (use Problem 6). An equivariant estimator of σ^2 must satisfy $t(a\Gamma x + b) = a^2 t(x)$. G acts transitively on $\mathcal{X}$ and $\overline{G}$ acts transitively on $(0, \infty)$ ($\mathcal{Y}$ for this case) so Proposition 7.8 and the argument given in Example 7.6 apply.
8. When $X \in \mathcal{X}$ with density $f(x'x)$, then $Y = X\Sigma^{1/2} = (I_n \otimes \Sigma^{1/2})X$ has density $f(\Sigma^{-1/2}x'x\Sigma^{-1/2})$ since $dx/|x'x|^{n/2}$ is invariant under $x \to xA$ for $A \in Gl_p$. Also, when X has density f, then $\mathcal{L}((\Gamma \otimes \Delta)X) = \mathcal{L}(X)$ for all $\Gamma \in \mathcal{O}_n$ and $\Delta \in \mathcal{O}_p$. This implies (see Proposition 2.19) that $\text{Cov}(X) = cI_n \otimes I_p$ for some $c > 0$. Hence $\text{Cov}((I_n \otimes \Sigma^{1/2})X) = cI_n \otimes \Sigma$. Part (ii) is clear and (iii) follows from Proposition 7.8 and Example 7.6. For (iv), the definition of C_0 and the assumption on f

imply $f(\Gamma C_0 \Gamma') = f(C_0 \Gamma' \Gamma) = f(C_0)$ for each $\Gamma \in \mathcal{O}_p$. The uniqueness of C_0 implies $C_0 = \alpha I_p$ for some $\alpha > 0$. Thus the maximum likelihood estimator of Σ must be $\alpha X'X$ (see Proposition 7.12 and Example 7.10).

9. If $\mathcal{L}(X) = P_0$, then $\mathcal{L}(\|X\|)$ is the same whenever $\mathcal{L}(X) \in \{P | P = gP_0, g \in \mathcal{O}(V)\}$ since $x \to \|x\|$ is a maximal invariant under the action of $\mathcal{O}(V)$ on V. For (ii), $\mathcal{L}(\|X\|)$ depends on μ through $\|\mu\|$.

10. Write $V = \omega \oplus (M - \omega) \oplus M^{\perp}$. Remove a set of Lebesgue measure zero from V and show the F ratio is a maximal invariant under the group action $x \to a\Gamma x + b$ where $a > 0$, $b \in \omega$, and $\Gamma \in \mathcal{O}(V)$ satisfies $\Gamma(\omega) \subseteq \omega$, $\Gamma(M - \omega) \subseteq (M - \omega)$. The group action on the parameter (μ, σ^2) is $\mu \to a\Gamma\mu + b$ and $\sigma^2 \to a^2\sigma^2$. A maximal invariant parameter is $\|P_{M-\omega}\mu\|^2/\sigma^2$, which is zero when $\mu \in \omega$.

11. The statistic V is invariant under $x_i \to Ax_i + b$, $i = 1, \ldots, n$, where $b \in R^p$, $A \in Gl_p$, and $\det A = 1$. The model is invariant under this group action where the induced group action on (μ, Σ) is $\mu \to A\mu + b$ and $\Sigma \to A\Sigma A'$. A direct calculation shows $\theta = \det(\Sigma)$ is a maximal invariant under the group action. Hence the distribution of V depends on (μ, Σ) only through θ.

12. For (i), if $h \in G$ and $B \in \mathcal{B}$, $(hP)(B) = P(h^{-1}B) = \int_G (g\overline{Q})(h^{-1}B) \mu(dg) = \int_G \overline{Q}(g^{-1}h^{-1}B)\mu(dg) = \int_G \overline{Q}((hg)^{-1}B)\mu(dg) = \int \overline{Q}(g^{-1}B)\mu(dg) = P(B)$, so $hP = P$ for $h \in G$ and P is G invariant. For (ii), let Q be the distribution described in Proposition 7.16 (ii), so if $\mathcal{L}(X) = P$, then $\mathcal{L}(X) = \mathcal{L}(UY)$ where U is uniform on G and is independent of Y. Thus for any bounded $\mathcal{B}$-measurable function f,

$$\int f(x)P(dx) = \int_G \int_{\mathcal{Y}} f(gy)\mu(dg)Q(dy) = \int_G \int_{\mathcal{X}} f(gx)\mu(dg)\overline{Q}(dx).$$

Set $f = I_B$ and we have $P(B) = \int_G \overline{Q}(g^{-1}B)\mu(dg)$ so (7.1) holds.

13. For $y \in \mathcal{Y}$ and $B \in \mathcal{B}$, define $R(B|y)$ by $R(B|y) = \int_G I_B(gy)\mu(dg)$. For each y, $R(\cdot|y)$ is a probability measure on $(\mathcal{X}, \mathcal{B})$ and for fixed B, $R(B|\cdot)$ is $(\mathcal{Y}, \mathcal{C})$ measurable. For $P \in \mathcal{P}$, (ii) of Proposition 7.16 shows that

$$\int h(x)P(dx) = \int_{\mathcal{Y}} \int_G h(gy)\mu(dg)Q(dy). \tag{7.2}$$

But by definition of $R(\cdot|\cdot)$, $\int_G h(gy)\mu(dg) = \int_{\mathcal{X}} h(x)R(dx|y)$, so (7.2) becomes

$$\int_{\mathcal{X}} h(x)P(dx) = \int_{\mathcal{Y}}\int_{\mathcal{X}} h(x)R(dx|y)Q(dy).$$

This shows that $R(\cdot|y)$ serves as a version of the conditional distribution of X given $\tau(X)$. Since R does not depend on $P \in \mathcal{P}$, $\tau(X)$ is sufficient.

14. For (i), that $t(gx) = g \circ t(x)$ is clear. Also, $X - \bar{X}e = Q_e X$, which is $N(0, Q_e)$ so is ancillary. For (ii), $\mathcal{E}(f(X_1)|\bar{X} = t) = \mathcal{E}(f(X_1 - \bar{X} + \bar{X})|\bar{X} = t) = \mathcal{E}(f(\varepsilon_1' Z(X) + \bar{X})|\bar{X} = t)$ since $Z(X)$ has coordinates $X_i - \bar{X}$, $i = 1, \ldots, n$. Since Z and $\bar{X}$ are independent, this last conditional expectation (given $\bar{X} = t$) is just the integral over the distribution of Z with $\bar{X} = t$. But $\varepsilon_1' Z(X) = X_1 - \bar{X}$ is $N(0, \delta^2)$ so the claimed integral expression holds. When $f(x) = 1$ for $x \leqslant u_0$ and 0 otherwise, the integral is just $\Phi((u_0 - t)/\delta)$ where Φ is the normal cumulative distribution function.
15. Let B be the set $(-\infty, u_0]$ so $I_B(X_1)$ is an unbiased estimator of $h(a, b)$ when $\mathcal{L}(X) = (a, b)P_0$. Thus $\hat{h}(t(X)) = \mathcal{E}(I_B(X_1)|t(X))$ is an unbiased estimator of $h(a, b)$ based on $t(X)$. To compute $\hat{h}$, we have $\mathcal{E}(I_B(X_1)|t(X)) = P\{X_1 \leqslant u_0|t(X)\} = P\{(X_1 - \bar{X})/s \leqslant (u_0 - \bar{X})/s|(s, \bar{X})\}$. But $(X_1 - \bar{X})/s \equiv Z_1$ is the first coordinate of $Z(X)$ so is independent of $(s, \bar{X})$. Thus $\hat{h}(s, \bar{X}) = P_{Z_1}\{Z_1 \leqslant (u_0 - \bar{X})/s\} = F((u_0 - \bar{X})/s)$ where F is the distribution function of the first coordinate of Z. To find F, first observe that Z takes values in $\mathcal{Z} = \{x|x \in R^n, x'e = 0, \|x\| = 1\}$ and the compact group $\mathcal{O}_n(e)$ acts transitively on $\mathcal{Z}$. Since $Z(\Gamma X) = \Gamma Z(X)$ for $\Gamma \in \mathcal{O}_n(e)$, it follows that Z has a uniform distribution on $\mathcal{Z}$ (see the argument in Example 7.19). Let U be $N(0, I_n)$ so Z has the same distribution as $Q_e U/\|Q_e U\|$ and $\mathcal{L}(Z_1) = \mathcal{L}(\varepsilon_1' Q_e U/\|Q_e U\|^2) = \mathcal{L}((Q_e\varepsilon_1)' Q_e U/\|Q_e U\|^2)$. Since $\|Q_e \varepsilon_1\|^2 = (n - 1)/n$ and $Q_e U$ is $N(0, Q_e)$, it follows that $\mathcal{L}(Z_1) = \mathcal{L}(((n - 1)/n)^{1/2} W_1)$ where $W_1 = U_1/(\Sigma_1^{n-1} U_i^2)^{1/2}$. The rest is a routine computation.
16. Part (i) is obvious and (ii) follows from

$$(7.3)\quad \mathcal{E}(f(X)|\tau(X) = g) = \mathcal{E}\Big(f\big(\tau(X)(\tau(X))^{-1}X\big)|\tau(X) = g\Big)$$

$$= \mathcal{E}(f(\tau(X)Z(X))|\tau(X) = g).$$

Since $Z(X)$ and $\tau(X)$ are independent and $\tau(X) = g$, the last member of (7.3) is just the expectation over Z of $f(gZ)$. Part (iii) is just an application and Q_0 is the uniform distribution on $\mathcal{F}_{p,n}$. For (iv), let B be a fixed Borel set in R^P and consider the parametric function

$h(\Sigma) = P_\Sigma(X_1 \in B) = \int I_B(x)(\sqrt{2\pi})^{-p}|\Sigma|^{-1/2}\exp[-\frac{1}{2}x'\Sigma^{-1}x]dx$, where X_1' is the first row of X. Since $\tau(X)$ is a complete sufficient statistic, the MVUE of $h(\Sigma)$ is

(7.4)

$$\hat{h}(T) = \mathcal{E}(I_B(X_1)|\tau(X) = T) = P\{T(\tau(X))^{-1}X_1 \in B|\tau(X) = T\}.$$

But $Z_1' = (\tau^{-1}(X)X_1)'$ is the first row of $Z(X)$ so is independent of $\tau(X)$. Hence $\hat{h}(T) = P_1\{Z_1 \in T^{-1}(B)\}$ where P_1 is the distribution of Z_1 when Z has a uniform distribution on $\mathcal{F}_{p,n}$. Since Z_1 is the first p coordinates of a random vector that is uniform on $\{x|\|x\| = 1, x \in R^n\}$, it follows that Z_1 has a density $\psi(\|u\|^2)$ for $u \in R^p$ where ψ is given by

$$\psi(v) = \begin{cases} c(1-v)^{(n-p-2)/2} & 0 < v < 1 \\ 0 & \text{otherwise} \end{cases}$$

where $c = \Gamma(n/2)/\pi^{p/2}\Gamma((n-p)/2)$. Therefore $\hat{h}(T) = \int_{R^p} I_B(Tu)\psi(\|u\|^2)du = (\det T)^{-1}\int_{R^p} I_B(u)\psi(\|T^{-1}u\|^2)du$. Now, let B shrink to the point u_0 to get that $(\det T)^{-1}\psi(\|T^{-1}u_0\|^2)$ is the MVUE for $(\sqrt{2\pi})^{-p}|\Sigma|^{-1/2}\exp[-\frac{1}{2}u_0'\Sigma^{-1}u_0]$.

CHAPTER 8

1. Make a change of variables to r, $x_1 = s_{11}/\sigma_{11}$ and $x_2 = s_{22}/\sigma_{22}$, and then integrate out x_1 and x_2. That $p(r|\rho)$ has the claimed form follows by inspection. Karlin's Lemma (see Appendix) implies that $\psi(\rho r)$ has a monotone likelihood ratio.
3. For $\alpha = 1/2, \ldots, (p-1)/2$, let $X_1, \ldots, X_r$ be i.i.d. $N(0, I_p)$ with $r = 2\alpha$. Then $S = X_i X_i'$ has ϕ_α as its characteristic function. For $\alpha > (p-1)/2$, the function $p_\alpha(s) = k(\alpha)|s|^\alpha \exp[-\frac{1}{2}\operatorname{tr} s]$ is a density with respect to $ds/|s|^{(p+1)/2}$ on $\mathcal{S}_p^+$. The characteristic function of p_α is ϕ_α. To show that $\phi_\alpha(\Sigma A)$ is a characteristic function, let S satisfy $\mathcal{E}\exp(i\langle A, S\rangle) = \phi_\alpha(A) = |I_p - 2iA|^\alpha$. Then $\Sigma^{1/2}S\Sigma^{1/2}$ has $\phi_\alpha(\Sigma A)$ as its characteristic function.
4. $\mathcal{L}(S) = \mathcal{L}(\Gamma S \Gamma')$ implies that $A = \mathcal{E}S$ satisfies $A = \Gamma A \Gamma'$ for all $\Gamma \in \mathcal{O}_p$. This implies $A = cI_p$ for some constant c. Obviously, $c = \mathcal{E}s_{11}$. For (ii) $\operatorname{var}(\operatorname{tr} DS) = \operatorname{var}(\Sigma_1^p d_i s_{ii}) = \Sigma_1^p d_i^2 \operatorname{var}(s_{ii}) + \Sigma\Sigma_{i \neq j} d_i d_j \operatorname{cov}(s_{ii}, s_{jj})$. Noting that $\mathcal{L}(S) = \mathcal{L}(\Gamma S \Gamma')$ for $\Gamma \in \mathcal{O}_p$, and in particular for permutation matrices, it follows that $\gamma = \operatorname{var}(s_{ii})$ does not depend on i and $\beta = \operatorname{cov}(s_{ii}, s_{jj})$ does not depend on i and j $(i \neq j)$. Thus $\operatorname{var}\langle D, S\rangle =$

$\gamma\Sigma_1^p d_i^2 + \beta\Sigma\Sigma_{i\neq j} d_i d_j = (\gamma - \beta)\Sigma_1^p d_i^2 + \beta(\Sigma_1^p d_i)^2$. For (iii), write $A \in \mathbb{S}_p$ as $\Gamma D \Gamma'$ so $\text{var}\langle A, S\rangle = \text{var}\langle \Gamma D \Gamma', S\rangle = \text{var}\langle D, \Gamma' S \Gamma\rangle = \text{var}\langle D, S\rangle = (\gamma - \beta)\Sigma_1^p d_i^2 + \beta(\Sigma_1^p d_i)^2 = (\gamma - \beta)\text{tr}\, A^2 + \beta(\text{tr}\, A)^2 = (\gamma - \beta)\langle A, A\rangle + \beta\langle I, A\rangle^2$. With $T = (\gamma - \beta)I_p \otimes I_p + \beta I_p \square I_p$, it follows that $\text{var}\langle A, S\rangle = \langle A, TA\rangle$, and since T is self-adjoint, this implies that $\text{Cov}(S) = T$.

5. Use Proposition 7.6.
6. Immediate from Problem 3.
7. For (i), it suffices to show that $\mathcal{L}((ASA')^{-1}) = W((A\Lambda A')^{-1}, r, \nu + r - 1)$. Since $\mathcal{L}(S^{-1}) = W(\Lambda^{-1}, p, \nu + p - 1)$, Proposition 8.9 implies that desired result. (ii) follows immediately from (i). For (iii), (i) implies $\tilde{S} = \Lambda^{-1/2} S \Lambda^{-1/2}$ is $IW(I_p, p, \nu)$ and $\mathcal{L}(\tilde{S}) = \mathcal{L}(\Gamma\tilde{S}\Gamma')$ for all $\Gamma \in \mathcal{O}_p$. Now, apply Problem 4 to conclude that $\mathcal{E}\tilde{S} = cI_p$ where $c = \mathcal{E}\tilde{s}_{11}$. That $c = (\nu - 2)^{-1}$ is an easy application of (i). Hence $(\nu - 2)^{-1} I_p = \mathcal{E}\tilde{S} = \Lambda^{-1/2}(\mathcal{E}S)\Lambda^{-1/2}$ so $\mathcal{E}S = (\nu - 2)^{-1}\Lambda$. Also, $\text{Cov}\,\tilde{S} = (\gamma - \beta)I_p \otimes I_p + \beta I_p \square I_p$ as in Problem 4. Thus $\text{Cov}(\tilde{S}) = (\Lambda^{1/2} \otimes \Lambda^{1/2})(\text{Cov}\,\tilde{S})(\Lambda^{1/2} \otimes \Lambda^{1/2}) = (\gamma - \beta)\Lambda \otimes \Lambda + \beta\Lambda \square \Lambda$. For (iv), that $\mathcal{L}(S_{11}) = IW(\Lambda_{11}, q, \nu)$, take $A = (I_q\ 0)$ in part (i). To show $\mathcal{L}(S_{22\cdot 1}^{-1}) = W(\Lambda_{22\cdot 1}^{-1}, r, \nu + q + r - 1)$, use Proposition 8.8 on S^{-1}, which is $W(\Lambda^{-1}, p, \nu + p - 1)$.
8. For (i), let $p_1(x)p_2(s)$ denote the joint density of X and S with respect to the measure $dx\, ds/|s|^{(p+1)/2}$. Setting $T = XS^{-1/2}$ and $V = S$, the joint density of T and V is $p_1(tv^{1/2})p_2(v)|v|^{r/2}$ with respect to $dt\, dv/|v|^{(p+1)/2}$—the Jacobian of $x \to tv^{1/2}$ is $|v|^{r/2}$—see Proposition 5.10. Now, integrate out v to get the claimed density. That $\mathcal{L}(T) = \mathcal{L}(\Gamma T \Delta')$ is clear from the form of the density (also from (ii) below). Use Proposition 2.19 to show $\text{Cov}(T) = c_1 I_r \otimes I_p$. Part (ii) follows by integrating out v from the conditional density of T to obtain the marginal density of T as given in (i). For (iii) represent T as: T given V is $N(0, I_r \otimes V)$ where V is $IW(I_p, p, \nu)$. Thus T_{11} given V is $N(0, I_k \otimes V_{11})$ where V_{11} is the $q \times q$ upper left-hand corner of V. Since $\mathcal{L}(V_{11}) = IW(I_q, q, \nu)$, the claimed result follows from (ii).
9. With $V = S_2^{-1/2} S_1 S_2^{-1/2}$ and $S = S_2^{-1}$, the conditional distribution of V given S is $W(S, p, m)$ and $\mathcal{L}(S) = IW(I_p, p, \nu)$. Since V is unconditionally $F(m, \nu, I_p)$, (i) follows. For (ii), $\mathcal{L}(T) = T(\nu, I_r, I_p)$ means that $\mathcal{L}(T) = \mathcal{L}(XS^{1/2})$ where $\mathcal{L}(X) = N(0, I_r \otimes I_p)$ and $\mathcal{L}(S) = IW(I_p, p, \nu)$. Thus $\mathcal{L}(T'T) = \mathcal{L}(S^{1/2}X'XS^{1/2})$. Since $\mathcal{L}(X'X) = W(I_p, p, r)$, (ii) follows by definition of $F(r, \nu, I_p)$. For (iii), write $F = T'T$ where $\mathcal{L}(T) = T(\nu, I_r, I_p)$, which has the density given in (i) of Problem 8. Since $r \geqslant p$, Proposition 7.6 is directly applicable to yield the density of F. To establish (iv), first note that $\mathcal{L}(F) = \mathcal{L}(\Gamma F \Gamma')$

for all $\Gamma \in \mathcal{O}_p$. Using Example 7.16, F has the same distributions as $\psi D \psi'$ where ψ is uniform on $\mathcal{O}_p$ and is independent of the diagonal matrix D whose diagonal elements $\lambda_1 \geqslant \cdots \geqslant \lambda_p$ are distributed as the eigenvalues of F. Thus $\lambda_1, \ldots, \lambda_p$ are distributed as the eigenvalues of $S_2^{-1}S_1$ where S_1 is $W(I_p, p, r)$ and S_2^{-1} is $IW(I_p, p, \nu)$. Hence $\mathcal{L}(F^{-1}) = \mathcal{L}(\psi D^{-1}\psi') = \mathcal{L}(\psi \tilde{D} \psi')$ where the diagonal elements of $\tilde{D}$, say $\lambda_p^{-1} \geqslant \cdots \geqslant \lambda_1^{-1}$, are the eigenvalues of $S_1^{-1}S_2$. Since S_2 is $W(I_p, p, \nu + p - 1)$, it follows that $\psi \tilde{D} \psi'$ has the same distribution as an $F(\nu + p - 1, r - p + 1, I_p)$ matrix by just repeating the orthogonal invariance argument given above. (v) is established by writing $F = T'T$ as in (ii) and partitioning T as $T_1 : r \times q$ and $T_2 : r \times (p - q)$ so

$$T'T = \begin{pmatrix} T_1'T_1 & T_1'T_2 \\ T_2'T_1 & T_2'T_2 \end{pmatrix}.$$

Since $\mathcal{L}(T_1) = T(\nu, I_r, I_q)$ and $F_{11} = T_1'T_1$, (ii) implies that $\mathcal{L}(F_{11}) = F(r, \nu, I_q)$. (vi) can be established by deriving the density of $XS^{-1}X'$ directly and using (iii), but an alternative argument is more instructive. First, apply Proposition 7.4 to X' and write $X = V^{1/2}\psi'$ where $V \in \mathcal{S}_r^+$, $V = XX'$ is $W(I_r, r, p)$ and is independent of $\psi : p \times r$, which is uniform on $\mathfrak{F}_{r,p}$. Then $XS^{-1}X' = V^{1/2}W^{-1}V^{1/2}$ where $W = (\psi' S^{-1} \psi)^{-1}$ and is independent of V. Proposition 8.1 implies that $\mathcal{L}(W) = W(I_r, r, m - p + r)$. Thus $\mathcal{L}(W^{-1}) = IW(I_r, r, m - p + 1)$. Now, use the orthogonal invariance of the distribution of $XS^{-1}X'$ to conclude that $\mathcal{L}(XS^{-1}X') = \mathcal{L}(\Gamma D \Gamma')$ where Γ and D are independent, Γ is uniform on $\mathcal{O}_r$, and the diagonal elements of D are distributed as the ordered eigenvalues of $W^{-1}V$. As in the proof of (iv), conclude that $\mathcal{L}(\Gamma D \Gamma') = F(p, m - p + 1, I_r)$.

10. The function $S \to S^{1/2}$ on $\mathcal{S}_p^+$ to $\mathcal{S}_p^+$ satisfies $(\Gamma S \Gamma')^{1/2} = \Gamma S^{1/2} \Gamma'$ for $\Gamma \in \mathcal{O}_p$. With $B(S_1, S_2) = (S_1 + S_2)^{-1/2} S_1 (S_1 + S_2)^{-1/2}$, it follows that $B(\Gamma S_1 \Gamma', \Gamma S_2 \Gamma') = \Gamma B(S_1, S_2) \Gamma'$. Since $\mathcal{L}(\Gamma S_i \Gamma') = \mathcal{L}(S_i)$, $i = 1, 2$, and S_1 and S_2 are independent, the above implies that $\mathcal{L}(B) = \mathcal{L}(\Gamma B \Gamma')$ for $\Gamma \in \mathcal{O}_p$. The rest of (i) is clear from Example 7.16. For (ii), let $B_1 = S_1^{1/2}(S_1 + S_2)^{-1}S_1^{1/2}$ so $\mathcal{L}(B_1) = \mathcal{L}(\Gamma B_1 \Gamma')$ for $\Gamma \in \mathcal{O}_p$. Thus $\mathcal{L}(B_1) = \mathcal{L}(\psi D \psi')$ where ψ and D are independent, ψ is uniform on $\mathcal{O}_p$. Also, the diagonal elements of D, say $\lambda_1 \geqslant \cdots \geqslant \lambda_p > 0$, are distributed as the ordered eigenvalues of $S_1(S_1 + S_2)^{-1}$ so B_1 is $B(m_1, m_2, I_p)$. (iii) is easy using (i) and (ii) and the fact that $F(I + F)^{-1}$ is symmetric. For (iv), let $B = X(S + X'X)^{-1}X'$ and observe that $\mathcal{L}(B) = \mathcal{L}(\Gamma B \Gamma')$, $\Gamma \in \mathcal{O}_p$. Since $m \geqslant p$, S^{-1} exists so $B = XS^{-1/2}(I_p + S^{-1/2}X'XS^{-1/2})^{-1}S^{-1/2}X'$. Hence $T = XS^{-1/2}$ is $T(m - p + 1, I_r, I_p)$. Thus $\mathcal{L}(B) = \mathcal{L}(\psi D \psi')$ where ψ is uniform on $\mathcal{O}_r$ and

is independent of D. The diagonal elements of D, say $\lambda_1, \ldots, \lambda_r$, are the eigenvalues of $T(I_p + T'T)^{-1}T'$. These are the same as the eigenvalues of $TT'(I_r + TT')^{-1}$ (use the singular value decomposition for T). But $\mathcal{L}(TT') = \mathcal{L}(XS^{-1}X') = F(p, m-p+1, I_r)$ by Problem 9 (vi). Now use (iii) above and the orthogonal invariance of $\mathcal{L}(B)$. (v) is trivial.

CHAPTER 9

1. Let B have rows $\nu_1', \ldots, \nu_k'$ and form X in the usual way (see Example 4.3) so $\mathcal{E}X = ZB$ with an appropriate $Z: n \times k$. Let $R: 1 \times k$ have entries $a_1, \ldots, a_k$. Then $RB = \Sigma_1^k a_i \mu_i'$ and H_0 holds iff $RB = 0$. Now apply the results in Section 9.1.
2. For (i), just do the algebra. For (ii), apply (i) with $S_1 = (Y - X\hat{B})'(Y - X\hat{B})$ and $S_2 = (X(B - \hat{B}))'(X(B - \hat{B}))$, so $\phi(S_1) \leqslant \phi(S_1 + S_2)$ for every B. Since $A \geqslant 0$, $\operatorname{tr} A(S_1 + S_2) = \operatorname{tr} AS_1 + \operatorname{tr} AS_2 \geqslant \operatorname{tr} AS_1$ since $\operatorname{tr} AS_2 \geqslant 0$ as $S_2 \geqslant 0$. To show $\det(A + S)$ is nondecreasing in $S \geqslant 0$, First note that $A + S_1 \leqslant A + S_1 + S_2$ in the sense of positive definiteness as $S_2 \geqslant 0$. Thus the ordered eigenvalues of $(A + S_1 + S_2)$, say $\lambda_1, \ldots, \lambda_p$, satisfy $\lambda_i \geqslant \mu_i$, $i = 1, \ldots, p$, where $\mu_1, \ldots, \mu_p$ are the ordered eigenvalues of $A + S_1$. Thus $\det(A + S_1 + S_2) \geqslant \det(A + S_1)$. This same argument solves (iv).
3. Since $\mathcal{L}(E\psi'A') = \mathcal{L}(EA')$ for $\psi \in \mathcal{O}_p$, the distribution of EA' depends only on a maximal invariant under the action $A \to A\psi$ of ψ on Gl_p. This maximal invariant is AA'. (ii) is clear and (iii) follows since the reduction to canonical form is achieved via an orthogonal transformation $\tilde{Y} = \Gamma Y$ where $\Gamma \in \mathcal{O}_n$. Thus $\tilde{Y} = \Gamma\mu + \Gamma EA'$. Γ is chosen so $\Gamma\mu$ has the claimed form and H_0 is $\tilde{B}_1 = 0$. Setting $\tilde{E} = \Gamma E$, the model has the claimed form and $\mathcal{L}(E) = \mathcal{L}(\tilde{E})$ by assumption. The arguments given in Section 9.1 show that the testing problem is invariant and a maximal invariant is the vector of the t largest eigenvalues of $Y_1(Y_3'Y_3)^{-1}Y_1'$. Under H_0, $Y_1 = E_1A'$, $Y_3 = E_3A'$ so $Y_1(Y_3'Y_3)^{-1}Y_1' = E_1(E_3'E_3)^{-1}E_1' \equiv W$. When $\mathcal{L}(\Gamma E) = \mathcal{L}(E)$ for all $\Gamma \in \mathcal{O}_n$, write $E = \psi U$ according to Proposition 7.3 where ψ and U are independent and ψ is uniform on $\mathcal{F}_{p,n}$. Partitioning ψ as E is partitioned, $E_i = \psi_i U$, $i = 1, 2, 3$, so $W = \psi_1 U((\psi_3 U)'\psi_3 U)^{-1}U'\psi_1' = \psi_1(\psi_3'\psi_3)^{-1}\psi_1'$. The rest is obvious as the distribution of W depends only on the distribution of ψ.
4. Use the independence of Y_1 and Y_3 and the fact that $\mathcal{E}(Y_3'Y_3)^{-1} = (m - p - 1)^{-1}\Sigma^{-1}$.

5. Let $\Gamma \in \mathcal{O}_2$ be given by

$$\Gamma = (\sqrt{2})^{-1}\begin{pmatrix} 1 & 1 \\ -1 & 1 \end{pmatrix}$$

and set $\tilde{Y} = Y\Gamma$. Then $\mathcal{L}(\tilde{Y}) = N(ZB\Gamma, I_n \otimes \Gamma'\Sigma\Gamma)$. Now, let $B\Gamma$ have columns β_1 and β_2. Then H_0 is that $\beta_1 = 0$. Also $\Gamma'\Sigma\Gamma$ is diagonal with unknown diagonal elements. The results of Section 9.2 apply directly to yield the likelihood ratio test. A standard invariance argument shows the test is UMP invariant.

6. For (i), look at the i, j elements of the equation for Y. To show $M_2 \perp M_3$, compute as follows: $\langle \alpha u_2', u_1\beta' \rangle = \text{tr}\, \alpha u_2'\beta u_1' = u_2'\beta u_1'\alpha = 0$ from the side conditions on α and β. The remaining relations $M_1 \perp M_2$ and $M_1 \perp M_2$ are verified similarly. For (iii) consider $(I_m \otimes A)(\mu u_1 u_2' + \alpha u_2' + u_1\beta') = \mu u_1(Au_2)' + \alpha(Au_2)' + u_1(A\beta)' = \mu\gamma u_1 u_2' + \gamma\alpha u_2' + \delta u_1\beta' \in M$ where the relations $Pu_2 = u_2$ and $Q\beta = \beta$ when $u_2'\beta = 0$ have been used. This shows that M is invariant under each $I_m \otimes A$. It is now readily verified that $\hat{\mu} = \overline{Y}_{..}$, $\hat{\alpha}_i = \hat{Y}_{i.} - \overline{Y}_{..}$ and $\hat{\beta}_j = \hat{Y}_{.j} - \overline{Y}_{..}$. For (iv), first note that the subspace $\omega = \{x | x \in M, \alpha = 0\}$ defined by H_0 is invariant under each $I_m \otimes A$. Obviously, $\omega = M_1 \oplus M_3$. Consider the group whose elements are $g = (c, \Gamma, b)$ where c is a positive scalar, $b \in M_1 \oplus M_3$, and Γ is an orthogonal transformation with invariant subspaces M_2, $M_1 \oplus M_3$, and $M^{\perp}$. The testing problem is invariant under $x \to c\Gamma x + b$ and a maximal invariant is W (up to a set a measure zero). Since W has a noncentral F-distribution, the test that rejects for large values of W is UMP invariant.

7. (i) is clear. The column space of W is contained in the column space of Z and has dimension r. Let $x_1, \ldots, x_r, x_{r+1}, \ldots, x_k, x_{k+1}, \ldots, x_n$ be an orthonormal basis for R^n such that $\text{span}\{x_1, \ldots, x_r\}$ = column space of W and $\text{span}\{x_1, \ldots, x_k\}$ = column space of Z. Also, let $y_1, \ldots, y_p$ be any orthonormal basis for R^p. Then $\{x_i \square y_j | i = 1, \ldots, r, j = 1, \ldots, p\}$ is a basis for $\mathcal{R}(P_W \otimes I_p)$, which has dimension rp. Obviously, $\mathcal{R}(P_W \otimes I_p) \subseteq M$. Consider $x \in \omega$ so $x = ZB$ with $RB = 0$. Thus $(P_W \otimes I_p)x = P_W ZB = W(W'W)^{-1}W'ZB = W(W'W)^{-1}R(Z'Z)^{-1}(ZZ)B = W(W'W)^{-1}RB = 0$. Thus $\mathcal{N}(P_W \otimes I_p) \supseteq \omega$, which implies $\mathcal{R}(P_W \otimes I_p) \subseteq \omega^{\perp}$. Hence $\mathcal{R}(P_W \otimes I_p) \subseteq M \cap \omega^{\perp}$. That $\dim \omega = (k - r)p$ can be shown by a reduction to canonical form as was done in Section 9.1. Since $\omega \subseteq M$, $\dim(M - \omega) = \dim M - \dim \omega = rp$, which entails $\mathcal{R}(P_W \otimes I_p) = M - \omega$. Hence $P_Z \otimes I_p - P_W \otimes I_p$ is the orthogonal projection onto ω.

8. Use the fact that $\Gamma\Sigma\Gamma$ is diagonal with diagonal entries $\alpha_1, \alpha_2, \alpha_3, \alpha_3, \alpha_2$ (see Proposition 9.13 ff.) so the maximum likelihood estimators α_1, α_2,

and α_3 are easy to find—just transform the data by Γ. Let $\hat{D}$ have diagonal entries $\hat{\alpha}_1$, $\hat{\alpha}_2$, $\hat{\alpha}_3$, $\hat{\alpha}_3$, $\hat{\alpha}_2$ so $\hat{\Sigma} = \Gamma\hat{D}\Gamma$ gives the maximum likelihood estimators of σ^2, ρ_1, and ρ_2.

9. Do the problems in the complex domain first to show that if $Z_1, \ldots, Z_n$ are i.i.d. $\mathbb{C}N(0, 2H)$, then $\hat{H} = (1/2n)\Sigma_1^n Z_j Z_j^*$. But if $Z_j = U_j + iV_j$ and

$$X_j = \begin{pmatrix} U_j \\ V_j \end{pmatrix},$$

then $\hat{H} = (1/2n)\Sigma_1^n(U_j + iV_j)(U_j - iV_j)' = (1/2n)[(S_{11} + S_{22}) + i(S_{12} - S_{21})]$ so $\hat{\psi} = \{\hat{H}\}$. This gives the desired result.

10. Write $R = M(I_r\ 0)\Gamma$ where M is $r \times r$ of rank r and $\Gamma \in \mathcal{O}_p$. With $\delta = \Gamma\mu$, the null hypothesis is $(I_r\ 0)\delta = 0$. Now, transform the data by Γ and proceed with the analysis as in the first testing problem considered in Section 9.6.

11. First write $P_Z = P_1 + P_2$ where P_1 is the orthogonal projection onto e and P_2 is the orthogonal projection onto (column space of Z) $\cap$ $\{\text{span}\, e\}^\perp$. Thus $P_M = P_1 \otimes I_p + P_2 \otimes I_p$. Also, write $A(\rho) = \gamma P_1 + \delta Q_1$ where $\gamma = 1 + (n-1)\rho$, $\delta = 1 - \rho$, and $Q_1 = I_n - P_1$. The relations $P_1 P_2 = 0 = Q_1 P_1$ and $P_2 Q_1 = Q_1 P_2 = P_2$ show that M is invariant under $A(\rho) \otimes \Sigma$ for each value of ρ and Σ. Write $ZB = eb_1' + \Sigma_2^k z_j b_j'$ so $Q_1 Y$ is $N(\Sigma_2^k(Q_1 z_j)b_j', (Q_1 A(\rho) Q_1) \otimes \Sigma)$. Now, $Q_1 A(\rho) Q_1 = \delta Q_1$ so $Q_1 Y$ is $N(\beta_2^k(Q_1 z_j)b_j', \delta Q_1 \otimes \Sigma)$. Also, $P_1 Y$ is $N(eb_1', \gamma P_1 \otimes \Sigma)$. Since hypotheses of the form $RB = 0$ involve only $b_2, \ldots, b_p$, an invariance argument shows that invariant tests of H_0 will not involve $P_1 Y$—so just ignore $P_1 Y$. But the model for $Q_1 Y$ is of the MANOVA type; change coordinates so

$$Q_1 = \begin{pmatrix} I_{n-1} & 0 \\ 0 & 0 \end{pmatrix}.$$

Now, the null hypothesis is of the type discussed in Section 9.1.

CHAPTER 10

1. Part (i) is clear since the number of nonzero canonical correlations is always the rank of Σ_{12} in the partitioned covariance of $\{X, Y\}$. For (ii), write

$$\mathrm{Cov}\{\tilde{X}, \tilde{Y}\} = \begin{pmatrix} \Sigma_{11} & \Sigma_{12} \\ \Sigma_{21} & \Sigma_{22} \end{pmatrix}$$

where Σ_{12} has rank t, and $\Sigma_{11} > 0$, $\Sigma_{22} > 0$. First, consider the case when $q \leqslant r$, $\Sigma_{11} = I_q$, $\Sigma_{22} = I_r$, and

$$\Sigma_{12} = \begin{pmatrix} D & 0 \\ 0 & 0 \end{pmatrix}$$

where $D > 0$ is $t \times t$ and diagonal. Set

$$A = \begin{pmatrix} D^{1/2} \\ 0 \end{pmatrix} : q \times t, \qquad B = \begin{pmatrix} D^{1/2} \\ 0 \end{pmatrix} : r \times t$$

so $AB' = \Sigma_{12}$. Now, set $\Lambda_{11} = I_q - AA'$, $\Lambda_{22} = I_r - BB'$, and the problem is solved for this case. The general case is solved by using Proposition 5.7 to reduce the problem to the case above.

2. That $\Sigma_{12} = \delta e_1 e_2'$ for some $\delta \in R^1$ is clear, and hence Σ_{12} has rank one–hence at most one nonzero canonical correlation. It is the square root of the largest eigenvalue of $\Sigma_{11}^{-1}\Sigma_{12}\Sigma_{22}^{-1}\Sigma_{21} = \delta^2\Sigma_{11}^{-1}e_1e_2'\Sigma_{22}^{-1}e_2e_1'$. The only nonzero (possibly) eigenvalue is $\delta^2 e_1'\Sigma_{11}^{-1}e_1e_2'\Sigma_{22}^{-1}e_2$. To describe canonical coordinates, let

$$\tilde{v}_1 = \frac{\Sigma_{11}^{-1/2}e_1}{\|\Sigma_{11}^{-1/2}e_1\|}, \qquad \tilde{w}_1 = \frac{\Sigma_{22}^{-1/2}e_2}{\|\Sigma_{22}^{-1/2}e_2\|}$$

and then form orthonormal bases $\{\tilde{v}_1, \tilde{v}_2, \ldots, \tilde{v}_q\}$ and $\{\tilde{w}_1, \ldots, \tilde{w}_r\}$ for R^q and R^r. Now, set $v_i = \Sigma_{11}^{-1/2}\tilde{v}_i$, $w_j = \Sigma_{22}^{-1/2}\tilde{w}_j$ for $i = 1, \ldots, q$, $j = 1, \ldots, r$. Then verify that $X_i = v_i'X$ and $Y_j = w_j'Y$ form a set of canonical coordinates for X and Y.

3. Part (i) follows immediately from Proposition 10.4 and the form of the covariance for $\{X, Y\}$. That $\delta(B) = \operatorname{tr} A(I - Q(B))$ is clear and the minimization of $\delta(B)$ follows from Proposition 1.44. To describe $\hat{B}$, let $\psi: p \times t$ have columns $a_1, \ldots, a_t$ so $\psi'\psi = I_t$ and $\hat{Q} = \psi\psi'$. Then show directly that $\hat{B} = \psi'\Sigma^{-1/2}$ is the minimizer and $\hat{C}\hat{B}X = \Sigma^{1/2}\hat{Q}\Sigma^{-1/2}X$ is the best predictor. (iii) is an immediate application of (ii).

4. Part (i) is easy. For (ii), with $u_i = x_i - a_0$,

$$\Delta(M, a_0) = \sum_1^n \|x_i - (P(x_i - a_0) + a_0)\|^2 = \sum_1^n \|u_i - Pu_i\|^2$$

$$= \sum_1^n \|Qu_i\|^2 = \sum_1^n \operatorname{tr} Qu_iu_i' = \operatorname{tr} Q\sum_1^n u_iu_i' = \operatorname{tr} S(a_0)Q.$$

Since $S(a_0) = S(\bar{x}) + n(\bar{x} - a_0)(\bar{x} - a_0)'$, (ii) follows. (iii) is an application of Proposition 1.44.

6. Part (i) follows from the singular value decomposition: For (ii), $\{x \in \mathcal{L}_{p,n} | x = \psi C, C \in \mathcal{L}_{p,k}\}$ is a linear subspace of $\mathcal{L}_{p,n}$ and the orthogonal projection onto this subspace is $(\psi\psi') \otimes I_p$. Thus the closest point to A is $((\psi\psi') \otimes I)A = \psi\psi' A$, and the C that achieves the minimum is $\hat{C} = \psi' A$. For $B \in \mathcal{B}_k$, write $B = \psi C$ as in (i). Then

$$\|A - B\|^2 \geqslant \inf_{\psi} \inf_{C} \|A - \psi C\|^2 = \inf_{\psi} \|A - \psi\psi' A\|^2 = \inf_{Q} \|AQ\|^2.$$

The last equality follows as each ψ determines a Q and conversely. Since $\|AQ\|^2 = \operatorname{tr} AQ(AQ)' = \operatorname{tr} AQ^2 A' = \operatorname{tr} QAA'$,

$$\|A - B\|^2 \geqslant \inf_{Q} \operatorname{tr} QAA'.$$

Writing $A = \Sigma_1^p \lambda_i u_i v_i'$ (the singular value decomposition for A), $AA' = \Sigma_1^p \lambda_i u_i u_i'$ is a spectral decomposition for AA'. Using Proposition 1.44, it follows easily that

$$\inf_{Q} \operatorname{tr} QAA' = \sum_{k+1}^{p} \lambda_i^2.$$

That $\hat{B}$ achieves the infimum is a routine calculation.

7. From Proposition 10.8, the density of W is

$$h(w|\theta) = \int_0^\infty p_{n-2}(w|\theta u^{1/2}) f(u)\, du$$

where p_{n-2} is the density of a noncentral t distribution and f is the density of a χ^2_{n-1} distribution. For $\theta > 0$, set $v = \theta u^{1/2}$ so

$$h(w|\theta) = \frac{2}{\theta^2} \int_0^\infty p_{n-2}(w|v) f\left(\frac{v^2}{\theta^2}\right) v\, dv.$$

Since $p_{n-2}(w|v)$ has a monotone likelihood ratio in w and v and $f(v^2/\theta^2)$ has a monotone likelihood ratio in v and θ, Karlin's Lemma implies that $h(w|\theta)$ has a monotone likelihood ratio. For $\theta < 0$, set $v = \theta u^{-1/2}$, change variables, and use Karlin's Lemma again. The last assertion is clear.

8. For U_2 fixed, the conditional distribution of W given U_2 can be described as the ratio of two independent random variables—the numerator has a χ^2_{r+2K} distribution (given K) and K is Poisson with parameter $\Delta/2$ where $\Delta = \rho^2(1-\rho^2)^{-1}U_2$ and the denominator is χ^2_{n-r-1}. Hence, given U_2, this ratio is $\mathcal{F}_{r+2K,\,n-r-1}$ with K described above, so the conditional density of W is

$$f_1(w|\rho, U_2) = \sum_{k=0}^{\infty} f_{r+2k,\,n-r-1}(w)\psi\left(k|\frac{\Delta}{2}\right)$$

where $\psi(\cdot|\Delta/2)$ is the Poisson probability function. Integrating out U_2 gives the unconditional density of W (at ρ). Thus it must be shown that $\mathcal{E}_{U_2}\psi(k|\Delta/2) = h(k|\rho)$—this is a calculation. That $f(\cdot|\rho)$ has a monotone likelihood ratio is a direct application of Karlin's Lemma.

9. Let M be the range of P. Each $R \in \mathcal{P}_s$ can be represented as $R = \psi\psi'$ where ψ is $n \times s$, $\psi'\psi = I_s$, and $P\psi = 0$. In other words, R corresponds to orthonormal vectors $\psi_1, \dots, \psi_s$ (the columns of ψ) and these vectors are in $M^{\perp}$ (of course, these vectors are not unique). But given any two such sets—say $\psi_1, \dots, \psi_s$ and $\delta_1, \dots, \delta_s$, there is a $\Gamma \in \mathcal{O}(P)$ such that $\Gamma\psi_i = \delta_i$, $i = 1, \dots, s$. This shows $\mathcal{O}(P)$ is compact and acts transitively on $\mathcal{P}_s$, so there is a unique $\mathcal{O}(P)$ invariant probability distribution on $\mathcal{P}_s$. For (iii), $\Delta R_0 \Delta'$ has an $\mathcal{O}(P)$ invariant distribution on $\mathcal{P}_s$—uniqueness does the rest.

10. For (i), use Proposition 7.3 to write $Z = \psi U$ with probability one where ψ and U are independent, ψ is uniform on $\mathcal{F}_{p,n}$, and $U \in G_U^+$. Thus with probability one, rank(QZ) = rank$(Q\psi)$. Let $S \geqslant 0$ be independent of ψ with $\mathcal{L}(S^2) = W(I_p, p, n)$ so S has rank p with probability one. Thus rank$(Q\psi)$ = rank$(Q\psi S)$ with probability one. But ψS is $N(0, I_n \otimes I_p)$, which implies that $Q\psi S$ has rank p. Part (ii) is a direct application of Problem 9.

12. That ψ is uniform follows from the uniformity of Γ on $\mathcal{O}_n$. For (ii), $\mathcal{L}(\psi) = \mathcal{L}(Z(Z'Z)^{-1/2})$ and $\Delta = (I_k\ 0)\psi$ implies that $\mathcal{L}(\psi) = \mathcal{L}(X(X'X + Y'Y)^{-1})$. (iii) is immediate from Problem 11, and (iv) is an application of Proposition 7.6. For (v), it suffices to show that $\int f(x)P_1(dx) = \int f(x)P_2(dx)$ for all bounded measurable f. The invariance of P_i implies that for $i = 1, 2$, $\int f(x)P_i(dx) = \int f(gx)P_i(dx)$, $g \in G$. Let ν be uniform probability measure on G and integrate the above to get $\int f(x)P_i(dx) = \int(\int_G f(gx)\nu(dg))P_i(dx)$. But the function $x \to \int_G f(gx)\nu(dg)$ is G-invariant and so can be written $\hat{f}(\tau(x))$ as τ is a maximal invariant. Since $P_1(\tau^{-1}(C)) = P_2(\tau^{-1}(C))$ for all measurable C, we have $\int k(\tau(x))P_1(dx) = \int k(\tau(x))P_2(dx)$ for all bounded

measurable k. Putting things together, we have $\int f(x)P_1(dx) = \int \hat{f}(\tau(x))P_1(dx) = \int \hat{f}(\tau(x))P_2(dx) = \int f(x)P_2(dx)$ so $P_1 = P_2$. Part (vi) is immediate from (v).

13. For (i), argue as in Example 4.4:

$$\begin{aligned}
&\operatorname{tr}(Z - TB)\Sigma^{-1}(Z - TB)' \\
&\quad = \operatorname{tr}(Z - T\hat{B} + T(\hat{B} - B))\Sigma^{-1}(Z - T\hat{B} + T(\hat{B} - B))' \\
&\quad = \operatorname{tr}(QZ + T(\hat{B} - B))\Sigma^{-1}(QZ + T(\hat{B} - B))' \\
&\quad = \operatorname{tr}(QZ)\Sigma^{-1}(QZ)' + \operatorname{tr} T(\hat{B} - B)\Sigma^{-1}(\hat{B} - B)'T' \\
&\quad \geqslant \operatorname{tr}(QZ)\Sigma^{-1}(QZ)' = \operatorname{tr} Z'QZ\Sigma^{-1}.
\end{aligned}$$

The third equality follows from the relation $QT = 0$ as in the normal case. Since h is nonincreasing, this shows that for each $\Sigma > 0$,

$$\sup_B f(Z|B, \Sigma) = f(Z|\hat{B}, \Sigma)$$

and it is obvious that $f(Z|\hat{B}, \Sigma) = |\Sigma|^{-n/2}h(\operatorname{tr} S\Sigma^{-1})$. For (ii), first note that $S > 0$ with probability one. Then, for $S > 0$,

$$\begin{aligned}
\sup_{H_1 \cup H_0} f(Z|B, \Sigma) &= \sup_{\Sigma > 0} f(Z|\hat{B}, \Sigma) \\
&= \sup_{\Sigma > 0} |\Sigma|^{-n/2}h(\operatorname{tr} S\Sigma^{-1}) \\
&= |S|^{-n/2} \sup_{C > 0} |C|^{n/2}h(\operatorname{tr} C).
\end{aligned}$$

Under H_0, we have

$$\begin{aligned}
&\sup_{H_0} f(Z|B, \Sigma) \\
&\quad = \sup_{\Sigma_{ii} > 0,\, i = 1,2} |\Sigma_{11}|^{-n/2}|\Sigma_{22}|^{-n/2}h(\operatorname{tr} \Sigma_{11}^{-1}S_{11} + \operatorname{tr} \Sigma_{22}^{-1}S_{22}) \\
&\quad = |S_{11}|^{-n/2}|S_{22}|^{-n/2} \sup_{C_{ii} > 0,\, i = 1,2} |C_{11}|^{n/2}|C_{22}|^{n/2}h(\operatorname{tr} C_{11} + \operatorname{tr} C_{22}).
\end{aligned}$$

This latter sup is bounded above by

$$\sup_{C > 0} |C|^{n/2}h(\operatorname{tr} C) \equiv k,$$

which is finite by assumption. Hence the likelihood ratio test rejects for small values of $k_1|S_{11}|^{-n/2}|S_{22}|^{-n/2}|S|^{n/2}$, which is equivalent to rejecting for small values of $\Lambda(Z)$. The identity of part (iii) follows from the equations relating the blocks of Σ to the blocks of Σ^{-1}. Partition B into $B_1 : k \times q$ and $B_2 : k \times r$ so $\mathcal{E}X = TB_1$ and $\mathcal{E}Y = TB_2$. Apply the identity with $U = X - TB_1$ and $V = Y - TB_2$ to give

$$f(Z|B, \Sigma) = |\Sigma_{11}|^{-n/2}|\Sigma_{22\cdot 1}|^{-n/2}$$
$$\times h\Big[\operatorname{tr}\big(Y - TB_2 - (X - TB_1)\Sigma_{11}^{-1}\Sigma_{12}\big)$$
$$\times \Sigma_{22\cdot 1}^{-1}\big(Y - TB_2 - (X - TB_1)\Sigma_{11}^{-1}\Sigma_{12}\big)'$$
$$+ \operatorname{tr}(X - TB_1)\Sigma_{11}^{-1}(X - TB_1)'\Big].$$

Using the notation of Section 10.5, write

$$f(X, Y|B, \Sigma) = |\Sigma_{11}|^{-n/2}|\Sigma_{22\cdot 1}|^{-n/2}$$
$$\times h\big[\operatorname{tr}(Y - WC)\Sigma_{22\cdot 1}^{-1}(Y - WC)'$$
$$+ \operatorname{tr}(X - TB_1)\Sigma_{11}^{-1}(X - TB_1)'\big].$$

Hence the conditional density of Y given X is

$$f_1(Y|C, B_1, \Sigma_{11}, \Sigma_{22\cdot 1}, X)$$
$$= |\Sigma_{22\cdot 1}|^{-n/2} h\big(\operatorname{tr}(Y - WC)\Sigma_{22\cdot 1}^{-1}(Y - WC)' + \eta\big)\phi(\eta)$$

where $\eta = \operatorname{tr}(X - TB_1)\Sigma_{11}^{-1}(X - TB_1)$ and $(\phi(\eta))^{-1} = \int_{\mathcal{L}_{r,n}} h(\operatorname{tr} uu' + \eta)\,du$. For (iv), argue as in (ii) and use the identities established in Proposition 10.17. Part (v) is easy, given the results of (iv)—just note that the sup over Σ_{11} and B_1 is equal to the sup over $\eta > 0$. Part (vi) is interesting—Proposition 10.13 is not applicable. Fix X, B_1, and Σ_{11} and note that under H_0, the conditional density of Y is

$$f_2(Y|C_2, \Sigma_{22\cdot 1}, \eta)$$
$$= |\Sigma_{22\cdot 1}|^{-n/2} h\big(\operatorname{tr}(Y - TC_2)\Sigma_{22\cdot 1}^{-1}(Y - TC_2) + \eta\big)\phi(\eta).$$

This shows that Y has the same distribution (conditionally) as $\tilde{Y} =$

$TC_2 + E\Sigma_{22\cdot 1}^{1/2}$ where $E \in \mathcal{L}_{r,n}$ has density $h(\operatorname{tr} EE' + \eta)\phi(\eta)$. Note that $\mathcal{L}(\Gamma E \Delta) = \mathcal{L}(E)$ for all $\Gamma \in \mathcal{O}_n$ and $\Delta \in \mathcal{O}_r$. Let $t = \min(q, r)$ and, given any $n \times n$ matrix A with real eigenvalues, let $\lambda(A)$ be the vector of the t largest eigenvalues of A. Thus the squares of the sample canonical correlations are the elements of the vector $\lambda(R_Y R_X)$ where $R_Y = (QY)(Y'QY)^{-1}(QY)$, $R_X = QX(X'QX)^{-1}QX$, since

$$S = \begin{pmatrix} X'QX & X'QY \\ Y'QX & Y'QY \end{pmatrix}.$$

(You may want to look at the discussion preceding Proposition 10.5.) Now, we use Problem 9 and the notation there—$P = I - Q$. First, $R_Y \in \mathcal{P}_r$, $R_X \in \mathcal{P}_q$, and $\mathcal{O}(P)$ acts transitively on $\mathcal{P}_r$ and $\mathcal{P}_q$. Under H_0 (and X fixed), $\mathcal{L}(QY) = \mathcal{L}(QE\Sigma_{22\cdot 1}^{1/2})$, which implies that $\mathcal{L}(\Gamma R_Y \Gamma') = \mathcal{L}(R_Y)$, $\Gamma \in \mathcal{O}(P)$. Hence R_Y is uniform on $\mathcal{P}_r$ for each X. Fix $R_0 \in \mathcal{R}_q$ and choose Γ_0 so that $\Gamma_0 R_0 \Gamma_0' = R_X$, Then, for each X,

$$\begin{aligned} \mathcal{L}(\lambda(R_Y R_0)) &= \mathcal{L}(\lambda(\Gamma_0 R_Y R_0 \Gamma_0')) = \mathcal{L}(\lambda(\Gamma_0 R_Y \Gamma_0' \Gamma_0 R_0 \Gamma_0')) \\ &= \mathcal{L}(\lambda(\Gamma_0 R_Y \Gamma_0' R_X) = \mathcal{L}(\lambda(R_Y R_X)). \end{aligned}$$

This shows that for each X, $\lambda(R_Y R_X)$ has the same distribution as $\lambda(R_Y R_0)$ for R_0 fixed where R_Y is uniform on $\mathcal{P}_r$. Since the distribution of $\lambda(R_Y R_0)$ does not depend on X and agrees with what we get in the normal case, the solution is complete.

BIBLIOGRAPHY

Anderson, T. W. (1946). The noncentral Wishart distribution and certain problems of multivariate statistics. *Ann. Math. Stat.*, **17**, 409–431.

Anderson, T. W. (1958). *An Introduction to Multivariate Statistical Analysis*. Wiley, New York.

Anderson, T. W., S. Das Gupta, and G. H. P. Styan (1972). *A Bibliography of Multivariate Analysis*. Oliver and Boyd, Edinburgh.

Andersson, S. A. (1975). Invariant normal models. *Ann. Stat.*, **3**, 132–154.

Andersson, S. A. (1982). Distributions of maximal invariants using quotient measures. *Ann. Stat.*, **10**, 955–961.

Arnold, S. F. (1973). Applications of the theory of products of problems to certain patterned covariance matrices. *Ann. Stat.*, **1**, 682–699.

Arnold, S. F. (1979). A coordinate free approach to finding optimal procedures for repeated measures designs. *Ann. Stat.*, **7**, 812–822.

Arnold, S. F. (1981). *The Theory of Linear Models and Multivariate Analysis*. Wiley, New York.

Basu, D. (1955). On statistics independent of a complete sufficient statistic. *Sankhyā*, **15**, 377–380.

Billingsley, P. (1979). *Probability and Measure*. Wiley, New York.

Björk, Å. (1967). Solving linear least squares problems by Gram–Schmidt orthogonalization. *BIT*, **7**, 1–21.

Blackwell, D. and M. Girshick (1954). *Theory of Games and Statistical Decision Functions*. Wiley, New York.

Bondesson, L. (1977). A note on sufficiency and independence. Preprint, University of Lund, Lund, Sweden.

Box, G. E. P. (1949). A general distribution theory for a class of likelihood criteria. *Biometrika*, **36**, 317–346.

Chung, K. L. (1974). *A Course in Probability Theory*, second edition. Academic Press, New York.

Cramér, H. (1946). *Mathematical Methods of Statistics*. Princeton University Press, Princeton, N.J.

Das Gupta, S. (1979). A note on anciliarity and independence via measure-preserving transformations. *Sankhyā*, **41**, Series A, 117–123.

Dawid, A. P. (1978). Spherical matrix distributions and a multivariate model. *J. Roy. Stat. Soc. B*, **39**, 254–261.

Dawid, A. P. (1981). Some matrix-variate distribution theory: Notational considerations and a Bayesian application. *Biometrika*, **68**, 265–274.

Deemer, W. L. and I. Olkin (1951). The Jacobians of certain matrix transformations useful in multivariate analysis. *Biometrika*, **38**, 345–367.

Dempster, A. P. (1969). *Elements of Continuous Multivariate Analysis*. Academic Press, Reading, Mass.

Eaton, M. L. (1970). Gauss–Markov estimation for multivariate linear models: A coordinate free approach. *Ann. Math. Stat.*, **41**, 528–538.

Eaton, M. L. (1972). *Multivariate Statistical Analysis*. Institute of Mathematical Statistics, University of Copenhagen, Copenhagen, Denmark.

Eaton, M. L. (1978). A note on the Gauss–Markov Theorem. *Ann. Inst. Stat. Math.*, **30**, 181–184.

Eaton, M. L. (1981). On the projections of isotropic distributions. *Ann. Stat.*, **9**, 391–400.

Eaton, M. L. and T. Kariya (1981). On a general condition for null robustness. University of Minnesota Technical Report No. 388, Minneapolis.

Eckhart, C. and G. Young (1936). The approximation of one matrix by another of lower rank. *Psychometrika*, **1**, 211–218.

Farrell, R. H. (1962). Representations of invariant measures. *Ill. J. Math.*, **6**, 447–467.

Farrell, R. H. (1976). *Techniques of Multivariate Calculation*. Lecture Notes in Mathematics #520. Springer-Verlag, Berlin.

Giri, N. (1964). On the likelihood ratio test of a normal multivariate testing problem. *Ann. Math. Stat.*, **35**, 181–190.

Giri, N. (1965a). On the complex analogues of T^2 and R^2-tests. *Ann. Math. Stat.*, **36**, 664–670.

Giri, N. (1965b). On the likelihood ratio test of a multivariate testing problem, II. *Ann. Math. Stat.*, **36**, 1061–1065.

Giri, N. (1975). *Invariance and Minimax Statistical Tests*. Hindustan Publishing Corporation, Dehli, India.

Giri, N. C. (1977). *Multivariate Statistical Inference*. Academic Press, New York.

Gnanadesikan, R. (1977). *Methods for Statistical Data Analysis of Multivariate Observations*. Wiley, New York.

Goodman, N. R. (1963). Statistical analysis based on a certain multivariate complex Gaussian distribution (An introduction). *Ann. Math. Stat.*, **34**, 152–177.

Hall, W. J., R. A. Wijsman, and J. K. Ghosh (1965). The relationship between sufficiency and invariance with applications in sequential analysis. *Ann. Math. Stat.*, **36**, 575–614.

Halmos, P. R. (1950). *Measure Theory*. D. Van Nostrand Company, Princeton, N.J.

Halmos, P. R. (1958). *Finite Dimensional Vector Spaces*. Undergraduate Texts in Mathematics, Springer-Verlag, New York.

Hoffman, K. and R. Kunze (1971). *Linear Algebra*, second edition. Prentice Hall, Englewood Cliffs, N.J.

Hotelling, H. (1931). The generalization of Student's ratio. *Ann. Math. Stat.*, **2**, 360–378.

Hotelling, H. (1935). The most predictable criterion. *J. Educ. Psych.*, **26**, 139–142.

Hotelling, H. (1936). Relations between two sets of variates. *Biometrika*, **28**, 321–377.

Hotelling, H. (1947). Multivariate quality control, illustrated by the air testing of sample bombsights, in *Techniques of Statistical Analysis*. McGraw-Hill, New York, pp. 111–184.

James, A. T. (1954). Normal multivariate analysis and the orthogonal group. *Ann. Math. Stat.*, **25**, 40–75.

Kariya, T. (1978). The general MANOVA problem. *Ann. Stat.*, **6**, 200–214.

Karlin, S. (1956). Decision theory for Polya-type distributions. Case of two actions, I, in *Proc. Third Berkeley Symp. Math. Stat. Prob.*, Vol. 1. University of California Press, Berkeley, pp. 115–129.

Karlin, S. and H. Rubin (1956). The theory of decision procedures for distributions with monotone likelihood ratio. *Ann. Math. Stat.* **27**, 272–299.

Kiefer, J. (1957). Invariance, minimax sequential estimation, and continuous time processes. *Ann. Math. Stat.*, **28**, 573–601.

Kiefer, J. (1966). Multivariate optimality results. In *Multivariate Analysis*, edited by P. R. Krishnaiah. Academic Press, New York.

Kruskal, W. (1961). The coordinate free approach to Gauss–Markov estimation and its application to missing and extra observations, in *Proc. Fourth Berkeley Symp. Math. Stat. Prob.*, Vol. 1. University of California Press, Berkeley, pp. 435–461.

Kruskal, W. (1968). When are Gauss–Markov and least squares estimators identical? A co-ordinate free approach. *Ann. Math. Stat.*, **39**, 70–75.

Kshirsagar, A. M. (1972). *Multivariate Analysis*. Marcel Dekker, New York.

Lang, S. (1969). *Analysis II*. Addison-Wesley, Reading, Massachusetts.

Lawley, D. N. (1938). A generalization of Fisher's *z*-test. *Biometrika*, **30**, 180–187.

Lehmann, E. L. (1959). *Testing Statistical Hypotheses*. Wiley, New York.

Mallows, C. L. (1960). Latent vectors of random symmetric matrices. *Biometrika*, **48**, 133–149.

Mardia, K. V., J. T. Kent, and J. M. Bibby (1979). *Multivariate Analysis*. Academic Press, New York.

Muirhead, R. J. (1982). *Aspects of Multivariate Statistical Theory*. Wiley, New York.

Nachbin, L. (1965). *The Haar Integral*. D. Van Nostrand Company, Princeton, N.J.

Noble, B. and J. Daniel (1977). *Applied Linear Algebra*, second edition. Prentice Hall, Englewood Cliffs, N.J.

Olkin, I. and S. J. Press (1969). Testing and estimation for a circular stationary model. *Ann. Math. Stat.*, **40**, 1358–1373.

Olkin, I. and H. Rubin (1964). Multivariate beta distributions and independence properties of the Wishart distribution. *Ann. Math. Stat.*, **35**, 261–269.

Olkin, I. and A. R. Sampson (1972). Jacobians of matrix transformations and induced functional equations. *Linear Algebra Appl.*, **5**, 257–276.

Peisakoff, M. (1950). Transformation parameters. Thesis, Princeton University, Princeton, N.J.

Pillai, K. C. S. (1955). Some new test criteria in multivariate analysis. *Ann. Math. Stat.*, **26**, 117–121.

Pitman, E. J. G. (1939). The estimation of location and scale parameters of a continuous population of any form. *Biometrika*, **30**, 391–421.

Potthoff, R. F. and S. N. Roy (1964). A generalized multivariate analysis of variance model useful especially for growth curve problems. *Biometrika*, **51**, 313–326.

Rao, C. R. (1973). *Linear Statistical Inference and Its Applications*, second edition. Wiley, New York.

Roy, S. N. (1953). On a heuristic method of test construction and its use in multivariate analysis. *Ann. Math. Stat.*, **24**, 220–238.

Roy, S. N. (1957). *Some Aspects of Multivariate Analysis*. Wiley, New York.

Rudin, W. (1953). *Principles of Mathematical Analysis*. McGraw-Hill, New York.

Scheffé, H. (1959). *The Analysis of Variance*. Wiley, New York.

Segal, I. E. and Kunze, R. A. (1978). *Integrals and Operators*, second revised and enlarged edition. Springer-Verlag, New York.

Serre, J. P. (1977). *Linear Representations of Finite Groups*. Springer-Verlag, New York.

Srivastava, M. S. and C. G. Khatri (1979). *An Introduction to Multivariate Statistics*. North Holland, Amsterdam.

Stein, C. (1956). Some problems in multivariate analysis, Part I. Stanford University Technical Report No. 6, Stanford, Calif.

Wijsman, R. A. (1957). Random orthogonal transformations and their use in some classical distribution problems in multivariate analysis. *Ann. Math. Statist.*, **28**, 415–423.

Wijsman, R. A. (1966). Cross-sections of orbits and their applications to densities of maximal invariants, in *Proc. Fifth Berkeley Symp. Math. Stat. Probl.*, Vol. 1. University of California Press, Berkeley, pp. 389–400.

Wilks, S. S. (1932). Certain generalizations in the analysis of variance. *Biometrika*, **24**, 471–494.

Wishart, J. (1928). The generalized product moment distribution in samples from a normal multivariate population. *Biometrika*, **20**, 32–52.

Index

This self-contained volume treats multivariate statistical theory using vector space and invariance methods, complementing the more traditional distribution theory-likelihood approach to the field. Emphasizing the application of these methods, the book contains an extensive set of problems with solutions, and chapter notes at the end of each chapter which provide a guide to the literature. The mathematical and statistical prerequisites include a knowledge of basic ideas in matrix and vector algebra in coordinates spaces, measure and integration theory, and the equivalent of a solid first-year graduate course in mathematical statistics.

MORRIS L. EATON is Professor of Theoretical Statistics at the University of Minnesota. A Fellow of the American Statistical Association and the Institute for Mathematical Statistics, Dr. Eaton is a former Associate Editor of the *Journal of the American Statistical Association* and *Annals of Statistics* and is the author of *Multivariate Statistical Analysis*, published by the Institute for Mathematical Statistics, University of Copenhagen, in 1972. He received his Ph.D. in statistics from Stanford University.

Applied Probability and Statistics (Continued)

ELANDT-JOHNSON and JOHNSON • Survival Models and Data Analysis

FLEISS • Statistical Methods for Rates and Proportions, *Second Edition*

FRANKEN, KÖNIG, ARNDT, and SCHMIDT • Queues and Point Processes

GALAMBOS • The Asymptotic Theory of Extreme Order Statistics

GIBBONS, OLKIN, and SOBEL • Selecting and Ordering Populations: A New Statistical Methodology

GNANADESIKAN • Methods for Statistical Data Analysis of Multivariate Observations

GOLDBERGER • Econometric Theory

GOLDSTEIN and DILLON • Discrete Discriminant Analysis

GREENBERG and WEBSTER • Advanced Econometrics: A Bridge to the Literature

GROSS and CLARK • Survival Distributions: Reliability Applications in the Biomedical Sciences

GROSS and HARRIS • Fundamentals of Queueing Theory

GUPTA and PANCHAPAKESAN • Multiple Decision Procedures: Theory and Methodology of Selecting and Ranking Populations

GUTTMAN, WILKS, and HUNTER • Introductory Engineering Statistics, *Third Edition*

HAHN and SHAPIRO • Statistical Models in Engineering

HALD • Statistical Tables and Formulas

HALD • Statistical Theory with Engineering Applications

HAND • Discrimination and Classification

HILDEBRAND, LAING, and ROSENTHAL • Prediction Analysis of Cross Classifications

HOAGLIN, MOSTELLER, and TUKEY • Understanding Robust and Exploratory Data Analysis

HOEL • Elementary Statistics, *Fourth Edition*

HOEL and JESSEN • Basic Statistics for Business and Economics, *Third Edition*

HOLLANDER and WOLFE • Nonparametric Statistical Methods

IMAN and CONOVER • Modern Business Statistics

JAGERS • Branching Processes with Biological Applications

JESSEN • Statistical Survey Techniques

JOHNSON and KOTZ • Distributions in Statistics
Discrete Distributions
Continuous Univariate Distributions—1
Continuous Univariate Distributions—2
Continuous Multivariate Distributions

JOHNSON and KOTZ • Urn Models and Their Application: An Approach to Modern Discrete Probability Theory

JOHNSON and LEONE • Statistics and Experimental Design in Engineering and the Physical Sciences, Volumes I and II, *Second Edition*

JUDGE, HILL, GRIFFITHS, LÜTKEPOHL and LEE • Introduction to the Theory and Practice of Econometrics

JUDGE, GRIFFITHS, HILL and LEE • The Theory and Practice of Econometrics

KALBFLEISCH and PRENTICE • The Statistical Analysis of Failure Time Data

KEENEY and RAIFFA • Decisions with Multiple Objectives

LAWLESS • Statistical Models and Methods for Lifetime Data

LEAMER • Specification Searches: Ad Hoc Inference with Nonexperimental Data

McNEIL • Interactive Data Analysis

MANN, SCHAFER and SINGPURWALLA • Methods for Statistical Analysis of Reliability and Life Data

MARTZ and WALLER • Bayesian Reliability Analysis

MEYER • Data Analysis for Scientists and Engineers

MIKE and STANLEY • Statistics in Medical Research: Methods and Issues with Applications in Cancer Research